Different Faces of Geometry

INTERNATIONAL MATHEMATICAL SERIES

Series Editor: Tamara Rozhkovskaya

Sobolev Institute of Mathematics of the Siberian Branch of the Russian Academy of Sciences, Novosibirsk, Russia

---

NONLINEAR PROBLEMS IN MATHEMATICAL PHYSICS AND RELATED TOPICS I: In Honor of Professor O. A. Ladyzhenskaya

Edited by M. Sh. Birman, S. Hildebrandt, V. A. Solonnikov, N. N. Uraltseva

NONLINEAR PROBLEMS IN MATHEMATICAL PHYSICS AND RELATED TOPICS II: In Honor of Professor O. A. Ladyzhenskaya

Edited by M. Sh. Birman, S. Hildebrandt, V. A. Solonnikov, N. N. Uraltseva

DIFFERENT FACES OF GEOMETRY

Edited by Simon Donaldson, Yakov Eliashberg, and Mikhael Gromov

---

Different Faces of Geometry

Edited by

Simon Donaldson
Imperial College of London
London, UK

Yakov Eliashberg
Stanford University
Stanford, California

and

Mikhael Gromov
Institut des Haute Etudes Scientifiques
Bures-sur-Yvette, France

Kluwer Academic / Plenum Publishers
New York, Boston, Dordrecht, London, Moscow

This series was founded by Kluwer Academic / Plenum Publishers and the Russian publisher Tamara Rozhkovskaya (Novosibirsk, Russia, rozhk@math.nsc.ru) in 2002. Each volume is simultaneously published in English and in Russian and presents contributions from volume editors and authors invited by the Series editor.

Series webpage: http://www.wkap.nl/prod/s/IMAT

The English camera-ready manuscript was prepared by Tamara Rozhkovskaya

ISSN 1571-5485

ISBN 0-306-47657-1

233 Spring Street, New York, N.Y. 10013

http://www.wkap.nl/

10 9 8 7 6 5 4 3 2 1

A C.I.P. record for this book is available from the Library of Congress

Printed in the United States of America

Preface

Modern information technology allows most mathematicians unprecedented access to the research literature. Published papers, reviews and preprints can be obtained in a few seconds. This mountain of information makes the need for articles which illuminate and survey important developments all the greater.

Our volume brings together articles by leading experts on a variety of different topics which are the scene of exciting current activity.

"Geometry" is famously hard to pin down, and means many different things to different mathematicians: it is probably best interpreted as a way of thought rather than a collection of specific subject areas-there is, perhaps, no branch of mathematics which cannot be considered a part of geometry, when approached in the right spirit.

Certainly we have not set out to cover all the main topics which would normally be denoted as geometry (for example, there is relatively little in the volume on algebraic geometry, nor on the interface between partial differential equations and Riemannian geometry); but we hope and believe that these articles will give a valuable picture of some major areas.

One can distinguish various themes running through the different contributions. There is some emphasis on invariants defined by elliptic equations and their applications in low-dimensional topology, symplectic and contact geometry (Bauer, Seidel, Ozsváth and Szabó). These ideas enter, more tangentially, in the articles of Joyce, Honda and LeBrun. Here and elsewhere, as well as explaining the rapid advances that have been made, the articles convey a wonderful sense of the vast areas lying beyond our current understanding.

Simpson's article emphasizes the need for interesting new constructions (in that case of Kähler and algebraic manifolds), a point which is also made by Bauer in the context of 4-manifolds and the "11/8 conjecture."

LeBrun's article gives another perspective on 4-manifold theory, via Riemannian geometry, and the challenging open questions involving the geometry of even "well-known" 4-manifolds.

There are also striking contrasts between the articles. The authors have taken different approaches: for example, the thoughtful essay of Simpson, the new research results of LeBrun and the thorough expositions with homework problems of Honda.

One can also ponder the differences in the style of mathematics. In the articles of Honda, Giannopoulos and Milman, and Mikhalkin, the "geometry" is present in a very vivid and tangible way; combining respectively with topology, analysis and algebra. The papers of Bauer and Seidel, on the other hand, makes the point that algebraic and algebro-topological abstraction (triangulated categories, spectra) can play an important role in very unexpected ways in concrete geometric problems.

Finally, we wish to thank all the authors for their splendid contributions and express the hope that the reader will find as much interest and excitement in the articles as we ourselves have done.

Simon Donaldson
Yakov Eliashberg
Mikhael Gromov

London–Stanford–Paris
March 2004

Main Topics

- Amoebas and Tropical Geometry
- Convex Geometry and Asymptotic Geometric Analysis
- Differential Topology of 4-Manifolds
- 3-Dimensional Contact Geometry
- Floer Homology and Low-Dimensional Topology
- Kähler Geometry
- Lagrangian and Special Lagrangian Submanifolds
- Refined Seiberg–Witten Invariants

Editors

Simon Donaldson (UK)

Department of Mathematics
Imperial College of London
Huxley Building, 180 Queen's Gate
London SW7 2BT, UK
s.donaldson@imperial.ac.uk

Yakov Eliashberg (USA)

Department of Mathematics
Stanford University
Stanford, CA 94305-2125, USA
eliash@math.stanford.edu

Mikhael Gromov (France)

Institut des Hautes Etudes Scientifiques
35, Route de Chartres
F-91440 Bures-sur-Yvette, France
gromov@ihes.fr

Authors

Stefan Bauer (Germany)

Department of Mathematics
University of Bielefeld
Postfach 100 131
D-33501 Bielefeld, Germany
bauer@mathematik.uni-bielefeld.de
http://www.mathematik.uni-bielefeld.de/~bauer

Ko Honda (USA)

Department of Mathematics
University of Southern California
3620 Vermont Ave.
Los Angeles, CA 90089, USA
khonda@math.usc.edu
http://math.usc.edu/~khonda

Apostolos A. Giannopoulos (Greece)

Department of Mathematics
University of Crete
Iraklion 714-09, Crete, Greece
giannop@fourier.math.uoc.gr
http://www.math.uoc.gr/~apostolo

Dominic Joyce (UK)

Lincoln College
Oxford, OX1 3DR, UK
dominic.joyce@lincoln.ox.ac.uk
http://www.maths.ox.ac.uk/~joyce

Claude LeBrun (USA)

Department of Mathematics
SUNY at Stony Brook
New York, NY 11794-3651, USA
claude@math.sunysb.edu
http://www.math.sunysb.edu/~claude

Grigory Mikhalkin (USA – Canada – Russia)

Department of Mathematics
University of Utah
Salt Lake City, Utah 84112 USA

Department of Mathematics
University of Toronto
Toronto, Ontario M5S 3G3 Canada

St. Petersburg Branch
of Steklov Mathematical Institute
Fontanka 27, St. Petersburg
191011 Russia

mikha@math.toronto.edu
http://www.math.utah.edu/~gmikhalk

Vitali Milman (Israel)

Department of Mathematics
Tel Aviv University
Tel Aviv 69978, Israel
milman@post.tau.ac.il
http://www.math.tau.ac.il/~milman

Peter Ozsváth (USA)

Math Department
Columbia University
2990, Broadway, 4406
New York, NY 10027, USA

Institute for Advanced Study
Princeton, New Jersey
08540 NJ, USA

petero@math.columbia.edu
http://www.math.columbia.edu/~petero

Paul Seidel (USA)

Department of Mathematics
5734 S. University Ave
Chicago, IL 60637, USA

seidel@math.uchicago.edu

Carlos Simpson (France)

CNRS, Laboratoire J.A. Dieudonné
Université de Nice-Sophia Antipolis
Parc Valrose, 06100 Nice, France

carlos@math.unice.fr
http://math.unice.fr/~carlos

Zoltán Szabó (USA)

Department of Mathematics
Princeton University
Fine Hall, Washington Road
Princeton, NJ 08544, USA

szabo@math.princeton.edu
http://www.math.princeton.edu/~szabo

Content

Apostolos A. Giannopoulos and Vitali D. Milman

Dominic Joyce

Grigory Mikhalkin

Different Faces of Geometry

Refined Seiberg–Witten Invariants

Stefan Bauer[†]

Universität Bielefeld
Bielefeld, Germany

In the past two decades, gauge theoretic methods became indispensable when considering manifolds in dimension four. Initially, research centered around the moduli spaces of Yang–Mills instantons. Simon Donaldson had introduced the instanton equations into the field. Using cohomological data of the corresponding moduli spaces, he defined invariants which could effectively distinguish differentiable structures on homeomorphic manifolds. Some years later, Nathan Seiberg and Ed Witten introduced the monopole equations. In a similar spirit as in Donaldson theory, cohomological data of the corresponding moduli spaces went into the definition of Seiberg–Witten invariants. These new invariants turned out to be far easier to compute, seemingly carrying the same information on differentiable structures as Donaldson's. His report

[†]This material is based upon work supported by the National Science Foundation under agreement No. DMS–0111298. Any opinions, findings and conclusions or recommendations expressed in this material are those of the author and do not necessarily reflect the views of the National Science Foundation.

[**14**] gives a glimpse of the wealth of insights in 4-manifold topology that could be extracted from these invariants.

However, there is more information in the monopole equations than is seen by the Seiberg–Witten invariants. The additional information is due to an interpretation of the monopole equations in terms of equivariant stable homotopy. The fact that certain partial differential equations admit a stable homotopy interpretation is not at all surprising. Indeed, this has been known for decades [**37**]. The good news is that in the case of the monopole equations it actually is possible to make effective use of this fact. The stable homotopy approach to the monopole equations does not only give a different view on known results, but also new insights.

This article is a mixture of a survey and a research article. It serves the multiple aims of introducing to this area of research, carefully outlining its foundations, presenting the known results in a unified framework and, last but not least, proving new results.

The new results concern various improvements to the definition of the refined invariants in [**7**]. Theorem 2.1, for example, specifies a class of nonlinear Fredholm maps between certain infinite dimensional manifolds and shows that the path connected components of the space of all such maps are naturally described by stable cohomotopy groups. This makes it possible to define the refined Seiberg–Witten invariant as the homotopy class of the monopole map in a precise way, clarifying a point left open in [**7**]. The proof also indicates how to avoid ad hoc arguments used in [**6**].

Another improvement is on the assumption $b^+ > b_1 + 1$, which had been necessary in [**7**] for a comparison with Seiberg–Witten invariants. The situation is now summarized in Theorem 4.5. The relation to Seiberg–Witten invariants is clarified without any restriction on b^+ or b_1. This includes in particular the wall-crossing phenomenon in the $b^+ = 1$ case, which had been missing in [**7**], and the case $b^+ = 0$.

1. The Monopole Map

The main part in the story to be told is figured by the monopole map

$$\mu : \mathcal{A} \to \mathcal{C},$$

which is defined for a closed Riemannian 4-manifold X after fixing a K-orientation, or equivalently both an orientation in the usual sense and a spin^c-structure $\mathfrak{s}$. In addition, also a background spin^c-connection has to be fixed. The monopole map then is a fiber preserving map between infinite dimensional vector bundles over the torus

$$\mathrm{Pic}^{\mathfrak{s}}(X) \cong H^1(X;\mathbb{R})/H^1(X;\mathbb{Z}).$$

The refined invariant by definition is the homotopy class of the monopole map in a sense to be made precise in the next chapter. This homotopy class does not depend on the chosen Riemannian metric or the chosen spin^c-connection as these choices are parameterized by connected (indeed contractible) spaces and so, indeed, becomes an invariant of the K-oriented differentiable manifold X (cf. 7.6).

Spinors are a main requisite in the definition of the monopole map. Let us start with the spinor group. The group $\mathrm{Spin}^c(4)$ consists of those pairs (u^+, u^-) of unitary rank two transformations which have the same determinant. If Δ^+ and Δ^- denote the two dimensional unitary representations on which the respective factors act, then the $\mathrm{Spin}^c(4)$-representation $\mathrm{Hom}_{\mathbb{C}}(\Delta^+, \Delta^-)$ admits a real structure. The choice of a basis for the real part H in this representation leads to a surjection $\mathrm{Spin}^c(4) \to SO(4)$ with kernel isomorphic to the group $\mathbb{T}$ of complex numbers of unit length. An element h of H has an adjoint h^* and acts on $\Delta = \Delta^+ \oplus \Delta^-$ via $h(\delta^+, \delta^-) = (-h^*(\delta^-), h(\delta^+))$. This action extends to an action of the Clifford algebra generated by H, resulting in an isomorphism $Cl(H) \otimes_{\mathbb{R}} \mathbb{C} \to \mathrm{End}_{\mathbb{C}}(\Delta)$ of $\mathrm{Spin}^c(4)$-representations. Combining with the complexified inverse to the isomorphism $Cl(H) \to \Lambda(H)$, which maps the product $h_1 h_2$ to $h_1 \wedge h_2 - \langle h_1, h_2 \rangle$, one obtains an isomorphism

$$\Lambda_{\mathbb{C}}(H) \to \mathrm{End}_{\mathbb{C}}(\Delta)$$

of $\mathrm{Spin}^c(4)$-representations. The decomposition $\Delta = \Delta^+ \oplus \Delta^-$ is preserved by elements of $\Lambda^2_{\mathbb{C}}(H)$. The kernel of the induced linear map

$$\rho : \Lambda^2_{\mathbb{C}}(H) \to \mathrm{End}_{\mathbb{C}}(\Delta^+)$$

consists of the anti-selfdual part $\Lambda^-_{\mathbb{C}}(H)$, its image of the traceless endomorphisms. The map ρ preserves the real structure, mapping the real selfdual part $\Lambda^+(H)$ isomorphically to the traceless skew Hermitian endomorphisms of Δ^+.

We may globalize the above identifications of $\mathrm{Spin}^c(4)$-representations to identifications of bundles by taking fibred products with a principal $\mathrm{Spin}^c(4)$-bundle. Particularly interesting are such principal bundles which arise as $\mathrm{Spin}^c(4)$-reductions of the orthonormal oriented frame bundle on an oriented Riemannian 4-manifold X. These are called spin^c-*structures.* In fact, the following data do characterize a spin^c-structure: Rank two Hermitian vector bundles S^+ and S^-, together with isomorphisms $\det(S^+) \cong \det(S^-)$ and $T^*_{\mathbb{C}}X \to \mathrm{Hom}_{\mathbb{C}}(S^+, S^-)$ of Hermitian bundles, the latter isomorphism preserving the real structures. In fact, such spin^c-structures always exist (cf. below). Taking tensor products with Hermitian line bundles results in a free and transitive action of $H^2(X;\mathbb{Z})$ on the set of all spin^c-structures.

There is an interpretation of spin^c-structures which is special to manifolds up to dimension 4: Choosing a spin^c-structure on X is equivalent to choosing a stably almost complex structure, i.e., an endomorphims I on the Whitney sum of the tangent bundle with a trivial rank two bundle over X satisfying $I^2 = -\mathrm{id}$. This is because the natural map

$$BU \to B\,\mathrm{Spin}^c$$

between the respective classifying spaces induces isomorphisms of homotopy groups up to dimension 5. By a theorem of Hirzebruch and Hopf [**26**], there always exists a stably almost complex structure on an oriented 4-manifold. If it comes from an (unstably) almost complex structure, then its second Chern class equals the Euler class of X. Using the equality $c_1^2 - 2c_2 = p_1$ of characteristic classes, we derive as a necessary condition for a stably almost complex structure to be almost complex that its first Chern class satisfies $c_1^2 = 3\,\mathrm{sign}(X) + 2\,e(X)$. If X is connected, this condition is also sufficient [**26**]. In this case the integer

$$k = \frac{c_1^2 - \mathrm{sign}(X)}{4} - (b^+ - b_1 + 1) \tag{1}$$

thus measures, how far a stably almost complex structure is away from being almost complex. Here b_1 denotes the first Betti number of X and $b^+ = \frac{1}{2}(b_2 + \mathrm{sign}(X))$ is the dimension of a maximal linear subspace of the second de Rham group of X on which the cup product pairing is positive definite.

After fixing a background spin^c-connection A, a spin^c-structure on X allows to define a Dirac operator

$$D_A : \Gamma(S^+) \to \Gamma(S^-)$$

mapping positive spinors, i.e., sections of the Hermitian vector bundle S^+, to negative spinors. The local model for the symbol of this operator over a point in X is obtained by identifying the cotangent space with the real part H of $\mathrm{Hom}_{\mathbb{C}}(\Delta^+, \Delta^-)$. At each point in X this symbol is the generator of Bott periodicity, so it provides a K-theory orientation class (cf. [**3**]) for the manifold X. Indeed, any K-theory orientation of X uniquely arises this way. The Dirac operator is complex elliptic. Its index is given by

$$\mathrm{ind}_{\mathbb{C}}(D_A) = \frac{c_1^2 - \mathrm{sign}(X)}{8}. \tag{2}$$

Now fix a spin^c-structure on the 4-manifold X, which from now on will be assumed to be connected unless explicitly stated differently. The gauge group $\mathcal{G} = \mathrm{map}(X, \mathbb{T})$ acts on spinors via multiplication with $u : X \to \mathbb{T}$, on spin^c-connections via addition of $u\,\mathrm{d}\,u^{-1}$. The map sending a pair (A, φ) consisting of a spin^c-connection and a positive spinor to $D_A(\varphi)$ is equivariant with respect to the gauge group. The action of the gauge group on the space of spin^c-connections is not free. However, restriction to the subgroup $\mathcal{G}_0$ consisting of functions which take value 1 at a chosen point in X results in a free action. In particular, the based gauge group $\mathcal{G}_0$ acts freely on the affine linear space $A + i\ \ker(\mathrm{d})$, where d denotes the de Rham differential on one-forms on X, with quotient

$$\mathrm{Pic}^{\mathfrak{s}}(X) \cong H^1(X;\mathbb{R})/H^1(X;\mathbb{Z}).$$

The gauge group acting trivially on forms, we obtain $\mathcal{G}$-spaces

$$\begin{aligned} \widetilde{\mathcal{A}} &= (A + i\ \ker(\mathrm{d})) \times \left(\Gamma(S^+) \oplus H^0(X;\mathbb{R}) \oplus \Omega^1(X)\right), \\ \widetilde{\mathcal{C}} &= (A + i\ \ker(\mathrm{d})) \times \left(\Gamma(S^-) \oplus \Omega^0(X) \oplus H^1(X;\mathbb{R}) \oplus \Omega^+(X)\right) \end{aligned}$$

consisting of spin^c-connections, spinors and forms on X.

Consider the map $\widetilde{\mu} : \widetilde{\mathcal{A}} \to \widetilde{\mathcal{C}}$ defined by

$$(A', \varphi, f, a) \mapsto \left(A', D_{A'}\varphi + ia\varphi, \mathrm{d}^* a + f, a_{harm}, \mathrm{d}^+ a + \sigma(\varphi)\right). \tag{3}$$

Here $\sigma(\varphi)$ denotes the trace free endomorphism $i(\varphi \otimes \varphi^* - \frac{1}{2}||\varphi||^2\ \mathrm{id})$ of S^+, considered via the map ρ as a selfdual 2-form on X. Restricted to forms, the map is familiar from Hodge theory: It is linear, injective with cokernel the space $H^+(X;\mathbb{R})$ of harmonic selfdual two-forms on X. The map $\widetilde{\mu}$ is equivariant with respect to the action of $\mathcal{G}$. Dividing by the free action of the pointed gauge group we obtain the monopole map

$$\mu = \widetilde{\mu}/\mathcal{G}_0 : \mathcal{A} \to \mathcal{C}$$

as a fiber preserving map between the bundles $\mathcal{A} = \widetilde{\mathcal{A}}/\mathcal{G}_0$ and $\mathcal{C} = \widetilde{\mathcal{C}}/\mathcal{G}_0$ over $\mathrm{Pic}^{\mathfrak{s}}(X)$. The preimage of the section $(A', 0, 0, 0, -F^+_{A'})$ of $\mathcal{C}$, divided by the residual $\mathbb{T}$-action, is called the *moduli space of monopoles.*

For a fixed $k > 2$, consider the fiberwise L^2_k Sobolev completion $\mathcal{A}_k$ and the fiberwise L^2_{k-1} Sobolev completion $\mathcal{C}_{k-1}$ of $\mathcal{A}$ and $\mathcal{C}$. The monopole map extends to a continuous map $\mathcal{A}_k \to \mathcal{C}_{k-1}$ over $\mathrm{Pic}^{\mathfrak{s}}(X)$, which will also be denoted by μ.

We will use the following properties of the monopole map:

1.1. *It is $\mathbb{T}$-equivariant.*

1.2. *Fiberwise, it is the sum $\mu = l + c$ of a linear Fredholm map l and a nonlinear compact operator c.*

1.3. *Preimages of bounded sets are bounded.*

Equivariance is immediate. The action is the residual action of the subgroup $\mathbb{T}$ of gauge transformations which are constant functions on X. This group acts by complex multiplication on the spaces $\Gamma(S^{\pm})$ of sections of complex vector bundles and trivially on forms.

Restricted to a fiber, the monopole map is a sum of the linear Fredholm operator l, consisting of the elliptic operators D_A and $\mathrm{d}^* + \mathrm{d}^+$, complemented by projections to and inclusions of harmonic forms. The nonlinear part of μ is built from the bilinear terms $a\varphi$ and $\sigma(\varphi)$. Multiplication $\mathcal{A}_k \times \mathcal{A}_k \to \mathcal{C}_k$ is continuous for $k > 2$. Combined with the compact restriction map $\mathcal{C}_k \to \mathcal{C}_{k-1}$ we gain the claimed compactness for c: Images of bounded sets are contained in compact sets.

Compact perturbations $l + c : \mathcal{U}' \to \mathcal{U}$ of linear Fredholm maps between Hilbert spaces enjoy a nice topological property: The restriction to any bounded, closed subset is proper. The argument is straightforward: Let p denote a projection to the kernel of l. Then the restriction of $l+c$ to a closed subset $A \subset \mathcal{U}'$ factors through an injective and closed and thus proper map $A \to \mathcal{U} \times \overline{c(A)} \times \overline{p(A)}$, $a \mapsto (l(a), c(a), p(a))$, a homeomorphism $(u, s, e) \mapsto (u + s, s, e)$ and the projection to $\mathcal{U}$, which is proper as the two other factors are compact.

If the bundles $\mathcal{A}$ and $\mathcal{C}$ were finite dimensional, then the boundedness property would be equivalent to properness. In this infinite dimensional setting, the argument above can be used the same way as Heine-Borel in the finite dimensional case to show that the boundedness condition implies properness. It turns out that the ingredients of the compactness proof for the moduli space [**44**] also prove the stronger boundedness property [**7**]: The Weitzenböck formula for the Dirac operator associated to the connection $A' = A + ib + ia$ reads

$$D^*_{A'} D_{A'} = \nabla^*_{A'} \nabla_{A'} + \frac{1}{4}s + \frac{1}{2}F^+_{A'}.$$

Applying the Laplacian $\Delta\,|\varphi|^2$ to the spinor part of an element $(A + ib, \varphi, f, a)$ in the preimage of μ leads to an estimate

$$\Delta|\varphi|^2 \;\leqslant\; 2\,\langle D^*_{A'}D_{A'}\varphi - \frac{s}{4}\varphi - \frac{1}{2}F^+_{A'}\varphi,\; \varphi\rangle.$$

The crucial point is that the term $F^+_{A'}$ can be replaced by an expression involving $\sigma(\varphi)$ and terms which are straightforward to estimate. The Laplacian at the maximum is nonnegative. Use of this fact and standard elliptic and Sobolev estimates then lead to an estimate

$$||\varphi||^4_\infty \leqslant P(||\varphi||_\infty)$$

with a polynomial P of order 3. The boundedness property follows easily from this.

2. Enter Stable Homotopy

In case the first Betti number of X vanishes, the monopole map is a map between Hilbert spaces. The boundedness property (1.3) of μ is equivalent to the statement that μ extends continuously to a map $S^{\mathcal{A}} \to S^{\mathcal{C}}$ between the one-point completions, where the neighborhoods of the points at infinity are the complements of bounded sets. As spaces, these one-point completions are infinite dimensional spheres. The monopole map thus rightly may be considered as a continuous map between spheres.

In the general case, we use a trivialization $\mathcal{C} \cong \mathrm{Pic}^s(X) \times \mathcal{U}$ of the bundle $\mathcal{C}$ to compose the monopole map with the projection p to the fiber $\mathcal{U}$. Now the boundedness property of μ translates as follows: The map $p \circ \mu$ extends continuously to a map $T\mathcal{A} \to S^{\mathcal{U}}$ from the Thom space of $\mathcal{A}$ to the sphere $S^{\mathcal{U}}$.

The idea of the refined invariant is to take the homotopy classes of these one-point completed maps. As it stands, this idea is of course nonsense: All the spaces involved are contractible, even equivariantly. So there is no interesting homotopy theory.

However, not all is lost. Restriction to maps satisfying not only (1.3), but also property (1.2) actually does the trick. We will consider the situation in a slightly more general setup.

Let $\mathcal{E}$ and $\mathcal{F}$ be infinite dimensional Hilbert space bundles over a compact base B. The structure group is the orthogonal group with its norm topology. Consider the set $\mathcal{P}(\mathcal{E}, \mathcal{F})$ of fiber-preserving continuous maps $\varphi : \mathcal{E} \to \mathcal{F}$ satisfying (1.2) and (1.3). Let us equip $\mathcal{P}(\mathcal{E}, \mathcal{F})$ with the topology induced by the metric

$$d(\varphi, \psi) = \sup_{e \in \mathcal{E}} \|j\varphi(e) - j\psi(e)\|,$$

where $j : \mathcal{F} \to \mathbb{R} \times \mathcal{F}$ denotes the embedding $f \mapsto (1 + f^2)^{-1}(1 - f^2, 2f)$ into the unit sphere bundle in $\mathbb{R} \times \mathcal{F}$ over B. (Actually, there are various topologies on $\mathcal{P}(\mathcal{E}, \mathcal{F})$ for which the following theorem is true; the choice made here is just to be definite.) Choosing a trivialization $\mathcal{F} \cong B \times \mathcal{U}$ of the bundle $\mathcal{F}$, the path components of $\mathcal{P}(\mathcal{E}, \mathcal{F})$ roughly can be described through a bijection

$$\pi_0(\mathcal{P}(\mathcal{E}, \mathcal{F})) \cong \coprod_{\alpha \in KO(B)} \pi^0_{\mathcal{U}}(B; \alpha).$$

This description uses stable cohomotopy groups of B with "twisted coefficients." These groups need some explanation and as it stands, the statement is rather imprecise. "For the purposes of planning strategy" ([**1**]) it is useful, to think of this decomposition as presented over the group $KO(B)$. For the purpose of rigorous definitions and proofs, much more care has to be taken.

Let us start from the beginning, from pointed spaces. The prototype of a topological space with a distinguished base point, usually denoted by $*$, is the one-point compactification S^U of a finite dimensional real vector space U with the point at infinity as base point. The smash product $A \wedge C$ of pointed spaces is the quotient of their product obtained by identifying $A \times \{*\} \cup \{*\} \times C$ to a point. In this way $S^U \wedge S^V$ is canonically homeomorphic to $S^{U \oplus V}$. The sphere $S^{\mathbb{R}^n}$ is usually denoted by S^n. The smash product with S^1 induces a functor from pointed spaces to pointed spaces, called *suspension*.

According to Freudenthal's suspension theorem, which holds for finite dimensional spaces, iterated suspensions eventually induce isomorphisms of sets of pointed homotopy classes

$$[S^n \wedge A, S^n \wedge C] \to [S^{n+1} \wedge A, S^{n+1} \wedge C].$$

The notion of a spectrum arose from the desire to define a category in which the elements of the resulting Abelian group

$$\operatorname*{colim}_{n \to \infty} [S^n \wedge A, S^n \wedge C]$$

appear as homotopy classes of maps between the objects. There are various ways to construct such categories. The situation suggests to use the Spanier–Whitehead category indexed by a universe: Objects and morphisms in this category are defined through colimit constructions. The index category consists of the finite dimensional linear subspaces of an infinite dimensional real Hilbert space $\mathfrak{U}$, called *universe*, with inclusions as morphisms. So an object A in the Spanier–Whitehead category associates to $U \subset \mathfrak{U}$ a pointed space A_U. To relate these spaces, we use the inclusion $U \subset W$ to identify W with $V \oplus U$, where V is the orthogonal complement to U in W. The collection of spaces A_U comes with identifications

$$\sigma_{U,W} : S^V \wedge A_U \to A_W \tag{4}$$

satisfying the obvious compatibility condition

$$\sigma_{U,W'} = \sigma_{W,W'} \circ (\mathrm{id}_{S^{V'}} \wedge \sigma_{U,W}) \tag{5}$$

for $W' = V' \oplus W \subset \mathfrak{U}$. The morphism set in the Spanier–Whitehead category is the colimit

$$\{A, C\}_{\mathfrak{U}} = \operatorname*{colim}_{U \subset \mathfrak{U}} [A_U, C_U].$$

over the maps

$$[A_U, C_U] \stackrel{\mathrm{id}_{S^V} \wedge -}{\to} [S^V \wedge A_U, S^V \wedge C_U] \leftrightarrow [A_W, C_W].$$

The latter identification is induced by the identifications $\sigma^A_{U,W}$ and $\sigma^C_{U,W}$.

Every (homotopy) category of spectra is supposed to contain some variant of the Spanier–Whitehead category as a full subcategory. So it should do no harm to call the objects spectra. It should, however, be pointed out that some authors reserve the name spectrum to objects in more elaborate categories.

Any pointed space A canonically defines its suspension spectrum, denoted by A as well, by setting $A_U = S^U \wedge A$.

To define objects in the Spanier–Whitehead category, it of course suffices to define them for a cofinal indexing category, as for example the subcategory of finite dimensional linear subspaces of $\mathfrak{U}$ containing a fixed subspace U. So associating for a given pointed space A to $W = V \oplus U \subset \mathfrak{U}$ the space $S^V \wedge A$ defines a spectrum different from A. We may safely denote this desuspension by $\Sigma^{-U} A$.

Let $p : \mathcal{F} \cong B \times \mathfrak{U} \to \mathfrak{U}$ be a trivialization and suppose $l : \mathcal{E} \to \mathcal{F}$ is a continuous, fiberwise linear Fredholm map. Let $U \subset \mathfrak{U}$ denote a finite dimensional linear subspace such that the index of l is represented by the difference $E - \underline{U}$ of finite dimensional vector bundles on B. Here

$\underline{U}$ denotes the trivial vector bundle $p^{-1}(U)$ and $E = l^{-1}(\underline{U})$. The one-point compactification TE of E is called *Thom space* of E. The Thom spectrum is defined as $T(\text{ind } l) = \Sigma^{-U}TE$. With this notation, stable cohomotopy with twisting ind l may be defined by

$$\pi^0_{\mathcal{U}}(B; \text{ind } l) := \{T(\text{ind } l), S^0\}_{\mathcal{U}}.$$

Such twisted cohomotopy groups are a natural habitat for Euler classes of vector bundles. To explain this, let F be a finite dimensional vector bundle over B. Choosing a bundle isomorphism $E \oplus F \cong \underline{U}$ and a section σ of F, this section and the projection to fibers together define a map $\sigma + \text{id}_E$ extending continuously to one-point compactifications $TE \to S^U$. This map then represents the stable cohomotopy Euler class $e(F) \in \pi^0_{\mathcal{U}}(B; -F)$.

The relation to the Euler class of a bundle in a multiplicative cohomology theory h is as follows [8]: A Thom class $u \in h^r(B; F) = h^r(TF, *)$ corresponds to an h-orientation of F. The h-theoretic Euler class is defined by $e_h(F) = \sigma^*(u) \in h^r(B)$. A generator $1 \in \widetilde{h}^0(S^0)$ gives rise to the Hurewicz map $\pi^0(B; -F) \to h^0(B; -F)$, which associates to a stable pointed map $\varphi : T(-F) \to S^0$ the element $h^0(\varphi)(1)$. Using the product pairing $h^0(B; -F) \times h^r(B; F) \to \widetilde{h}^r(B)$, the h-theoretic Euler class and the stable cohomotopy one are related by

$$e_h(F) = h^0(e(F))(1) \cdot u.$$

To formulate the theorem, let us introduce for a fixed fiberwise linear Fredholm operator $l : \mathcal{E} \to \mathcal{F}$ the subspace $\mathcal{P}_l(\mathcal{E}, \mathcal{F})$ of $\mathcal{P}(\mathcal{E}, \mathcal{F})$ consisting of elements φ such that $\varphi - l$ is fiberwise compact.

Theorem 2.1. *A projection $p : \mathcal{F} \cong B \times \mathcal{U} \to \mathcal{U}$ induces a natural bijection*

$$\pi_0(\mathcal{P}_l(\mathcal{E}, \mathcal{F})) \cong \pi^0_{\mathcal{U}}(B; \text{ind } l).$$

The theorem also handles homotopies by applying it to the base space $B \times [0, 1]$. Note that the restriction maps

$$\pi^0_{\mathcal{U}}(B \times [0, 1]; \text{ind } l) \to \pi^0_{\mathcal{U}}(B \times \{i\}; \text{ind } l|_{B \times \{i\}})$$

are isomorphisms. So if for example $\varphi = l + c = l' + c'$ are two different presentations as a sum, then the constant homotopy $\varphi = \varphi_t = (1-t)(l + c) + t(l' + c')$ can be used to identify $\pi^0_{\mathcal{U}}(B; \text{ind } l)$ with $\pi^0_{\mathcal{U}}(B; \text{ind } l')$. Under this identification, the element associated to the decomposition $l + c$ of φ is mapped to the element associated to the decomposition $l' + c'$.

PROOF OF THEOREM 2.1. Let us briefly sketch a proof of the theorem: An element $\xi \in \pi^0_{\mathcal{U}}(B; \text{ind}\, l)$ is represented by a virtual bundle $E - \underline{U}$ over B, together with a map $TE \to S^U$. It may be necessary to suspend the given map in order that it can be replaced by a homotopic map for which the preimage of the base point consists only of the base point. In particular, ξ then is represented by a proper map $E \to \underline{U}$. The given embedding of E into $\mathcal{E}$ and an identification of the orthogonal complements $E^\perp \subset \mathcal{E}$ and $\underline{U}^\perp \subset \mathcal{F}$, results in an element of $\mathcal{P}_l(\mathcal{E}, \mathcal{F})$.

On the other hand, for a given element $\varphi \in \mathcal{P}_l(\mathcal{E}, \mathcal{F})$, choose a real number $R > 0$ and an ε with $0 < \varepsilon < R$. The boundedness property (1.3) of φ implies that the preimage under $p\varphi$ of the ball of radius R in $\mathcal{U}$ is bounded in $\mathcal{E}$. Using compactness of B, this bounded preimage is mapped by the fiberwise compact operator $p \circ (\varphi - l)$ into a compact subset of $\mathcal{U}$. We may cover this image with finitely many ε-balls, the centers of which generate a finite dimensional vector space $U \subset \mathcal{U}$. After possibly enlarging U, we can assume that the virtual bundle $E - \underline{U}$ with $E = (pl)^{-1}(U)$ represents ind l. The restriction $p\varphi|_E$ by construction misses the sphere $S_R(U^\perp)$ of radius R in the orthogonal complement of $U \subset \mathcal{U}$. This map $p\varphi|_E$ extends to the one-point completions to give a continuous map

$$TE \to S^{\mathcal{U}} \setminus S_R(U^\perp).$$

Composition with a homotopy inverse to the inclusion $S^U \to S^{\mathcal{U}} \setminus S_R(U^\perp)$ defines an element of $\{T(\text{ind}\, l), S^0\}$. It remains of course to be checked that the two constructions lead to well defined maps between the sets in the theorem which are inverse to each other. This is straightforward, but a little tedious. Well-definedness uses the discussion in [7] and in particular lemma 2.3 there. The second construction obviously is left inverse to the first. To show that it is right inverse, one has to construct paths in $\mathcal{P}_l(\mathcal{E}, \mathcal{F})$ from an arbitrary element to an element, which can be "projected" onto the image of the first construction. Such a path is made explicit through the following homotopy φ_t which starts from $\varphi = \varphi_0$ and ends at φ_1. It is constant on a disk bundle of radius Q in $\mathcal{E}$, which contains the preimage of an R-disk bundle in $\mathcal{F}$. Outside it is defined by

$$\varphi_t(e) = \left(\frac{|e|}{Q}\right)^t \varphi\left(\left(\frac{|e|}{Q}\right)^{-t} e\right). \tag{6}$$

The theorem is proved. □

The theorem describes the path-connected components of these mapping spaces in terms of a disjoint union of algebraic objects. So one can hardly expect the algebraic structure to be reflected in the world of Fredholm maps by natural constructions. In particular, addition of elements in the respective cohomotopy groups seems to be difficult to describe in terms of Fredholm maps without making use of the theorem. However, two aspects are inherent in the Fredholm setup.

Remark 2.2. A Fredholm map $\varphi \in \mathcal{P}_l(\mathcal{E}, \mathcal{F})$, for which $p\varphi$ is not surjective, describes the zero element in the stable cohomotopy group associated to its linearization l. To see this, recall from the previous section that φ is proper. In particular, the image of $p\varphi$ is a closed subset in U. If a point $u \in \mathcal{U}$ is in the complement of the image, then so is a whole ε-neighborhood of u. Now apply the construction above for some $R > |u| + \varepsilon$ and replace φ by the homotopic φ_1. We choose U so that it also contains u. The map $p\varphi_1|_E$, followed by the orthogonal projection to U is proper and by construction misses u. So its one-point compactification is null homotopic.

Remark 2.3. The bijection (2.1) respects products: If $\mathcal{E}_i$ and $\mathcal{F}_i$ denote Hilbert space bundles over a base space B_i, then taking products of maps results in a product

$$\mathcal{P}_{l_1}(\mathcal{E}_1, \mathcal{F}_1) \times \mathcal{P}_{l_2}(\mathcal{E}_2, \mathcal{F}_2) \to \mathcal{P}_l(\mathcal{E}_1 \times \mathcal{E}_2, \mathcal{F}_1 \times \mathcal{F}_2)$$

with $l = \mathrm{pr}_1^* l_1 \times \mathrm{pr}_2^* l_2$. This product structure is reflected in the stable cohomotopy counterpart. The natural smash product of objects in the Spanier–Whitehead category is not defined within one universe, but comes with a change of universes: Let A and C be objects in the Spanier–Whitehead categories indexed by universes $\mathcal{U}$ and $\mathcal{V}$. The smash product $A \wedge C$ then is an object in the Spanier–Whitehead category indexed by $\mathcal{U} \oplus \mathcal{V}$. It is defined by

$$(A \wedge C)_{U \oplus V} := A_U \wedge C_V,$$

whenever both sides make sense, which is at least for a cofinal subcategory of the indexing category.

If Fredholm maps $\varphi_i \in \mathcal{P}_{l_i}(\mathcal{E}_i, \mathcal{F}_i)$ represent elements $\xi_i \in \pi^0_{\mathcal{U}_i}(B_i; \mathrm{ind}\, l_i)$, then the product $\varphi_1 \times \varphi_2$ represents the cohomotopy class $\xi_1 \wedge xi_2$, which is an element of the group $\pi^0_{\mathcal{U}_1 \oplus \mathcal{U}_2}(B_1 \times B_2; \mathrm{ind}\, l)$.

3. Some Equivariant Topology

Using equivariant spaces and maps throughout, the above concepts carry over to an equivariant setting in a straightforward manner. An appropriate reference is [**1**].

The action of a compact Lie group G on a pointed G-space A fixes the distinguished base point. If A and C are pointed G-spaces, then the smash product $A \wedge C$ obtains a G-action by restricting the natural $\mathrm{G} \times \mathrm{G}$-action to the diagonal subgroup. A G-universe $\mathfrak{U}$ is a Hilbert space on which G acts via isometries in such a way that an irreducible G-representation, if contained in $\mathfrak{U}$, is so with infinite multiplicity. A complete G-universe contains all irreducible representations.

An object of the G-Spanier–Whitehead category indexed by $\mathfrak{U}$ associates to a finite dimensional representation $U \subset \mathfrak{U}$ a pointed G-space A_U. The morphism set is the colimit

$$\{A, C\}^{\mathrm{G}}_{\mathfrak{U}} = \operatorname*{colim}_{U \subset \mathfrak{U}} [A_U, C_U]^{\mathrm{G}}$$

of G-homotopy classes of equivariant maps. This morphism set is a group if the G-universe $\mathfrak{U}$ contains trivial G-representations.

An equivariant projection $p : \mathcal{F} \cong B \times \mathfrak{U} \to \mathfrak{U}$ need not exist. This may happen for example, if the G-action on the (unpointed) base space B of the G-Hilbert bundle $\mathcal{F}$ is nontrivial or if the fiber over a fixed point in B does not qualify as a universe. If such a projection p exists, it induces a natural bijection

$$\pi_0(\mathcal{P}_l(\mathcal{E}, \mathcal{F})^{\mathrm{G}}) \cong \pi^0_{\mathrm{G},\mathfrak{U}}(B; \mathrm{ind}\ l)$$

as before.

The stable cohomotopy groups in 2.1 and in particular their equivariant counterparts are barely known. To get a rough impression, consider the case of a point $B = \{P\}$. If we choose a universe with a trivial G-action, then the twist ind l is characterized by an integer i and the group $\pi^0_{\mathrm{G},\mathfrak{U}}(P; \mathrm{ind}\ l)$ can be identified with the i-th stable stem $\pi_i^{st}(S^0)$. On the other extreme, if we choose $\mathfrak{U}$ to be a complete universe, then the isomorphism class of ind l gives an element in the representation ring $RO(\mathrm{G})$. In the case where l is an isomorphism, $\pi^0_{\mathrm{G},\mathfrak{U}}(P; 0)$ is isomorphic to the Burnside ring $A(\mathrm{G})$ [**10**, II.8.4]. If G is a finite group, $A(G)$ is the Grothendieck ring of finite G-sets with addition given by disjoint union and multiplication given by product. A point with the trivial G-action on it represents 1.

Understanding the group $\pi^0_{G,\mathcal{U}}(P; \text{ind } l)$ for a virtual representation ind $l = V - W$ of G in any universe $\mathcal{U}$ whatsoever boils down to understanding the homotopy classes of G-maps $f : S^V \to S^W$. Equivariant K-theory provides some information in case both V and W are complex representations. The method is explained in [**10**, II.5]:

Let $K_G(B)$ be the Grothendieck group of equivariant complex vector bundles over the G-space B. For a pointed space B as usual $\widetilde{K}_G(B)$ denotes the kernel of the restriction homomorphism $K_G(B) \to K_G(*) \cong R(G)$. If V is a complex G-representation, then $K_G(V) := \widetilde{K}_G(S^V)$ is a free $R(G)$-module generated by a Bott class $b(V)$ [**2**]. The image of the Bott class $b(U \oplus V)$ under the restriction homomorphism $\widetilde{K}_G(S^{U\oplus V}) \to \widetilde{K}_G(S^U)$ is $e_{K_G}(V)b(U)$. This defines the Euler class, which was determined by Segal [**38**] to be the element

$$e_{K_G}(V) = \sum_{i=0}^{dim V} (-1)^i \Lambda^i(V) \in R(G). \tag{7}$$

A pointed G-map $f : S^V \to S^W$ induces a $R(G)$-linear homomorphism in K_G-theory. The image of the Bott class of W is a multiple $a_G(f)b(V)$ of the Bott class of V. The K_G-theory degree $a_G(f)$ is an element of the complex representation ring $R(G)$.

To determine the K_G-degree as a character on G, we have to evaluate it at elements $g \in G$. Let C denote the closure of the subgroup generated by g. Decompose $V = V_C \oplus V^C$ into the C-fixed point set V^C and its orthogonal complement. The inclusions of the fixed point sets induce a commuting diagram

$$\begin{array}{ccc} K_C(W) & \stackrel{f^*}{\to} & K_C(V) \\ \downarrow & & \downarrow \\ K_C(W^C) & \stackrel{f^{C*}}{\to} & K_C(V^C). \end{array}$$

The lower map is multiplication by the degree of the map f^C as a map between oriented spheres, if V^C and W^C both have the same dimension. Otherwise it is zero. This is because $\widetilde{K}(S^{2n}) \cong \mathbb{Z}$ and $\pi^{st}_{2n-2m}(S^0)$ is torsion for $n \neq m$.

Commutativity of the diagram relates Euler classes and degrees by

$$e_{K_C}(W_C)d(f^C) = a_C(f)e_{K_C}(V_C). \tag{8}$$

The representation V_C does not contain a trivial summand. So the character $e_{K_C}(V_C)$ does not vanish at g. In particular, the K_G-degree can be computed from the ordinary degrees of the restrictions of f to fixed points.

Let us apply this concept to a simple example: Consider the group $\mathbb{T}$ of complex numbers of unit length acting on the representation V with character $z \mapsto n + mz$,

Proposition 3.1. *Let $f : S^{2n} \wedge S^{\mathbb{C}^m} \to S^{2n} \wedge S^{\mathbb{C}^{m+l}}$ be a $\mathbb{T}$ equivariant map such that the restricted map on the fixed points has degree $d \neq 0$. Then $l \geqslant 0$ and in case $l = 0$ the degree of f nonequivariantly on the total space is d as well.*

PROOF. For $z \neq 1$, the above equation (8) reads as follows:

$$(1 - z)^{m+l} d = a_{\mathbb{T}}(f)(z)(1 - z)^m.$$

The function $d(1 - z)^l = a_{\mathbb{T}}(f)(z)$ is a character in $R(\mathrm{G})$ only if $l \geqslant 0$. In case $l = 0$, the $K_{\mathbb{T}}$-degree and hence the (K-)degree equals d. □

Algebraic topology provides quite heavy machinery for the equivariant world. Basic equipment can be found in [**1**], [**10**]. Here is a survival kit:

3.2. *An equivariant isometry $\mathcal{U} \hookrightarrow \mathcal{V}$ induces a change-of-universe morphism $\{\,\_, \_\,\}^G_{\mathcal{U}} \to \{\,\_, \_\,\}^G_{\mathcal{V}}$. It is bijective, if both universes are built from the same irreducible representations.*

3.3. *A cofiber sequence $A' \to A \to A''$ of pointed* G*-spaces induces long exact sequences*

$$\ldots \leftarrow \{S^i \wedge A', C\}^{\mathrm{G}}_{\mathcal{U}} \leftarrow \{S^i \wedge A, C\}^{\mathrm{G}}_{\mathcal{U}} \leftarrow \{S^i \wedge A'', C\}^{\mathrm{G}}_{\mathcal{U}} \leftarrow \{S^{i+1} \wedge A', C\}^{\mathrm{G}}_{\mathcal{U}} \leftarrow \ldots$$

$$\ldots \to \{C, S^i \wedge A'\}^{\mathrm{G}}_{\mathcal{U}} \to \{C, S^i \wedge A\}^{\mathrm{G}}_{\mathcal{U}} \to \{C, S^i \wedge A''\}^{\mathrm{G}}_{\mathcal{U}} \to \{C, S^{i+1} \wedge A'\}^{\mathrm{G}}_{\mathcal{U}} \to \ldots$$

3.4. *Let* $\mathrm{H} < \mathrm{G}$ *be a subgroup of finite index. Then there are natural bijections*

$$\{A, \mathrm{res}^{\mathrm{G}}_{\mathrm{H}} C\}^{\mathrm{H}}_{\mathrm{res}^{\mathrm{G}}_{\mathrm{H}} \mathcal{U}} \leftrightarrow \{(\mathrm{G} \sqcup *) \wedge_{\mathrm{H}} A, C\}^{\mathrm{G}}_{\mathcal{U}},$$

$$\{\mathrm{res}^{\mathrm{G}}_{\mathrm{H}} C, A\}^{\mathrm{H}}_{\mathrm{res}^{\mathrm{G}}_{\mathrm{H}} \mathcal{U}} \leftrightarrow \{C, (\mathrm{G} \sqcup *) \wedge_{\mathrm{H}} A\}^{\mathrm{G}}_{\mathcal{U}}.$$

Here $\mathrm{G} \sqcup *$ *denotes the group* G *with a disjoint base point and* $(\mathrm{G} \sqcup *) \wedge_{\mathrm{H}} A$ *is the orbit space of* $(\mathrm{G} \sqcup *) \wedge A$ *by the action* $(h, (g, a)) \mapsto (gh^{-1}, ha)$. *Here A is a Spanier–Whitehead spectrum for the group* H *and C one for the group* G. *The first adjointness property follows from a corresponding property on the space level. The second Wirthmüller isomorphism can be found in* [**10**, II.6.14]

3.5. *For spaces with free* G*-action, the equivariant cohomotopy is naturally isomorphic to the nonequivariant cohomotopy of the quotient space.*

In more exact diction, this reads: Let A *be finite dimensional* G*-space with a free* G*-action away from the base point and let* C *be a nonequivariant spectrum, indexed by the fixed point universe* $\mathfrak{U}^{\mathrm{G}}$ *of the universe* $\mathfrak{U}$ *indexing the suspension spectrum of* A*. For such objects, there is a natural bijection*

$$\{A, j^*C\}^{\mathrm{G}}_{\mathfrak{U}} \leftrightarrow \{A/\mathrm{G}, C\}_{\mathfrak{U}^{\mathrm{G}}}.$$

Here j^*C *denotes the spectrum obtained from* C*, considered as a spectrum with trivial* G*-action, by change of universe* $\mathfrak{U}^{\mathrm{G}} \hookrightarrow \mathfrak{U}$*. This property is obvious for spaces. The fact that it carries over to the equivariantly stable world follows from a careful analysis of the equivariant suspension theorem: In this situation, it suffices to suspend with trivial representations only to get into the stable range.*

4. Topology of the Monopole Map

The first statement in the following theorem summarizes the previous discussion.

Theorem 4.1 ([7]). *The monopole map* $\mu : \mathcal{A} \to \mathcal{C}$ *defines an element in the equivariant stable cohomotopy group*

$$\pi^0_{\mathbb{T},\mathfrak{U}}\left(\mathrm{Pic}^{\mathfrak{s}}(X); \mathrm{ind}(D) - \underline{H^+(X;\mathbb{R})}\right).$$

For $b^+ > \dim(\mathrm{Pic}^{\mathfrak{s}}(X)) + 1$*, a homology orientation determines a homomorphism of this stable cohomotopy group to* $\mathbb{Z}$*, which maps* $[\mu]$ *to the integer valued Seiberg–Witten invariant.*

Proof. The universe in this statement is explicitly given as the fiber

$$\mathfrak{U} = \Gamma(S^-) \oplus \Omega^0(X) \oplus H^1(X;\mathbb{R}) \oplus \Omega^+(X)$$

of the bundle $\mathcal{C}$. The index of the linearization of the monopole map consists of two summands. The Dirac operator associated to $\mathfrak{s}$ defines a virtual complex index bundle $\mathrm{ind}(D)$ over the Picard torus. The second bundle is the trivial bundle with fiber the b^+-dimensional space of self-dual harmonic forms $H^+(X;\mathbb{R})$. An orientation of $\mathrm{Pic}^{\mathfrak{s}}(X) \times H^+(X;\mathbb{R})$ is called a *homology orientation*.

Let us define the homomorphism of the stable cohomotopy group to $\mathbb{Z}$. An element of the stable cohomotopy group is represented by an equivariant map $f : TE \to S^U$ from the Thom space of a bundle E over $\mathrm{Pic}^{\mathfrak{s}}(X)$, where $E - \underline{U} = \mathrm{ind}\, l$ is the index of the linearization l of μ. Let Ci denote the mapping cone of the inclusion $i : TE^{\mathbb{T}} \to TE$ of the $\mathbb{T}$-fixed point set. In the long exact sequence associated to the cofiber sequence $TE^{\mathbb{T}} \to TE \to Ci$,

$$\pi^{-1}_{\mathbb{T},\mathcal{U}}(\Sigma^{-U}(TE^{\mathbb{T}})) \to \pi^{0}_{\mathbb{T},\mathcal{U}}(\Sigma^{-U}Ci) \to \pi^{0}_{\mathbb{T},\mathcal{U}}(\Sigma^{-U}TE) \to \pi^{0}_{\mathbb{T},\mathcal{U}}(\Sigma^{-U}(TE^{\mathbb{T}})) \tag{9}$$

the first and last term are vanishing because by assumption the dimension of the space $S^1 \wedge TE^{\mathbb{T}}$ is less than the dimension of the $\mathbb{T}$-fixed point sphere in S^U, the difference in dimension being $b^+ - b_1 - 1$. So the map μ can be described by a cohomotopy element of $\Sigma^{-U}Ci$. The Hurewicz image $h(\mu)$ of this element in equivariant Borel-cohomology lies in the relative group $H^0_{\mathbb{T}}(\Sigma^{-U}TE, \Sigma^{-U}TE^{\mathbb{T}})$. The $\mathbb{T}$-action on the pair of spaces $(TE, TE^{\mathbb{T}})$ is relatively free. So its equivariant cohomology group identifies with the singular cohomology $H^*(TE/\mathbb{T}, TE^{\mathbb{T}})$ of the quotient. After replacing $TE^{\mathbb{T}}$ by a tubular neighborhood, this is the singular cohomology of a connected manifold relative to its boundary. An orientation of $\mathrm{Pic}^{\mathfrak{s}}(X)$ together with the standard orientation of complex vector bundles defines an orientation class $[TE]_{\mathbb{T}}$ in the top cohomology of this manifold. Similarly, the chosen homology orientation of X and the orientation of $\mathrm{Pic}^{\mathfrak{s}}(X)$ determine the orientation of U and thus a generator $\Sigma^{-U}[TE]_{\mathbb{T}}$ of the graded cohomology group $H^*_{\mathbb{T}}(\Sigma^{-U}TE, \Sigma^{-U}TE^{\mathbb{T}})$ in its top grading $* = k$. This cohomology group is a graded module over the polynomial ring $H^*_{\mathbb{T}}(*) \cong \mathbb{Z}[t]$ in one variable t of degree 2. The homomorphism sought for is zero if k is odd or negative. Otherwise $t^{\frac{k}{2}}h(\mu)$ is a multiple of the generator $\Sigma^{-U}[TE]_{\mathbb{T}}$. This multiplicity is the Seiberg–Witten invariant. □

To see what happens in the cases $b^+ \leqslant b_1 + 1$ not covered by this theorem we have to take a closer look at the monopole map and distinguish different cases.

4.1. The case $b^+ = 0$. The choice of a point $P \in \mathrm{Pic}^{\mathfrak{s}}(X)$ induces a restriction map

$$\pi^0_{\mathbb{T},\mathcal{U}}(\mathrm{Pic}^{\mathfrak{s}}(X); \mathrm{ind}\, l) \to \pi^0_{\mathbb{T},\mathcal{U}}(P; \mathrm{ind}\, l) \cong \{S^{\mathrm{ind}(D)}, S^{H^+(X;\mathbb{R})}\}^{\mathbb{T}}_{\mathcal{U}}.$$

The index of the Dirac operator ind(D) consists of $d = \frac{1}{8}(c^2 - \text{sign}(X))$ copies of the tautological complex $\mathbb{T}$-representation. The restriction to the $\mathbb{T}$-fixed point set of an element in this group is an element in the stable stem $\pi^{st}_{-b^+}(S^0)$, which is trivial except in the case $b^+ = 0$. In this case the restriction of the monopole map is a linear isomorphism on the fixed point set. Here is an immediate consequence, well known from Seiberg–Witten theory:

Proposition 4.2. *Let X be an oriented 4-manifold with $b^+ = 0$. Then the first Chern class of any K-orientation on X satisfies*

$$c^2 \leqslant \text{sign}(X).$$

PROOF. Otherwise the monopole map represented an element in $\{S^{\mathbb{C}^d}, S^0\}^{\mathbb{T}}_{\mathcal{U}}$ for some $d > 0$ which is of degree 1 on the $\mathbb{T}$-fixed point set. The existence of such an element contradicts 3.1. □

Applying Elkies' theorem [**17**], we obtain as a corollary Donaldson's theorem:

Theorem 4.3 ([**11**], [**12**]). *Let X be a closed oriented four-manifold with negative definite intersection form. Then the intersection pairing on $H_2(X;\mathbb{Z})/$ Torsion is diagonal.* □

4.2. The case $b^+ = 1$, $b_1 = 0$. In the case $b^+ = 1$, the Seiberg–Witten invariants depend in a well understood manner ([**44**], [**30**]) on the Riemannian metric and on an additional perturbation parameter. To understand the phenomenon, let us illustrate it in a characteristic example. This example describes the situation in the case of an almost complex manifold with $b_1(X) = 0$, cf. [**7**]:

View the spinning globe as a two-sphere with an $\mathbb{T}$-action and choose the north pole as a base point. As a target space, take a one-sphere with trivial action and choose two points on this one-sphere as "poles," the north pole again as base point. Based equivariant maps from the spinning globe to the one-sphere are determined by their restriction to a latitude, which as an arc is a contractible space. So there is only the trivial homotopy class of equivariant such maps.

In contrast, consider equivariant maps, which take north and south pole to north and south pole respectively. The monopole maps for all choices of metrics and background connections actually are of this type. Such a map basically wraps a latitude $n + 1/2$ times around the one-sphere. Choosing a generic point in the one-sphere, the oriented count of preimages in a fixed latitude defines in a natural way a map of the set of relative homotopy classes to the integers. This oriented count, however, depends on the choice of the generic point. It changes by ± 1, if the generic point is chosen in the "other half" of the one-sphere.

There are two ways to deal with the problem. If one prefers to have homotopy classes, one may consider the monopole map up to equivariant homotopy relative to the fixed point. The monopole map then is a well defined element in the set of all such homotopy classes. However, this set will not be a group anymore. There is a comparison map to the integers depending upon the choice of a "chamber." An alternative is described below.

4.3. The case $b^+ > 1$, $b_1 \neq 0$. The restriction of the monopole map to the $\mathbb{T}$-fixed point set $\mathcal{A}^{\mathbb{T}} = \mathrm{Pic}^{\mathfrak{s}}(X) \times (H^0(X;\mathbb{R}) \oplus \Omega^1(X))$ is a product map. On the factor $\mathrm{Pic}^{\mathfrak{s}}(X)$ it is the identity, on the second factor it is a linear embedding with cokernel $H^+(X;\mathbb{R})$. To take this into account, we would like to construct an equivariant spectrum Q encoding this information. Ideally, this spectrum would be obtained by a push-out of two maps. The one map describes the inclusion $\mathcal{A}^{\mathbb{T}} \to \mathcal{A}$ of the fixed point set, the other the projection $\mathcal{A}^{\mathbb{T}} \to H^0(X;\mathbb{R}) \oplus \Omega^1(X)$. Such a push-out seems not available in the category we are working in. Let us try to define a substitute. Suppose $E - \underline{U}$ for $U \subset \Gamma(S^-)$ represents the index of the Dirac operator as a virtual complex bundle over $\mathrm{Pic}^{\mathfrak{s}}(X)$. Then let $TE/\mathrm{Pic}^{\mathfrak{s}}(X)$ denote the quotient of the Thom space TE, where the subspace $\mathrm{Pic}^{\mathfrak{s}}(X)$, the image of the zero section in E, is identified to a point. Alternatively, $TE/\mathrm{Pic}^{\mathfrak{s}}(X)$ is described as the unreduced suspension of the unit sphere bundle in E. As a $\mathbb{T}$-space it has two fixed points. The spectrum $Q(X, \mathfrak{s}, U)$ then is defined by

$$Q(X,\mathfrak{s},U) = \Sigma^{-U-H^+(X;\mathbb{R})}(TE/\mathrm{Pic}^{\mathfrak{s}}(X)) .$$

Using this spectrum, we obtain a straightforward sharpening of 4.1:

Proposition 4.4. *For sufficiently large $U \subset \Gamma(S^-)$, the monopole map $\mu : \mathcal{A} \to \mathcal{C}$ defines an element in the equivariant stable cohomotopy group $\pi^0_{\mathbb{T},\mathcal{U}}(Q(X,\mathfrak{s},U))$. For $b^+ > 1$, a homology orientation determines a homomorphism of this stable cohomotopy group to $\mathbb{Z}$, which maps $[\mu]$ to the integer valued Seiberg–Witten invariant.*

Proof is a slight variation of that of 4.1. To construct the homomorphism to $\mathbb{Z}$, use the cofiber sequence $S^0 \to TE/\mathrm{Pic}^{\mathfrak{s}}(X) \to Ci$ of spaces.

The outer terms in the analogue of sequence (9) now are vanishing for $b^+ > 1$ for dimension reasons. $\square$

One can prove that for U big enough the groups $\pi^0_{\mathbb{T},\mathcal{U}}(Q(X,\mathfrak{s},U))$ become isomorphic. The description, however, still does not look satisfactory.

4.4. The case $b^+ = 1$, $b_1 \neq 0$. Consider the cofiber sequence $S^0 \to TE/\mathrm{Pic}^{\mathfrak{s}}(X) \to Ci$ in the proof of 4.4 and set $W = U + H^+(X;\mathbb{R}) \subset \mathcal{U}$. This leads to the analogue of the exact sequence (9)

$$\pi^{-1}_{\mathbb{T},\mathcal{U}}(\Sigma^{-W}S^0) \to \pi^0_{\mathbb{T},\mathcal{U}}(\Sigma^{-W}Ci) \to \pi^0_{\mathbb{T},\mathcal{U}}(Q(U)) \to \pi^0_{\mathbb{T},\mathcal{U}}(\Sigma^{-W}S^0)$$

The last term in this sequence vanishes, but the first term is isomorphic to $\mathbb{Z}$. As in the proof of 4.4, the Seiberg–Witten construction describes a homomorphism $\pi^0_{\mathbb{T},\mathcal{U}}(\Sigma^{-W}Ci) \to \mathbb{Z}$. The choice of a "chamber" in computing the Seiberg–Witten invariant amounts to the choice of a null homotopy of the restriction $S^0 \to S^{H^+(X;\mathbb{R})}$ of the monopole map to the fixed point set. Such a null homotopy gives rise to a lift of the class of the monopole map to an element in $\pi^0_{\mathbb{T},\mathcal{U}}(\Sigma^{-W}Ci)$. The wall-crossing formulas mentioned above can be understood as describing the degree of the composite map

$$\mathbb{Z} \cong \pi^{-1}_{\mathbb{T},\mathcal{U}}(\Sigma^{-W}S^0) \to \pi^0_{\mathbb{T},\mathcal{U}}(\Sigma^{-W}Ci) \to \mathbb{Z}.$$

4.5. Summary. To summarize the preceding discussion, let $\pi^0_{\mathbb{T},\mathcal{U}}(Q,Q^{\mathbb{T}})$ denote the group $\mathrm{colim}_{U\subset\Gamma(S^-)}\,\pi^0_{\mathbb{T},\mathcal{U}}(\Sigma^{-W}Ci)$. Note that the spectrum $\Sigma^{-W}Ci$ depends on the chosen presentation $E - \underline{U}$ for the virtual index bundle over $\mathrm{Pic}^{\mathfrak{s}}(X)$. However, the group above by construction is independent of the chosen linear subspace $U \subset \mathcal{U}$. The groups $\pi^0_{\mathbb{T},\mathcal{U}}(Q(U))$ for big enough U become isomorphic, but not in a natural way. When writing $\pi^0_{\mathbb{T},\mathcal{U}}(Q)$, we tacitly fix some large $U \subset \Gamma(S^-)$.

Theorem 4.5. *The monopole map $\mu : \mathcal{A} \to \mathcal{C}$ for an oriented 4-manifold X with* spin^c*-structure $\mathfrak{s}$ defines an element in the equivariant stable cohomotopy group $\pi^0_{\mathbb{T},\mathcal{U}}(Q(X,\mathfrak{s}))$, which fits into an exact sequence*

$$\pi_1^{st}(S^{H^+(X;\mathbb{R})}) \xrightarrow{\alpha} \pi^0_{\mathbb{T},\mathcal{U}}(Q(X,\mathfrak{s}), Q(X,\mathfrak{s})^{\mathbb{T}}) \xrightarrow{\beta} \pi^0_{\mathbb{T},\mathcal{U}}(Q(X,\mathfrak{s})) \xrightarrow{\gamma} \pi_0^{st}(S^{H^+(X;\mathbb{R})}).$$

The Seiberg–Witten homomorphism $h : \pi^0_{\mathbb{T},\mathcal{U}}(Q(X,\mathfrak{s}), Q(X,\mathfrak{s})^{\mathbb{T}}) \to \mathbb{Z}$ is determined by the choice of a homology orientation and relates the *monopole class to the integer valued Seiberg–Witten invariant* $h\beta^{-1}([\mu])$ *in case* $b^+ > 1$. *For* $b^+ = 1$, *the choice of a chamber determines a lift* $[\mu]^{rel} \in \beta^{-1}([\mu])$ *and* $h([\mu]^{rel})$ *is the corresponding Seiberg–Witten invariant. The degree of the map* $h\alpha$ *describes the effect of wall-crossing on the Seiberg–Witten invariant. In case* $b^+ = 0$, *finally,* $\gamma([\mu]) = 1$.

5. Kähler, Symplectic, and Almost Complex Manifolds

The current knowledge about differentiable structures on four-dimensional manifolds builds on the fact that the gauge theoretic invariants are closely related to the Cauchy-Riemann equations. Witten explained how in the case of Kähler surfaces Seiberg–Witten invariants can be determined by complex analytic methods. Taubes modified the arguments for the case of symplectic manifolds. Various mathematicians consequently studied Seiberg–Witten invariants for Kähler and symplectic manifolds. Cutting-and-pasting methods were developed to transfer these computations to other almost complex manifolds. These efforts resulted in a diverse and fascinating picture.

The refined invariants have little to add to this direction in four-manifold theory. This section intends to explain why. For the sake of brevity, let us focus on central aspects and let us assume $b^+ > 1$ in this section. As noted in the first section, a spin^c-structure is equivalent to a stably almost complex structure on the tangent bundle of a four-manifold. In particular, an almost complex manifold comes with a canonical spin^c-structure $\mathfrak{s}_{can}$. Any other spin^c-structure on the underlying oriented 4-manifold is of the form $\mathfrak{s}_{can} \otimes L$ for some $L \in H^2(X;\mathbb{Z})$, represented by a line bundle on X. With this convention the first Chern class of $\mathfrak{s}_{can}$ is minus the first Chern class K_X of the cotangent bundle.

Theorem 5.1 ([44], [40]). *Let X be a symplectic four-manifold with* $b^+ > 1$. *The Seiberg–Witten invariant for the canonical* spin^c*-structure* $\mathfrak{s}_{can}$ *is* ± 1. *Furthermore, Serre-duality holds in the following form:*

$$SW(\mathfrak{s}_{can} \otimes L) = \pm SW(\mathfrak{s}_{can} \otimes (K_X - L)).$$

Theorem 5.2 ([44], [41]). *Let X be a symplectic four-manifold with* $b^+ > 1$. *If for some* $L \in H^2(X;\mathbb{Z})$ *the Seiberg–Witten invariant of*

$\mathfrak{s}_{can} \otimes L$ is nonvanishing, then this spin^c*-structure corresponds to an almost complex structure.*

Witten and Taubes actually prove more than is stated in these theorems: The monopole map is not surjective, unless there is a pseudoholomorphic curve in X which is Poincaré dual to the class L. The result follows by the application of adjunction inequalities [**34**]. By Remark 2.2, we get as an immediate consequence:

Corollary 5.3. *Let X be an oriented four-manifold with $b^+ > 1$, which admits a symplectic structure. If the stable cohomotopy invariant $[\mu] \in \pi^0_{\mathbb{T},\mathcal{U}}(Q)$ in 4.4 is nonvanishing for some* spin^c*-structure $\mathfrak{s}$ on X, then $\mathfrak{s}$ describes an almost complex structure on X.*

The refined invariants, when applied to symplectic manifolds, carry exactly the same information as the Seiberg–Witten invariants. This is a consequence of 5.3 and the following statement.

Proposition 5.4. *Let X be an almost complex four-manifold with $b^+ > 1$. Then the homomorphism $\pi^0_{\mathbb{T},\mathcal{U}}(Q) \to \mathbb{Z}$ in 4.4 comparing the Seiberg–Witten invariant with its refinement is an isomorphism.*

PROOF. For an almost complex 4-manifold, the "virtual dimension of the moduli space" k is zero (1). The construction of the comparison homomorphism in 4.1 and 4.4 considers a map from a pair $(TE/\mathbb{T}, TE^{\mathbb{T}})$ of spaces to a sphere. The integer k is exactly the difference of the dimensions of $TE/\mathbb{T}$ and the sphere. The dimensions being equal and $(TE/\mathbb{T}, TE^{\mathbb{T}})$ being a connected and oriented manifold relative to its boundary, one can apply a classical theorem of Hopf. It states that the homotopy classes of such maps are classified by their degree. □

So in order to test, whether the refined invariants are of any use, we have to leave the by now familiar world of symplectic or at least almost complex 4-manifolds and enter the jungle.

6. Some Stable Cohomotopy Groups

The groups $\pi^0_{\mathbb{T},\mathcal{U}}(\mathrm{Pic}^{\mathfrak{s}}(X); \mathrm{ind}\, \mathrm{l})$ seem to be at least as hard to compute as the stable homotopy groups of spheres. Let us restrict to the simplest cases. In particular, let us only consider 4-manifolds X with vanishing first Betti number and $b^+ > 1$. The groups then are then determined by the index of the linearization l of the monopole map. We will write $\pi^0_{\mathbb{T},\mathcal{U}}(\operatorname{ind} l)$ for short. The index of the Dirac operator is denoted by $d = \operatorname{ind}_{\mathbb{C}}(D) = \dfrac{c^2 - \operatorname{sign}(X)}{8}$. The virtual dimension (1) of the moduli space is $k = 2d - b^+ - 1$.

Proposition 6.1 ([7]). *Let X be a K-oriented, closed 4-manifold with vanishing first Betti number and $b^+ > 1$. The stable equivariant cohomotopy group $\pi^0_{\mathbb{T},\mathcal{U}}(\operatorname{ind} l)$ is isomorphic to the nonequivariant stable cohomotopy group $\pi^{b^+-1}(P(\mathbb{C}^d))$ of the complex $(d-1)$-dimensional projective space. This group vanishes for $k < 0$. It is isomorphic to $\mathbb{Z} \oplus A(k,d)$, if $k \geqslant 0$ is even, and to $A(k,d)$ otherwise. Here $A(k,d)$ denotes a finite Abelian group. For any prime p, the p-primary part of $A(k,d)$ vanishes for $k < 2p-3$. For $k \leqslant 4$, the groups $A(k,d)$ can be described as follows:*

- *$A(0,d) \cong A(4,d) = 0$.*
- *$A(1,d) \cong A(2,d)$. For even d these groups are isomorphic to $\mathbb{Z}/2$, otherwise they vanish.*
- *The 2-primary part of $A(3,d)$ is a cyclic group, depending on the congruence class of d modulo 8. The order of the group is 8, 0, 2, 4, 4, 0, 2, 2 for the congruence classes 0, 1, 2, ...*
- *The 3-primary part of $A(3,d)$ is of order 3 if d is divisible by 3 and else vanishes.*

The proof of the first statement uses the sequence (9), which in this situation by excision is a part of the long exact cohomotopy sequence for the pair $(D(\mathbb{C}^d)\sqcup *, S(\mathbb{C}^d)\sqcup *)$ consisting of the unit ball and its bounding sphere in the complex vector space $\mathbb{C}^d$ with an extra base point added. The $\mathbb{T}$-action on the sphere is free, so we may apply (3.5) to get the result. The Atiyah–Hirzebruch spectral sequence accounts for the rest of the statement.

Instead of chasing through technicalities, let us try to understand in an informal way, how to represent elements in these groups for small k. Recall the structure of the stable homotopy groups of spheres in low dimensions. The group $\pi^{st}_n(S^0)$ is cyclic for $n \leqslant 5$. It is infinite for $n = 0$, of order 2 for $n = 1$ or 2, of order 24 for $n = 3$ and zero else in this range. For $n = 1$ and 3, these groups are generated by Hopf maps $S(\mathbb{F}^2) \to P(\mathbb{F}^2)$ for $\mathbb{F} = \mathbb{C}$ and $\mathbb{F} = \mathbb{H}$, denoted by η and ν. These generators satisfy the relation $\eta^3 = 12\nu$.

First consider the map obtained by forgetting the $\mathbb{T}$-action. This homomorphism

$$f : \pi^0_{\mathbb{T},\mathcal{U}}(\operatorname{ind} l) \to \pi_{k+1}(S^0)$$

associates to a $\mathbb{T}$-equivariant map between $\mathbb{T}$-representation spheres its underlying nonequivariant map. In the case $k = 0$, $d = 2$, the group

$$\pi^0_{\mathbb{T},\mathcal{U}}(\operatorname{ind} l) \cong \{S^{\mathbb{C}^2}, S^3\}^{\mathbb{T}}_{\mathcal{U}} \cong \pi^2(P(\mathbb{C}^2)) \cong \mathbb{Z}$$

is generated by the unreduced suspension of the Hopf map η. For $k = 0$ and general d, the generator of $\pi_{\mathbb{T},\mathcal{U}}(Q) \cong \mathbb{Z}$ is mapped to $(d-1)\eta$.

The collapsing map $P(\mathbb{C}^d) \to P(\mathbb{C}^d)/P(\mathbb{C}^{d-1}) \cong S^{2d-2}$ induces a homomorphism

$$c : \pi_k(S^0) = \pi^{b^+-1}(S^{2d-2}) \to \pi^{b^+-1}(P(\mathbb{C}^d)) \cong \pi_{\mathbb{T},\mathcal{U}}(\operatorname{ind} l).$$

This map turns out to be an isomorphism for $k = 0$ and surjective onto the torsion subgroup $A(k,d)$ for $0 < k \leqslant 4$. The composite map $f \circ c : \pi_k(S^0) \to \pi_{k+1}(S^0)$ is multiplication by $(d-1)\eta$.

Finally consider the Hurewicz map

$$\pi^{b^+-1}(P(\mathbb{C}^d)) \to H^{b^+-1}(P(\mathbb{C}^d)).$$

If an element in $\pi^{b^+-1}(P(\mathbb{C}^d))$ represents a monopole map, then the image of this element under the Hurewicz map is a multiple of the generator in the cohomology group in the respective dimension. This multiplicity is the Seiberg–Witten invariant.

The Hurewicz map is neither surjective nor injective, the kernel being torsion. The noninjectivity issue makes the stable cohomotopy invariant a true refinement of Seiberg–Witten invariants. This will be addressed in the next sections. Nonsurjectivity implies that, depending on k and d, the Seiberg–Witten invariants automatically satisfy certain divisibility conditions. The index of the image of the Hurewicz map $\pi^{2m}(P(\mathbb{C}^{m+n})) \to H^{2m}(P(\mathbb{C}^{m+n}))$ for $m, n \geqslant 0$ is known to be the stable James number $U(-m,n)$ (cf. [**9**, Remark 2.7]). These James numbers can be defined in a more general setup and appear in various geometric situations. K-theory methods provide an estimate for them, which conjecturally is sharp:

Theorem 6.2 ([9]). *The power series in z with rational coefficients*

$$\left(\frac{z}{\log(1+z)}\right)^m,$$

when multiplied with $U(m,n)$, becomes integral modulo z^n.

7. Intermezzo

This chapter aims at sensitizing for some snags one should be aware of when working in this field. One concerns a misinterpretation of the Pontrijagin–Thom construction, another the proper use of homotopy categories.

The main difference between the familiar approach to gauge theory and the homotopy approach is the replacement of spaces by maps. The Pontrijagin–Thom construction provides a perfect and well-known duality between the concepts "stable homotopy classes of maps between spheres" and "bordism classes of framed manifolds." At first glance, this duality suggests stable maps to contain equivalent information as localized data in the form of moduli spaces together with suitably specified normal bundle data. This idea is particularly appealing to anybody working in gauge theory, since the use of localized data — often in form of characteristic classes — is a main trick of the trade. I propose to dispose of this idea as quickly as possible, since it is prone to deception and self-deception. Here is a much too long discussion; for related topics cf. [**1**, Chapter 6].

7.1. Equivariant transversality. One minor reason is due to the fact that the Pontrijagin–Thom correspondence fails in general in an equivariant setting due to the fact that transversality arguments do not work in sufficient generality.

7.2. A variant on the Eilenberg-swindle. More seriously, the information cannot possibly localize as suggested above. The reason is as puzzling as it is simple: Any two framings on a bundle by their very definition are isomorphic. Framings can be distinguished only embedded in a surrounding space.

But how to keep control over framings when changing the surrounding space? The default surrounding space we are dealing with is a Hilbert space. In order to get into business, we have to reduce to finite dimensions. And, to get this straight, the only natural way is by linear projection. Indeed, such projections are used in the proof of 2.1. Now comes the point: Embedded framed manifolds are extremely ill behaved under projections.

Let us look at this in more detail. Embed S^1 as the $\mathbb{T}$-orbit of a nonzero element in $\mathbb{C}^d$ and fix a framing of an affine normal disk to a given point in S^1. Using the $\mathbb{T}$-action on $\mathbb{C}^d$, this framing extends to a framing of the normal bundle of S^1. By equivariant (here it is okay) Pontrijagin–Thom, this framing corresponds to a generator in the corresponding equivariant stable homotopy group, which happens to be isomorphic to $\mathbb{Z}$, as we have seen in the preceding chapter. Now consider a generic projection $\mathbb{C}^d \to \mathbb{C}^{d-1}$. This is a $\mathbb{T}$-equivariant map and the $\mathbb{T}$-equivariant normal framing in d complex dimensions is equivariantly projected to one constructed the same way in complex $d-1$ dimensions, which also represents a generator in the corresponding group. The disastrous effect on the framing becomes apparent only after forgetting the $\mathbb{T}$-action. As explained in the preceding chapter, nonequivariantly the constructed framing of the embedding in $\mathbb{C}^d$ is $(d-1)\eta \in \pi_1^{st}(S^0) \cong (\mathbb{Z}/2)\eta$. So it is trivial for odd d and nontrivial for even. In particular, when projecting along an infinite dimensional Hilbert space in an uncontrolled manner, we systematically do Eilenberg-swindles. There are several ways to deal with this. I'll explain some commonly used ones.

7.3. Equivariance to the rescue. The way to gain control is by the use of the stable map representing the framing. Let us do that. This is an equivariant map $S^{\mathbb{C}^d} \to S^{2d-1}$. From the equivariant picture it is clear that this map has nothing to do with equivariant maps $S^{\mathbb{C}^{d-1}} \to S^{2d-3}$: Projection should correspond to desuspension. But considering source and target, we immediately realize that if our example were a desuspension, then it were along different $\mathbb{T}$-representations on either side of the map. The lesson should be that only by holding to the map as a double-entry book-keeping device, we can tell legal and harmless projections (desuspensions) from the illegal and harmful. But actually, in our case this is not enough.

7.4. Universes to the rescue. Let us take a closer look at the example just discussed and let us forget that there was a $\mathbb{T}$-action. As pointed out, linear projections should correspond to desuspensions. But if we forget the $\mathbb{T}$-action, the linear projections in the example on both sides are real linear along an $\mathbb{R}^2$. As we have seen, they cannot correspond to desuspensions. Intuitively, the problem is easy to understand: In the source, we are trying to desuspend a "moving frame," whereas in the target, we want to desuspend a "fixed frame." Now that we have excluded representation theory to act as a savior, we need a replacement to convey that idea. The notion of a universe, which seems to go back to

Peter May, is such a replacement. The point here is that the projection above along $\mathbb{R}^2$ does not factor through a projection along $\mathbb{R}^1$ as it should. If one uses universes, this feature is built in.

7.5. On the usage of spectra I. I want to present a way how not to define the refined invariants: This uses the spectrum of a selfadjoint elliptic operator, acting on a Hilbert space $\mathcal{U}$. After choosing an ordered basis for eigenspaces, we get a canonical embedding $\mathbb{R}^\infty \to \mathcal{U}$, which we may use to make suspensions ordered by the integers instead of finite dimensional linear subspaces of $\mathcal{U}$. This is okay if one does not change the operator.

The snag appears if one wants to change the operator. Let us do that, say by changing a metric used to define it. At first sight this looks controllable: A small change of the operator will result in a small change of the eigenvalues, so locally, up to "canonical" homotopy, this should define a "canonical" homotopy equivalence between the sphere spectra indexed by the integers.

Will this stand up to scrutiny? Assume we have a closed path of operators such that the eigenspaces for eigenvalues in a fixed interval constitute a bundle over $B = S^1$. It may happen that the bundle for a chosen set of eigenvalues is not orientable. Following the "canonical" homotopy equivalences of the sphere along the circle, we obtain that the identity map over the base point is "canonically" homotopic to minus the identity map. This is not what we want.

But, the space of metrics is contractible. So we may always extend the operator to an operator parameterized by a disk. In the critical cases this will involve other eigenspaces than the ones we started with. So only very special arrangements of eigenspaces will be "admissible" for the argument. And which arrangements are "admissible" may depend heavily on the chosen extension of the operator on the disk. There may exist no "admissible" arrangement that works in all situations.

Orientation is governed by a determinant line bundle, which exists in the Fredholm setting. So, indeed, there may be a way to coherently enforce all such bundles over B to be orientable. I do not know any, but let us suppose we found one. Then, as bundles over B, they are trivial. However, there are two trivializations up to homotopy to choose from. If we pick the wrong one, we will have the following phenomenon: Using the trivialization, we may parallel transport an embedded S^1 with framed normal bundle in the fiber over a point in B once around the loop B. This parallel transport changes the framing.

But, the space of metrics is contractible. Indeed, if the operator is such that the bundle over B extends to a bundle of eigenspaces over the disc, then this would pick a trivialization. However, there may be a different extension of the operator to the disk such that we get a trivialization only if we add a 2-dimensional eigenspace. The two trivializations obtained that way need not be the same, as the example (7.2) shows. Which to choose?

7.6. Why do universes work? The discussion above shows well-definedness of the refined invariants to be a nontrivial issue. The following argument is not based on the contractibility of some parameter space, but on the contractibility of the orthogonal group of Hilbert space. If we take a path in our parameter space (metrics, spin^c-connections), then we will get a bundle of universes over that path. The theorem of Kuiper [**29**] shows that this is a trivial bundle and has a unique trivialization up to homotopy. A trivialization identifies the universes defined for different parameters. Such an identification of universes provides for a change-of-universe isomorphism of the stable cohomotopy groups defined with respect to the respective universes. A trivialization homotopic modulo end points to the one chosen will induce the same isomorphism of stable cohomotopy groups. This uses 2.1 for the parameter space $B \times [0,1] \times [0,1]$. In particular, by taking closed paths, we get that the invariants are well defined.

7.7. On the usage of spectra II. Finally, I want to point out a reliable avenue to create nonsense. Is it possible to construct (homotopy types of) spectra out of spaces, which themselves are only defined up to homotopy? That means, all spaces are defined up to homotopy, the suspensions are defined up to homotopy, the compatibility condition (5) holds only up to homotopy. The answer in general is: No. This would amount to a lift from the homotopy category of topological spaces to the category of topological spaces. This problem has been addressed in work of Dwyer and Kan, cf., for example, [**15**], [**16**]. To see the problems, just assume for each $U \subset \mathcal{U}$, the space A_U to be a sphere homotopy equivalent to S^U. When trying to prove well-definedness of the identity map, not only similar problems as above turn up, but also higher dimensional phenomena. There is no magic to cure this problem.

Since I am using [**10**] as a reference, I should point out that his definition of spectra looks similar to the one I am criticizing. It actually is different: The author wisely only uses complex representations as suspension coordinates. Because of the implicit $\mathbb{T}$-equivariance (7.3) this gets rid of all the complications I lamented about. Moreover, the author is only interested in spectra as realizing equivariant homology and cohomology functors on spaces. He does not define a category of spectra and in particular he does not define maps of spectra, thus avoiding any discussion about the indicated higher dimensional phenomena.

Not all authors have taken this problem seriously. Sadly enough, it renders a considerable part of the homotopy theory literature in this subject useless.

8. Gluing Results

8.1. Gluing along positive curvature. Connected sums of oriented 4-manifolds have vanishing Seiberg–Witten invariants, unless one of the summands has negative definite intersection form. The same statement holds for Donaldson invariants. This fits very well with known stability results on simply connected 4-manifolds: A theorem of Wall [**43**] states that if any two differentiable 4-manifolds are homotopy equivalent, then after taking connected sum with sufficiently many copies of $S^2 \times S^2$, the resulting manifolds will be diffeomorphic. In many cases it is known that already one copy is "sufficiently many." For example, complete intersections or elliptic surfaces are almost completely decomposable [**31**]. That means, the result of taking connected sum with a single complex projective plane is diffeomorphic to a connected sum of projective planes, taken with both standard and reversed orientation. S. Donaldson defined in [**13**] mod 2-polynomial invariants, which potentially could distinguish different structures on connected sums. However, no examples were found.

The stable cohomotopy invariants do not vanish in general for connected sums. This shows that they are true refinements of Seiberg–Witten invariants. The connected sum theorem [**6**] states that for a connected sum $X_0 \# X_1$ of 4-manifolds, the stable equivariant cohomotopy invariant is the smash product of the invariants of its summands. It is straightforward to compute explicit examples.

A precise statement of the theorem constitutes already a major part of its proof. We will discuss a slightly more general setup. Let X be the disjoint union of a finite number, say n, of closed connected Riemannian 4-manifolds X_i, each equipped with a K-theory orientation. Suppose each component contains a separating neck

$$N_i \cong Y \times [-L, L].$$

So it is a union

$$X_i = X_i^- \cup X_i^+$$

of closed submanifolds with common boundary

$$\partial X_i^\pm = Y \times \{0\}.$$

Here Y denotes a 3-manifold with a fixed Riemannian structure. The length $2L > 2$ of the neck is considered a variable. For an even permutation τ of the indices, let X^τ be the manifold obtained from X by interchanging the positive parts of its components, i.e.,

$$X_i^\tau = X_i^- \cup X_{\tau(i)}^+.$$

Next comes the question of whether and how K-orientations glue. In order to be able to glue, we of course need the following

ASSUMPTION. The K-orientations on all components X_i, when pulled back along the inclusion

$$Y \times [-L, L] \xrightarrow{\sim} N_i \hookrightarrow X_i,$$

lead to the same K-orientation.

This assumption is automatically satisfied in case Y is an integral homology sphere. In general, in order to get a well-defined K-orientation on the manifold X^τ, it does not suffice to fix an isomorphism class, but we also have to fix identifications. Note that the gauge group map$(Y \times [-L, L], \mathbb{T})$ acts freely on the set of all such identifications. If the gauge group is connected, any such identification will give the same K-orientation on X^τ. We can enforce connectedness by the following

ASSUMPTION. Let Y have vanishing first Betti number.

It turns out that we will have to put much stronger assumptions on the geometry of Y in order to prove the gluing theorem. So we need not discuss this tricky issue at this point. Under these assumptions, a

K-theory orientation of X uniquely induces by gluing one on X^τ. A main ingredient for the gluing setup is a change of universe isomorphism $V_Y : \mathcal{U} \to \mathcal{U}^\tau$. Its explicit construction uses a smooth path

$$\psi : [-1, 1] \to SO(n)$$

starting from the unit, i.e., $\psi(-1) = \mathrm{id}$, and ending at τ, considered as the permutation matrix $(\delta_{i,\tau(j)})_{i,j} \in SO(n)$. Suppose we are given a bundle over X such that the restrictions over the necks are identified with a bundle F over $Y \times [-L, L]$. Using these identifications, the restrictions of the bundle to $X_i^{\pm}$ glue together to a bundle over X^τ. Sections of the given bundle, when restricted over the neck, can be viewed as a section of the bundle $\oplus_{i=1}^n F$ over $Y \times [-L, L]$. Consider the path ψ as rotation of the components of this bundle. Rotating via ψ a given section of a bundle over X results in a section of the glued bundle over X^τ. This gluing construction, applied to forms and spinors on X, defines fiberwise linear bundle isomorphisms $V_Y : \mathcal{A} \to \mathcal{A}^\tau$ and $V_Y : \mathcal{C} \to \mathcal{C}^\tau$ of the Hilbert space bundles over a suitably defined identification $\mathrm{Pic}^{\mathfrak{s}}(X) \xrightarrow{\cong} \mathrm{Pic}^{\mathfrak{s}}(X^\tau)$. The following theorem is formulated in [**6**] only for the cohomotopy groups in 4.1 and the case $Y = S^3$. The proof extends without further changes to the version in 4.4 and to positively curved manifolds Y, i.e., quotients of the sphere.

Theorem 8.1. *Let Y be a manifold with positive Ricci and in particular scalar curvature. Then the change of universe isomorphism*

$$V_Y : \pi^0_{\mathbb{T},\mathcal{U}}(\mathrm{Pic}^{\mathfrak{s}}(X); \mathrm{ind}\, l) \to \pi^0_{\mathbb{T},\mathcal{U}^\tau}(\mathrm{Pic}^{\mathfrak{s}}(X^\tau); \mathrm{ind}\, l^\tau)$$

identifies the monopole classes of X and X^τ for corresponding K-theory orientations.

The theorem claims the diagram

$$\begin{array}{ccc} \mathcal{A} & \xrightarrow{\mu} & \mathcal{C} \\ V \downarrow & & \downarrow V \\ \mathcal{A}^\tau & \xrightarrow{\mu^\tau} & \mathcal{C}^\tau \end{array}$$

to commute up to homotopy, i.e., there is a path in $\mathcal{P}_l(\mathcal{A}, \mathcal{C})$ connecting the maps μ and $V^{-1}\mu^\tau V$. The difference between the two maps is a compact operator. So the homotopy need only change the compact summand in the monopole map such that at any time during the homotopy the boundedness condition (1.3) remains satisfied. Control is achieved by the use of Weitzenböck formulas for both the Dirac operator and the covariant derivative. Positivity of scalar and Ricci curvature, respectively, along the neck provide the necessary estimates on the spinor and form components during the homotopy. The estimates on spinor and forms finally are tuned by neckstretching. So the theorem holds for sufficiently large L and hence for any $L > 1$. The proof in [**6**] actually constructs a path in a slightly bigger space than $\mathcal{P}_l(\mathcal{A}, \mathcal{C})$. This can be avoided by the use of the homotopy (6).

To apply this theorem, let us spell out the following elementary observation.

Proposition 8.2 ([**6**]). *Let X be the disjoint union of a finite number of K-oriented 4-manifolds X_i. Then the Thom spectrum $T(\mathrm{ind}\, \mathrm{l})$ of the index bundle over $\mathrm{Pic}^{\mathfrak{s}}(X)$ is the smash product of the corresponding spectra $T(\mathrm{ind}\, \mathrm{l}_i)$ associated to the components and the stable cohomotopy class of the monopole map of X is the smash product*

$$[\mu(X, \mathfrak{s})] = \wedge_{i=1}^{n} [\mu(X_i, \mathfrak{s}_i] \in \pi^0_{\mathbb{T}^n, \oplus \mathcal{U}_i}(\mathrm{Pic}^{\mathfrak{s}}(X); \mathrm{ind}\, \mathrm{l})$$

of the stable cohomotopy classes associated to the respective components. The action of the torus $\mathbb{T}^n$ on the sum $\oplus_{i=1}^{n} \mathcal{U}_i$ is factorwise.

Note that the $\mathbb{T}$-action in 8.1 on these spectra is the diagonal one.

The proof of the connected sum formula follows from applying this theorem to the case $Y = S^3$ when X is the disjoint union of a connected sum $X_0 \# X_1$ and two copies of the 4-sphere. The manifold X^τ then will be the disjoint union of X_0, X_1 and one further copy of the 4-sphere. Using (4.5) and (3.1) it is immediate to recognize $\mu(S^4)$ as homotopic to the identity map on the sphere spectrum. In particular, we may identify the monopole class $[\mu(X_0 \# X_1, \mathfrak{s}_0 \# \mathfrak{s}_1)]$, with the monopole class

$$[\mu(X_0 \# X_1, \mathfrak{s}_0 \# \mathfrak{s}_1) \wedge \mu(S^4) \wedge \mu(S^4)] = [\mu(X_0 \# X_1, \mathfrak{s}_0 \# \mathfrak{s}_1) \wedge \mathrm{id}_{S^0} \wedge \mathrm{id}_{S^0}],$$

via some obvious change-of-universe identifications. So as a corollary to 8.1 we obtain the connected sum theorem.

Theorem 8.3 ([**6**]). *The gluing map V_{S^3} identifies the class*

$$[\mu(X_0 \# X_1, \mathfrak{s}_0 \# \mathfrak{s}_1)]$$

of the monopole map of the connected sum of two K-oriented 4-manifolds with the smash product $[\mu(X_0, \mathfrak{s}_0)] \wedge [\mu(X_1, \mathfrak{s}_1)]$ of the monopole classes of the summands.

The gluing theorem applies to a further construction, which is discussed in [**25**, p. 411]: Suppose, the K-oriented 4-manifolds X_0 and

X_1 both contain a -2-curve, i.e., a smoothly embedded 2-sphere with self-intersection number -2. Cutting out tubular neighborhoods of these -2-curves, we obtain manifolds with real projective 3-space as boundary. Using an orientation reversing diffeomorphism of the boundaries, we may glue the manifolds along their boundaries. Let $X_0\#_2X_1$ denote the resulting manifold.

The orientation reversing diffeomorphism permutes the two spin-structures on the real projective 3-space $P(\mathbb{R}^4)$. One of these two spin-structures extends as a spin-structure to the tubular neighborhood of a -2-curve. This property distinguishes the two spin-structures. The other spin structure extends to a spin^c-structure on the tubular neighborhood, the determinant line bundle of which has degree congruent 2 mod 4, when restricted to the -2-curve.

Gluing two copies of tubular neighborhoods of -2-curves by the use of an orientation reversing diffeomorphism of the boundaries, results in a manifold N. This manifold can also be recognized as the manifold $N = \overline{P(\mathbb{C}^3)\#P(\mathbb{C}^3)}$ obtained by reversing the orientation on the connected sum of two copies of the complex projective plane. There are four spin^c-structures on N for which the monopole map is homotopic to the identity map on the sphere spectrum. This again is immediate from (4.5) and (3.1). Exactly the same argument as in 8.3, with S^4 replaced by N, thus proves:

Theorem 8.4. *Let $X_0\#_2X_1$ be the sum of two 4-manifolds along -2-curves with* spin^c*-structure $\mathfrak{s}$. Then there are* spin^c*-structures $\mathfrak{s}_0$ and $\mathfrak{s}_1$ on X_0 and X_1, respectively, such that one of the associated first Chern classes evaluates at the corresponding -2-curve with 2, the other with 0 and $\mathfrak{s} = \mathfrak{s}_0\#_2\mathfrak{s}_1$. The gluing map $V_{P(\mathbb{R}^4)}$ identifies the class of the monopole map $[\mu(X_0\#_2X_1,\mathfrak{s})]$ with $[\mu(X_0,\mathfrak{s}_0)] \wedge [\mu(X_1,\mathfrak{s}_1)]$.*

Obviously, the range of applications of 8.1 is rather limited. It would be desirable to extend the stable cohomotopy approach in a well-defined manner (cf. 7.7) to manifolds with boundary explaining the behavior under cutting and pasting.

8.2. Applications to 4-manifolds. The computations of the stable cohomotopy groups in (6.1) can now be combined with known results on Seiberg–Witten invariants (5.4). Most of the following statements are immediate.

Theorem 8.5 (vanishing results for connected sums, [6]). *Let X be a connected sum of oriented 4-manifolds. Then the refined invariants vanish for any* spin^c*-structure on X in the following cases:*

1. *The refined invariants vanish for any* spin^c*-structure on one of the summands.*
2. *There are two or more summands which are symplectic and have vanishing first Betti numbers. Furthermore, one symplectic summand X_0 satisfies $b^+(X_0) \equiv 1 \bmod 4$.*
3. *The manifold X has vanishing first Betti number, $b^+(X) \not\equiv 4 \bmod 8$ and is a connected sum of 4 symplectic manifolds.*
4. *The manifold X has vanishing first Betti number and is a connected sum of 5 symplectic manifolds.*

The theorem remains true, if one replaces "symplectic" by the weaker assumption "all spin^c-structures with nontrivial refined invariants are almost complex."

Theorem 8.6 (vanishing results for sums along -2-spheres). *Let X_0 and X_1 be oriented 4-manifolds containing -2-spheres C_0 and C_1 respectively. Then the refined invariants vanish for any* spin^c*-structure on the sum $X_0 \#_2 X_1$ along these spheres in the following cases:*

1. *The refined invariants vanish for any* spin^c*-structure on X_0.*
2. *The first Chern class of any* spin^c*-structure on X_i, for which the refined invariant is nonvanishing, gives the same number modulo 4 when evaluated on C_i (for both $i = 0, 1$).*
3. *The first Chern classes of those* spin^c*-structures on X_i, which have nonvanishing refined invariants, span a linear subspace of $H^2(X_i; \mathbb{Q})$ on which the cup-product is positive semi-definite.*
4. *Both X_0 and X_1 can be equipped with the structure of a minimal Kähler surface with $b^+(X_i) > 1$.*

Proof. The first two statements are immediate from 8.4: The assumptions imply one of the two factors in the smash product to vanish. The fourth statement is a special case of the third. The proof of 3 uses the following, well-known fact: Complex conjugation in a small tubular neighborhood of a -2-sphere extends to an automorphism of the 4-manifold X_i which is constant outside a larger tubular neighborhood. The effect in second cohomology is a reflection on the hyperplane perpendicular to the Poincaré dual $PD(C_i)$ of the -2-curve. If there was a spin^c-structure with nonvanishing refined invariant, whose first Chern class is not perpendicular to $PD(C_i)$, then $PD(C_i)$ were a linear combination of the first Chern classes of this spin^c-structure and its reflected spin^c-structure. This would contradict the assumption. As a consequence, the second condition is satisfied, the number modulo 4 being 0. □

Question 8.7. *Is there a minimal symplectic 4-manifold with $b^+ > 1$ for which the first Chern classes of* spin^c*-structures with nonvanishing Seiberg–Witten invariants span a linear subspace of $H^2(X;\mathbb{Q})$ which is not positive semidefinite?*

Here are some general nonvanishing results. Of course, no manifold can be on both a vanishing list as above and a nonvanishing list. This has some nontrivial implications. Note that the assumptions are met by symplectic manifolds.

Theorem 8.8 (nonvanishing results for connected sums, [**6**]).
There is a spin^c*-structure on the oriented 4-manifold X for which the associated refined invariant is nonvanishing, if one of the following holds:*

1. *The manifold X is a connected sum $X = X_0 \# X_1$ of a manifold X_0, which admits a* spin^c*-structure with nonvanishing refined invariant, and a manifold X_1 with $b^+(X_1) = 0$.*
2. *The manifold X has vanishing first Betti number and is a connected sum with two or three summands. For every summand X_i there is an almost complex structure for which the integer Seiberg–Witten invariant is odd and $b^+(X_i) \equiv 3 \bmod 4$.*
3. *The manifold X is a connected sum with four summands, has vanishing first Betti number and $b^+(X) \equiv 4 \bmod 8$. For every summand X_i there is an almost complex structure for which the integer Seiberg–Witten invariant is odd and $b^+(X_i) \equiv 3 \bmod 4$.*

PROOF. Only the first statement is not discussed in [**6**]. Using Donaldson's theorem 4.3 we can find a spin^c-structure on X_1 such that the virtual index bundle of the Dirac operator over $\mathrm{Pic}^{\mathfrak{s}}(X_1)$ has rank 0. The inclusion of a point in $\mathrm{Pic}^{\mathfrak{s}}(X_1)$ induces a restriction map

$$\pi^0_{\mathbb{T},\mathcal{U}}(\mathrm{Pic}^{\mathfrak{s}}(X_1);\mathrm{ind}\,\mathrm{l}) \to \pi^0_{\mathbb{T},\mathcal{U}}(S^0).$$

The image of the monopole class is the identity map. □

The information retained in the refined invariants of connected sums is much more detailed than these sweeping vanishing and nonvanishing theorems might suggest. To get an impression, let us consider connected sums of certain elliptic surfaces which had been classified [**5**] up to diffeomorphism with methods from Donaldson theory. Note that in each of the two homeomorphism classes of such elliptic surfaces there are infinitely many diffeomorphism classes.

Corollary 8.9 ([6]). *Suppose the connected sum $\#_{i=1}^4 E_i$ of simply connected minimal elliptic surfaces of geometric genus one is diffeomorphic to a connected sum $\#_{j=1}^n F_j$ of elliptic surfaces. Then $n = 4$ and the F_j and the E_i are diffeomorphic up to permutation.*

Ishida and LeBrun [**27**], [**28**] pointed out some differential geometric applications of the connected sum theorem. In particular they proved nonexistence statements for Einstein metrics on connected sums of algebraic surfaces.

9. Additional Symmetries

9.1. Spin structures. The case of spin structures was pioneered by Furuta [**20**]. The key observation is that for a spin 4-manifold X the monopole map is actually Pin(2)-equivariant, where $\mathrm{Pin}(2) \subset Sp(1) \subset \mathbb{H}$ is the normalizer of the maximal torus $\mathbb{T} \subset \mathbb{C} \subset \mathbb{H}$ in $Sp(1)$. This subgroup is generated by $\mathbb{T}$ and an additional element $j \in \mathbb{H}$ satisfying $j^2 = -1$ and $ij + ji = 0$.

The group Spin(4) is isomorphic to the product of two copies of $Sp(1) \cong SU(2)$ and embeds as a subgroup in $\mathrm{Spin}^c(4)$. So the $\mathrm{Spin}^c(4)$-representations used in the definition of the monopole map naturally restrict to Spin(4)-representations. Considered this way as Spin(4)-representations, Δ^+ and Δ^- admit quaternionic structures. The Dirac operator, therefore, is $\mathbb{H}$-linear.

This additional structure is not preserved by the monopole map: Consider the induced action of $Sp(1)$ on the space of all Spin(4)-equivariant quadratic maps $\Delta^+ \to \Lambda^+$. The isotropy group of the term σ in the definition of the monopole map is $\mathbb{T}$. The normalizer of the torus interchanges σ and $-\sigma$. This indicates, for which action and which group the monopole map can be made equivariant.

Taking the *spin*-connection A as the background spin^c-connection, we can define a Pin(2)-action on the spaces $\mathcal{A}$ and $\mathcal{C}$ used in the definition of the monopole map: The group acts via the quaternionic structure on the sections of the quaternionic bundles S^+ and S^-. The element j acts via multiplication by -1 on both forms and spin^c-connections (after identifying the space of connections with $A + i\Omega^1(X)$. The monopole map (3) indeed is equivariant with respect to this Pin(2)-action.

In this setup, our standard universe

$$\mathcal{U} = \Gamma(S^-) \oplus \Omega^0(X) \oplus H^1(X;\mathbb{R}) \oplus \Omega^+(X)$$

will not contain trivial Pin(2)-representations. As a consequence, equivariant cohomotopy $\pi^0_{\mathrm{Pin}(2),\mathcal{U}}(\mathrm{Pic}^{\mathfrak{s}}(X);\mathrm{ind}\,\mathrm{l})$ in general does not carry a group structure; it is just a set which even may be empty. Indeed, the main result in **[20]** in effect proves emptiness of this set in certain cases.

In order to get groups, we may simply enlarge the universe $\mathcal{V} = \mathcal{U} \oplus H$ by adding an infinite dimensional Hilbert space H with trivial Pin(2)-action. The change-of-universe map

$$\pi^0_{\mathrm{Pin}(2),\mathcal{U}}(\mathrm{Pin}^{\mathfrak{s}}(X);\mathrm{ind}\,\mathrm{l}) \to \pi^0_{\mathrm{Pin}(2),\mathcal{V}}(\mathrm{Pin}^{\mathfrak{s}}(X);\mathrm{ind}\,\mathrm{l})$$

can be viewed as induced by smash product with the identity element in $\pi^0_{\mathrm{Pin}(2),H}(S^0)$ and turns out **[36]** to be injective in case $b_1 = 0$; the image is characterized in algebraic terms. In particular, there is no loss of information by this change-of-universe, but a gain of convenient algebraic structure.

Theorem 9.1 ([**20**]). *Let X be a spin 4-manifold with* $\mathrm{sign}(X) < 0$. *Then the second Betti number of X satisfies the inequality*

$$b_2(X) \geqslant 2 - \frac{10}{8}\,\mathrm{sign}(X).$$

PROOF. The inclusion $A \hookrightarrow \mathrm{Pic}^0(X)$ of the *spin*-connection induces a restriction map $\pi^0_{\mathrm{Pin}(2),\mathcal{V}}(\mathrm{Pic}^{\mathfrak{s}}(X);\mathrm{ind}\,\mathrm{l}) \to \pi^0_{\mathrm{Pin}(2),\mathcal{V}}(\mathrm{ind}\; l)$. The index of l is $\dfrac{-\,\mathrm{sign}(X)}{16}\mathbb{H} - H^+(X;\mathbb{R})$. In order to apply the K-theory degree formula, we need to complexify these Pin(2)-representations. So, consider the square of the monopole map

$$\nu = [\mu(X)] \wedge [i\mu(X)] \in \pi^0_{\mathrm{Pin}(2),\mathcal{V}+i\mathcal{V}}\left(\Sigma^{-H^+(X;\mathbb{R})\otimes\mathbb{C}}(S^{\mathbb{H}^d})\right).$$

The element $j \in \mathrm{Pin}(2)$ acts by multiplication with -1 on the Pin(2)-representation $H^+(X;\mathbb{R}) \otimes \mathbb{C}$. We would like to compute the $K_{\mathrm{Pin}(2)}$-mapping degree $a_{\mathrm{Pin}(2)}(\nu)$ of ν via formula (8)

$$e_{K_{\mathrm{Pin}(2)}}(H^+(X;\mathbb{R}) \otimes \mathbb{C}) \cdot d_{\mathrm{Pin}(2)} = a_{\mathrm{Pin}(2)}(\nu) \cdot e_{K_{\mathrm{Pin}(2)}}(\mathbb{H}^d),$$

which takes place in the representation ring $R(\mathrm{Pin}(2)) \cong \mathbb{Z}[\lambda, h]/(\lambda^2 - 1, \lambda h - h)$. Here λ stands for the one-dimensional representation on which j acts by multiplication with -1 and h stands for the quaternions, viewed as a Pin(2)-representation. The singular cohomology mapping degree $d_{\mathrm{Pin}(2)}$ can be computed by considering for each element in Pin(2) the cohomology degree on the fixed point spheres of that element. By dimension reasons, this vanishes except for the conjugates of j. It is 1 for j itself, as by construction ν is the identity on the fixed point set. So we get $d_{\mathrm{Pin}(2)} = \frac{1}{2}(1-\lambda)$. The $K_{\mathrm{Pin}(2)}$-Euler classes are computed via (7) to be $e_{K_{\mathrm{Pin}(2)}}(H^+(X;\mathbb{R})\otimes\mathbb{C}) = (1-\lambda)^{b^+}$ and $e_{K_{\mathrm{Pin}(2)}}(\mathbb{H}^d) = (2-h)^d$. The mapping degree formula thus reads

$$\frac{1}{2}(1-\lambda)^{b^+ + 1} = a_{\mathrm{Pin}(2)}(2-h)^d.$$

In the representation ring, this equality can only be satisfied, if $a_{\mathrm{Pin}(2)}$ is of the form $a(1-\lambda)$ for some integer a (the character on the left hand side is zero on $\mathbb{T}$!). So we are left with the equation

$$2^{b^+ - 1}(1-\lambda) = a2^d(1-\lambda),$$

which can be satisfied only for $d \leqslant b^+ - 1$, or equivalently, $b^+ \geqslant 1 - \dfrac{\mathrm{sign}(X)}{8}$. □

This theorem can be sharpened a little:

Theorem 9.2 ([**35**], [**36**]). *Let X be a spin 4-manifold with* $\mathrm{sign}(X) < 0$. *Then the second Betti number of X satisfies the inequality*

$$b_2(X) \geqslant 2a - \frac{10}{8}\,\mathrm{sign}(X),$$

with $a = 2$, if $\mathrm{sign}(X) \equiv 32 \bmod 64$ *and $a = 3$, if* $|\mathrm{sign}(X)| \equiv 48 \bmod 64$ *and $a = 1$ else. Moreover, in the case* $\mathrm{sign}(X) = -64$, *one has*

$$b_2(X) \geqslant 88.$$

The two references correspond to two different proofs. The first one relies on results of S. Stolz and M. Crabb in $\mathbb{Z}/4$-equivariant stable homotopy. The second one imitates in principle the proof above, using more refined $KO_{\mathrm{Pin}(2)}$-mapping degrees instead. Furuta [**21**] announced that by the same methods one can prove $a = 3$ if $\mathrm{sign}(X) \equiv 0 \bmod 64$.

The following example shows that these methods cannot be carried through to prove the so-called $\frac{11}{8}$-conjecture stating $b_2(X) \geqslant -\frac{11}{8}\,\mathrm{sign}(X)$.

Theorem 9.3 ([**36**]). *There is an element in* $\{S^{\mathbb{H}^5}, S^{V^{12}}\}_{\mathcal{U}}^{\mathrm{Pin}(2)}$, *where* V *is the real* 1*-dimensional nontrivial* Pin(2)*-representation.*

So the lowest rank of a potential counterexample to the $\frac{11}{8}$-conjecture, at least according to current knowledge, is $b_2 = 104$.

The connected sum theorem also works in the Pin(2)-equivariant setting for *spin*-manifolds. Taking connected sum of a *spin*-manifold X with $S^2 \times S^2$ amounts for the Pin(2)-equivariant monopole classes to multiplication with the stable cohomotopy Euler class $e(V) : S^0 \hookrightarrow S^V$ of the Pin(2)-representation V.

The long exact sequence for the pair of spaces $(D(V) \sqcup *, S(V) \sqcup *)$, together with the adjunction 3.3 leads for a Pin(2)-spectrum A to a long exact Gysin sequence

$$\ldots \to \pi^{-1}_{\mathbb{T},\mathcal{V}}(S^V \wedge A) \to \pi^0_{\mathrm{Pin}(2),\mathcal{V}}(S^V \wedge A) \to \pi^0_{\mathrm{Pin}(2),\mathcal{V}+V}(A) \to \pi^0_{\mathbb{T},\mathcal{V}+V}(A) \to \ldots$$

The map in the middle is multiplication with the Euler class $e(V)$. The next map restricts the group action. Application to the Thom spectrum $T(\mathrm{ind}\, l)$ of the index bundle over $\mathrm{Pic}^{\mathfrak{s}}(X)$ gives as an immediate consequence:

Theorem 9.4 ([**36**]). *Suppose that the* Pin(2)*-equivariant monopole class of a spin* 4*-manifold* X *with* $\mathrm{sign}(X) < 0$ *is not divisible by the Euler class* $e(V)$*. Then the refined Seiberg–Witten invariant of* X *is nonzero. In particular, if* X *has second Betti number*

$$b_2(X) = -\frac{11}{8}\,\mathrm{sign}(X)$$

for $|\,\mathrm{sign}(X)| \leqslant 64$ *(or* $b_2(X) = 104$ *and* $\mathrm{sign}(X) = -80$*), then* X *has nonvanishing refined invariants.*

The special cases of vanishing first Betti numbers and $\mathrm{sign}(X) = -16$, $\mathrm{sign}(X) = -32$, and $\mathrm{sign}(X) = -48$ were obtained in [**33**], [**22**], and [**23**] respectively.

9.2. Symplectic structures with $c_1 = 0$. Let X be a K-oriented 4-manifold, which is both symplectic and spin. This means the canonical spin^c-structure coming with the symplectic structure has vanishing integral first Chern class. For such a manifold one can combine the considerations above with Taubes' result (5.1).

According to the Kodaira classification of complex surfaces [**4**] there are only three families of complex surfaces with $c_1 = 0$. The obvious families are the simply connected $K3$-surfaces and the tori. Furthermore, there are primary Kodaira surfaces with first Betti number 3. The first two families are Kähler, hence symplectic by default. Other symplectic, non-Kähler and even noncomplex manifolds with $c_1 = 0$ and Betti numbers 2 and 3, were constructed [**42**], [**18**], [**24**].

Theorem 9.5 ([**33**]). *Let X be a symplectic 4-manifold with vanishing first Betti number and with trivial canonical line bundle. Then* $\text{sign}(X) = -16$.

PROOF. The vanishing of c_1 forces X to be spin with $\text{sign}(X) = -16n$ and $b^+ = 4n - 1$. The monopole map is Pin(2)-equivariant, and after moding out the $\mathbb{T}$-action as in 6.1, we obtain a stable $\mathbb{Z}/2$-equivariant map in $\{P(\mathbb{C}^{2n}), S(V^{4n-1})\}^{\mathbb{Z}/2}$. The action is free on both spaces and the spaces are of the same dimension. This allows to apply a $\mathbb{Z}/2$-equivariant version of the Hopf theorem [**10**, p. 126].

According to this theorem, the (nonequivariant) degree of such a map is determined modulo 2. As we will see, this degree, which is the Seiberg–Witten invariant, can be an odd number only in the case $\text{sign}(X) = -16$. Taubes' theorem 5.1 then completes the argument. To show that the degree is even for $n > 1$, it suffices because of Hopf's theorem to exhibit an element in $\{P(\mathbb{C}^{2n}), S(V^{4n-1})\}^{\mathbb{Z}/2}$ which has even degree. Here it is: The n-th power η^n of the Pin(2)-equivariant Hopf map induces a $\mathbb{Z}/2$-equivariant map $P(\mathbb{C}^{2n}) \to S(V^{3n})$. Composed with the inclusion $S(V^{3n}) \to S(V^{4n-1})$ we get a map of degree zero for $n > 1$. □

Here is an immediate corollary:

Corollary 9.6. *A symplectic 4-manifold with finite fundamental group and with trivial canonical line bundle is homeomorphic to a $K3$-surface.*

It is well known that there are infinitely many different smooth structures on the topological 4-manifold underlying a $K3$-surface. The infinitely many smooth structures which come from complex Kähler structures were classified in [**5**]. There are infinitely many more smooth structures which come from symplectic ones, cf. [**25**, p. 396f]. And there are again infinitely many more smooth structures which do not allow for a symplectic structure at all, cf. [**19**]. Nevertheless, amongst all these smooth structures only the $K3$-surface seems to be known to carry a symplectic structure with $c_1 = 0$. The analogy to the Kodaira-dimension zero case in the Kodaira-classification therefore is tantalizing:

Question 9.7. *Are symplectic 4-manifolds with $c_1 = 0$ necessarily either parallelizable or $K3$-surfaces?*

9.3. Group actions. The stable cohomotopy approach does not rely on transversality results and therefore seems suitable for considering group actions on 4-manifolds. In the discussion below, which closely follows [**39**], the first Betti number of the manifolds will always be zero. A compact Lie group acting on a 4-manifold X is supposed to preserve its K-orientation.

Theorem 9.8 ([**39**]). *Let G act on the K-oriented 4-manifold X. There is a central extension $\mathbb{G}$ of the group G by the torus $\mathbb{T}$, such that the monopole map $\mu : \mathcal{A} \to \mathcal{C}$ is $\mathbb{G}$-equivariant. The associated element $[\mu]_G \in \pi^0_{\mathbb{G},\mathcal{U}}(\mathrm{ind}\, \mathrm{l})$ restricts to the $\mathbb{T}$-equivariant stable cohomotopy invariant.*

When considering free actions of finite groups, one gets into a situation which very much resembles Galois theory. The quotients X/H by the various subgroups $H < G$ are 4-manifolds carrying a residual action of the Weyl group $WH = N_G H/H$.

Theorem 9.9 ([**39**]). *Let X be a K-oriented 4-manifold with a free action of a finite group G and $H < G$ a subgroup. The set $J(H, \mathfrak{s}_X)$ of* spin^c*-structures on the quotient X/H which pull back to the given* spin^c*-structure $\mathfrak{s}_X$ on X can be canonically identified with the set of subgroups of $\mathbb{G}$ which map isomorphically to H under the projection to G. For $j \in J(H, \mathfrak{s}_X)$ the invariant $[\mu(X/H, \mathfrak{s}_j)]_{WH}$ can be identified with the restriction of $[\mu(X, \mathfrak{s})]_G$ to the fixed points of $H(j) < \mathbb{G}$.*

In particular, stable cohomotopy invariants of oriented 4-manifolds with finite fundamental group are determined by equivariant stable cohomotopy invariants of simply connected 4-manifolds.

One can combine all the restrictions to fixed points into a comparison map

$$\pi^0_{\mathbb{G},\mathcal{U}}(\mathrm{ind}\, \mathrm{l}(X, \mathfrak{s})) \to \bigoplus_{(H) \leqslant G} \; \bigoplus_{J(H, \mathfrak{s}_X)} H^0(WH; \pi^0_{\mathbb{T}, \mathcal{U}^{H(j)}}(\mathrm{ind}\, \mathrm{l}(X/H; \mathfrak{s}_j)))).$$

Under certain conditions a general splitting result in equivariant homotopy theory implies that this comparison map is an isomorphism after localization away from the order of the group. This splitting theorem can be applied for example if both $b^+(X/H) > 1$ holds for any subgroup of G and the index of the Dirac operator can be represented by an actual representation. So in this case kernel and cokernel are torsion groups with nonzero p-primary parts only for those primes which do divide the order of G.

Finally, let us restrict to the case of a group of prime order p. Again the case where the K-orientation on X comes from an almost complex structure is easy to handle.

Theorem 9.10 ([39]). *If the group G of prime order p acts freely on the almost complex manifold X, then the invariant $[\mu(X,\mathfrak{s})]_G$ is completely determined by the nonequivariant invariants for X and for X/G. Among the latter, the relation*

$$[\mu(X,\mathfrak{s})] \equiv \sum_{J(G,\mathfrak{s})} [\mu(X/G,\mathfrak{s}_j)] \bmod p$$

is satisfied.

The comparison map, however, is not injective in general. This is proved in [**39**] using Adams spectral sequence calculations. To find geometrical applications for these homotopy theoretical computations looks like a challenging problem.

10. Final Remarks

There is no chance to determine stable cohomotopy invariants by direct computation. This seems obvious. So the only way to get further information out of the monopole map is through a better conceptual understanding.

Any improvement in our knowledge about the groups which arise as equivariant stable cohomotopy groups in this field could help as a guideline to computing invariants as well as to constructing 4-manifolds. We know disturbingly few examples of nonvanishing refined invariants. All examples known at the moment are powers of the Hopf map η. Actually this reflects the fact that η by Pontrjiagin–Thom describes the Lie group framing of the group $\mathbb{T}$ acting.

A hypothetical way to realizing other stable cohomotopy elements was pointed out in 9.4: Construct a minimal counterexample to the $\frac{11}{8}$-conjecture! Now we know, where to start the search (9.3). It does not look like that hopeless an enterprise anymore.

It were symmetry considerations which lead to 9.4. Indeed, symmetry considerations may be a key to further progress. Let us dwell upon it a little more. One can consider the monopole map as a map between infinite dimensional bundles over some configuration space Conf(X) consisting of all the choices made: metrics, spin^c-connections, harmonic 1-forms. There is a symmetry group $\mathbb{G}$ acting: It is an extension of the subgroup of the diffeomorphism group preserving the K-orientation by some gauge group. Ideally, the monopole map can be understood as an Euler class of the virtual index bundle in a "proper stable $\mathbb{G}$-equivariant cohomotopy group"

$$\pi^0_{\mathbb{G}}(\mathrm{Conf}(X); \mathrm{ind}\ l)$$

with twisting in an element of "proper $\mathbb{G}$-equivariant KO-theory." The space Conf(X) is the classifying space for proper $\mathbb{G}$-actions. The obvious map $E\mathbb{G} \to \mathrm{Conf}(X)$ from the classifying space of free actions induces a "Segal map"

$$\pi^0_{\mathbb{G}}(\mathrm{Conf}(X); \mathrm{ind}\ l) \to \pi^0_{\mathbb{G}}(E\mathbb{G}; \mathrm{ind}\ l).$$

In analogy to the compact Lie group case one would expect the latter group to be isomorphic (or at least related) to nonequivariant stable cohomotopy $\pi^0(B\mathbb{G}; \mathrm{ind}\ l)$. Now the classifying space $B\mathbb{G}$ of the group $\mathbb{G}$ indeed classifies parameterized families of K-oriented 4-manifolds. The image of the monopole class in this last group therefore is the universal parameterized stable cohomotopy invariant. Of course, everything here is ill defined and probably cannot be made precise at all. However, it can be made precise for compact approximations, i.e., for compact subgroups of $\mathbb{G}$ or for finite equivariant subcomplexes of Conf(X). This might lead to information on the diffeomorphism groups of 4-manifolds. Already the case of the four-dimensional sphere looks interesting.

Interesting first results in this direction, relating diffeomorphisms of 4-manifolds to parameterized stable cohomotopy invariants over the 1-sphere, can be found in a recent preprint [**32**].

The space Conf(X) might also be of interest for considering the behavior of the stable cohomotopy invariants at its "boundary," i.e., study the behaviour of the maps under degeneration of the manifolds.

Another challenging direction of research is to find homotopy interpretations of Donaldson invariants and of Gromov–Witten invariants and to relate these concepts. At the moment these seem to be totally out of reach.

Most urgently needed, however, are more general concepts of gluing. Ideally, there should be relative invariants for manifolds with boundaries defining a "stable homotopy" quantum field theory.

Indeed, there are papers in the literature claiming such a construction. The proposal is to use Conley indices of finite dimensional approximations to define homotopy types of "Seiberg-Witten-Floer"-spectra and related relative invariants. I do see serious problems in this approach, concerning well-definedness.

Here is a brief discussion of the main point. The choices, which go into the construction of the Conley index of a given flow, can be put together to define a colimit in the homotopy category of spaces. More precisely, they form a "connected simple system", a concept which also carries over to parameterized situations as long as the parameter space is simply connected. Amongst many other choices, the proposed construction uses the choice of a homotopy equivalence of the suspension of the Conley index for a given approximation with the Conley index of a higher dimensional approximation. The space of such choices, however, is not simply connected in general. Indeed, let us consider the special case where all the relevant Conley indices are represented by spheres. The space of stable homotopy equivalences of spheres has fundamental group $\mathbb{Z}/2$.

Acknowledgement: I am grateful to M. Szymik for critical and helpful comments. This material was prepared during a stay at the Institute for Advanced Study, which I thank for its great hospitality.

References

1. J. F. Adams, *Prerequisites (on equivariant stable homotopy) for Carlsson's lecture*, in: Algebraic Topology (Aarhus 1982), Springer LNM 1051, Berlin (1982), 483–532.
2. M. Atiyah, *Bott periodicity and the index of elliptic operators*, Quart. J. Math. Oxford (2) **19** (1968), 113-140.
3. M. Atiyah, R. Bott, and A. Shapiro, *Clifford modules*, Topology **3**, Suppl. 1 (1964), 3–38.
4. W. Barth, C. Peters, and A. van de Ven, *Compact complex surfaces*, Ergebnisse der Mathematik, Springer-Verlag Berlin, 1984.
5. S. Bauer, *Diffeomorphism types of elliptic surfaces with $p_g = 1$*, J. Reine Angew. Math. **451** (1994), 89–148.
6. S. Bauer, *A stable cohomotopy refinement of Seiberg–Witten invariants: II*, Invent. Math. **155** (2004), 21–40.
7. S. Bauer and M. Furuta, *A stable cohomotopy refinement of Seiberg–Witten invariants: I*, Invent. Math. **155** (2004), 1–19.

8. M. C. Crabb and K. H. Knapp, *On the codegree of negative multiples of the Hopf map*, Proc. R. Soc. Edinb., Sect. A, Math. **107** (1987), 87–107.
9. M. C. Crabb and K. H. Knapp, *James numbers*, Math. Ann. **282** (1988), 395–422.
10. T. tom Dieck, *Transformation Groups*, de Gruyter, Berlin, 1987.
11. S. K. Donaldson, *An application of gauge theory to four-dimensional topology*, J. Differ. Geom. **18** (1983), 279–315.
12. S. K. Donaldson, *The orientation of Yang-Mills moduli spaces and 4-dimensional topology*, J. Differ. Geom. **26** (1987), 397–428.
13. S. K. Donaldson, *Yang-Mills invariants of four-manifolds*, in: Geometry of Low-Dimensional Manifolds, 1 (Durham 1989), Lond. Math. Soc. Lect. Note Ser. **150** (1990), 5–40.
14. S. K. Donaldson, *The Seiberg-Witten equations and 4-manifold topology*, Bull. Am. Math. Soc. **33** (1996), 45–70.
15. W. G. Dwyer and D. M. Kan, *An obstruction theory for diagrams of simplicial sets*, Nederl. Akad. Wetensch. Indag. Math. **46** (1984), no. 2, 139–146.
16. W. G. Dwyer and D. M. Kan, *Singular functors and realization functors*, Nederl. Akad. Wetensch. Indag. Math. **46** (1984), no.2, 147–153.
17. N. Elkies, *A characterization of the $\mathbb{Z}^n$-lattice*, Math. Res. Lett. **2** (1995), 321–326.
18. M. Fernández, M. Gotay, and A. Gray, *Compact parallelizable 4-dimensional symplectic and complex manifolds*, Proc. Am. Math. Soc. **103** (1988), 1209–1212.
19. R. Fintushel and R. Stern, *Knots, links and 4-manifolds*, Invent. Math. **138** (1998), 363–400.
20. M. Furuta, *Monopole equation and the 11/8-conjecture*, Math. Res. Lett. **8** (2001), 279–291.
21. M. Furuta, *Finite dimensional approximations in geometry*, Proc. ICM 2002, Vol. II, 395–403.
22. M. Furuta, Y. Kametani, and H. Matsue, *Spin 4-manifolds with signature = −32*, Math. Res. Lett. **8** (2001), 293–301.
23. M. Furuta, Y. Kametani, H. Matsue, and N. Minami, *Stable-homotopy Seiberg-Witten invariants and Pin bordisms*, Preprint (UTMS 2000-46, University of Tokyo)
24. H.-J. Geiges, *Symplectic structures on T^2-bundles over T^2*, Duke Math. J. **67** (1992), 539–555.
25. R. E. Gompf and A. I. Stipsicz, *4-Manifolds and Kirby Calculus*, Graduate Studies in Mathematics **20**, AMS, Providence, 1999.
26. F. Hirzebruch and H. Hopf, *Felder von Flächenelementen in 4-dimensionalen Mannigfaltigkeiten*, Math. Ann. **136** (1958), 156–172.
27. M. Ishida and C. LeBrun, *Spin manifolds, Einstein metrics and differential topology*, Math. Res. Lett. **9** (2002), 229–240.
28. M. Ishida and C. LeBrun, *Curvature, connected aums and Seiberg-Witten theory*, Comm. An. Geom. [to appear].
29. N. H. Kuiper, *The homotopy type of the unitary group of Hilbert space*, Topology **3** (1965), 19–30.
30. T. J. Li and A. Liu, *General wall crossing formula*, Math. Res. Lett. **2** (1995), 797–810.

31. R. Mandelbaum and B. Moishezon, *On the topology of simply connected algebraic surfaces*, Transl. Am. Math. Soc. **260** (1980), 195–222.
32. Ming Xu, 1-*Parameter Bauer–Furuta Invariant*, Preprint (2003).
33. J. Morgan and Z. Szabo, *Homotopy K3-surfaces and mod 2 Seiberg–Witten invariants*, Math. Res. Lett. **4** (1997), 17–21.
34. J. Morgan, Z. Szabo, and C. H. Taubes, *A product formula for the Seiberg–Witten invariants and the generalized Thom conjecture*, J. Differ. Geom. **44** (1996), 706–788.
35. B. Schmidt, *Ein Kriterium für die Existenz äquivarianter Abbildungen zwischen reellen Darstellungssphären der Gruppe* Pin(2), Diplomarbeit, Universität Bielefeld (1997).
36. B. Schmidt, *Spin* 4-*manifolds and* Pin(2)-*equivariant homotopy theory*, Dissertation, Bielefeld (2003). urn:nbn:hbz:361–3165, electronically available at http://bieson.ub.uni-bielefeld.de/volltexte/2003/316/
37. A. S. Schwarz, Dokl. Akad. Nauk USSR **154**, 61–63. English translation: *The homotopic Topology of Banach Spaces* AMS Transl. **5** (1964), 57–59.
38. G. B. Segal, *Equivariant K-Theory*, Publ. Math. Inst. Hautes Études Sci. **34** (1968), 129–151.
39. M. Szymik *Galois Symmetries of* 4-*Manifolds and Equivariant Stable Homotopy Theory*, Dissertation, Bielefeld (2002).
40. C. H. Taubes, *The Seiberg–Witten invariants and symplectic forms*, Math. Res. Lett. **1** (1994), 809–822.
41. C. H. Taubes, *SW ⇒ Gr: from the Seiberg–Witten equations to pseudoholomorphic curves*, J. Am. Math. Soc. **9** (1996), 819–918.
42. W. Thurston, *Some simple examples of symplectic manifolds*, Proc. Am. Math. Soc. **55** (1976), 467–468.
43. C. T. C. Wall, *On simply-connected* 4-*manifolds*, J. Lond. Math. Soc. **39** (1964), 141–149.
44. E. Witten, *Monopoles and four-manifolds*, Math. Res. Lett. **1** (1994), 769–796.

3–Dimensional Methods in Contact Geometry

Ko Honda†

University of Southern California
Los Angeles, USA

A *contact manifold* (M, ξ) is a $(2n+1)$-dimensional manifold M equipped with a smooth maximally nonintegrable hyperplane field $\xi \subset TM$, i.e., locally $\xi = \ker \alpha$, where α is a 1-form which satisfies $\alpha \wedge (d\alpha)^n \neq 0$. Since $d\alpha$ is a nondegenerate 2-form when restricted to ξ, contact geometry is customarily viewed as the odd-dimensional sibling of symplectic geometry. Although contact geometry in dimensions $\geqslant 5$ is still in an incipient state, contact structures in dimension 3 are much better understood, largely due to the fact that symplectic geometry in two dimensions is just the study of area. The goal of this article is to explain some of the recent developments in 3-dimensional contact geometry, with an emphasis on methods from 3-dimensional topology. Basic references include [**1, 7, 13, 18**]. The article [**31**] is similar in spirit to ours.

Three-dimensional contact geometry lies at the interface between 3- and 4-manifold geometries, and has been an essential part of the flurry in low-dimensional geometry and topology over the last 20 years. In dimension 3, it relates to foliation theory and knot theory; in dimension

†Supported by an Alfred P. Sloan Fellowship and an NSF grant.

4, there are rich interactions with symplectic geometry. In both dimensions, there are relations with gauge theories such as Seiberg–Witten theory and Heegaard Floer homology.

1. Introduction

From now on we will restrict our attention to contact structures on 3-manifolds. We will implicitly assume that our contact structures ξ on M satisfy the following:

1. ξ is oriented, and hence given as the kernel of a global 1-form α.
2. $\alpha \wedge d\alpha > 0$, i.e., the contact structure is *positive.*

Such contact structures are often said to be *cooriented.*

HW 1. *Show that if ξ is a smooth oriented 2-plane field, then ξ can be written as the kernel of a global 1-form α.*

1.1. First examples.

EXAMPLE 1: $(\mathbf{R}^3, \xi_0)$, where $\mathbf{R}^3$ has coordinates (x, y, z), and ξ_0 is given by $\alpha_0 = dz - ydx$. Then

$$\xi_0 = \ker \alpha_0 = \mathbf{R}\left\{ \frac{\partial}{\partial y}, \frac{\partial}{\partial x} + y\frac{\partial}{\partial z} \right\}.$$

According to the standard "propeller picture" (see Figure 1), all the straight lines parallel to the y-axis are everywhere tangent to ξ_0, and the 2-planes rotate in unison along these straight lines.

EXAMPLE 2: (T^3, ξ_n). Here $T^3 \simeq \mathbf{R}^3/\mathbf{Z}^3$, with coordinates (x, y, z), and $n \in \mathbf{Z}^+$. Then ξ_n is given by $\alpha_n = \sin(2\pi nz)dx + \cos(2\pi nz)dy$. We have

$$\xi_n = \mathbf{R}\left\{ \frac{\partial}{\partial z}, \cos(2\pi nz)\frac{\partial}{\partial x} - \sin(2\pi nz)\frac{\partial}{\partial y} \right\}.$$

This time, the circles $x = y = const$ (parallel to the z-axis) are everywhere tangent to ξ_n, and the contact structure makes n full twists along such circles.

HW 2. *Verify that $(\mathbf{R}^3, \xi_0)$ and (T^3, ξ_n) are indeed contact manifolds.*

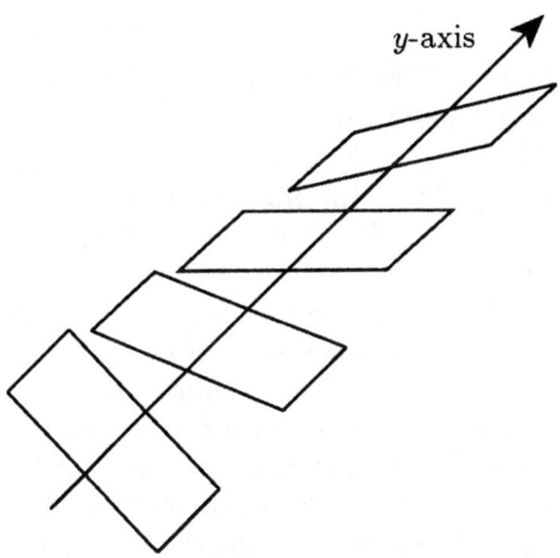

FIGURE 1. The propeller picture

The significance of Example 1 is the following:

Theorem 1.1 (Pfaff). *Every contact 3-manifold* (M, ξ) *locally looks like* $(\mathbf{R}^3, \xi_0)$, *i.e., for all* $p \in M$ *there is an open set* $U \supset p$ *such that* $(U, \xi) \simeq (\mathbf{R}^3, \xi_0)$.

Note that an isomorphism in the contact category (usually called a *contactomorphism*) is a diffeomorphism $\phi : (M_1, \xi_1) \xrightarrow{\sim} (M_2, \xi_2)$ which maps $\phi_* \xi_1 = \xi_2$. Pfaff's theorem says that there are no local invariants in contact geometry.

REMARK. A contactomorphism usually does not preserve the contact 1-form.

HW 3. *Prove Pfaff's theorem in dimension 3. Then generalize it to higher dimensions.*

EXAMPLE 3: (S^3, ξ), the standard contact structure on S^3. Consider $B^4 = \{|z_1|^2 + |z_2|^2 \leqslant 1\} \subset \mathbf{C}^2$. Then take $S^3 = \partial B^4$. The contact structure ξ is defined as follows: for all $p \in S^3$, ξ_p is the unique complex line $\subset T_p S^3$ (the unique 2-plane invariant under the complex structure J).

HW 4. *Write down a contact 1-form α for (S^3, ξ) and verify that $\alpha \wedge d\alpha > 0$.*

1.2. Legendrian knots. Given a contact manifold (M, ξ), a curve $L \subset M$ is *Legendrian* if L is everywhere tangent to ξ, i.e., $\dot{L}(p) \in \xi_p$ at every point $p \in L$. In this section we describe the invariants that can be assigned to a Legendrian knot (= embedded closed curve) L. For a more thorough discussion, see the survey article [**15**].

Twisting number/Thurston–Bennequin invariant. Our first invariant is the *relative Thurston–Bennequin invariant* $t(L, \mathcal{F})$, also known as the *twisting number*, where $\mathcal{F}$ is some fixed framing for L. Although $t(L, \mathcal{F})$ is an invariant of the *unoriented* knot L, for convenience pick one orientation of L. L has a natural framing called the *normal framing*, induced from ξ by taking $v_p \in \xi_p$ so that $(v_p, \dot{L}(p))$ form an oriented basis for ξ_p. We then define $t(L, \mathcal{F})$ to be the integer difference in the number of twists between the normal framing and $\mathcal{F}$. By convention, left twists are negative. Now, the framing $\mathcal{F}$ that we choose is often dictated by the topology. For example, if $[L] = 0 \in H_1(M; \mathbf{Z})$ (which is the case when $M = S^3$), then there is a compact surface $\Sigma \subset M$ with $\partial\Sigma = L$, i.e., a *Seifert surface*. Now Σ induces a framing $\mathcal{F}_\Sigma$, which is the normal framing to the 2-plane field $T\Sigma$ along L, and the *Thurston–Bennequin invariant* $tb(L)$ is given by:

$$tb(L) = t(L, \mathcal{F}_\Sigma).$$

HW 5. *Show that $tb(L)$ does not depend on the choice of Seifert surface Σ.*

In Example 2, if $L = \{x = y = const\}$, then a convenient framing $\mathcal{F}$ is induced from tori $x = const$ (or equivalently from $y = const$). We have $t(L, \mathcal{F}) = -n$.

Rotation number. Given an oriented Legendrian knot L in S^3, we define the *rotation number* $r(L)$ as follows: Choose a Seifert surface Σ and trivialize $\xi|_\Sigma$. Then $r(L)$ is the winding number of $\dot{L}$ along L with respect to the trivialization.

HW 6. *Show that $r(L)$ does not depend on the choice of trivialization or Seifert surface.*

Front projection. We now consider Legendrian knots in the standard contact $(\mathbf{R}^3, \xi_0)$ given by $dz - ydx = 0$. Consider the *front projection* $\pi : \mathbf{R}^3 \to \mathbf{R}^2$, where $(x, y, z) \mapsto (x, z)$. Generic Legendrian knots L (the genericity can be achieved by applying a small contact isotopy) can be projected to closed curves in $\mathbf{R}^2$ with cusps and ordinary double points but no vertical tangencies. Conversely, such a closed curve in $\mathbf{R}^2$ can be lifted to a Legendrian knot in $\mathbf{R}^3$ by setting y to be the slope of the curve at (x, z). (Observe that if $dz - ydx = 0$, then $\frac{dz}{dx} = y$.) The Thurston–Bennequin invariant and rotation number of a Legendrian knot L can be computed in the front projection using the following formula:

$$\begin{aligned} tb(L) &= -\frac{1}{2}(\#\text{cusps}) + \#\text{positive crossings} \\ &\quad - \#\text{negative crossings}. \\ r(L) &= \frac{1}{2}(\#\text{downward cusps} - \#\text{upward cusps}) \end{aligned}$$

HW 7. *Prove the above formulas for tb and r in the front projection.*

Stabilization. Given an oriented Legendrian knot L, its *positive stabilization* (respectively, *negative stabilization*) $S_+(L)$ (respectively, $S_-(L)$) is an operation that decreases tb by adding a zigzag in the front projection as in Figure 2.

We have $tb(S_\pm(L)) = tb(L) - 1$ and $r(S_\pm(L)) = r(L) \pm 1$.

HW 8. *Prove that the stabilization operation is well-defined (independent of the location where the zigzag is added).*

The following theorem of Eliashberg–Fraser [**10**] enumerates all the Legendrian unknots:

Theorem 1.2 (Eliashberg–Fraser). *Legendrian unknots in the standard contact* $\mathbf{R}^3$ *(or* S^3*) are completely determined by tb and r.*

In fact, all the Legendrian unknots are stabilizations $S_+^{k_1} S_-^{k_2}(L_0)$ of the unique maximal tb Legendrian unknot L_0 with $tb(L_0) = -1$ and $r(L_0) = 0$, given on the left-hand side of Figure 3. The right-hand picture is $S_+^2 S_-^1(L_0)$.

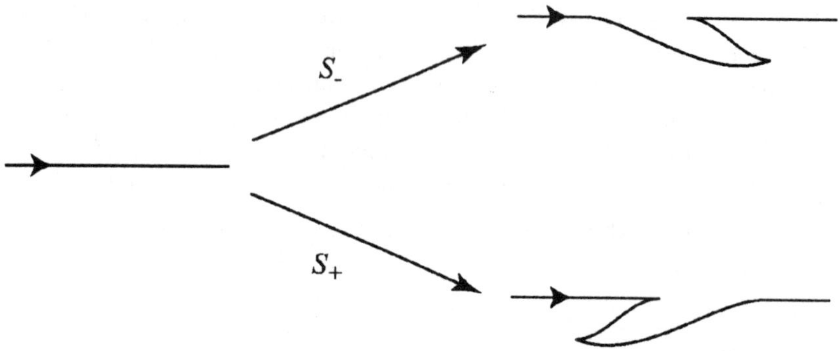

FIGURE 2. Positive and negative stabilizations

For an oriented Legendrian knot in $\mathbf{R}^3$ or S^3, the topological knot type, the Thurston–Bennequin invariant, and the rotation number are called the *classical* invariants. Although Legendrian unknots are completely determined by their classical invariants according to Theorem 1.2, Legendrian knots in general are not completely classified by the classical invariants. One way of distinguishing two Legendrian knots with the same classical invariants is through *contact homology*. (See [**3, 11**] for more details.)

1.3. Tight vs. overtwisted. In the 1970's, Lutz [**36**] and Martinet [**37**] proved the following:

Theorem 1.3 (Lutz, Martinet). *Let M be a closed oriented 3-manifold,* $\text{Dist}(M)$ *be the set of smooth 2-plane field distributions on M, and* $\text{Cont}(M)$ *be the set of smooth contact 2-plane field distributions on M. Then* $\pi_0(\text{Cont}(M)) \to \pi_0(\text{Dist}(M))$ *is surjective.*

STRATEGY OF PROOF. 1. Start with a 2-plane field ξ. Take a fine enough triangulation of M so that on each 3-simplex Δ, ξ is close to a linear foliation by planes.

2. It is easy to homotop ξ near the 2-skeleton so it becomes contact. Now we have an extension problem to the interior of each 3-simplex.

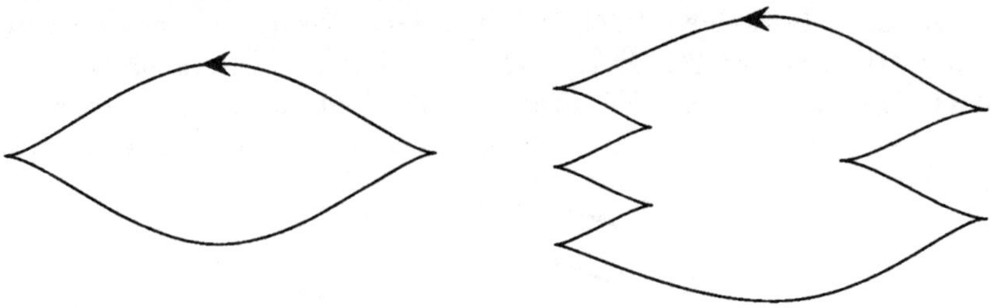

$tb(L_0) = -1$
$r(L_0) = 0$

$tb(L) = -4$
$r(L) = 1$

FIGURE 3. Legendrian unknots in the front projection

3. Insert a Lutz tube. A *Lutz tube* is a contact structure on $S^1 \times D^2$ (with cylindrical coordinates (z, r, θ), where $D^2 = \{(r, \theta) | r \leqslant 1\}$) given by the 1-form

$$\alpha = \cos(2\pi r)dz + r \sin(2\pi r)d\theta. \qquad \square$$

HW 9. *Think about how to use a Lutz tube ("perform a Lutz twist") to finish the construction. Keep in mind that the homotopy class of the 2-plane field needs to be preserved.*

Having introduced Lutz twists, we can now write down more contact structures on $\mathbf{R}^3$:

EXAMPLE 1_R: $(\mathbf{R}^3, \zeta_R)$, where $\mathbf{R}^3$ has cylindrical coordinates (r, θ, z), R is a positive real number, and ζ_R is given by $\alpha_R = \cos f_R(r)dz + r \sin f_R(r)d\theta$. Here $f_R(r)$ is a function with positive derivative satisfying $f_R(r) = r$ near $r = 0$ and $\lim_{r \to +\infty} f_R(r) = R$.

HW 10. *Show that* $(\mathbf{R}^3, \xi_0) \simeq (\mathbf{R}^3, \zeta_R)$ *for all* $R \leqslant \pi$.

However, we have the following key result of Bennequin [**2**]:

Theorem 1.4 (Bennequin). $(\mathbf{R}^3, \xi_0) \not\simeq (\mathbf{R}^3, \zeta_R)$ *if* $R > \pi$.

The distinguishing feature is the existence of an *overtwisted (OT) disk*, i.e., an embedded disk $D \subset (M, \xi)$ such that $\xi_p = T_p D$ at all

$p \in \partial D$. A typical OT disk looks like $\{pt\} \times D^2$ in the Lutz tube $S^1 \times D^2$ described above (also see Figure 4). While it is not hard to see that $(\mathbf{R}^3, \zeta_R)$ has OT disks if $R > \pi$, what Bennequin proved was that $(\mathbf{R}^3, \xi_0)$ contains no OT disks. It turns out that the existence of an OT disk is equivalent to the existence of a Legendrian unknot L with $tb(L) = 0$.

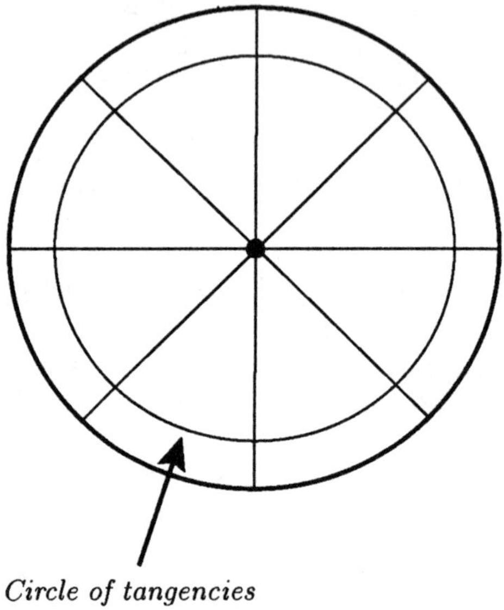

FIGURE 4. An overtwisted disk D. (Precisely speaking, the disk should end at the circle of tangencies.) The straight lines represent the singular (characteristic) foliation that $\xi \cap TD$ traces on D, and the circle is the set of points where $\xi = TD$. There is also an elliptic tangency at the center.

HW 11 (Hard). *Try to prove that* $(\mathbf{R}^3, \xi_0)$ *has no overtwisted disks.*

It is not an exaggeration to say that modern contact geometry has its beginnings in Bennequin's theorem. There is a dichotomy in the world of contact structures, those that contain OT disks (called *overtwisted* contact structures) and those that do not (called *tight* contact structures). In view of Theorem 1.1, every contact structure is *locally tight*, and therefore the question of overtwistedness is a global one.

The following is an important inequality for knots in tight contact manifolds.

Theorem 1.5 (Bennequin inequality). *Let L be nullhomologous Legendrian knot in a tight (M,ξ). If Σ is a Seifert surface for L with Euler characteristic $\chi(\Sigma)$, then*

$$tb(L) \pm r(L) \leqslant -\chi(\Sigma).$$

1.4. Classification of contact structures. When discussing the classification of contact structures, it is important to keep in mind the following theorem:

Theorem 1.6 (Gray). *Let ξ_t, $t \in [0,1]$, be a 1-parameter family of contact structures on a closed manifold M. Then there is a 1-parameter family of diffeomorphisms φ_t such that $\varphi_0 = id$ and $\varphi_t^* \xi_t = \xi_0$.*

In other words, a *homotopy* of contact structures gives rise to a contact *isotopy*.

The *overtwisted* classification (on closed 3-manifolds) was shown by Eliashberg [**8**] to be essentially the same as the homotopy classification of 2-plane fields. (The result is quite striking, especially when contrasted with the tight classification on T^3 below.)

Theorem 1.7 (Eliashberg). *Let M be a closed oriented 3-manifold, and $\mathrm{Cont}^{OT}(M) \subset \mathrm{Dist}(M)$ be the overtwisted 2-plane field distributions. Then*

$$\pi_0(\mathrm{Cont}^{OT}(M)) \simeq \pi_0(\mathrm{Dist}(M)).$$

On the other hand, tight contact structures tend to reflect the underlying topology of the manifold, and are more difficult to understand. The goal of this article is to introduce techniques which enable us to better understand tight contact structures. In the meantime, we list a couple of examples:

1. S^3. Eliashberg [**7**] proved that there is a unique tight contact structure up to isotopy. It is the one given in Example 3.

2. T^3. Giroux [**20**] and Kanda [**30**] independently proved that
   (a) every tight contact structure is isomorphic to some ξ_n,
   (b) $(T^3, \xi_m) \not\simeq (T^3, \xi_n)$ if $m \neq n$.

HW 12. *Try to prove that $(T^3, \xi_m) \not\simeq (T^3, \xi_n)$ if $m \neq n$.*

In Section 4 we will give a classification of tight contact structures for the lens spaces $L(p,q)$.

1.5. A criterion for tightness. A contact structure (M,ξ) is *symplectically fillable* if there exists a compact symplectic 4-manifold (X,ω) such that $\partial X = M$ and $\omega|_\xi > 0$. (X,ω) is said to be a *symplectic filling* of (M,ξ). (Technically speaking, what we are calling "symplectically fillable" is usually called "weakly symplectically fillable," but since we have no need of such taxonomy in this article, we will stick to "symplectically fillable" or even just "fillable." For more information, refer to [**16**].)

HW 13. *Show that* (S^3,ξ) *in Example* 3 *is symplectically fillable.*

HW 14. *Show* (T^3,ξ_n) *in Example* 2 *is symplectically fillable.*
Hint: first modify $\alpha_n \mapsto dz + t\alpha_n$ with t small.

A powerful general method for producing tight contact structures is the following theorem of Gromov and Eliashberg [**6, 23**]:

Theorem 1.8 (Gromov–Eliashberg). *A symplectically fillable contact structure is tight.*

It immediately follows from the symplectic filling theorem that the standard (S^3,ξ) from Example 3 and the contact structures (T^3,ξ_n) from Example 2 are tight.

Symplectic filling is a 4-dimensional way of checking whether (M,ξ) is tight. We will discuss other methods (including a purely 3-dimensional one) of proving tightness in Section 5.

1.6. Relationship with foliation theory. Foliations are the other type of locally homogeneous 2-plane field distributions. The following table is a brief list of analogous objects from both worlds (note that the analogies are not precise):

Foliations	Contact Structures
$\alpha \wedge d\alpha = 0$ integrable	$\alpha \wedge d\alpha > 0$ nonintegrable
$\alpha = dz$ Frobenius	$\alpha = dz - ydx$ Pfaff
Reeb components	Overtwisted disks
Taut	Tight

A (rank 2) *foliation* ξ is an integrable 2-plane field distribution, i.e., locally given as the kernel of a 1-form α with $\alpha \wedge d\alpha = 0$. According to Frobenius' theorem, ξ can locally be written as the kernel of $\alpha = dz$. The world of foliations also breaks up into the topologically significant *taut* foliations (i.e., foliations for which there is a closed transversal curve through each leaf), and the foliations with *generalized Reeb components*, which exist on every 3-manifold. A generalized Reeb component is a compact submanifold $N \subset M$ whose boundary ∂N is a union of torus leaves, and such that there are no transversal arcs which begin and end on ∂N. The primary example of a generalized Reeb component is a *Reeb component*, i.e., a foliation of the solid torus $S^1 \times D^2$ whose boundary $S^1 \times S^1$ is a leaf and whose interior is foliated by planes as in Figure 5.

FIGURE 5. A Reeb component. Here the top and bottom are identified.

The following is a key theorem which allows us to transfer information from foliation theory to contact geometry.

Theorem 1.9 (Eliashberg–Thurston). *Let M be a closed, oriented 3-manifold $\neq S^1 \times S^2$. Then every taut foliation admits a C^0-small perturbation into a tight contact structure.*

For a thorough treatment of the relationship with foliation theory, see [**12**]. In Section 2.3, we will discuss one aspect, namely the relationship with Gabai's *sutured manifold* theory.

2. Convex Surfaces

In this section, we investigate embedded surfaces Σ in the contact manifold (M, ξ). The principal notion is that of *convexity*. For the time being, ξ may be tight or overtwisted.

2.1. Characteristic foliations.

Before discussing convexity, we first examine how ξ traces a singular line field on an embedded surface Σ.

Definition 2.1. The *characteristic foliation* Σ_ξ is the singular foliation induced on Σ from ξ, where $\Sigma_\xi(p) = \xi_p \cap T_p\Sigma$. The *singular points* (or tangencies) are points $p \in \Sigma$ where $\xi_p = T_p\Sigma$.

Lemma 2.2. *A C^∞-generic characteristic foliation Σ_ξ is of Morse–Smale type, i.e., satisfies the following:*

(1) *the singularities and closed orbits are dynamically hyperbolic, i.e, hyperbolic in the dynamical systems sense,*
(2) *there are no saddle-saddle connections,*
(3) *every point $p \in \Sigma$ limits to some isolated singularity or closed orbit in forward time and likewise in backward time.*

The proof of Lemma 2.2 uses the fact that a C^∞-small perturbation of ξ is still contact. We choose the perturbation of ξ to be compactly supported near Σ, and hence the isotopy in Gray's theorem is compactly supported near Σ. Therefore, generic properties of 1-forms (in particular the Morse–Smale condition) are satisfied.

HW 15. *Show that if α is a contact 1-form and β is any 1-form, then $\alpha + t\beta$ is contact for sufficiently small t.*

There are two types of dynamically hyperbolic singularities: *elliptic* and *hyperbolic* (not in the dynamical systems sense). Choose coordinates

(x, y) on Σ and let the origin be the singular point. If we write

$$\alpha = dz + f dx + g dy,$$

then

$$X = g\frac{\partial}{\partial x} - f\frac{\partial}{\partial y}$$

is a vector field for the characteristic foliation near the origin. If the determinant of the matrix

$$\begin{pmatrix} \frac{\partial g}{\partial x} & \frac{\partial g}{\partial y} \\ -\frac{\partial f}{\partial x} & -\frac{\partial f}{\partial y} \end{pmatrix}$$

is positive (respectively, negative), then the singular point is elliptic (respectively, hyperbolic). An example of an elliptic singularity is $\alpha = dz + (xdy - ydx)$, and an example of a hyperbolic singularity is $\alpha = dz + (2xdy + ydx)$.

Next we discuss signs. Assume Σ and ξ are both oriented. Then a singular point p is *positive* (respectively, *negative*) if $T_p\Sigma$ and ξ_p have the same orientation (respectively, opposite orientations).

The characteristic foliation Σ_ξ is oriented.

We use the convention that positive elliptic points are sources and negative elliptic points are sinks. If p is a nonsingular point of a leaf L of the characteristic foliation, then we choose $v \in T_pL$ so that (v, n) is an oriented basis for $T_p\Sigma$. Here $n \in T_p\Sigma$ is an oriented normal vector to ξ_p.

EXAMPLES OF CHARACTERISTIC FOLIATIONS.

1. Consider $S^2 = \{x^2 + y^2 + z^2 = 1\} \subset (\mathbf{R}^3, \zeta_{\pi/2})$. Then S^2 will have two singular points, the positive elliptic point $(0, 0, 1)$ and the negative elliptic point $(0, 0, -1)$, and the leaves spiral downward from $(0, 0, 1)$ to $(0, 0, -1)$.

2. An example of an overtwisted disk D is one which has a positive elliptic point at the center and radial leaves emanating from the center, such that ∂D is a circle of singularities. Often in the literature one sees overtwisted disks whose boundary is transverse to ξ and whose leaves emanating from the center spiral towards the limit cycle ∂D. (Strictly speaking, such a D with a limit cycle is not an OT disk according to our definition, but can easily be modified to fit our definition.)

The importance of the characteristic foliation Σ_ξ comes from the following proposition:

Proposition 2.3. *Let ξ_0 and ξ_1 be two contact structures which induce the same characteristic foliation on Σ. Then there is an isotopy φ_t, $t \in [0,1]$, rel Σ, with $\varphi_0 = id$ and $(\varphi_1)_*\xi_0 = \xi_1$.*

2.2. Convexity. The notion of a *convex surface*, introduced by Giroux in **[19]** and extended to the case of a compact surface with Legendrian boundary by Kanda in **[30]**, is the key ingredient in the cut-and-paste theory of contact structures.

Definition 2.4. A properly embedded oriented surface Σ is *convex* if there exists a contact vector field $v \pitchfork \Sigma$. Here, a *contact vector field* is a vector field whose corresponding flow preserves the contact structure ξ. In this article we assume that our convex surfaces are either *closed* or *compact with Legendrian boundary.*

If $\Sigma = \Sigma \times \{0\}$ is convex, then there is an invariant neighborhood $\Sigma \times [-\varepsilon, \varepsilon] \subset M$. We usually assume that v agrees with the normal orientation to Σ.

PROPERTIES OF CONVEX SURFACES.

1. A C^∞-generic closed embedded surface Σ is convex. This is because an embedded surface Σ with a Morse–Smale characteristic foliation is convex. (The same is almost true for compact surfaces with Legendrian boundary, but more care is needed along the boundary.)

2. To a convex surface Σ we may associate a multicurve (i.e., a properly embedded (smooth) 1-manifold, possibly disconnected and possibly with boundary)

$$\Gamma_\Sigma = \{x \in \Sigma | v(x) \in \xi_x\},$$

called the *dividing set*. It satisfies the following:

(a) $\Gamma_\Sigma \pitchfork \Sigma_\xi$,
(b) the isotopy class of Γ_Σ does not depend on the choice of v,
(c) $\Sigma \setminus \Gamma_\Sigma = R_+(\Gamma_\Sigma) \sqcup R_-(\Gamma_\Sigma)$, where $R_+(\Gamma_\Sigma) \subset \Sigma$ (respectively, $R_-(\Gamma_\Sigma)$) is the set of points x where the normal orientation to Σ given by $v(x)$ agrees with (respectively, is opposite to) the normal orientation to ξ_x.

REMARK. We may think of Γ_Σ as the set of points where $\xi \perp \Sigma$, where $\perp$ is measured with respect to v.

Write $\#\Gamma_\Sigma$ for the number of connected components of Γ_Σ.

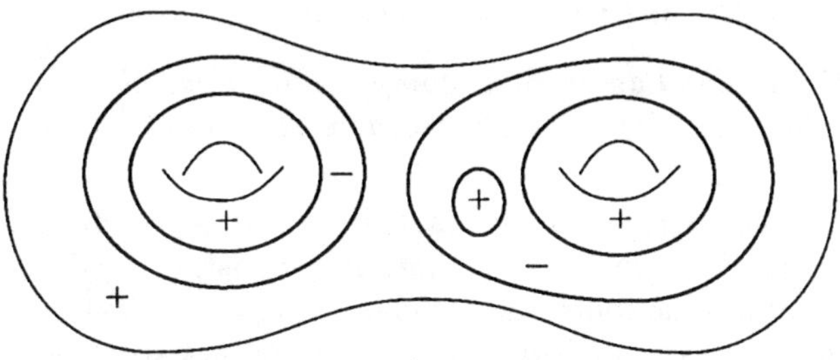

FIGURE 6. A sample dividing set

The usefulness of the dividing set Γ_Σ comes from the following:

Theorem 2.5 (Giroux's flexibility theorem). *Assume Σ is convex with characteristic foliation Σ_ξ, contact vector field v, and dividing set Γ_Σ. Let $\mathcal{F}$ be another singular foliation on Σ which is* adapted to Γ_Σ *(i.e., there is a contact structure ξ' in a neighborhood of Σ such that $\Sigma_{\xi'} = \mathcal{F}$ and Γ_Σ is also a dividing set for ξ'). Then there is an isotopy φ_t, $t \in [0,1]$, of Σ in (M,ξ) such that*
(1) *$\varphi_0 = id$ and $\varphi_t|_{\Gamma_\Sigma} = id$ for all t,*
(2) *$\varphi_t(\Sigma) \pitchfork v$ for all t,*
(3) *$\varphi_1(\Sigma)$ has characteristic foliation $\mathcal{F}$.*

In essence, Γ_Σ encodes ALL of the essential contact-topological information in a neighborhood of Σ. Therefore, having discussed characteristic foliations in Section 2.1, we may proceed to discard them and simply remember the dividing set.

HW 16. *Prove Giroux Flexibility.*

EXAMPLES ON T^2.
There are two common characteristic foliations on T^2:

1. *Nonsingular Morse–Smale.* This is when the characteristic foliation is nonsingular and has exactly $2n$ closed orbits, n of which are sources (repelling periodic orbits) and the other n are sinks (attracting

periodic orbits). Γ_{T^2} consists of $2n$ closed curves parallel to the closed orbits. Each dividing curve lies inbetween two periodic orbits.

2. *Standard form.* An example is $x = const$ inside (T^3, ξ_n). The torus is fibered by closed Legendrian fibers, called *ruling curves*, and the singular set consists of $2n$ closed curves, called *Legendrian divides.* The $2n$ curves of Γ_{T^2} lie between the Legendrian divides.

HW 17. *Find an explicit example of a T^2 inside a contact manifold with nonsingular Morse–Smale characteristic foliation.*

What Giroux Flexibility tells us is that it is easy to switch between the two types of characteristic foliations – nonsingular Morse–Smale and standard form. The following corollary of Giroux Flexibility is a crucial ingredient in the cut-and-paste theory of contact structures.

Corollary 2.6 (Legendrian realization principle (LeRP)). *Let Σ be a convex surface and C be a multicurve on Σ. Assume $C \pitchfork \Gamma_\Sigma$ and C is nonisolating, i.e., each connected component of $\Sigma \setminus C$ nontrivially intersects Γ_Σ. Then there is an isotopy (as in the Giroux Flexibility Theorem) such that $\varphi_1(C)$ is Legendrian.*

HW 18. *Try to prove LeRP, assuming Giroux Flexibility.*

REMARK. C may have extraneous intersections with Γ_Σ, i.e., the actual number of intersections $\#(C \cap \Gamma_\Sigma)$ is allowed to be larger than the geometric intersection number.

FACT. If C is a Legendrian curve on the convex surface Σ, then the twisting number $t(C, \mathcal{F}_\Sigma)$ relative to the framing $\mathcal{F}_\Sigma$ from Σ is $-\frac{1}{2}\#(C \cap \Gamma_\Sigma)$. Here $\#(\cdot)$ represents cardinality, not geometric intersection.

Now we present the criterion (see **[21]**) for determining when a convex surface has a tight neighborhood.

Proposition 2.7 (Giroux's criterion). *A convex surface $\Sigma \neq S^2$ has a tight neighborhood if and only if Γ_Σ has no homotopically trivial dividing curves. If $\Sigma = S^2$, then there is a tight neighborhood if and only if $\#\Gamma_\Sigma = 1$.*

HW 19. *Prove that if Γ_Σ has a homotopically trivial dividing curve, then there exists an overtwisted disk in a neighborhood of Σ, provided we are not in the situation where $\Sigma = S^2$ and $\#\Gamma_\Sigma = 1$.*

Hint: Use LeRP, together with a trick when Γ_Σ has no other components besides the homotopically trivial curve.

The "only if" direction in Giroux's Criterion follows from HW 19. The "if" direction follows from constructing an explicit model inside a tight 3-ball or gluing (for the latter, see [**4**]).

Suppose that (M, ξ) is tight. If $\Sigma = S^2$ is a convex surface in (M, ξ), then Γ_Σ is unique up to isotopy, consisting on one (homotopically trivial) circle. If $\Sigma = T^2$ is convex, then it consists of $2n$ parallel, homotopically essential curves. Therefore Γ_{T^2} is determined by $\#\Gamma_{T^2}$ and the slope, once a trivialization $T^2 \simeq \mathbf{R}^2/\mathbf{Z}^2$ is fixed.

2.3. Convex decomposition theory. The reader may have already noticed certain similarities between convex surfaces and the theory of sutured manifolds due to Gabai [**17**].

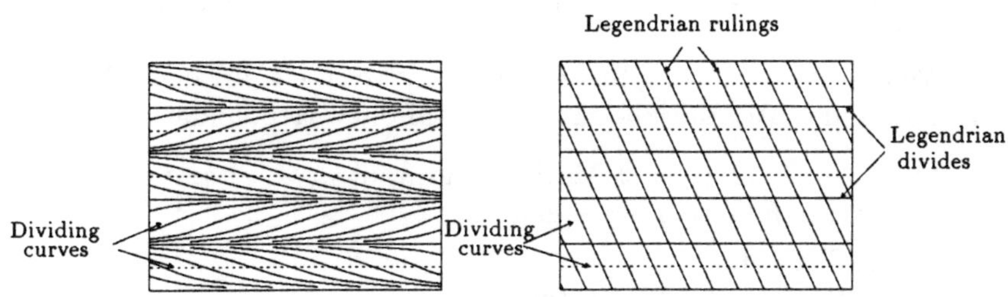

FIGURE 7. The left-hand side is a torus with nonsingular Morse–Smale characteristic foliation. The right-hand side is a torus in standard form. Here the sides are identified and the top and bottom are identified.

Definition 2.8. A *sutured manifold* (M, Γ) consists of the following data:

1) M is a compact, oriented, irreducible 3-manifold; each component of M has nonempty boundary,

2) Γ is a multicurve on ∂M which has nonempty intersection with each component of ∂M, and
3) Γ divides ∂M into positive and negative regions, whose sign changes every time Γ is crossed. We write $\partial M \setminus \Gamma = R_+(\Gamma) \sqcup R_-(\Gamma)$.

Here, a 3-manifold M is *irreducible* if every embedded 2-sphere S^2 bounds a 3-ball B^3.

Note that our definition of a sutured manifold, chosen to simplify the exposition in this paper, is slightly different from that of Gabai [**17**].

Definition 2.9. Let S be a compact oriented surface with connected components $S_1, \ldots, S_n$. The *Thurston norm* of S is:

$$x(S) = \sum_{i \text{ such that } \chi(S_i) < 0} |\chi(S_i)|.$$

Definition 2.10. A sutured manifold (M, Γ) is *taut* if $R_\pm(\Gamma)$ are incompressible in M and minimize the Thurston norm in $H_2(M, \Gamma)$. Here, a surface $S \subset M$ is *incompressible* if for every embedded disk $D \subset M$ with $D \cap S = \partial D$, there is a disk $D' \subset S$ such that $\partial D = \partial D'$.

Roughly speaking, (M, Γ) is taut if $R_\pm(\Gamma)$ attain the minimum genus amongst all the embedded representatives in the relative homology class $H_2(M, \Gamma)$.

We have the following theorem which gives the equivalence between tightness and tautness in the case of a manifold with boundary (see [**26**]):

Theorem 2.11 (Kazez–Matić–Honda). *Let (M, Γ) be a sutured manifold. Then the following are equivalent:*

1. *(M, Γ) is taut.*
2. *(M, Γ) carries a taut foliation.*
3. *(M, Γ) carries a universally tight contact structure.*
4. *(M, Γ) carries a tight contact structure.*

A contact structure ξ on M is *carried by* (M, Γ) if ∂M is a convex surface for ξ with dividing set Γ. A transversely oriented foliation ξ on M is *carried by* (M, Γ) if there exists a thickening of Γ to a union $\gamma \subset \partial M$ of annuli, so that $\partial M \setminus \gamma$ is a union of leaves of ξ, ξ is transverse to γ, and the orientations of $R_\pm(\Gamma)$ and ξ agree. (Strictly speaking, in this case M is a manifold with corners.) A tight contact structure is *universally tight* if it remains tight when pulled back to the universal cover of M.

In the rest of this section, we explain how *sutured manifold decompositions* have an analog in the contact world, namely the theory of *convex decompositions*. Using it we outline the proof of (1)⇒(4).

Definition 2.12. Let S be an oriented, properly embedded surface in (M, Γ) which intersects Γ transversely. Then a *sutured manifold splitting* $(M, \Gamma) \overset{S}{\rightsquigarrow} (M', \Gamma')$ is given as follows (see Figure 8 for an illustration): Define $M' = M \setminus S$, and let S_+ (respectively, S_-) be the copy of S on $\partial M'$ where the orientation inherited from S and the outward normal agree (are opposite). Then set $R_\pm(\Gamma') = (R_\pm(\Gamma) \setminus S) \cup S_\pm$. The new suture Γ' forms the boundary between the regions $R_+(\Gamma')$ and $R_-(\Gamma')$.

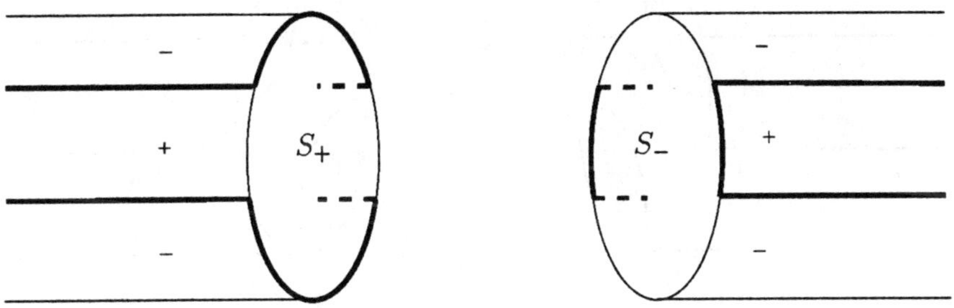

FIGURE 8

A sutured manifold (M, Γ) is *decomposable*, if there is a sequence of sutured manifold splittings:

$$(M, \Gamma) \overset{S_1}{\rightsquigarrow} (M_1, \Gamma_1) \overset{S_2}{\rightsquigarrow} \cdots \overset{S_n}{\rightsquigarrow} (M_n, \Gamma_n) = \sqcup(B^3, S^1).$$

Gabai, in [**17**], proved the following theorem:

Theorem 2.13 (Gabai).

1. Decomposition: *If (M, Γ) is taut, then it is decomposable.*
2. Reconstruction: *Given a sutured manifold decomposition, we can backtrack and construct a taut foliation which is carried by (M, Γ).*

Now, in the contact category, we choose a dividing set Γ_S so that every component of Γ_S is *∂-parallel*, i.e., cuts off a half-disk of S which does not intersect any other component of Γ_S. Such a dividing set Γ_S is also called *∂-parallel*.

If there is an invariant contact structure defined in a neighborhood of ∂M with dividing set $\Gamma = \Gamma_{\partial M}$, then by an application of LeRP, we may take ∂S to be Legendrian. (There are some exceptional cases, but we will not worry about them here.) Extend the contact structure to be an invariant contact structure in a neighborhood of S with ∂-parallel dividing set Γ_S. Now, if we cut M along S, we obtain a manifold with corners. To smooth the corners, we apply *edge-rounding*. This is given in Figures 9 and 10. Figure 9 gives the surface S before rounding, and Figure 10 after rounding. Notice that we may think of S as a lid of a jar, and the edge-rounding operation as twisting to close the jar.

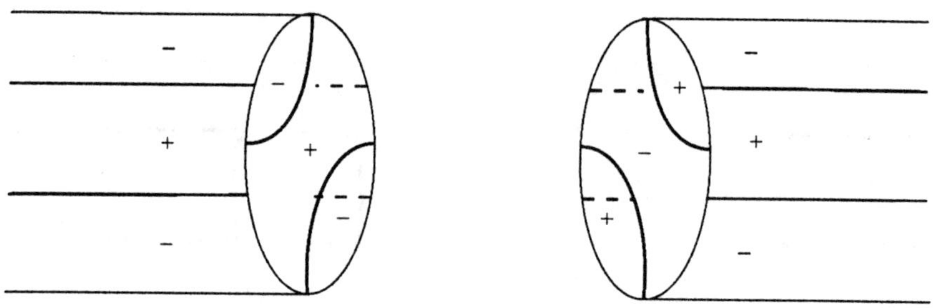

FIGURE 9

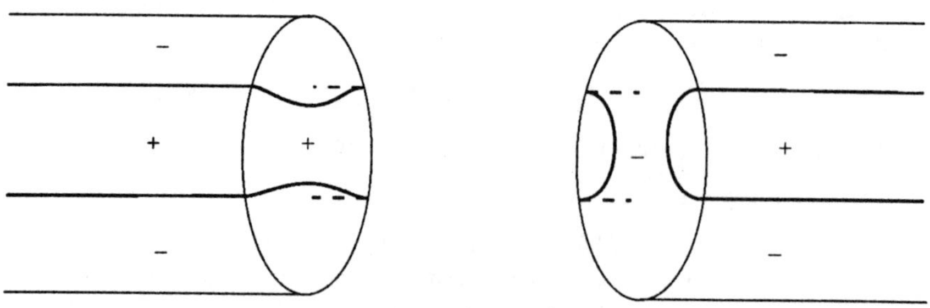

FIGURE 10

HW 20. *Explain why edge-rounding works as in Figures* 9 *and* 10.

Observe that the dividing set in Figure 10 is isotopic to the sutures in Figure 8. Therefore, given a sutured manifold splitting $(M,\Gamma) \overset{S}{\rightsquigarrow} (M',\Gamma')$, there is a corresponding convex splitting $(M,\Gamma) \overset{(S,\Gamma_S)}{\rightsquigarrow} (M',\Gamma')$, with a ∂-parallel dividing set Γ_S. Using the decomposition theorem of Gabai, if (M,Γ) is taut, then there exist s a convex decomposition:

$$(M,\Gamma) \overset{(S_1,\Gamma_{S_1})}{\rightsquigarrow} (M_1,\Gamma_1) \overset{(S_2,\Gamma_{S_2})}{\rightsquigarrow} \cdots \overset{(S_n,\Gamma_{S_n})}{\rightsquigarrow} (M_n,\Gamma_n) = \sqcup(B^3,S^1).$$

We now work backwards, starting with the following theorem of Eliashberg [**7**]:

Theorem 2.14 (Eliashberg). *Fix a characteristic foliation $\mathcal{F}$ adapted to $\Gamma_{\partial B^3} = S^1$. Then there is a unique tight contact structure on B^3 up to isotopy relative to ∂B^3.*

The following gluing theorem of Colin [**4**] allows us to inductively build a universally tight contact structure carried by (M,Γ).

Theorem 2.15 (Colin). *Let Σ be an incompressible surface with $\partial\Sigma \neq \varnothing$. If Γ_Σ is ∂-parallel and $(M \setminus \Sigma, \xi|_{M\setminus\Sigma})$ is universally tight, then (M,ξ) is also universally tight.*

This theorem and other similar theorems will be discussed in Section 5. Theorem 2.11 is a refinement, in the case of manifolds with boundary, of the following theorem:

Theorem 2.16 (Gabai–Eliashberg–Thurston). *Let M be a oriented, closed, irreducible 3-manifold with $H_2(M;\mathbf{Z}) \neq 0$. Then M carries a universally tight contact structure.*

The Gabai–Eliashberg–Thurston theorem was originally proved in two parts: Gabai [**17**] proved that such an M carries a taut foliation, and Eliashberg–Thurston [**12**] proved that the taut foliation can be perturbed into a universally tight contact structure. There is also an alternate, purely 3-dimensional method for proving this theorem [**27, 28, 29**].

3. Bypasses

In this section, we introduce the other chief ingredient in the cut-and-paste theory of tight contact structures: the *bypass*. As a surface is isotoped inside the ambient tight contact manifold (M,ξ), the dividing set changes in discrete units, and the fundamental unit of change is effected by the bypass. Bypasses would be quite useless if they were difficult to find. For the cases we examine in Section 4, namely solid tori, $T^2 \times I$, and lens spaces, they can be found relatively easily by examining the next step in the Haken hierarchy. This will be explained in Section 3.2. For more information on bypasses, refer to [**24**].

3.1. Definition and examples.

Definition 3.1. Let Σ be a convex surface and α be a Legendrian arc in Σ which intersects Γ_Σ in three points p_1, p_2, p_3, where p_1 and p_3 are endpoints of α. A *bypass half-disk* is a convex half-disk D with Legendrian boundary, where $D \cap \Sigma = \alpha$ and $tb(\partial D) = -1$. α is called the *arc of attachment* of the bypass, and D is said to be a bypass *along* α or Σ.

REMARK. Most bypasses do not come for free. Finding a bypass is equivalent to raising the twisting number (or Thurston–Bennequin invariant) by 1. Although it is easy to lower the twisting number by attaching "zigzags" in a front projection, raising the twisting number is usually a nontrivial operation.

Lemma 3.2 (Bypass Attachment Lemma). *Let D be a bypass for Σ. If Σ is isotoped across D, then we obtain a new convex surface Σ' whose dividing set is obtained from Γ_Σ via the move in Figure 12.*

Note that this is reasonable because a bypass attachment increases the twisting number along the arc of attachment by 1.

EXAMPLE: T^2. Let us enumerate the possible bypass attachments – see Figure 13.

(a) is the case where $\#\Gamma_{T^2} = 2n > 2$, and the bypass reduced $\#\Gamma$ by two, while keeping the slope fixed.

(b) is the case where $\#\Gamma_{T^2} = 2$, and the slope is modified.

In addition, there also are trivial and disallowed moves, which are moves locally given in Figure 14. It turns out that the trivial move always exists inside a tight contact manifold, whereas the disallowed move can never exist inside a tight contact manifold.

HW 21. *Is there a bypass attachment which increases $\#\Gamma$?*

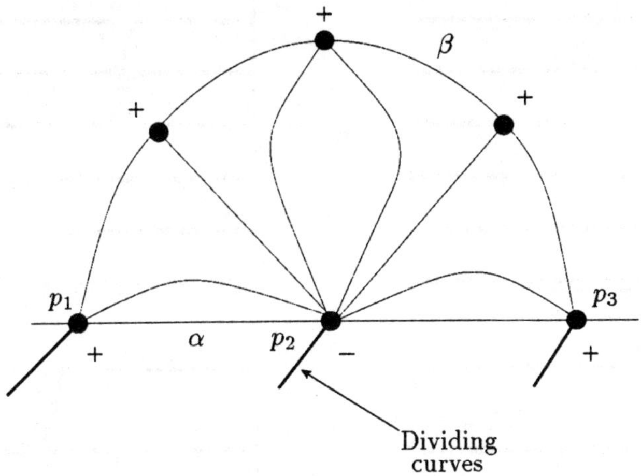

FIGURE 11. A bypass

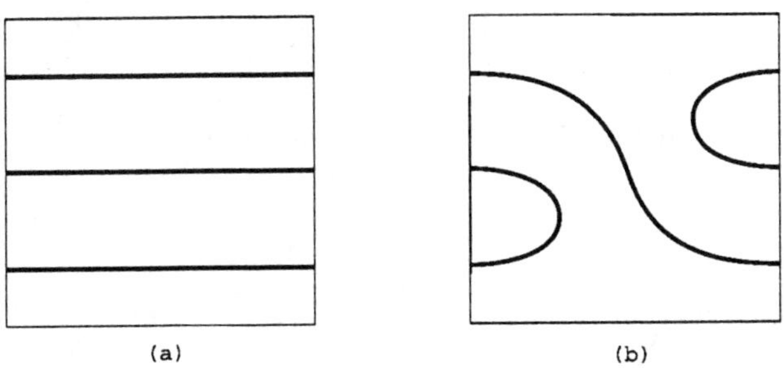

FIGURE 12. The effect of attaching a bypass from the front. Γ_Σ is (a) and $\Gamma_{\Sigma'}$ is (b)

Intrinsic interpretation. Observe that, in case (b), the bypass move is equivalent to performing a *positive Dehn twist* along a particular curve. We can therefore reformulate this bypass move and give an *intrinsic interpretation* in terms of the Farey tessellation of the hyperbolic unit disk **H** (Figure 15). The set of vertices of the Farey tessellation

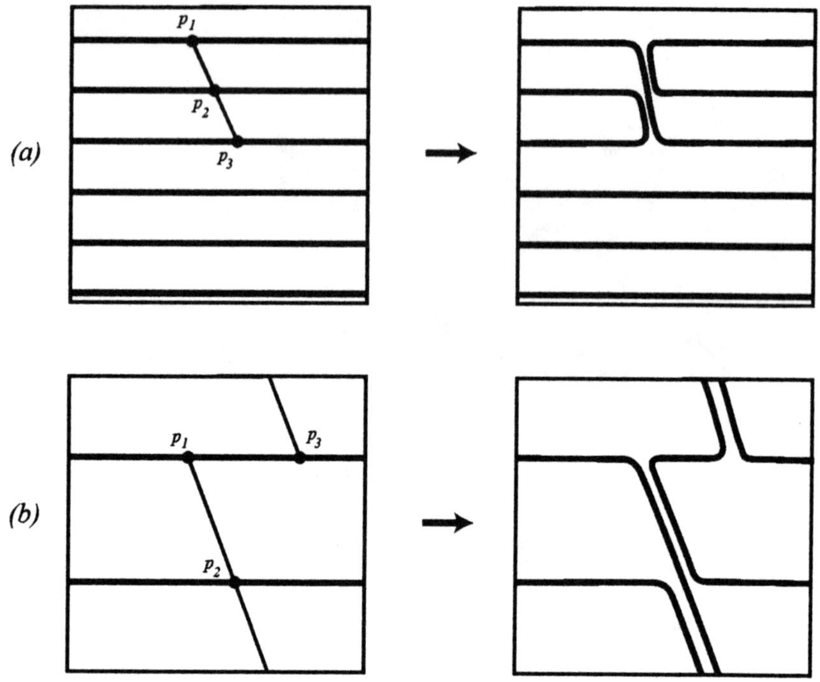

FIGURE 13. Possible bypasses on tori

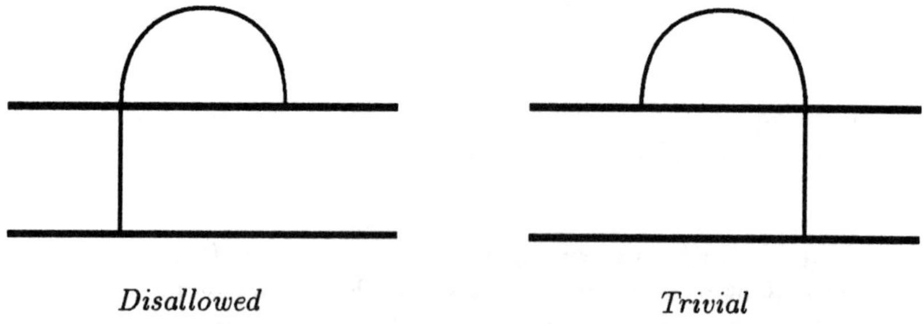

FIGURE 14. A disallowed bypass attachment and a trivial bypass attachment

is $\mathbf{Q} \cup \{\infty\}$ on $\partial\mathbf{H}$. (More precisely, fix a fractional linear transformation f from the upper half-plane model of hyperbolic space to the unit disk model $\mathbf{H}$. Then the set of vertices is the image of $\mathbf{Q} \cup \{\infty\}$ under f.) There is a unique edge between $\frac{p}{q}$ and $\frac{p'}{q'}$ if and only if the corresponding shortest integer vectors form an integral basis for $\mathbf{Z}^2$. (The edge is usually taken to be a geodesic in $\mathbf{H}$.)

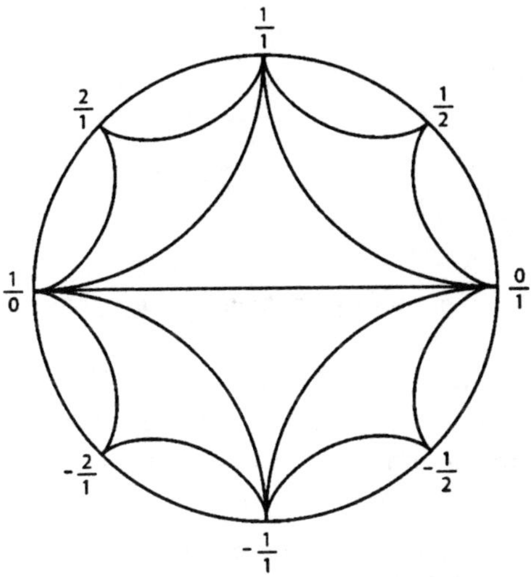

FIGURE 15. The Farey tessellation. The spacing between vertices are not drawn to scale

Proposition 3.3. *Let* $s = \text{slope}(\Gamma_{T^2})$. *If a bypass is attached along a closed Legendrian curve of slope* s', *then the resulting slope* s'' *is obtained as follows: Let* $(s', s) \subset \partial\mathbf{H}$ *be the counterclockwise interval from* s' *to* s. *Then* s'' *is the point on* (s', s) *which is closest to* s' *and has an edge to* s.

See Figure 16 for an illustration.

HW 22. *Prove Proposition 3.3.*

3.2. Finding bypasses. We now explain how to find bypasses. Let M be a closed manifold and $\Sigma \subset M$ be a closed surface. In order to find a

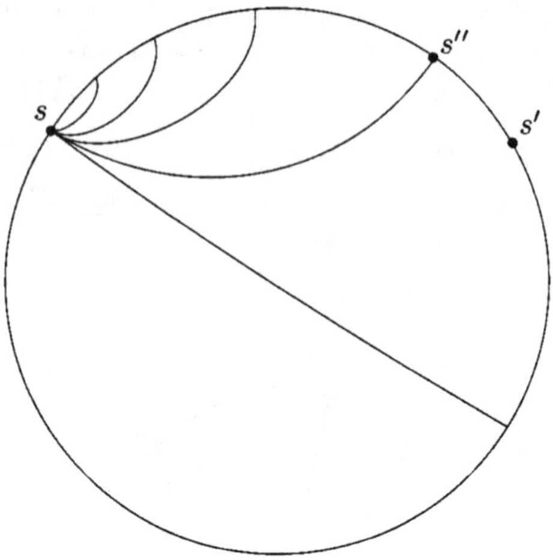

FIGURE 16. Intrinsic interpretation of the bypass attachment

bypass along Σ, we consider $M \setminus \Sigma$. Let $S \subset M \setminus \Sigma$ be an incompressible surface with nonempty boundary, for example the next cutting surface in the Haken hierarchy. Under mild conditions on ∂S, we can take S to be a convex surface with nonempty Legendrian boundary.

Lemma 3.4. *Suppose that Γ_S has a ∂-parallel component and either $S \neq D^2$ or else if $S = D^2$ then $tb(\partial S) < -1$. Then there exists a bypass along ∂S and hence along Σ.*

PROOF. Draw an arc $\delta' \subset S$ so that δ' cuts off a half-disk with only the ∂-parallel arc δ on it. The condition on S is needed to ensure that we can use LeRP to find a Legendrian arc δ''. The half-disk cut off by δ'' (and containing a copy of δ) is the bypass for Σ. □

Corollary 3.5. *Let $S = D^2$ be a convex disk with Legendrian boundary so that $tb(\partial S) < -1$. Then there exists a bypass along ∂S.*

Corollary 3.5 follows from Lemma 3.4, by observing that all components of Γ_{D^2} cut off half-disks of D^2 and that a ∂-parallel component is simply an outermost arc of Γ_{D^2}.

REMARK. Corollary 3.5 does not work when $tb(\partial D) = -1$.

Similarly, we can prove the following:

Corollary 3.6 (imbalance principle). *Let $S = S^1 \times [0,1]$ be a convex annulus. If $t(S^1 \times \{1\}, \mathcal{F}_S) < t(S^1 \times \{0\}, \mathcal{F}_S)$, then there is a ∂-parallel arc and hence a bypass along $S^1 \times \{1\}$. Here $\mathcal{F}_S$ is the framing induced from the surface S.*

Figure 17 gives an example of a convex annulus with $t(S^1 \times \{1\}, \mathcal{F}_S) < t(S^1 \times \{0\}, \mathcal{F}_S)$. There is necessarily a bypass along $S^1 \times \{1\}$.

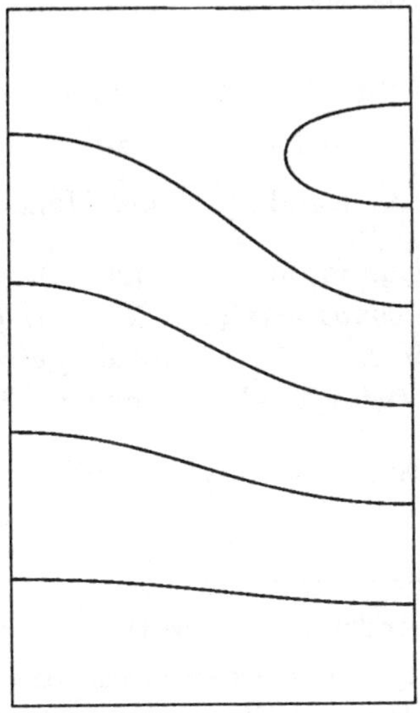

FIGURE 17. One possible dividing set for the annulus. Here the top and the bottom are identified

4. Classification of Tight Contact Structures on Lens Spaces

As an illustration of the technology introduced in the previous two sections, we give a complete classification of tight contact structures on the lens spaces $L(p,q)$. This classification was obtained independently by Giroux [**20**] and Honda [**24**]; partial results had been obtained previously by Etnyre [**14**]. In this article, we follow the method of [**24**].

4.1. The standard neighborhood of a Legendrian curve. Consider a (closed) Legendrian curve L with $t(L, \mathcal{F}) = -n < 0$, $n \in \mathbf{Z}^+$. (Pick some framing $\mathcal{F}$ for which the twisting number is negative.) Then a *standard neighborhood* $S^1 \times D^2 = \mathbf{R}/\mathbf{Z} \times \{x^2 + y^2 \leqslant \varepsilon\}$ (with coordinates z, x, y) of the Legendrian curve $L = S^1 \times \{(0,0)\}$ is given by

$$\alpha = \sin(2\pi nz)dx + \cos(2\pi nz)dy,$$

and satisfies the following:

1. $T^2 = \partial(S^1 \times D^2)$ is convex.
2. $\#\Gamma_{T^2} = 2$.
3. $\text{slope}(\Gamma_{T^2}) = -\frac{1}{n}$ if the meridian has zero slope and the longitude given by $x = y = const$ has slope ∞.

The following is due to Kanda [**30**] and Makar-Limanov [**38**].

Proposition 4.1 (Kanda, Makar-Limanov). *Given a solid torus $S^1 \times D^2$ and boundary conditions* (1), (2), (3), *there exists a unique tight contact structure on $S^1 \times D^2$ up to isotopy rel boundary, provided we have fixed a characteristic foliation $\mathcal{F}$ adapted to $\Gamma_{\partial(S^1 \times T^2)}$.*

REMARK. The precise characteristic foliation is irrelevant in view of Giroux Flexibility.

PROOF. 1. $L \subset T^2$ be a curve which bounds the meridian D. Using LeRP, realize it as a Legendrian curve with $tb(L) = -1$.

2. Using the genericity of convex surfaces, realize the surface D with $\partial D = L$ as a convex surface with Legendrian boundary. Since $tb(L) = -1$, there is only one possibility for Γ_D, up to isotopy.

3. Next, using Giroux Flexibility, fix some characteristic foliation on D adapted to Γ_D. Note that any two tight contact structures on $S^1 \times D$ with boundary condition $\mathcal{F}$ can be isotoped to agree on $T^2 \cup D$.

4. The rest is a 3-ball B^3. Use Eliashberg's uniqueness theorem for tight contact structures on B^3. □

HW 23. *Try to prove Eliashberg's theorem, using convex surfaces.*

4.2. Lens spaces. Let $p > q > 0$ be relatively prime integers. The *lens space* $L(p, q)$ is obtained by gluing $V_1 = S^1 \times D^2$ and $V_2 = S^1 \times D^2$ together via $A : \partial V_2 \xrightarrow{\sim} \partial V_1$, where $A = \begin{pmatrix} -q & q' \\ p & -p' \end{pmatrix} \in -SL(2, \mathbf{Z})$. Here we are making an oriented identification $\partial V_i \simeq \mathbf{R}^2/\mathbf{Z}^2$, where the meridian of V_i is mapped to $\pm(1,0)$, and some chosen longitude is mapped to $\pm(0,1)$.

Continued fractions. Let $-\frac{p}{q}$ have a continued fraction expansion

$$-\frac{p}{q} = r_0 - \frac{1}{r_1 - \frac{1}{r_2 \ldots -1/r_k}}, \quad \text{where } r_i \leqslant -2.$$

EXAMPLE:

$$-\frac{14}{5} = -3 - \frac{1}{-5}.$$

We write

$$-\frac{14}{5} \leftrightarrow (-3, -5).$$

Theorem 4.2 (Giroux, Honda). *On $L(p,q)$, there are exactly $|(r_0+1)(r_1+1)\ldots(r_k+1)|$ tight contact structures up to isotopy. They are all holomorphically fillable.*

A surgery presentation for $L(p,q)$ is given as follows:

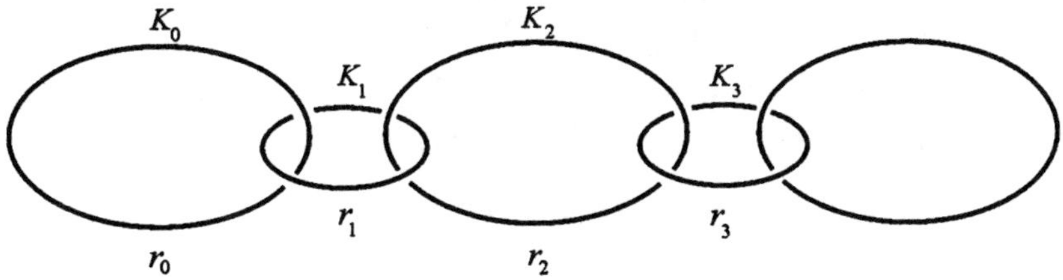

FIGURE 18

Legendrian surgery. Given a Legendrian knot $K = K_0$ or link $L = \sqcup_{i=0}^{k} K_i$ in a contact manifold (M,ξ), we can perform a surgery along the K_i with coefficient $tb(K_i) - 1$. At the 4-dimensional level, if $M = S^3$, then we start with a Stein domain B^4 with $\partial B^4 = S^3$, and attach 2-handles in a way which makes the resulting 4-manifold X^4 a Stein domain (and in particular symplectic). The resulting contact 3-manifold (M', ξ') with $\partial X = M'$ is said to be *holomorphically fillable*. Similarly, if (M,ξ) is symplectically fillable, then (M',ξ') obtained by Legendrian surgery is also symplectically fillable. The Stein construction was done by Eliashberg in [**9**] and the symplectic construction by Weinstein [**43**].

Suppose K_i is a Legendrian unknot with $tb(K_i) = r_i + 1$ and $r(K_i) =$ one of $r_i + 2, r_i + 4, \ldots, -(r_i + 2)$. There are precisely $|r_i + 1|$ choices for the rotation number $r(K_i)$. (In fact, these are all the Legendrian unknots with $tb(K_i) = r_i + 1$ by Theorem 1.2.)

HW 24. *Show that the* $|r_0 + 1||r_1 + 1|...|r_k + 1|$ *holomorphically fillable contact structures are distinct.*

Therefore, we have the lower bound:

$$\# \text{Tight}(L(p,q)) \geqslant |(r_0 + 1)(r_1 + 1) \ldots (r_k + 1)|.$$

Here Tight(M) refers to the set of isotopy classes of tight contact structures on M. In order to prove Theorem 4.2, it remains to show the reverse inequality.

4.3. Solid tori. We now consider tight contact structures on the solid torus $S^1 \times D^2$ with the following conditions on the boundary $T = S^1 \times D^2$:

(1) $\#\Gamma_T = 2$.

(2) slope$(\Gamma_T) = -\dfrac{p}{q}$, where $-\infty < -\frac{p}{q} \leqslant -1$. (After performing Dehn twists, we can normalize the slope as such.)

(3) The fixed characteristic foliation $\mathcal{F}$ is adapted to Γ_T.

Theorem 4.3. *There are exactly* $|(r_0 + 1)(r_1 + 1) \ldots (r_{k-1} + 1)r_k|$ *tight contact structures on* $S^1 \times D^2$ *with this boundary condition.*

STEP 1. In this step we factor $S^1 \times D^2$ into a union of $T^2 \times I$ layers and a standard neighborhood of a Legendrian curve isotopic to the core curve of $S^1 \times D^2$. Assume $-\dfrac{p}{q} < -1$ since $-\dfrac{p}{q} = -1$ has already been treated.

Let D be a meridional disk with ∂D Legendrian and $tb(\partial D) = -p < -1$. Then by Lemma 3.5 there is at least one bypass along ∂D. Attach the bypass to T from the interior and apply the Bypass Attachment Lemma. We obtain a convex torus T' isotopic to T, such that T and T' cobound a $T^2 \times I$. Denote

$$\text{slope}(\Gamma_{T'}) = -\frac{p'}{q'}.$$

HW 25. *If* $-\dfrac{p}{q} \leftrightarrow (r_0, r_1, \ldots, r_{k-1}, r_k)$, *then*

$$-\frac{p'}{q'} \leftrightarrow (r_0, r_1, \ldots, r_{k-1}, r_k + 1).$$

We successively peel off $T^2 \times I$ layers according to the Farey tessellation. The sequence of slopes is given by the continued fraction expansion, or, equivalently, by the shortest sequence of counterclockwise arcs in the Farey tessellation from $-\frac{p}{q}$ to -1. Once slope -1 is reached, $S^1 \times D^2$ with boundary slope -1 is the standard neighborhood of a Legendrian core curve with twisting number -1 (with respect to the fibration induced from the S^1-fibers $S^1 \times \{pt\}$).

STEP 2 (analysis of each $T^2 \times I$ layer).

FACT. Consider $T^2 \times [0,1]$ with convex boundary conditions $\#\Gamma_0 = \#\Gamma_1 = 2$, $s_0 = \infty$, and $s_1 = 0$. Here we write $\Gamma_i = \Gamma_{T^2 \times \{i\}}$ and $s_i = \text{slope}(\Gamma_i)$. (More invariantly, the shortest integers corresponding s_0, s_1 form an integral basis for $\mathbf{Z}^2$.) Then there are exactly two tight contact structures (up to isotopy rel boundary) which are *minimally twisting*, i.e., every convex torus T' isotopic to $T^2 \times \{i\}$ has slope$(\Gamma_{T'})$ in the interval $(0, +\infty)$. They are distinguished by the Poincaré duals of the *relative half-Euler class*, which are computed to be $\pm((1,0)-(0,1)) \in H_1(T^2 \times [0,1]; \mathbf{Z})$. We call these $T^2 \times [0,1]$ layers *basic slices*.

The proof of the fact will be omitted, but one of the key elements in the proof is the following lemma:

HW 26. *Prove, using the Imbalance Principle, that for any tight contact structure on $T^2 \times [0,1]$ with boundary slopes $s_0 \neq s_1$ and any rational slope s in the interval (s_1, s_0), there exists a convex surface $T' \subset T^2 \times [0,1]$, which is parallel to $T^2 \times \{pt\}$ and has slope s. Here, if $s_0 < s_1$, (s_1, s_0) means $(s_1, +\infty] \cup [-\infty, s_0)$.*

STEP 3 (shuffling). Consider the example of the solid torus where $-\frac{p}{q} = -\frac{14}{5}$. We have the following factorization:

$$\begin{aligned} -\frac{14}{5} &\leftrightarrow (-3,-5) \\ -\frac{11}{4} &\leftrightarrow (-3,-4) \\ -\frac{8}{3} &\leftrightarrow (-3,-3) \end{aligned}$$

$$\begin{aligned} -\frac{5}{2} &\leftrightarrow (-3,-2) \\ -2 &\leftrightarrow (-3,-1)=(-2) \\ -1 &\leftrightarrow (-1) \end{aligned}$$

We group the basic slices into *continued fraction blocks*. Each block consists of all the slopes whose continued fraction representations are of the same length. In the example, we have two blocks: slope $-\frac{14}{5}$ to -2, and slope -2 to -1. All the relative half-Euler classes of the basic slices in the first block are $\pm(-1,3)$; for the second block, they are $\pm(0,1)$. Therefore, a naive upper bound for the number of tight contact structures would be 2 to the power #(basic slices).

A closer inspection however reveals that we may *shuffle* basic slices which are in the same continued fraction block. More precisely, if $T^2 \times [0,2]$ admits a factoring into basic slices $T^2 \times [0,1]$ and $T^2 \times [1,2]$ with relative half-Euler classes (a,b) and $-(a,b)$, then it also admits a factoring into basic slices where the relative half-Euler classes are $-(a,b)$ and (a,b), i.e., the order is reversed.

Shuffling is (more or less) equivalent to the following proposition:

Lemma 4.4. *Let L be a Legendrian knot. Then*

$$S_+S_-(L) = S_-S_+(L).$$

HW 27. *Prove Lemma* 4.4. (*Observe that the ambient contact manifold is irrelevant and that the commutation can be done in a standard tubular neighborhood of L.*)

Returning to the example at hand, the first continued fraction block has at most $|-5| = 4+1$ tight contact structures (distinguished by the relative half-Euler class), and the second has at most $|-3+1| = 2$ tight contact structures. We compute $\#\,\text{Tight} \leqslant 2 \cdot 5$.

In general, for the solid torus with slope $-\frac{p}{q} \leftrightarrow (r_0, r_1, \ldots, r_k)$ we have

$$\#\,\text{Tight} \leqslant |(r_0+1)(r_1+1)\ldots(r_{k-1}+1)r_k|. \tag{1}$$

4.4. Completion of the proof of Theorems 4.2 and 4.3. We prove the following, which instantaneously completes the proof of both theorems.

$$\#\,\text{Tight}(L(p,q)) \leqslant |(r_0+1)(r_1+1)\ldots(r_k+1)|. \tag{2}$$

Recall that on ∂V_1, the meridian of V_2 has slope

$$-\frac{p}{q} \leftrightarrow (r_0, r_1, \ldots, r_{k-1}, r_k).$$

First, take a Legendrian curve γ isotopic to the core curve of V_2 with largest twisting number. (Such a Legendrian curve exists since any closed curve admits a C^0-small approximation by a Legendrian curve; the upper bound exists by the Thurston–Bennequin inequality.) We may assume V_2 is the standard neighborhood of γ; the tight contact structure on V_2 is then unique up to isotopy. Next,

$$\mathrm{slope}(\Gamma_{\partial V_1}) = -\frac{p'}{q'} \leftrightarrow (r_0, \ldots, r_{k-1}, r_k + 1),$$

and we have already computed the upper bound for $\#\,\mathrm{Tight}(V_2)$ to be $|(r_0+1)\ldots(r_{k-1}+1)(r_k+1)|$ by Equation 1. This completes the proof of Equation 2 and hence of Theorems 4.2 and 4.3.

Open question. Give a complete classification of tight contact structures on $T^2 \times [0,1]$ when $\#\Gamma_{T^2\times\{i\}} > 2$, $i = 0, 1$. (Contrary to what is claimed in [**24**], the general answer is not yet known.)

5. Gluing

There are three general methods for proving tightness:

1. Symplectic filling.
2. Gauge theory (in particular Heegaard Floer homology).
3. Gluing (state traversal).

Symplectic filling was already discussed in Section 1.5. We briefly explain the relationship between contact structures and the *Heegaard Floer homology* of Ozsvath and Szabo [**40, 41**]. To an oriented closed 3-manifold M one can assign a *Heegaard Floer homology group* $\widehat{HF}(M)$, constructed out of the Heegaard decomposition of M. In [**42**], Ozsvath and Szabo assigned a class $c(\xi) \in \widehat{HF}(-M)$ to every contact structure (M, ξ) (tight or overtwisted). This was done via the work of Giroux [**22**] in which it was shown that every contact structure (M, ξ) corresponds to an equivalence class of open book decompositions of M (and hence an equivalence class of fibered knots). Lisca and Stipsicz [**35**] showed that large families of contact structures are tight (but not fillable) by showing that their Heegaard Floer homology class is nonzero. The Heegaard

Floer homology approach appears to be very promising at the time of the writing of this article.

In this section we focus on the last technique, namely gluing. Many of the key ideas in gluing were introduced by Colin [**4**, **5**] and Makar-Limanov [**39**], and subsequently enhanced by Honda [**25**] who combined them with the bypass technology.

Let us start by asking the following question:

Question 5.1. *Let* Σ *be a convex surface in* (M,ξ). *If* $(M\backslash\Sigma, \xi|_{M\backslash\Sigma})$ *is tight, then is* (M,ξ) *tight?*

Answer: This is usually not true. Our goal is to understand to what extent it is true.

HW 28. *Give an example of an overtwisted* $T^2\times[0,1]$ *which is tight when restricted to* $T^2\times[0,1/2]$ *and to* $T^2\times[1/2,1]$.

5.1. Basic examples with trivial state transitions.

EXAMPLE A (Colin [**5**], Makar-Limanov [**39**]). Suppose $\Sigma = S^2$. If $(M\setminus\Sigma, \xi|_{M\backslash\Sigma})$ is tight, then (M,ξ) is tight.

PROOF. Recall that there is only one possibility for Γ_{S^2} inside a tight contact manifold. We argue by contradiction. Suppose there is an OT disk $D\subset M$. A priori, the OT disk D can intersect Σ in a very complicated manner. We obtain a contradiction as follows:

1. Isotop Σ to Σ' so that $\Sigma'\cap D=\varnothing$.
2. *Discretize* the isotopy $\Sigma_0=\Sigma\to\Sigma_1\to\cdots\to\Sigma_n=\Sigma'$, so that each step is obtained by attaching a *bypass*.
3. If $(M\backslash\Sigma_i, \xi|_{M\backslash\Sigma_i})$ is tight, then $\Gamma_{\Sigma_i}=\Gamma_{\Sigma_{i+1}}=S^1$ and the bypass must be *trivial*. Hence

$$(M\setminus\Sigma_i, \xi|_{M\backslash\Sigma_i})\simeq(M\setminus\Sigma_{i+1}, \xi|_{M\backslash\Sigma_{i+1}}).$$

We have proved inductively that $(M\setminus\Sigma', \xi|_{(M\backslash\Sigma')})$ is tight, a contradiction. □

More generally, one can prove:

Theorem 5.2 (Colin [**5**]). *If* $M=M_1\#M_2$, *then*

$$\text{Tight}(M)\simeq\text{Tight}(M_1)\times\text{Tight}(M_2).$$

HW 29. *Classify tight contact structures on $S^1 \times S^2$.*

EXAMPLE B (Colin [4]). If $\Sigma = D^2$ and Γ_Σ is ∂-*parallel*, then $(M \setminus \Sigma, \xi|_{M\setminus\Sigma})$ tight $\Rightarrow (M, \xi)$ tight.

EXAMPLE C (Colin [4]). Let Σ be an incompressible surface with $\partial\Sigma \neq \varnothing$. If Γ_Σ is ∂-parallel and $(M \setminus \Sigma, \xi|_{M\setminus\Sigma})$ is universally tight, then (M, ξ) is universally tight. (This is Theorem 2.15 above.)

Question 5.3. *In Example* C, *does $(M\setminus\Sigma, \xi|_{M\setminus\Sigma})$ tight imply (M, ξ) tight? In other words, can universal tightness be avoided?*

All of the above examples can be characterized by the fact that the *state transitions* are trivial. However, to create more interesting examples, we need to "traverse all states."

5.2. More complicated example.

EXAMPLE D (Honda [25]). Let H be a handlebody of genus g and $D_1, \ldots, D_g$ be compressing disks so that $H \setminus (D_1 \sqcup \cdots \sqcup D_g) = B^3$. Fix $\Gamma_{\partial H}$ (and a compatible characteristic foliation). Note that we need $tb(D_i) \leqslant -1$ since otherwise we can find an OT disk using LeRP.

Let $\mathcal{C}$ be the *configuration space*, i.e., the set of all possible $C = (\Gamma_{D_1}, \ldots, \Gamma_{D_g})$, where each Γ_{D_i} has no closed curves. The cardinality of $\mathcal{C}$ is *finite*. If we cut H along $\Sigma = D_1 \cup \cdots \cup D_g$, then we obtain a 3-ball with corners. Given a configuration C, we can round the corners, as previously explained in Section 2.3. Now, if $\Gamma_{\partial(H\setminus\Sigma)} = S^1$ after rounding, then C is said to be *potentially allowable.*

State transitions. The smallest unit of isotopy (in the contact world) is a bypass attachment. Therefore we examine the effect of one bypass attachment onto D_i. First we need to ascertain whether a candidate bypass exists.

Criterion for existence of state transition. The candidate bypass exists if and only if attaching the bypass from the interio r of $B^3 = H \setminus \Sigma$ does not increase $\#\Gamma_{\partial B^3}$.

We construct a graph Γ with $\mathcal{C}$ as the vertices. We assign an edge from $(\Gamma_{D_1}, \ldots, \Gamma_{D_i}, \ldots, \Gamma_{D_g})$ to $(\Gamma_{D_1}, \ldots, \Gamma_{D_i'}, \ldots, \Gamma_{D_g})$ if there is a state transition $D_i \to D_i'$ given by a single bypass move. Note that the bypass may be from either side of D_i. Then we have:

Theorem 5.4. Tight$(H, \Gamma_{\partial H})$ *is in* 1-1 *correspondence with the connected components of* Γ, *all of whose vertices* C *are potentially allowable.*

HW 30. *Explain why* Tight$(H, \Gamma_{\partial H})$ *is finite.*

REMARK. Since $\mathcal{C}$ is a finite graph, in theory we can compute Tight$(H, \Gamma_{\partial H})$ for any handlebody H with a fixed boundary $\Gamma_{\partial H}$. Tanya Cofer, a (former) graduate student at the University of Georgia, has programmed this for $g = 1$, and the experiment agrees with the theoretical number from Theorem 4.2, in case $\#\Gamma_{\partial H} = 2$ and the slope is $-\frac{p}{q}$ with $p \leqslant 10$.

HW 31. *Using the state transition technique, analyze tight contact structures on* $S^1 \times D^2$, *where* Γ_{T^2}, $T^2 = \partial(S^1 \times D^2)$, *satisfies the following:*

1. $\#\Gamma_{T^2} = 2$ *and* slope$(\Gamma_{T^2}) = -2$.
2. $\#\Gamma_{T^2} = 2$ *and* slope$(\Gamma_{T^2}) = -3$.
3. $\#\Gamma_{T^2} = 4$ *and* slope$(\Gamma_{T^2}) = \infty$.

Here the slope of the meridian is 0 *and the slope of some preferred longitude is* ∞.

5.3. Tightness and fillability. We present two examples which show that the world of tight contact structures is larger than the world of symplectically fillable contact structures.

EXAMPLE E (Honda [**25**]). We present a tight handlebody H of genus 4 which becomes OT after a Legendrian surgery. Since Legendrian surgery preserves fillability, the tight handlebody cannot be embedded in side any closed fillable contact 3-manifold.

We take the union $H = M_1 \cup M_2$, where $M_1 = S^1 \times D^2$ is the standard tubular neighborhood of a Legendrian curve and M_2 is an I-invariant neighborhood of a convex disk S with 4 holes. Here $\partial S = \gamma - \bigcup_{i=1}^{4} \gamma_i$ and Γ_S consists of 4 arcs, one each from γ_i to γ_{i+1} (i mod 4). The gluing is presented in Figure 19, where $T^2 = \partial(S^1 \times D^2)$ is drawn so that Γ_{T^2} has slope ∞, the γ_i have slope 0, and the meridian of M_1 has slope 1.

A Legendrian surgery along the core curve of M_1 yields a new meridional slope of 0 along T^2, and hence allows S to be completed to an OT disk. Using the state transition method, one can prove that the contact structure is tight.

HW 32. *Verify the tightness.*

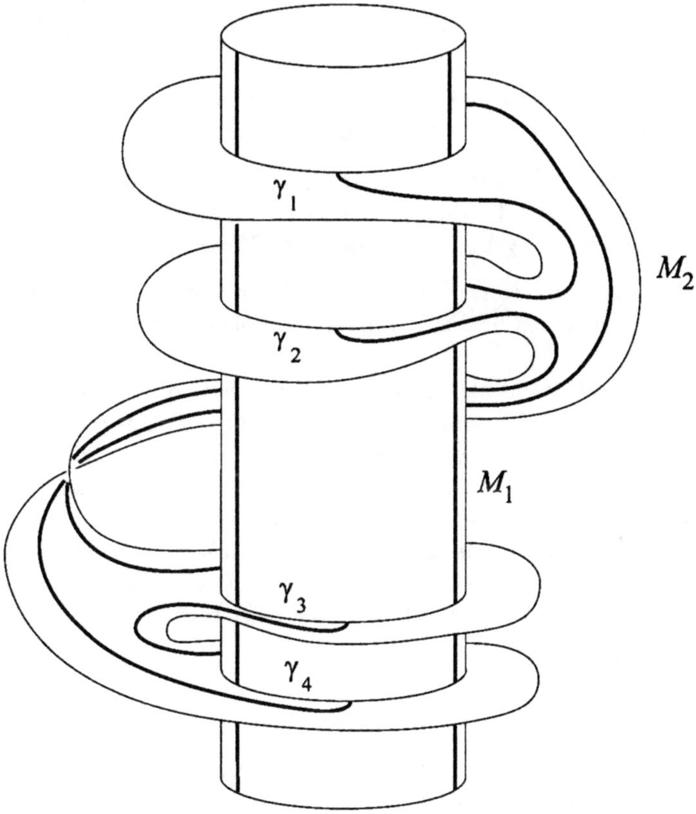

FIGURE 19. The top and bottom are identified

EXAMPLE F (Etnyre–Honda [**16**]). Consider the torus bundle $M = (T^2\times[0,1])/\sim$, where $(x,1)\sim(Ax,0)$, $T^2 = \mathbf{R}^2/\mathbf{Z}^2$, and $A = \begin{pmatrix} 0 & 1 \\ -1 & 0 \end{pmatrix}$. Let $T^2\times[0,1]$ be a basic slice with boundary slopes $s_0 = \infty$ and $s_1 = 0$. The glued-up contact structure ξ is proved to be tight using state traversal. However, ξ is not symplectically fillable by the following contradiction argument:

(1) M is a Seifert fibered space over S^2 with Seifert invariants $(-\frac{1}{2}, \frac{1}{4}, \frac{1}{4})$.

(2) There exists a Legendrian surgery taking (M, ξ) to (M', ξ'), where M' is a Seifert fibered space over S^2 with invariants $(-\frac{1}{2}, \frac{1}{3}, \frac{1}{4})$. Since Legendrian surgery preserves fillability, if ξ is fillable, ξ' is also fillable.

(3) A theorem of Lisca [**32**], proved using Seiberg–Witten theory, states that there are no fillable contact structures on M'.

REMARK. Example F was the first example of a tight contact structure which is not fillable. Since then, numerous other examples have been discovered by Lisca and Stipsicz [**33, 34, 35**].

Open question. Elucidate the difference between the world of tight contact structures and the world of fillable contact structures.

Acknowledgements. This manuscript grew out of a lecture series given at the Winter School in Contact Geometry in München in February 2003 and a minicourse given at the *Geometry and Foliations 2003* conference, held at Ryokoku University in Kyoto in September 2003. I would like to thank Kai Cieliebak and Dieter Kotschick for the former, and Takashi Tsuboi for the latter, as well as for his hospitality during my visit to the University of Tokyo and the Tokyo Institute of Technology during the summer and fall of 2003. Much of the actual writing took place during this visit.

References

1. B. Aebischer, et. al., *Symplectic Geometry*, Progress in Math. **124**, Birkhäuser, Basel, Boston and Berlin, 1994.
2. D. Bennequin, *Entrelacements et équations de Pfaff*, Astérisque **107–108** (1983), 87–161.
3. Y. Chekanov, *Differential algebra of Legendrian links*, Invent. Math. **150** (2002), 441–483.
4. V. Colin, *Recollement de variétés de contact tendues*, Bull. Soc. Math. France **127** (1999), 43–69.
5. V. Colin, *Chirurgies d'indice un et isotopies de sphères dans les variétés de contact tendues*, C. R. Acad. Sci., Paris, Ser. I, Math. **324** (1997), 659–663.
6. Ya. Eliashberg, *Filling by holomorphic discs and its applications*, Lond. Math. Soc. Lect. Note Ser. **151** (1991), 45–67.
7. Ya. Eliashberg, *Contact 3-manifolds twenty years since J. Martinet's work*, Ann. Inst. Fourier (Grenoble) **42** (1992), 165–192.

8. Ya. Eliashberg, *Classification of overtwisted contact structures on 3-manifolds*, Invent. Math. **98** (1989), 623–637.
9. Ya. Eliashberg, *Topological characterization of Stein manifolds of dimension > 2*, Internat. J. Math. **1** (1990), 29–46.
10. Ya. Eliashberg and M. Fraser, *Classification of topologically trivial Legendrian knots*, Geometry, topology, and dynamics (Montreal, PQ, 1995), 17–51, CRM Proc. Lecture Notes **15**, Am. Math. Soc., Providence, RI, 1998.
11. Ya. Eliashberg, A. Givental, and H. Hofer, *Introduction to symplectic field theory*, GAFA 2000 (Tel Aviv, 1999), Geom. Funct. Anal. 2000, Special Volume, Part II, 560–673.
12. Ya. Eliashberg and W. Thurston, *Confoliations*, University Lecture Series **13**, Am. Math. Soc., Providence, 1998.
13. J. Etnyre, *Introductory lectures on contact geometry*, Topology and Geometry of Manifolds (Athens, GA, 2001), 81–107, Proc. Sympos. Pure Math. **67**, Am. Math. Soc., Providence, RI, 2003.
14. J. Etnyre, *Tight contact structures on lens spaces*, Commun. Contemp. Math. **2** (2000), 559–577.
15. J. Etnyre, *Legendrian and transversal knots*, ArXiv: math.SG/0306256, to appear in the Handbook of Knot Theory.
16. J. Etnyre and K. Honda, *Tight contact structures with no symplectic fillings*, Invent. Math. **148** (2002), 609–626.
17. D. Gabai, *Foliations and the topology of* 3*-manifolds*, J. Differ. Geom. **18** (1983), 445–503.
18. H. Geiges, *Contact geometry*, ArXiv: math.SG/0307242, to appear in the Handbook of Differential Geometry, Vol. 2.
19. E. Giroux, *Convexité en topologie de contact*, Comment. Math. Helv. **66** (1991), 637–677.
20. E. Giroux, *Structures de contact en dimension trois et bifurcations des feuilletages de surfaces*, Invent. Math. **141** (2000), 615–689.
21. E. Giroux, *Structures de contact sur les variétés fibrées en cercles au-dessus d'une surface*, Comment. Math. Helv. **76** (2001), 218–262.
22. E. Giroux, *Géométrie de contact: de la dimension trois vers les dimensions supérieures*, Proceedings of the International Congress of Mathematicians, Vol. II (Beijing, 2002), 405–414, Higher Ed. Press, Beijing, 2002
23. M. Gromov, *Pseudo-holomorphic curves in symplectic manifolds*, Invent. Math. **82** (1985), 307–347.
24. K. Honda, *On the classification of tight contact structures I*, Geom. Topol. **4** (2000), 309–368.
25. K. Honda, *Gluing tight contact structures*, Duke Math. J. **115** (2002), 435–478.
26. K. Honda, W. Kazez, and G. Matić, *Tight contact structures and taut foliations*, Geom. Topol. **4** (2000), 219–242.
27. K. Honda, W. Kazez, and G. Matić, *Convex decomposition theory*, Internat. Math. Res. Notices **2002**, 55–88.
28. K. Honda, W. Kazez, and G. Matić, *Tight contact structures on fibered hyperbolic 3-manifolds*, J. Differ. Geom. **64** (2003), 305–358.

29. K. Honda, W. Kazez, and G. Matić, *On the Gabai–Eliashberg–Thurston theorem*, ArXiv: math.GT/0111123, to appear in Comment. Math. Helv.
30. Y. Kanda, *The classification of tight contact structures on the 3-torus*, Comm. Anal. Geom. **5** (1997), 413–438.
31. W. Kazez, *A cut-and-paste approach to contact topology*, ArXiv: math.GT/0210091.
32. P. Lisca, *Symplectic fillings and positive scalar curvature*, Geom. Topol. **2** (1998), 103–116.
33. P. Lisca and A. Stipsicz, *An infinite family of tight, not semi-fillable contact three-manifolds*, ArXiv: math.SG/0208063.
34. P. Lisca and A. Stipsicz, *Tight, not semi-fillable contact circle bundles*, ArXiv: math.SG/0211429.
35. P. Lisca and A. Stipsicz, *Heegaard Floer invariants and tight contact 3-manifolds*, ArXiv: math.SG/0303280.
36. R. Lutz, *Sur quelques propriétes des formes différentielles en dimension 3*, Ph. D. Thesis, Strasbourg, 1971.
37. J. Martinet, *Formes de contact sur les variétés de dimension 3*, Springer Lecture Notes in Math **209**, 142–163.
38. S. Makar-Limanov, *Tight contact structures on solid tori*, Trans. Am. Math. Soc. **350** (1998), 1045–1078.
39. S. Makar-Limanov, *Morse Surgeries of Index 0 on Tight Manifolds*, preprint 1997.
40. P. Ozsvath and Z. Szabo, *Holomorphic disks and topological invariants for closed three-manifolds*, ArXiv: math.SG/0101206, to appear in Ann. Math.
41. P. Ozsvath and Z. Szabo, *Holomorphic disks and three-manifold invariants: properties and applications*, ArXiv: math.SG/0105202, to appear in Ann. Math.
42. P. Ozsvath and Z. Szabo, *Heegaard Floer homologies and contact structures*, ArXiv: math.SG/0210127.
43. A. Weinstein, *Contact surgery and symplectic handlebodies*, Hokkaido Math. J. **20** (1991), 241–251.

Asymptotic Convex Geometry Short Overview

A. A. Giannopoulos
University of Crete
Iraklion, Greece

V. D. Milman†
Tel Aviv University
Tel Aviv, Israel

We outline the development of Asymptotic Convex Geometry, a relatively recent theory having its origin in Convex Geometry and Geometric Functional Analysis, which is concerned with the geometric and linear properties of high dimensional normed spaces and convex bodies. The asymptotic theory has succeeded to reveal the underlying order and structures which accompany high dimensional convex bodies. In many important respects, convex bodies exhibit less and less diversity as the dimension increases; all bodies share important common features and most bodies "look similar." We demonstrate this unified behavior of the family of "all convex bodies" through classical results which form the main body of the theory as well as important very recent discoveries.

---

†The work was supported in part by the Israel Science Foundation and by the Minkowski Center for Geometry.

***Different Faces of Geometry*, Edited by Donaldson *et al.*,**
Kluwer Academic/Plenum Publishers, New York, 2004

1. Introduction

In this article, we outline a rapidly developing theory of high dimensional normed spaces and convex bodies. The classical Convex Geometry, sometimes called Brunn–Minkowski theory, studies the geometry of convex bodies and related geometric inequalities in Euclidean space of a fixed dimension (because of this, it is an *isometric* theory). The classical Functional Analysis is standardly understood as the theory of infinite dimensional spaces. However, it is a relatively recent discovery that there is a theory "in between," which is concerned with the geometric and linear properties of finite dimensional normed spaces or convex bodies, the emphasis now being on the asymptotic behavior of various quantitative parameters as the dimension grows to infinity. We call it Asymptotic Geometric Analysis, but also Asymptotic Convex Geometry (actually, more names are associated to it: history has not yet selected the right one). In the framework of this theory, very unexpected phenomena, hidden structures and forms of behavior were discovered, new intuition was built and many new tools were developed. It is now clear that the theory provides the right questions to reveal the underlying "order" and structures which accompany high dimensional spaces.

The quantitative study of high dimensional normed spaces used many of the tools of convex geometry. However, these tools were now used under a different point of view. The isometric questions which were typical in classical convexity were replaced by isomorphic ones, which were most natural for functional analysis but alien to convexity. Isoperimetric type problems provide a bold example of this transformation. Isomorphic versions of such problems, which make sense only from the asymptotic point of view, led to the discovery of the concentration of measure phenomenon, which plays a crucial role in the proof of Dvoretzky type theorems. Later, the method spread and influenced the development of other "asymptotic" theories in Probability, Asymptotic Combinatorics and Complexity, where much more general high parametric systems arise.

After this major step on the conceptual level, many unsolved problems of classical convexity were put in asymptotic form and were studied systematically. In this way, the two theories started to interact with many deep consequences in both analysis and geometry. Typical examples are the reverse Brunn–Minkowski inequality and the reverse Santaló inequality, which provides an affirmative answer — at least in the asymptotic sense — to a classical conjecture of Mahler.

The article is organized as follows: Section 2 gives a brief synopsis of the major results of asymptotic convex geometry (the concept of concentration, Dvoretzky type theorems, Pisier's inequality on the Rademacher projection, Milman's low M^*-estimate and the quotient of subspace theorem, entropy estimates) and of some more recent important directions (global theory, asymptotic formulas and phase transition behavior, "coordinate theory").

Section 3 contains background material from classical convexity: the Brunn–Minkowski inequality and its functional forms, the Alexandrov–Fenchel inequality and related geometric inequalities about mixed volumes of convex bodies, volume preserving transformations (Knöthe and Brenier maps).

Section 4 describes classical positions of convex bodies such as John's position, the minimal surface area and the minimal mean width positions. They are all characterized as isotropic ones, an observation which relates them to the Brascamp–Lieb inequality and its reverse form. Some sharp geometric inequalities are applications of this point of view, a fact which was first observed and successfully exploited by Ball. We also give a short account on the challenging slicing problem.

Section 5 gives some classical and recent examples of the interaction between the asymptotic convex geometry point of view and classical convexity. The reverse Santaló inequality and the reverse Brunn–Minkowski inequality are proved with the method of isomorphic symmetrization. This discussion introduces M-ellipsoids and their basic properties. Recent results of Klartag and Milman on the number of Minkowski or Steiner symmetrizations that are needed in order to bring an arbitrary convex body close to a ball give another example of use of the asymptotic theory in questions with classical convexity flavor.

Section 6, which is closer to the spirit of geometric functional analysis, is devoted to the geometry of the Banach–Mazur compactum and some questions on the local structure of high-dimensional normed spaces. Random spaces, which were first introduced by Gluskin, play an important role in this discussion.

A number of surveys on different aspects of the theory were recently published (see [**14, 86, 87, 109, 110**]). In particular, [**54**] gives a more geometrically directed point of view on the theory. However, this article was written before 1999 and a new stream of results is now available.

We cannot avoid repeating the very basic and already classical line of development we described, but we refer to [54] for many proofs which are outlined there in a very condense form. General references on the Brunn–Minkowski theory and geometric inequalities are the books of Schneider [134] and Burago–Zalgaller [34]. The reader may consult the books of Milman–Schechtman [113], Pisier [121] and Tomczak-Jaegermann [147] for various aspects of the asymptotic theory of finite dimensional normed spaces.

2. Asymptotic Convex Geometry

We study finite-dimensional real normed spaces $X = (\mathbb{R}^n, \|\cdot\|)$. The unit ball K_X of such a space is a symmetric (with respect to the origin) convex body in $\mathbb{R}^n$. Conversely, if K is a symmetric convex body, then $\|x\|_K = \min\{\lambda \geqslant 0 : x \in \lambda K\}$ is a norm defining a space X_K with K as its unit ball. If K_1 and K_2 are symmetric convex bodies in $\mathbb{R}^n$, their geometric distance $d(K_1, K_2)$ is defined by

$$d(K_1, K_2) = \inf\{ab : a, b > 0, K_1 \subseteq bK_2, K_2 \subseteq aK_1\}.$$

The natural distance between the n-dimensional spaces X_{K_1} and X_{K_2} is the *Banach–Mazur distance*

$$d(X_{K_1}, X_{K_2}) = \inf\{d(K_1, T(K_2)) : T \in GL(n)\}.$$

Note that $d(X_{K_1}, X_{K_2})$ is the smallest positive number d for which we can find $T \in GL(n)$ such that $K_1 \subseteq T(K_2) \subseteq dK_1$. In the language of geometric functional analysis, if X and Y are two n-dimensional normed spaces, then

$$d(X, Y) = \min\{\|T\| \, \|T^{-1}\| \ : \ T : X \to Y \text{ is an isomorphism}\}.$$

We assume that $\mathbb{R}^n$ is equipped with a Euclidean structure $\langle \cdot, \cdot \rangle$ and denote the corresponding Euclidean norm by $|\cdot|$. B_2^n is the Euclidean unit ball and S^{n-1} is the unit sphere. The rotationally invariant probability measure on S^{n-1} will be denoted by σ. The unit ball of ℓ_p^n is denoted by B_p^n. By a classical theorem of John [73], one has $d(X, \ell_2^n) \leqslant \sqrt{n}$ for every n-dimensional normed space X (see also § 4.1).

If K is a symmetric convex body in $\mathbb{R}^n$, its polar body is defined by $\|y\|_{K^\circ} = \max_{x \in K} |\langle x, y \rangle|$. Note that $X_{K^\circ} = X_K^*$: K° is the unit ball of the dual space of X.

Let K be a convex body in $\mathbb{R}^n$ with $0 \in \mathrm{int}(K)$. The radial function $\rho_K : \mathbb{R}^n \setminus \{0\} \to \mathbb{R}^+$ of K is defined by $\rho_K(x) = \max\{\lambda > 0 : \lambda x \in K\}$. The support function $h_K : \mathbb{R}^n \to \mathbb{R}$ of K is defined by $h_K(x) = \max\{\langle x, y\rangle : y \in K\}$. The width of K in the direction of $\theta \in S^{n-1}$ is the quantity $w(K,\theta) = h_K(\theta) + h_K(-\theta)$, and the mean width of K is defined by

$$w(K) = \frac{1}{2} \int_{S^{n-1}} w(K,\theta)\sigma(d\theta) = \int_{S^{n-1}} h_K(\theta)\sigma(d\theta).$$

Note that if K is symmetric then $\rho_K(x) = 1/\|x\|_K$ and $h_K(x) = \|x\|_{K^\circ}$.

2.1. Isomorphic isoperimetric inequalities and concentration of measure. Concentration of measure was understood and developed as a method for the goals of geometric functional analysis, but it was soon realized that it was very well adapted to the needs of probability theory, asymptotic combinatorics and complexity. General references on concentration, from various viewpoints, are the following surveys and books: [**14, 68, 69, 80, 81, 104, 110, 132**].

The general framework is a probability space $(X, \mathcal{A}, d, \mu)$, where $\mathcal{A}$ is the Borel σ-algebra with respect to a given metric d on X. For every $A \in \mathcal{A}$, we consider the t-extension $A_t = \{x \in X : d(x, A) \leqslant t\}$ of A. One can then formulate the abstract isoperimetric problem for metric probability spaces as follows: Given $0 < \alpha < 1$ and $t > 0$, find $\inf\{\mu(A_t) : A \in \mathcal{A}, \mu(A) \geqslant \alpha\}$ and describe the sets A on which this infimum is possibly attained. The complete answer to the isoperimetric problem is available for a few but very important geometric examples.

Spherical isoperimetric inequality. Consider the sphere S^{n-1} as a metric probability space, with the geodesic distance ρ and the $O(n)$-invariant probability measure σ. The spherical isoperimetric inequality states that spherical caps of the form $B(x,r)$ are the extremal sets: if A is a Borel subset of S^{n-1} and $\sigma(A) = \sigma(B(x_0,r))$ for some $x_0 \in S^{n-1}$ and $r > 0$, then

$$\sigma(A_t) \geqslant \sigma(B(x_0, r+t)) \tag{2.1.1}$$

for every $t > 0$. This is proved by spherical symmetrization (see, for example, [**44**]). Since spherical caps are easy to work with, one can use (2.1.1) to obtain a good lower bound for the measure of the t-extension of an arbitrary subset of the sphere in terms of its measure. The most important case is when $\sigma(A) = 1/2$ (see [**113**]).

Theorem 2.1. *If A is a Borel subset of S^{n+1} and $\sigma(A) = 1/2$, then*

$$\sigma(A_t) \geqslant 1 - \sqrt{\pi/8}\exp(-t^2 n/2) \qquad (2.1.2)$$

for every $t > 0$. □

Isoperimetric inequality in Gauss space. Consider $\mathbb{R}^n$ as a metric probability space, with the Euclidean distance $|\cdot|$ and the standard Gaussian probability measure γ_n. The isoperimetric inequality in Gauss space (proved by Borell and Sudakov–Tsirelson, see [**80**] or [**81**] for references) states that halfspaces are the extremal sets: if $\alpha \in (0,1)$, $\theta \in S^{n-1}$ and $H = \{x \in \mathbb{R}^n : \langle x, \theta\rangle \leqslant s\}$ is a halfspace in $\mathbb{R}^n$ with $\gamma_n(H) = \alpha$, then, for every $t > 0$ and every Borel subset A of $\mathbb{R}^n$ with $\gamma_n(A) = \alpha$, one has

$$\gamma_n(A_t) \geqslant \gamma_n(H_t). \qquad (2.1.3)$$

A direct computation shows the following.

Theorem 2.2. *If $\gamma_n(A) \geqslant 1/2$ then for every $t > 0$*

$$\gamma_n(A_t) \geqslant 1 - \frac{1}{2}\exp(-t^2/2). \qquad (2.1.4)$$

□

These examples lead to the definition of the concentration function of a metric probability space. For every $t \geqslant 0$ we set

$$\alpha(X,t) := 1 - \inf\{\mu(A_t) : \mu(A) \geqslant 1/2\}. \qquad (2.1.5)$$

P. Lévy [**82**] realized the role of the dimension in the spherical isoperimetric inequality (2.1.2): if we fix $\alpha = 1/2$ and $t > 0$, as the dimension n increases to infinity the measure of the complement of A_t decreases exponentially to zero for every subset A of S^{n-1} with $\sigma(A) = 1/2$. Following this basic example, we say that a sequence $(X_n, \mathcal{A}_n, d_n, \mu_n)$ of metric probability spaces is a normal Lévy family with constants (c_1, c_2) if

$$\alpha(X_n, t) \leqslant c_1 \exp(-c_2 t^2 n). \qquad (2.1.6)$$

There are many examples of normal Lévy families which have found applications in the asymptotic theory of finite dimensional normed spaces. For some important metric probability spaces X, the exact solution to the isoperimetric problem was (and still is) unknown: new and very interesting techniques were invented in order to estimate the concentration function $\alpha(X,t)$. Natural families of obvious geometric importance are the following.

1. The family of the orthogonal groups $(SO(n), \rho_n, \mu_n)$ equipped with the Hilbert–Schmidt metric and the Haar probability measure is a Lévy family with constants $c_1 = \sqrt{\pi/8}$ and $c_2 = 1/8$.

2. The family $X_n = \prod_{i=1}^{m_n} S^n$ with the natural Riemannian metric and the product probability measure is a Lévy family with constants $c_1 = \sqrt{\pi/8}$ and $c_2 = 1/2$.

3. All homogeneous spaces of $SO(n)$ inherit the property of forming Lévy families. In particular, any family of Stiefel manifolds W_{n,k_n} or any family of Grassman manifolds G_{n,k_n} is a Lévy family with the same constants as in 1. These first three examples appear in [**70**].

4. The spaces $E_2^n = \{-1,1\}^n$ with the normalized Hamming distance $d(\eta,\eta') = \#\{i \leqslant n : \eta_i \neq \eta_i'\}/n$ and the normalized counting measure form a Lévy family with constants $c_1 = 1/2$ and $c_2 = 2$. This follows from an isoperimetric inequality of Harper [**72**] and it was first stated in this form and used in [**5**].

5. The group Π_n of permutations of $\{1,\dots,n\}$ equipped with the normalized Hamming distance $d(\sigma,\tau) = \#\{i \leqslant n : \sigma(i) \neq \tau(i)\}/n$ and the normalized counting measure satisfies $\alpha(\Pi_n,t) \leqslant 2\exp(-t^2 n/64)$. This was proved by Maurey [**94**] with a martingale method, which was further developed by Schechtman [**131**].

An equivalent way to express concentration is by means of Lipschitz functions (see [**80**] or [**113**]).

Theorem 2.3. *Let $(X, \mathcal{A}, d, \mu)$ be a metric probability space. If $f : X \to \mathbb{R}$ is a Lipschitz function with constant 1, then*

$$\mu(\{x \in X : |f(x) - M_f| > t\}) \leqslant 2\alpha(X,t), \tag{2.1.7}$$

where M_f is the Lévy median of f. □

Therefore, if the concentration function of X is small, Lipschitz functions are almost constant on almost all space. This observation has very important applications to the study of the normal Lévy families above.

Many problems which arise in the asymptotic geometric analysis require the proof of the existence of some geometric structure with prescribed behavior. The basic idea of the probabilistic method is to show that a random element of a suitable metric probability space has the required properties. The method (which was first used in combinatorial geometry and graph theory) works because the desirable structure is quite often the typical one. The concentration phenomenon provides a powerful tool for the probabilistic method, since it enables us to identify the typical structure in many situations. The first appearance of this idea in Analysis was in the proof of Dvoretzky's theorem in [**97**], which we discuss in the next subsection.

2.2. Dvoretzky type theorems. Dvoretzky's theorem [**40, 41**] states that every high-dimensional normed space has a subspace of "large dimension" which is well isomorphic to the Euclidean space. We use the terminology "Dvoretzky type theorems" for a wide family of results which exhibit large nice substructures inside normed spaces of sufficiently high dimension. The concrete estimates regarding the different parameters which enter in this type of results have become a crucial and important topic in the theory. There are many theorems which provide such estimates and even asymptotic formulas depending on different parameters.

The starting point for Dvoretzky's original theorem is a lemma of Dvoretzky and Rogers [**42**], which shows that for every symmetric convex body K whose maximal volume ellipsoid is B_2^n (see § 4.1), there exist $k \simeq \sqrt{n}$ and a k-dimensional subspace E_k of $\mathbb{R}^n$ such that $B_2^n \cap E_k \subseteq K \cap E_k \subseteq 2Q_k$, where Q_k is the unit cube in E_k with respect to a suitable coordinate system. Grothendieck asked whether it is possible to replace Q_k by $B_2^n \cap E_k$ in this statement, so that k will be still increasing to infinity with n. Dvoretzky's theorem provides an affirmative answer to this question. The best known version can be stated in the language of geometric functional analysis as follows.

Theorem 2.4. *Let X be an n-dimensional normed space and $\varepsilon > 0$. There exist an integer $k \geqslant c\varepsilon^2 \log n$ and a k-dimensional subspace E_k of X which satisfies $d(E_k, \ell_2^k) \leqslant 1 + \varepsilon$.* □

The example of ℓ_∞^n shows that the logarithmic dependence of k on n is best possible for small values of ε. The exact relation between n, ε and k has not been settled. It seems reasonable that ℓ_∞^n represents the worst case. This would mean that, for fixed k and ε, every n-dimensional normed space has a k-dimensional subspace which is $(1+\varepsilon)$-isomorphic to ℓ_2^k, provided that $n \geqslant c(k)\varepsilon^{-\frac{k-1}{2}}$. The problem is very interesting even for small values of k (actually, it is completely understood only in the case $k = 2$) and has connections with other branches of mathematics (algebraic topology, number theory, harmonic analysis, see [**103**] for a discussion).

The proof of Theorem 2.4 given in [**97**] (with a slightly worse dependence on ε) uses the concentration of measure on S^{n-1}. We start with an n-dimensional normed space X, and we may assume that B_2^n is the ellipsoid of maximal volume inscribed in the unit ball K of X. Then, the function $r : S^{n-1} \to \mathbb{R}$ defined by $r(x) = \|x\|$ is Lipschitz continuous with constant 1. If L_r is the Lévy median of r, Theorems 2.1 and 2.3 imply that for every $t \in (0,1)$,

$$\sigma(x \in S^{n-1} : |r(x) - L_r| \geqslant tL_r) \leqslant 2c_1 \exp(-c_2 t^2 L_r^2 n), \tag{2.2.1}$$

where $c_1, c_2 > 0$ are absolute constants. Since the function $r(x) = \|x\|$ is almost constant and equal to L_r on a subset of the sphere whose measure is practically equal to 1, one can extract a subsphere on which r is almost constant. This is done by a discretization argument via nets of spheres (see [**54**] for an outline of the argument).

Theorem 2.5. *Let $X = (\mathbb{R}^n, \|\cdot\|)$ and assume that $\|x\| \leqslant |x|$ for all $x \in \mathbb{R}^n$. For every $\varepsilon \in (0,1)$ we can find $k \geqslant c_3\varepsilon^2 L_r^2 n$ and a k-dimensional subspace F of $\mathbb{R}^n$ such that*

$$(1+\varepsilon)^{-1/2} L_r |x| \leqslant \|x\| \leqslant L_r (1+\varepsilon)^{1/2} |x| \tag{2.2.3}$$

for every $x \in F$. □

If $Y = (F, \|\cdot\|)$, it is clear that $d(Y, \ell_2^k) \leqslant 1 + \varepsilon$, and what remains is to give a lower bound for L_r. It is easier to work with the expectation

$$M = M(X) = \int_{S^{n-1}} \|x\| \,\sigma(dx), \tag{2.2.4}$$

of the norm on the sphere, and a simple computation shows that $L_r \simeq M$.

We now make full use of the fact that B_2^n is the ellipsoid of maximal volume inscribed in K. By the Dvoretzky–Rogers lemma (see [**42**]), we can find an orthonormal basis $\{v_1, \ldots, v_n\}$ with $\|v_i\| \geqslant 1/2$ for all

$i \leqslant n/2$. One may check that

$$M = \int_{S^{n-1}} \Big\| \sum_{i=1}^{n} a_i v_i \Big\| \sigma(da) \geqslant \frac{1}{2} \int_{S^{n-1}} \max_{1 \leqslant i \leqslant n/2} |a_i| \sigma(da)$$
$$\geqslant c\sqrt{\log n / n}, \tag{2.2.5}$$

where $c > 0$ is an absolute constant. Going back to Theorem 2.5 we conclude the proof of Theorem 2.4. □

Let $X = (\mathbb{R}^n, \|\cdot\|)$ be an n-dimensional normed space. We denote by b the smallest constant for which $\|x\| \leqslant b|x|$ holds for every $x \in \mathbb{R}^n$. Let $k(X)$ be the largest positive integer $k \leqslant n$ for which $E_k \in G_{n,k}$ satisfies

$$(M/2)|x| \leqslant \|x\| \leqslant (2M)|x|, \qquad x \in E_k \tag{2.2.6}$$

with probability greater than $1 - e^{-k}$. The proof of Theorem 2.4 shows that there exists $k \geqslant c_1 n (M/b)^2$ such that a random k-dimensional subspace E_k of X has this property. In other words, $k(X) \geqslant c_1 n (M/b)^2$. It was observed in [**115**] that this inequality is in fact an "asymptotic formula": for every n-dimensional normed space X one has $k(X) \leqslant Cn(M/b)^2$.

Theorem 2.6. *Let X be an n-dimensional normed space. Then, $k(X) \simeq n(M/b)^2$.* □

The estimate $k(X) \geqslant cn(M/b)^2$ allows one to check that in several situations the dimension of "spherical sections" of high-dimensional convex bodies may be much larger than logarithmic in the dimension. For example, one has $k(\ell_p^n) \simeq n$ if $1 < p < 2$ and $k(\ell_q^n) \simeq \sqrt{q} n^{2/q}$ if $q > 2$ (see [**44**] or [**113**]).

It is interesting to check the strength of Theorem 2.5 in the particular example of ℓ_1^n. For every $\varepsilon \in (0,1)$ there exists $c(\varepsilon) > 0$ such that ℓ_1^n has a subspace E_k of dimension $k \geqslant c(\varepsilon) n$ with $d(E_k, \ell_2^k) \leqslant 1 + \varepsilon$. Because of the nature of the argument, we have subspaces of ℓ_1^n of *some* dimension proportional to n which are "almost isometric" to Euclidean, but no information on $d(E_k, \ell_2^k)$ if k exceeds some fixed proportion of n. An isomorphic Dvoretzky type theorem for ℓ_1^n was proved by Kashin [**74**]: there exist $c(\alpha)$-Euclidean subspaces of ℓ_1^n of dimension $[\alpha n]$, for every $\alpha \in (0,1)$. Szarek realized that this property of ℓ_1^n is a consequence of the fact that its unit ball has bounded "volume ratio." This notion was formally introduced in [**143**]: The volume ratio of a symmetric convex body K in $\mathbb{R}^n$ is the parameter

$$\mathrm{vr}(K) = \inf\left\{ \left(\frac{|K|}{|E|}\right)^{1/n} : E \subseteq K \right\}, \tag{2.2.7}$$

where the inf is taken over all ellipsoids E contained in K. A simple computation shows that $\mathrm{vr}(B_1^n) \leqslant C$ for some absolute constant $C > 0$. Then, Kashin's theorem admits the following generalization [**136, 143**].

Theorem 2.7. *Let K be a symmetric convex body in $\mathbb{R}^n$ with $\mathrm{vr}(K) = A$. For every $k \leqslant n$ there exists a k-dimensional subspace E_k of X_K such that*

$$d(E_k, \ell_2^k) \leqslant (cA)^{\frac{n}{n-k}}, \tag{2.2.8}$$

where $c > 0$ is an absolute constant. □

Isomorphic versions of Dvoretzky's theorem for arbitrary n-dimensional normed spaces were studied by Milman and Schechtman [**114**]. There exists an absolute constant $C > 0$ such that if $\dim X = n$ and $C \log n \leqslant k < n$, then X has a k-dimensional subspace E_k with $d(E_k, \ell_2^k) \leqslant C\sqrt{k/\log(n/k)}$.

We close this subsection with a recent result of Rudelson and Vershynin [**129**], which is different in nature but very close in spirit to the Dvoretzky type theorems we discussed. Let (T, μ, d) be a metric probability space whose concentration function satisfies the "normal Lévy estimate"

$$\alpha(T, t) \leqslant c_1 \exp(-c_2 t^2 n)$$

for some n and all $t > 0$. In order to avoid degenerate cases we also assume that there exist $\varepsilon, \delta > 0$ such that the ε-neighborhood of any point in T has measure smaller than $1 - \delta$ (T is (ε, δ)-regular). We say that (T, d) is K-Lipschitz embedded into a normed space X if there exists $F : T \to X$ such that $d(x, y) \leqslant \|F(x) - F(y)\| \leqslant K \cdot d(x, y)$ for all $x, y \in T$. Assume that X is n-dimensional. If an (ε, δ)-regular metric probability space as above is K-Lipschitz embedded into X, then $k(X) \geqslant c\left(\frac{\varepsilon\delta}{K}\right)^4 n$. In other words, X must have Euclidean subspaces of proportional dimension. This fact underlines the importance of the concentration of measure phenomenon on the sphere: if some metric probability space with a normal concentration function well embeds into a normed space, this must be also true for the Euclidean space.

2.3. The ℓ-position and Pisier's inequality. One of the fundamental facts in the local theory of normed spaces is Pisier's estimate on the K-convexity constant. Combined with important results of Lewis, Figiel and Tomczak-Jaegermann, it leads to the following geometric statement: every convex body K in $\mathbb{R}^n$ has an affine image TK of volume 1 whose mean width satisfies the "reverse Urysohn inequality"

$$w(TK) \leqslant c\sqrt{n}\log n, \tag{2.3.1}$$

where $c > 0$ is an absolute constant. In this subsection we give a very concise description of this circle of ideas.

Let X be an n-dimensional normed space, and let α be a norm on $L(\ell_2^n, X)$. The trace dual norm α^* of α is defined on $L(X, \ell_2^n)$ by

$$\alpha^*(v) = \sup\{\mathrm{tr}(vu) : \alpha(u) \leqslant 1\}. \tag{2.3.2}$$

The lemma of Lewis [**83**] applies to any pair of trace dual norms.

Theorem 2.8. *For any norm α on $L(\ell_2^n, X)$, there exists $u : \ell_2^n \to X$ such that $\alpha(u) = 1$ and $\alpha^*(u^{-1}) = n$.* □

The ℓ-norm on $L(\ell_2^n, X)$ was defined by Figiel and Tomczak-Jaegermann in [**45**]: Let $\{g_1, \dots, g_n\}$ be a sequence of independent standard Gaussian random variables on some probability space, and let $\{e_1, ..., e_n\}$ be the standard orthonormal basis of $\mathbb{R}^n$. If $u : \ell_2^n \to X$, the ℓ-norm of u is defined by

$$\ell(u) = \left(\mathbb{E}\left\|\sum_{i=1}^n g_i u(e_i)\right\|^2\right)^{1/2}. \tag{2.3.3}$$

A standard computation gives

$$\ell(u) \simeq \sqrt{n}w((u^{-1})^*(K^\circ)), \tag{2.3.4}$$

where K is the unit ball of X. This formula connects the ℓ-norm to the mean width. It is more instructive to replace the Gaussians by the Rademacher functions $r_i : E_2^n \to \{-1, 1\}$ defined by $r_i(\varepsilon) = \varepsilon_i$, where $E_2^n = \{-1, 1\}^n$ is viewed as a probability space with the uniform measure. An inequality of Maurey and Pisier (see [**113**] or [**147**]) shows that

$$\ell(u) \simeq \left(\int_{E_2^n}\left\|\sum_{i=1}^n r_i(\varepsilon)u(e_i)\right\|^2 d\varepsilon\right)^{1/2} \tag{2.3.5}$$

up to a $\sqrt{\log n}$-term.

Consider the Walsh functions

$$w_A(\varepsilon) = \prod_{i \in A} r_i(\varepsilon),$$

where $A \subseteq \{1, \dots, n\}$. It is not hard to see that every function $f : E_2^n \to X$ is uniquely represented in the form

$$f(\varepsilon) = \sum_A w_A(\varepsilon) x_A \tag{2.3.6}$$

for some $x_A \in X$. The space of all functions $f : E_2^n \to X$ becomes a Banach space with the norm

$$\|f\|_{L_2(X)} = \left(\int_{E_2^n} \|f(\varepsilon)\|^2 d\varepsilon \right)^{1/2} \tag{2.3.7}$$

The Rademacher projection $R_n : L_2(X) \to L_2(X)$ is the operator sending $f = \sum w_A x_A$ to the function $R_n f := \sum_{i=1}^{n} r_i x_{\{i\}}$. Denote by $\mathrm{Rad}(X)$ the norm of this projection. Pisier **[119]** gave a sharp estimate in terms of the Banach–Mazur distance $d(X, \ell_2^n)$.

Theorem 2.9. *Let X be an n-dimensional normed space. Then,*

$$\mathrm{Rad}(X) \leqslant c \log[d(X, \ell_2^n) + 1], \tag{2.3.8}$$

where $c > 0$ is an absolute constant. □

Figiel and Tomczak-Jaegermann **[45]** had previously shown the relevance of this estimate to the study of the ℓ-norm.

Theorem 2.10. *Let X be an n-dimensional normed space. There exists $u : \ell_2^n \to X$ such that*

$$\ell(u)\ell((u^{-1})^*) \leqslant n \mathrm{Rad}(X). \tag{2.3.9}$$

Let us briefly sketch the proof. From Theorem 2.8 we can find an isomorphism $u : \ell_2^n \to X$ such that $\ell(u)\ell^*(u^{-1}) = n$. On the other hand,

$$\ell((u^{-1})^*) = \left(\int_{E_2^n} \left\| \sum_{i=1}^{n} r_i(\varepsilon)(u^{-1})^*(e_i) \right\|_*^2 d\varepsilon \right)^{1/2}. \tag{2.3.10}$$

There exists a function $\varphi : E_2^n \to X$, which can be represented in the form $\varphi = \sum_A w_A x_A$ and has norm $\|\varphi\|_{L_2(X)} = 1$, such that

$$\ell((u^{-1})^*) = \Big\langle \sum_{i=1}^n r_i (u^{-1})^*(e_i), \varphi \Big\rangle = \sum_{i=1}^n \langle (u^{-1})^*(e_i), x_{\{i\}} \rangle. \tag{2.3.11}$$

If we define $v : \ell_2^n \to X$ by $v(e_i) = x_{\{i\}}$, we easily check that

$$\ell((u^{-1})^*) = \mathrm{tr}(u^{-1}v) \leqslant \ell^*(u^{-1})\ell(v). \tag{2.3.12}$$

On observing that

$$\ell(v) = \|R_n(\varphi)\|_{L_2(X)} \leqslant \mathrm{Rad}(X)\|\varphi\|_{L_2(X)} = \mathrm{Rad}(X), \tag{2.3.13}$$

we get

$$\ell(u)\ell((u^{-1})^*) \leqslant \ell(u)\ell^*(u^{-1})\mathrm{Rad}(X) = n\mathrm{Rad}(X). \tag{2.3.14}$$

This concludes the proof. □

Combining the above with John's estimate $d(X, \ell_2^n) \leqslant \sqrt{n}$ [**73**], we can give an upper bound for the "minimal mean width" of a symmetric convex body (see § 4.1 for a discussion on different "positions" of convex bodies).

Theorem 2.11. *If K is a symmetric convex body in $\mathbb{R}^n$, there exists a linear image $\widetilde{K}$ of K with volume $|\widetilde{K}| = 1$ and mean width*

$$w(\widetilde{K}) \leqslant c\sqrt{n}\log n, \tag{2.3.15}$$

where $c > 0$ is an absolute constant.

For the proof, consider the operator $u : \ell_2^n \to X_K$ in Theorem 2.10 and set $\widetilde{K} = (u^{-1})^*(K)$. In view of (2.3.4), John's theorem and Theorem 2.9, we have

$$w(\widetilde{K})w(\widetilde{K}^\circ) \leqslant c_1 \log n. \tag{2.3.16}$$

Computing the volume of $\widetilde{K}$ in polar coordinates and using Hölder's inequality, we check that $w(\widetilde{K}^\circ)^{-1} \leqslant c_2\sqrt{n}|\widetilde{K}|^{1/n}$. It follows that

$$w(\widetilde{K}) \leqslant c_3\sqrt{n}\log n|\widetilde{K}|^{1/n}. \tag{2.3.17}$$

Normalizing the volume we obtain the assertion of the theorem. A simple argument based on the Rogers–Shephard inequality [**125**] shows that the symmetry of K is not necessary.

2.4. Low M^*-estimate and the quotient of subspace theorem. The Low M^*-estimate is the first step towards a general theory of sections and projections of symmetric convex bodies in $\mathbb{R}^n$ with dimension proportional to n. In geometric terms, it says that for fixed $\lambda \in (0,1)$, the diameter of a random $[\lambda n]$-dimensional section of the body K is controlled by its mean width

$$M^* := M(X^*) = \int_{S^{n-1}} \|x\|_* \sigma(dx) \tag{2.4.1}$$

up to a function depending only on λ.

Theorem 2.12. (Milman, [**98**, **99**]) *There exists a function $f : (0,1) \to \mathbb{R}^+$ with the following property: for every $\lambda \in (0,1)$ and every n-dimensional normed space X, a random subspace $H \in G_{n,[\lambda n]}$ satisfies*

$$\frac{f(\lambda)}{M^*}|x| \leqslant \|x\| \tag{2.4.2}$$

for every $x \in H$.

The precise dependence on λ was established in a series of papers. Theorem 2.12 was originally proved in [**98**] and a second proof using the isoperimetric inequality on S^{n-1} was given in [**99**], with a bound of the form $f(\lambda) \geqslant c(1-\lambda)$. Pajor and Tomczak-Jaegermann [**123**] later showed that one can take $f(\lambda) \geqslant c\sqrt{1-\lambda}$ (see also [**106**] for a different proof with this dependence on λ). Finally, Gordon [**64**] proved that the theorem holds true with

$$f(\lambda) \geqslant \sqrt{1-\lambda}\left(1 + O(\frac{1}{(1-\lambda)n})\right). \tag{2.4.3}$$

If we dualize the statement of the theorem, we get that a random $[\lambda n]$-dimensional projection of K_X contains a ball whose radius is of the order of $1/M$. For a random $H \in G_{n,[\lambda n]}$ we have

$$P_H(K_X) \supseteq \frac{f(\lambda)}{M} B_2^n \cap H. \tag{2.4.4}$$

The next step is the quotient of subspace theorem (Milman, [**100**]). In geometric terms, it says that for every symmetric convex body K in $\mathbb{R}^n$ and any $\alpha \in [1/2, 1)$, we can find subspaces $G \subset H$ with $\dim G \geqslant \alpha n$ and an ellipsoid $\mathcal{E}$ in G such that

$$\mathcal{E} \subset P_G(K \cap H) \subset c(1-\alpha)^{-1}|\log(1-\alpha)|\mathcal{E}. \tag{2.4.5}$$

Theorem 2.13. [100] *Let X be an n-dimensional normed space and let $\alpha \in [1/2, 1)$. Then, there exist subspaces $H \supset G$ of X such that $k = \dim(H/G) \geqslant \alpha n$ and*

$$d(H/G, \ell_2^k) \leqslant c(1-\alpha)^{-1}|\log(1-\alpha)|. \tag{2.4.6}$$

The proof of the theorem is based on the Low M^*-estimate and an iteration procedure in which Pisier's inequality plays a crucial role. We show the idea by describing the first step. We may assume that K_X satisfies the assertion of Theorem 2.10: because of (2.3.4) this can be written in the form $M(X)M^*(X) \leqslant c\log[d(X, \ell_2^n) + 1]$.

Let $\lambda \in (0, 1)$. Theorem 2.12 shows that on a random $[\lambda n]$-dimensional subspace H of X we have

$$\frac{c_1\sqrt{1-\lambda}}{M^*(X)}|x| \leqslant \|x\| \leqslant b|x|. \tag{2.4.7}$$

It is easy to check that for most $H \in G_{n,[\lambda n]}$ we have

$$M(H) \leqslant c_2 M(X). \tag{2.4.8}$$

If H satisfies both conditions, repeating the same argument for H^*, we may find a subspace G of H^* with $\dim G = k \geqslant \lambda^2 n$ and

$$\frac{c_3\sqrt{1-\lambda}}{M(X)}|x| \leqslant \frac{c_1\sqrt{1-\lambda}}{M^*(H^*)}|x| \leqslant \|x\|_{H^*} \leqslant \frac{M^*(X)}{c_1\sqrt{1-\lambda}}|x| \tag{2.4.9}$$

for every $x \in G$. In other words, $F := H/G$ satisfies

$$\begin{aligned} d(F, \ell_2^k) &\leqslant c_4(1-\lambda)^{-1}M(X)M^*(X) \\ &\leqslant c(1-\lambda)^{-1}\log[d(X, \ell_2^n) + 1]. \end{aligned} \tag{2.4.10}$$

To set up the iteration, we write $QS(X)$ for the class of all quotient spaces of a subspace of X, and define a function $f : (0, 1) \to \mathbb{R}^+$ by

$$f(\alpha) = \inf\{d(F, \ell_2^k) : F \in QS(X), \dim F \geqslant \alpha n\}. \tag{2.4.11}$$

The argument we have just described proves that

$$f(\lambda^2\alpha) \leqslant c(1-\lambda)^{-1}\log f(\alpha). \tag{2.4.12}$$

This is enough to estimate the function f as in Theorem 2.13. □

It is natural to ask whether the estimate on the diameter of proportional dimensional sections given by Theorem 2.12 is precise in some sense. From the computational geometry point of view it would be desirable to have a simple way to determine the diameter of a random section of fixed proportion. One can easily rephrase the Low M^*-estimate as follows [**108**]: If r_1 is the solution of the equation

$$M^*(K \cap sB_2^n) = f(\lambda)s, \tag{2.4.13}$$

then for a random $[\lambda n]$-dimensional section $K \cap H$ of K we have

$$\mathrm{diam}(K \cap H) \leqslant 2r_1. \tag{2.4.14}$$

In view of Gordon's proof of Theorem 2.12, we can choose $f(\lambda) = (1 - \varepsilon)\sqrt{1-\lambda}$ for any $\varepsilon \in (0,1)$, and then (2.4.14) is satisfied for all H in a subset of $G_{n,[\lambda n]}$ of measure greater than $1 - c_1 \exp(-c_2\varepsilon^2(1-\lambda)n)$. It turns out that the function $s \mapsto M^*(K \cap sB_2^n)$ can be used for a dual estimate [**52**]. There exists a second function $g : (0,1) \to \mathbb{R}$ with the following property: if $\lambda \in (1/2, 1)$ and if r_2 is the solution of the equation $M^*(K \cap sB_2^n) = g(\lambda)s$, then a random $[\lambda n]$-dimensional section $K \cap H$ of K satisfies $\mathrm{diam}(K \cap H) \geqslant 2r_2$, This gives a "confidence interval" $[r_2, r_1]$ for $\mathrm{diam}(K \cap H)$, which may be viewed as an asymptotic formula. What is essential is of course that the functions f and g can be described analytically and they do not depend on the dimension n or on the body K.

Another consequence of the Low M^*-estimate is that very accurate linear relations hold true in full generality for the diameter of sections of a body and its polar. This fact can be made precise in the following way [**107**]. Let $t(r) = t(X_K; r)$ be the greatest integer k for which a random subspace $H \in G_{n,k}$ satisfies $\mathrm{diam}(K \cap H) \leqslant 2r$. If $t^*(r) = t(X_K^*; r)$, then for any $\zeta > 0$ and any $r > 0$ we have

$$t(r) + t^*\left(\frac{1}{\zeta r}\right) \geqslant (1-\zeta)n - C, \tag{2.4.15}$$

where $C > 0$ is an absolute constant.

2.5. Coordinate theory. We fix an orthonormal basis $\{e_1, \dots, e_n\}$ of $\mathbb{R}^n$ and for every non empty $\sigma \subseteq \{1, \dots, n\}$ we consider the coordinate subspace $\mathbb{R}^\sigma = \mathrm{span}\{e_j : j \in \sigma\}$. The following coordinate version of the Low M^*-estimate was established by Giannopoulos and Milman in [**51**]: If K is an ellipsoid in $\mathbb{R}^n$, then for every $\lambda \in (0,1)$ we can find $\sigma \subseteq \{1, \dots, n\}$ of cardinality $|\sigma| \geqslant (1-\lambda)n$ such that

$$P_{\mathbb{R}^\sigma}(K) \supseteq \frac{[\lambda/\log(1/\lambda)]^{1/2}}{M(K)} B_2^n \cap \mathbb{R}^\sigma. \tag{2.5.1}$$

This observation (which has its origin in [**48, 49**]) has consequences for the question of the maximal Banach–Mazur distance to the cube (see also the proportional Dvoretzky–Rogers factorization theorem in § 6.1). The proof has its roots in an isomorphic version of the Sauer–Shelah lemma from Combinatorics, which was proved by Szarek and Talagrand [**141**] (see also [**3, 142**]), and is close in spirit to the theory of restricted invertibility of operators which was developed by Bourgain and Tzafriri [**30**].

As the example of the cube shows, one cannot have a coordinate low M^*-estimate for an arbitrary convex body. Under assumptions which guarantee the existence of "large ellipsoids" of any proportional dimension inside the body, one can use the above ellipsoidal result and obtain analogues of (2.5.1). This is done in [**51**] for bodies whose volume ratio or cotype-2 constant is well-bounded. These results can be applied to give estimates on the number of points with "many" integer coordinates inside a given convex body.

Very recently, Rudelson and Vershynin [**130**] obtained a new family of coordinate results. Assume that K is a symmetric convex body in $\mathbb{R}^n$ such that the norm $\|\cdot\|$ induced by K satisfies the conditions $\|x\| \leqslant |x|$ for all x and $M = M(K) \geqslant \delta$ for some positive constant $\delta > 0$. Then, there exist two positive numbers s and t with $c\delta \leqslant t \leqslant 1$ and $st \geqslant \delta/\log^{3/2}(2/\delta)$ and a subset σ of $\{1, \dots, n\}$ with cardinality $|\sigma| \geqslant s^2 n$, such that

$$\left\|\sum_{i\in\sigma} a_i e_i\right\| \geqslant \frac{ct}{\sqrt{n}} \sum_{i\in\sigma} |a_i| \tag{2.5.2}$$

for all choices of reals a_i, $i \in \sigma$. From this statement one can recover Elton's theorem about spaces which contain large dimensional copies of ℓ_1's [**43**] in an optimal form.

Note that the space $X = (\mathbb{R}^n, \|\cdot\|)$ satisfies $k(X) \simeq n(M/b)^2 \geqslant \delta n$. In other words, the result concerns spaces which have Euclidean subspaces of some dimension proportional to n (depending on δ). The estimate in (2.5.2) shows that

$$K \cap \mathbb{R}^\sigma \subseteq c(\delta)\sqrt{n} B_1^\sigma. \tag{2.5.3}$$

This may be viewed as a coordinate version of the low M^*-estimate for this class of bodies. The formulation is dual to the one in (2.5.1): one now considers sections instead of projections. The condition $k(X) \simeq n$ is in some sense dual to the assumptions on the volume ratio or the cotype-2 constant in [**51**].

To feel the analogy even more, we state the following "condition-free" version of the result in **[130]**: Let K be a symmetric convex body in $\mathbb{R}^n$ with $B_2^n \subseteq K$. There exists a subset σ of $\{1, \dots, n\}$ with cardinality $|\sigma| \geqslant cf(M)n$, such that

$$M \cdot (K \cap \mathbb{R}^\sigma) \subseteq \sqrt{|\sigma|} B_1^\sigma, \tag{2.5.4}$$

where $f(x) = x \log^{-3/2}(2/x)$. Compare with the low M^*-estimate: one has sections of the body inside an appropriate ℓ_1-ball on coordinate subspaces (this is weaker, but the example of ℓ_1^n shows that it is natural). Also, the parameter $1/M^*$ is replaced by M (which is stronger). However, the estimates hold for some proportional dimensions and not for any proportion.

All these are still preliminary but interesting results which show that a coordinate theory may be further developed in the future. This would have several consequences for the theory.

2.6. Covering results. Let K_1 and K_2 be two convex bodies in $\mathbb{R}^n$. The covering number $N(K_1, K_2)$ is the minimal cardinality of a finite subset A of $\mathbb{R}^n$ with the property

$$K_1 \subseteq A + K_2 = \bigcup_{x \in A} (x + K_2). \tag{2.6.1}$$

Note the multiplicative inequality $N(K_1, stK_3) \leqslant N(K_1, sK_2)N(K_2, tK_3)$ for all $t, s > 0$.

If we require $A \subset K_1$ we get the variant $\widetilde{N}(K_1, K_2)$. If K_2 is symmetric, it is easy to see that

$$\widetilde{N}(K_1, 2sK_2) \leqslant N(K_1, sK_2) \leqslant \widetilde{N}(K_1, sK_2)$$

for every $s > 0$. The standard way to estimate $\widetilde{N}(K_1, K_2)$ is to consider a maximal subset $\{x_1, \dots, x_N\}$ of K_1 any two points of which are at distance greater than or equal to 1 with respect to $\|\cdot\|_{K_2}$. Then, $K_1 \subseteq \cup(x_i + K_2)$ and this shows that $\widetilde{N}(K_1, K_2) \leqslant N$.

The most classical estimate on covering numbers is Sudakov's inequality which gives a bound on $N(K, tB_2^n)$ in terms of the mean width of K.

Theorem 2.14. *Let K be a convex body in $\mathbb{R}^n$. For every $t > 0$,*

$$\log N(K, tB_2^n) \leqslant cn\,(w(K)/t)^2\,, \tag{2.6.2}$$

where $c > 0$ is an absolute constant.

This fact is an immediate translation of an inequality of Sudakov [**135**] on the expectation of the supremum of a Gaussian process (this in turn follows from Slepian's lemma). Let $\mathcal{Y} = (Y_x)_{x\in A}$ be a Gaussian process and let ρ denote the induced semimetric on T. If $M(A,t)$ is the largest possible number of elements of A which are t-separated, then

$$\mathbb{E}\sup_{x\in A} Y_x \geqslant 2^{-1/2}\nu(M(A,t))\log^{1/2}(M(A,t))t, \tag{2.6.3}$$

where $\nu(n) = 0.648$ for $1 \leqslant n \leqslant 23$ and $\nu(n) = 2^{1/2} - \log n^{-1/2}$ for $24 \leqslant n$ (see [**85**, Section 14]). Actually, the inequality is true for the sequence $\nu(n) = 2^{1/2} - \log\log n/(2^{3/2}\log n) + O(1/\log n)$ as $n \to \infty$ (see [**47**]).

Let $g_1, \dots, g_n$ be independent standard Gaussian random variables on some probability space and let $\{e_1, \dots, e_n\}$ be an orthonormal basis in $\mathbb{R}^n$. If we consider the Gaussian process $Y_x = \langle \sum g_i e_i, x\rangle$, $x \in K$, then the induced metric on K is the Euclidean one and the estimates above show that, asymptotically,

$$\log^{1/2}(N(K, tB_2^n))t \leqslant \mathbb{E}\left\|\sum g_i e_i\right\|_*, \tag{2.6.4}$$

which gives (2.6.2) with a constant $c = c_n \to 1$ as $n \to \infty$.

A dual inequality was proved by Pajor and Tomczak-Jaegermann [**123**].

Theorem 2.15. *Let K be a symmetric convex body in $\mathbb{R}^n$. For every $t > 0$,*

$$\log N(B_2^n, tK) \leqslant cn\,(w(K^\circ)/t)^2, \tag{2.6.5}$$

where $c > 0$ is an absolute constant.

A simple proof of this fact was given by Talagrand (see [**81**] or [**54**]). From Theorem 2.15 one can deduce Sudakov's inequality with a duality argument of Tomczak-Jaegermann [**146**].

We close this subsection with some information on the duality conjecture for the entropy numbers of operators. The conjecture, which was stated by Pietsch [**118**], asserts that if X, Y are Banach spaces, if $T : X \to Y$ is a compact operator and if $N(T,\varepsilon)$ denotes the covering number $N(T(B_X), \varepsilon B_Y)$, then

$$b^{-1}\log N(T, a^{-1}\varepsilon) \leqslant \log N(T^*, \varepsilon) \leqslant b\log N(T, a\varepsilon) \tag{2.6.6}$$

for every $\varepsilon > 0$, where $a, b > 0$ are absolute constants, and T^* is the adjoint operator of T. Until recently, this conjecture had been verified only under strong assumptions for both spaces X and Y (see [**65**] and [**123**]). In the case where one of the two spaces is a Hilbert space, the conjecture is equivalent to the following statement about covering numbers of convex bodies: There exist two constants $a, b > 0$ such that

$$\frac{1}{b}\log N(B_2^n, a^{-1}K^\circ) \leqslant \log N(K, B_2^n) \leqslant b\log N(B_2^n, aK^\circ) \qquad (2.6.7)$$

for every symmetric convex body K in $\mathbb{R}^n$.

A weaker but general duality inequality was proved by König and Milman [**79**]. Using the reverse Santaló and Brunn–Minkowski inequalities (see § 5.2) they showed that

$$c^{-1}N(K_2^\circ, K_1^\circ)^{1/n} \leqslant N(K_1, K_2)^{1/n} \leqslant cN(K_2^\circ, K_1^\circ)^{1/n} \qquad (2.6.8)$$

for every pair of symmetric convex bodies K_1 and K_2 in $\mathbb{R}^n$. Note that this inequality proves the duality conjecture in the case where the logarithm of the covering numbers is large enough with respect to the dimension n.

Very recently, Artstein, Milman and Szarek [**7**, **8**] proved (2.6.7) in full generality. This settles the duality conjecture in the case either X or $Y = H$ (a Hilbert space). The proof consists of three steps: Given a symmetric convex body K in $\mathbb{R}^n$, in the first step one shows that there exists a parameter γ depending on K such that $N(K, B_2^n) \leqslant N(B_2^n, \gamma^{-1}K^\circ)^3$ and $N(B_2^n, \gamma K^\circ) \leqslant N(K, B_2^n)^2$, which is "the conjecture up to γ." The idea is to project onto a random k-dimensional subspace: one knows that c-separated sets of points are mapped onto $c\sqrt{k/n}$-separated sets under such random projections, so the information on covering numbers is kept during this process (with the cost of γ). The dimension k is chosen so that the result of [**79**] will be enough to give duality for the projected bodies.

This step can be iterated, each time applied to an intersection of some multiple of K with a ball of suitable radius (here, a variant of Tomczak's duality argument is used). As a result, $N(K, B_2^n)$ and $N(B_2^n, K^\circ)$ are bounded by products of covering numbers of polar bodies. In the last step, each product can be "telescoped" to a product of only two or three terms, which establishes duality.

2.7. Global theory and asymptotic formulas. Let K be a (symmetric) convex body in $\mathbb{R}^n$. For a fixed dimension $1 \leqslant l \leqslant n$ consider the expected value

$$D_l(K) = \int_{G_{n,l}} \mathrm{diam}(P_E(K))\nu_{n,l}(dE) \tag{2.7.1}$$

of the diameter of the orthogonal projection $P_E(K)$ onto $E \in G_{n,l}$. Theorem 2.5 shows that there is a critical value $k^* = n\big(w(K)/\mathrm{diam}(K)\big)^2$ such that: if $1 \leqslant l \leqslant k^*$ then

$$cw(K) \leqslant D_l(K) \leqslant Cw(K), \tag{2.7.2}$$

while if $k^* \leqslant l \leqslant n$, then

$$c\sqrt{l/n}\,\mathrm{diam}(K) \leqslant D_l(K) \leqslant C\sqrt{l/n}\,\mathrm{diam}(K). \tag{2.7.3}$$

Observe the phase transition at k^*: the random diameter of l-dimensional projections is stabilized since below the critical dimension k^* maximal symmetry has been achieved: most projections of the body have become isomorphic Euclidean balls of radius $w(K)/2$.

The same situation appears if one considers a dual "global problem." We want to approximate a Euclidean ball by Minkowski averages of rotations

$$K_t = \frac{1}{t}(u_1(K) + \cdots + u_t(K)) \tag{2.7.4}$$

of the body K. One way is to fix an integer $t \geqslant 2$ and ask for the infimum of $\mathrm{diam}(K_t)$ or the expected value $\mathbb{E}\mathrm{diam}(K_t)$ over all choices of $u_1, \dots, u_t \in O(n)$. It turns out (see **[115]**) that both quantities are of the same order, and

$$\mathbb{E}\,\mathrm{diam}(K_t) \simeq \frac{\mathrm{diam}(K)}{\sqrt{t}} \tag{2.7.5}$$

if $1 \leqslant t \leqslant t^* = \big[(\mathrm{diam}(K)/w(K))^2\big]$, while

$$\mathbb{E}\,\mathrm{diam}(K_t) \simeq w(K) \tag{2.7.6}$$

if $t^* \leqslant t \leqslant n$. Again, observe the phase transition at t^*. Stabilization occurs at $t \simeq t^*$ because above this integer $K_t \simeq w(K)B_2^n$ with very high probability: the norm of a random K_t has already become roughly Euclidean. Note also that, in this global process of forming averages of rotations, the "best possibility" (infimum of the diameter) coincides with the random one (expectation of the diameter).

The fact that the "asymptotic formula" $k^*t^* \simeq n$ holds true for every convex body K is only one instance of a remarkable duality. Local statements can be translated to global ones, and a very useful intuition can be developed through their comparison. However, the proofs of dual statements are not "direct translations" of each other, and they should often be invented from the start.

We proceed to another example of phase transition in which the stabilized behavior is of a different nature. Let $\|\cdot\|$ be a norm on $\mathbb{R}^n$, and let a, b be the smallest positive constants for which $(1/a)|x| \leqslant \|x\| \leqslant b|x|$ is satisfied for every $x \in \mathbb{R}^n$. For every $q \geqslant 1$ consider the parameter

$$M_q = \left(\int_{S^{n-1}} \|x\|^q \sigma(dx) \right)^{1/q}. \tag{2.7.7}$$

Then, if $k(X) = n(M_1/b)^2$ one has the following behavior of M_q (see [**89**]):

(a) $M_q \simeq M_1$ if $1 \leqslant q \leqslant k(X)$.

(b) $M_q \simeq b\sqrt{q/n}$ if $k(X) \leqslant q \leqslant n$.

(c) $M_q \simeq b$ if $q > n$.

The global q-approximation results are as follows: write

$$\|x\|_{q,t} = \left(\frac{1}{t} \sum_{i=1}^{t} \|u_i x\|^q \right)^{1/q}, \tag{2.7.8}$$

where $u_1, \dots, u_t \in O(n)$, and let t_q be the smallest integer for which there exist $u_1, \dots, u_t \in O(n)$ such that

$$(M_q/2)|x| \leqslant \|x\|_{q,t} \leqslant (2M_q)|x|. \tag{2.7.9}$$

Then, for the optimal value of t_q a random choice of $u_1, \dots, u_t$ satisfies (2.7.9) up to some universal constants, and $t_q \simeq t_1$ for $1 \leqslant q \leqslant 2$, while $t_q^{2/q} \simeq t_1(M_1/M_q)^2$ for $q \geqslant 2$. If we insert the formulas for M_q in the above relations, we check that there are two phase transitions which occur on the interval $(1, n)$ at the values $q = k(X)$ and $q = 2$.

In this more complicated example of process, the initial "constant behavior" of M_q may be viewed as a concentration phenomenon: the norm is almost constant on the sphere and this creates "inertia" in the behavior of M_q for small values of q.

Our next example is a problem of approximation: write $I = [-x, x]$ for an interval, where $x \in S^{n-1}$. We would like to approximate the Euclidean ball B_2^n by zonotopes $K_N = \frac{1}{N} \sum_{i=1}^{N} I_i$. If we fix the cardinality

N of summands and ask for the best approximation

$$A(N,n) := \inf\{d(K_N, B_2^n) : x_1, \dots, x_N \in S^{n-1}\},$$

then we have $A(N,n) = \infty$ if $N < n$, $A(N,n) = \sqrt{n}$ if $N = n$, and $A(N,n) = C(\lambda)$ if $N = [\lambda n]$ for some $\lambda > 1$ (see Kashin [**74**]). The behavior of $C(\lambda)$ (say, for $\lambda < 2$) was determined by Gluskin [**62**]:

$$C(\lambda) \simeq \min\left\{\sqrt{n}, \sqrt{(\log(1/(\lambda-1)))/(\lambda-1)}\right\}.$$

Observe that we have a sharp threshold at the value $N = n$.

The same problem can be generalized as follows: let $\|\cdot\|$ be the norm defined by a symmetric convex body K on $\mathbb{R}^n$. Consider bodies of the form $K_N = \frac{1}{N}\sum_{i=1}^{N} u_i(K)$, where $u_i \in O(n)$. The question is what is the minimal value of N for which there exist $u_1, \dots, u_N \in O(n)$ such that, for example, $d(K_N, B_2^n) \leqslant 4$. The answer is $N_0 \simeq t^* = (\mathrm{diam}(K)/w(K))^2$, and typically we have a sharp threshold for $\inf d(K_N, B_2^n)$ at this point. So, changing our parameter of study from "minimal diameter of K_N" to "geometric distance from the Euclidean ball," we often observe a phase transition behavior being replaced by a threshold type one. Again, optimal and random behaviors are equivalent: if $N \geqslant ct^*/\varepsilon^2$ then a random choice of $u_1, \dots, u_N \in O(n)$ satisfies $d(K_N, B_2^n) \leqslant 1 + \varepsilon$.

3. Classical Convexity Connected to the Asymptotic Theory

3.1. Brunn–Minkowski inequality: classical proofs and functional forms. The fundamental Brunn–Minkowski inequality states that if K and T are two nonempty compact subsets of $\mathbb{R}^n$, then

$$|K+T|^{1/n} \geqslant |K|^{1/n} + |T|^{1/n}. \tag{3.1.1}$$

If we make the additional hypothesis that K and T are convex bodies, then we can have equality only if K and T are homothetical.

The inequality expresses in a sense the fact that volume is an "n-concave" function with respect to Minkowski addition. For this reason, it is often written in the following form: If K, T are nonempty compact subsets of $\mathbb{R}^n$ and $\lambda \in (0,1)$, then

$$|\lambda K + (1-\lambda)T|^{1/n} \geqslant \lambda |K|^{1/n} + (1-\lambda)|T|^{1/n}. \tag{3.1.2}$$

Using (3.1.2) and the arithmetic-geometric means inequality we can also write

$$|\lambda K + (1-\lambda)T| \geqslant |K|^{\lambda} |T|^{1-\lambda}. \tag{3.1.3}$$

This weaker, but actually equivalent, form of the Brunn–Minkowski inequality has the advantage (or disadvantage) of being dimension free.

There are many interesting proofs of the Brunn–Minkowski inequality, all of them related to important ideas. Historically, the first proof of the Brunn–Minkowski inequality was based on Brunn's concavity principle:

Theorem 3.1. *Let K be a convex body in $\mathbb{R}^n$ and let F be a k-dimensional subspace of $\mathbb{R}^n$, $1 \leqslant k \leqslant n$. Then, the function $f : F^{\perp} \to \mathbb{R}$ defined by $f(x) = |K \cap (F+x)|^{1/k}$ is concave on its support.*

The proof goes by symmetrization. The Steiner symmetrization of K in the direction of $\theta \in S^{n-1}$ is the set $S_\theta(K)$ consisting of all points of the form $x + \lambda\theta$, where x is in the projection $P_{\theta^{\perp}}(K)$ of K onto $\theta^{\perp}$ and $|\lambda| \leqslant \frac{1}{2} \times \text{length}(x + \mathbb{R}\theta) \cap K$. Steiner symmetrization preserves convexity and volume: if K is a convex body then $S_\theta(K)$ is also a convex body, and $|S_\theta(K)| = |K|$. A well known fact which goes back to Steiner and Schwarz is that for every convex body K one can find a sequence of successive Steiner symmetrizations in directions $\theta \in F$ so that the limiting convex body $\widetilde{K}$ has the following property:

> For every $x \in F^{\perp}$, $\widetilde{K} \cap (F+x)$ is a ball with center at x and radius $r(x)$ such that $|\widetilde{K} \cap (F+x)| = |K \cap (F+x)|$.

Now, the proof of the theorem is immediate. Convexity of $\widetilde{K}$ implies that r is concave on its support, and this shows that f is also concave.

Brunn's concavity principle implies the Brunn–Minkowski inequality as follows. If K and T are convex bodies in $\mathbb{R}^n$, we define $K_1 = K \times \{0\}$ and $T_1 = T \times \{1\}$ in $\mathbb{R}^{n+1}$ and consider their convex hull L. If we set $L(t) = \{x \in \mathbb{R}^n : (x,t) \in L\}$ for all $t \in [0,1]$, we easily check that $L(0) = K$, $L(1) = T$ and $L(1/2) = \frac{K+T}{2}$. Then, Brunn's concavity principle for $F = \mathbb{R}^n$ shows that

$$\left|\frac{K+T}{2}\right|^{1/n} \geqslant \frac{1}{2}|K|^{1/n} + \frac{1}{2}|T|^{1/n}. \tag{3.1.4}$$

A functional form of the Brunn–Minkowski inequality is an integral inequality which reduces to (3.1.1) by appropriate choice of the functions involved. The advantage of such functional inequalities is that they can be applied in many other contexts: an example is given by the Prékopa-Leindler inequality (see **[121]** or **[14]**) which is stated below: it can be applied to yield the logarithmic Sobolev inequality and several important concentration results in Gauss space.

Theorem 3.2. *Let $f,g,h:\mathbb{R}^n\to\mathbb{R}^+$ be measurable functions, and let $\lambda\in(0,1)$. We assume that f and g are integrable, and for every $x,y\in\mathbb{R}^n$*

$$h(\lambda x+(1-\lambda)y)\geqslant f(x)^{\lambda}g(y)^{1-\lambda}. \tag{3.1.5}$$

Then,

$$\int_{\mathbb{R}^n} h\geqslant\left(\int_{\mathbb{R}^n} f\right)^{\lambda}\left(\int_{\mathbb{R}^n} g\right)^{1-\lambda}. \tag{3.1.6}$$

We shall only sketch the case $n=1$. We may assume that f and g are continuous and strictly positive and then define $x,y:(0,1)\to\mathbb{R}$ by the equations

$$\int_{-\infty}^{x(t)} f=t\int f \quad\text{and}\quad \int_{-\infty}^{y(t)} g=t\int g. \tag{3.1.7}$$

Then, x and y are differentiable, and for every $t\in(0,1)$ we have

$$x'(t)f(x(t))=\int f \quad\text{and}\quad y'(t)g(y(t))=\int g. \tag{3.1.8}$$

We now define $z:(0,1)\to\mathbb{R}$ by $z(t)=\lambda x(t)+(1-\lambda)y(t)$. Since x and y are strictly increasing, z is also strictly increasing, and the arithmetic-geometric means inequality shows that

$$z'(t)=\lambda x'(t)+(1-\lambda)y'(t)\geqslant(x'(t))^{\lambda}(y'(t))^{1-\lambda}. \tag{3.1.9}$$

Hence, we can estimate the integral of h making the change of variables $s=z(t)$:

$$\begin{aligned}\int h&=\int_0^1 h(z(t))z'(t)dt\\&\geqslant\int_0^1 h(\lambda x(t)+(1-\lambda)y(t))(x'(t))^{\lambda}(y'(t))^{1-\lambda}dt\end{aligned}$$

$$\geqslant \int_0^1 f^{\lambda}(x(t))g^{1-\lambda}(y(t))\left(\frac{\int f}{f(x(t))}\right)^{\lambda}\left(\frac{\int g}{g(y(t))}\right)^{1-\lambda} dt$$

$$= \left(\int f\right)^{\lambda}\left(\int g\right)^{1-\lambda}.$$

Induction on the dimension completes the proof.

The Brunn–Minkowski inequality is a simple consequence of Theorem 3.2. Let K and T be nonempty compact subsets of $\mathbb{R}^n$, and let $\lambda \in (0,1)$. We define $f = \chi_K$, $g = \chi_T$, and $h = \chi_{\lambda K+(1-\lambda)T}$. It is easily checked that the assumptions of Theorem 3.2 are satisfied. Therefore,

$$|\lambda K + (1-\lambda)T| = \int h \geqslant \left(\int f\right)^{\lambda}\left(\int g\right)^{1-\lambda} = |K|^{\lambda}|T|^{1-\lambda}. \quad (3.1.10)$$

There are many variants of the Prékopa–Leindler inequality. All of them can be proved by a "transportation of measure" argument similar to the one used above. We shall state one of them and use it to give a functional version of a proof of Brunn's principle which was given by Gromov and Milman [**71**].

We first introduce some notation: If $p > 0$ and $\lambda \in (0,1)$, for all $x, y > 0$ we set

$$M_p^{\lambda}(x,y) = (\lambda x^p + (1-\lambda)y^p)^{1/p}.$$

If $x, y \geqslant 0$ and $xy = 0$, we set $M_p^{\lambda}(x,y) = 0$. Observe that

$$\lim_{p\to 0+} M_p^{\lambda}(x,y) = x^{\lambda}y^{1-\lambda}.$$

Statement. *Suppose that $f, g, h : \mathbb{R}^n \to \mathbb{R}^+$ are measurable functions, and let $p > 0$, $\lambda \in (0,1)$. We assume that f and g are integrable, and for every $x, y \in \mathbb{R}^n$*

$$h(\lambda x + (1-\lambda)y) \geqslant M_p^{\lambda}(f(x), g(y)). \quad (3.1.11)$$

Then,

$$\int_{\mathbb{R}^n} h \geqslant M_{p/(pn+1)}^{\lambda}\left(\int_{\mathbb{R}^n} f, \int_{\mathbb{R}^n} g\right). \quad (3.1.12)$$

The proof of the statement is quite similar to the proof of the Prékopa–Leindler inequality given above.

We need a few more definitions: Let K be a convex set in $\mathbb{R}^n$ and let $f : K \to \mathbb{R}^+$. We say that f is α-concave for some $\alpha > 0$, if $f^{1/\alpha}$ is concave on K. It is easy to see that if $f, g : K \to \mathbb{R}^+$ and if f is α-concave and g is β-concave, then fg is $(\alpha + \beta)$-concave.

Let now K be a convex body in $\mathbb{R}^n$ and let $\theta \in S^{n-1}$. For every $y \in P_{\theta^\perp}(K)$ we write I_y for the interval $\{t \in \mathbb{R} : y + t\theta \in K\}$. For every continuous function $f : K \to \mathbb{R}^+$ we define the projection $P_\theta f$ of f with respect to θ by

$$(P_\theta f)(y) := \int_{I_y} f(y + t\theta)dt, \quad y \in P_{\theta^\perp}(K). \tag{3.1.13}$$

If we define $F_y(t) = \chi_K(y + t\theta)f(y + t\theta)$ for $y \in P_{\theta^\perp}(K)$, then by the α-concavity of f and the convexity of K we easily check that

$$F_{\lambda y+(1-\lambda)w}(\lambda t + (1 - \lambda)s) \geqslant M^{\lambda}_{1/\alpha}(F_y(t), F_w(s)) \tag{3.1.14}$$

for all $y, w \in P_{\theta^\perp}(K)$. Applying the statement, we immediately get:

Claim. *If f is α-concave, then $P_\theta f$ is $(1 + \alpha)$-concave.*

We now finish the proof of Brunn's principle as follows. Let F be a k-dimensional subspace of $\mathbb{R}^n$. The indicator function of K is constant on K, and hence it is α-concave for every $\alpha > 0$. We choose an orthonormal basis $\{\theta_1, \dots, \theta_k\}$ of F and perform successive projections in the directions of θ_i. The claim shows that the function $x \mapsto |K \cap (F + x)|$ is $(\alpha + k)$-concave on $P_{F^\perp}(K)$, for every $\alpha > 0$. It follows that $f(x) = |K \cap (F + x)|^{1/k}$ is concave.

The Prékopa–Leindler inequality and the statement above, have recently been extended to Riemannian manifolds [**37**]. There, the curvature plays an essential role (through the Ricci curvature, in particular) and a *distortion* coefficient has to be added to the condition (3.1.5). We will state the spherical extension of the Prékopa–Leindler inequality obtained in [**36**]. Let ρ denote the (geodesic) distance on the sphere S^n and σ the usual rotationally invariant measure on S^n. For $x, y \in S^n$ with $x \neq -y$, introduce the geodesic analogue of the point $tx + (1 - t)y$, namely the point $z = \gamma_t(x, y) \in S^n$ verifying

$$\rho(x, z) = (1 - t)\rho(x, y) \quad \text{and} \quad \rho(z, y) = t\rho(x, y). \tag{3.1.15}$$

If $x = \cos(\theta)y + \sin(\theta)v$ with $\theta \in [0, \pi)$ and $v \in S^n$ orthogonal to y, then $\gamma_t(x, y) = \cos(t\theta)y + \sin(t\theta)v$. For $t \in (0, 1)$ and $d \in [0, \pi]$, set

$S(d) := d^{-1} \sin d$ and

$$L_t(d) := \big(S(d)/S(td)\big)^t \big(S(d)/S((1-t)d)\big)^{1-t}. \tag{3.1.16}$$

Theorem 3.3. *Let $f, g, h : S^n \to R^+$ be Borel functions and $t \in (0,1)$. We assume that for every $x \neq -y \in S^n$,*

$$h(\gamma_t(x,y)) \geqslant L_t(\rho(x,y))^{n-1} f(x)^t g(y)^{1-t}. \tag{3.1.17}$$

Then

$$\int h\, d\sigma \geqslant \left(\int f\, d\sigma\right)^t \left(\int g\, d\sigma\right)^{1-t}. \tag{3.1.18}$$

Since $L_t(\pi) = 0$, the condition (3.1.17) is always satisfied when $x = -y$. From $L_t(d) \leqslant 1$ we deduce in particular that the Brunn–Minkowski inequality holds on the sphere for the geodesic midsum of two sets, say. It is known that $L_t(d) \leqslant e^{-t(1-t)d^2/2}$ and thus the coefficient $L_t(\rho(x,y))^{n-1}$ in (3.1.17) can be replaced by the coefficient

$$e^{-(n-1)t(1-t)\rho^2(x,y)/2}.$$

With this form, one can recover, as in **[95]**, the classical concentration results for the sphere.

3.2. Geometric inequalities of hyperbolic type. We write $\mathcal{K}_n$ for the class of nonempty, compact convex subsets of $\mathbb{R}^n$. Minkowski's fundamental theorem states that if $K_1, \dots, K_m \in \mathcal{K}_n$, $m \in \mathbb{N}$, there exist coefficients $V(K_{i_1}, \dots, K_{i_n})$, $1 \leqslant i_1, \dots, i_n \leqslant m$ which are invariant under permutations of their arguments, such that

$$|t_1 K_1 + \cdots + t_m K_m| = \sum_{1 \leqslant i_1, \dots, i_n \leqslant m} V(K_{i_1}, \dots, K_{i_n}) t_{i_1} \dots t_{i_n} \tag{3.2.1}$$

for every choice of nonnegative real numbers t_i (see **[134]** or **[34]**). The coefficient $V(A_1, \dots, A_n)$ is called the mixed volume of the compact convex sets $A_1, \dots, A_n$. A special case of Minkowski's theorem is Steiner's formula. If $K \in \mathcal{K}_n$, then

$$|K + tB_2^n| = \sum_{i=0}^{n} \binom{n}{i} V_{n-i}(K) t^i \tag{3.2.2}$$

for all $t > 0$, where $V_{n-i}(K) = V(K; n-i, B_2^n; i)$ is the ith quermassintegral of K.

A very deep and strong generalization of the Brunn–Minkowski inequality is the Alexandrov–Fenchel inequality **[1, 2]** (see **[134]**):

If $K, T, A_3, \dots, A_n \in \mathcal{K}_n$, then

$$V(K, T, A_3, \dots, A_n)^2 \geqslant V(K, K, A_3, \dots, A_n) V(T, T, A_3, \dots, A_n). \tag{3.2.3}$$

Among many consequences of (3.2.3), one should mention the inequalities

$$V_i(K+T)^{1/i} \geqslant V_i(K)^{1/i} + V_i(T)^{1/i} \tag{3.2.4}$$

which hold true for all convex bodies K, T in $\mathbb{R}^n$ and all $i \in \{1, \dots, n\}$, and the Alexandrov inequalities

$$\left(\frac{V_i(K)}{|B_2^n|}\right)^{1/i} \geqslant \left(\frac{V_j(K)}{|B_2^n|}\right)^{1/j}, \tag{3.2.5}$$

where $1 \leqslant i < j \leqslant n$. Note that the Brunn–Minkowski inequality and the isoperimetric inequality are special cases of (3.2.4) and (3.2.5) respectively.

Going back in time, we locate numerical inequalities which are surprisingly similar to the ones above (see [**18**]). Let $\overline{x} = (x_1, \dots, x_n)$ be an n-tuple of positive real numbers, and consider the normalized elementary symmetric functions $E_0(\overline{x}) \equiv 1$ and

$$E_i(x_1, \dots, x_n) = \frac{1}{\binom{n}{i}} \sum_{1 \leqslant j_1 < \dots, j_i \leqslant n} x_{j_1} x_{j_2} \dots x_{j_i} \tag{3.2.6}$$

for $i = 1, \dots, n$. With this definition, $E_1(\overline{x})$ and $E_n^{1/n}(\overline{x})$ correspond to the arithmetic and geometric means of $x_1, \dots, x_n$. Newton proved that

$$E_k^2(\overline{x}) \geqslant E_{k-1}(\overline{x}) E_{k+1}(\overline{x}) \tag{3.2.7}$$

for all $k = 1, \dots, n-1$, with equality if and only if all the x_i's are equal. Maclaurin observed that

$$E_1(\overline{x}) \geqslant E_2^{1/2}(\overline{x}) \geqslant \dots \geqslant E_n^{1/n}(\overline{x}). \tag{3.2.8}$$

These inequalities follow immediately from Newton's inequality (3.2.7) and they generalize the arithmetic-geometric means inequality.

One can feel the analogy with the Alexandrov–Fenchel inequalities even more, by considering the more recent Marcus–Lopes inequality

$$\frac{E_k(\overline{x}+\overline{y})}{E_{k-1}(\overline{x}+\overline{y})} \geqslant \frac{E_k(\overline{x})}{E_{k-1}(\overline{x})} + \frac{E_k(\overline{y})}{E_{k-1}(\overline{y})}, \tag{3.2.9}$$

which holds true for all $k = 1, \dots, n$. As a formal consequence one gets

$$[E_k(\overline{x}+\overline{y})]^{1/k} \geqslant [E_k(\overline{x})]^{1/k} + [E_k(\overline{y})]^{1/k}. \tag{3.2.10}$$

We now pass to the multidimensional case: let S_n^+ be the space of real positive symmetric $n \times n$ matrices. If $t_1, \ldots, t_m > 0$ and $A_1, \ldots, A_m \in S_n^+$, then

$$\det(t_1A_1 + \cdots + t_mA_m) = \sum_{1 \leqslant i_1 \leqslant \ldots \leqslant i_n \leqslant m} n! D(A_{i_1}, \ldots, A_{i_n}) t_{i_1} \ldots t_{i_n}, \tag{3.2.11}$$

where the coefficient $D(B_1, \ldots, B_n)$ is invariant under permutations of its arguments and is called the mixed discriminant of $B_1, \ldots, B_n$. Based on the fact that $P(t) = \det(A + tI)$ has only real roots for any $A \in S_n^+$ one can prove some very interesting inequalities about mixed discriminants, which are completely analogous to Newton's inequalities, and were discovered by Alexandrov. Examples are the inequalities

$$D(A, B, C_3, \ldots, C_n)^2 \geqslant D(A, A, C_3, \ldots, C_n) D(B, B, C_3, \ldots, C_n) \tag{3.2.12}$$

for all $A, B, C_3, \ldots, C_n \in S_n^+$ and

$$D(A_1, A_2, \ldots, A_n) \geqslant \prod_{i=1}^{n} [\det A_i]^{1/n}. \tag{3.2.13}$$

There are many other inequalities on positive symmetric matrices, and one is tempted to look for their analogues in the setting of convex geometry. An inequality of Bergstrom (see [**18**]), which is the matrix analogue of (3.2.9), states that if A and B are symmetric positive definite matrices and if A_i, B_i denote the submatrices obtained by deleting the ith row and column, then

$$\frac{\det(A+B)}{\det(A_i+B_i)} \geqslant \frac{\det(A)}{\det(A_i)} + \frac{\det(B)}{\det(B_i)}. \tag{3.2.14}$$

This is generalized by Ky Fan in the form

$$\left(\frac{\det(A+B)}{\det(A_k+B_k)}\right)^{1/k} \geqslant \left(\frac{\det(A)}{\det(A_k)}\right)^{1/k} + \left(\frac{\det(B)}{\det(B_k)}\right)^{1/k}, \tag{3.2.15}$$

where A_k is the submatrix of A we obtain if we delete k rows and the corresponding columns of A. For $k = n$ this reduces to Minkowski's inequality $[\det(A+B)]^{1/n} \geqslant [\det A]^{1/n} + [\det B]^{1/n}$. For related inequalities about mixed volumes see [**50, 46**].

One last comment is that behind all these numerical or convex geometric inequalities there is a unified principle: "the minimum of certain functionals is achieved on equal objects." Statements like the Brunn–Minkowski or the Alexandrov–Fenchel inequality may be equivalently expressed in the form

$$f(A,B) \geqslant \min\{f(A,A), f(B,B)\}. \tag{3.2.16}$$

The Brunn–Minkowski inequality can be rederived from its simple consequence $|aK + bT| \geqslant \min\{|(a+b)K|, |(a+b)T|\}$. Likewise, the Alexandrov–Fenchel inequality is equivalent to the inequality

$$V(K,T,A_3,\ldots,A_n)^2 \geqslant \min\{V(K,K,A_3,\ldots,A_n), V(T,T,A_3,\ldots,A_n)\}. \tag{3.2.17}$$

The same principle applies to all the hyperbolic type inequalities we discussed in this subsection. In contrast, "elliptic type" inequalities like the triangle inequality and the Cauchy–Schwarz inequality obey a "maximum principle": for example, the latter inequality is equivalent to the statement

$$\int |f\cdot g|d\mu \leqslant \max\left\{\int |f|^2 d\mu, \int |g|^2 d\mu\right\}. \tag{3.2.18}$$

The maximum of the functional $(f,g) \mapsto \int |f \cdot g| d\mu$ is "achieved on equal objects." Hölder's inequality is also a consequence of such an "elliptic" principle, which should however be correctly applied so that the functions f and g involved stay in "correct" spaces. If p and q are conjugate exponents, then the inequality

$$\int |f\cdot g|d\mu \leqslant \max\left\{\left(\int |f|^p d\mu\right)^{1/(p-1)}, \left(\int |g|^q d\mu\right)^{1/(q-1)}\right\} \tag{3.2.19}$$

for all $f \in L^p$ and $g \in L^q$, is equivalent to the classical Hölder's inequality.

3.3. Volume preserving transformations. Let K and T be two open convex bodies in $\mathbb{R}^n$. A volume preserving transformation from K onto T is a map $\varphi : K \to T$ which is one to one, onto and has a Jacobian with constant determinant equal to $|K|/|T|$. In this section we describe two such maps, the Knöthe map and the Brenier map. Applying each one of them we may obtain alternative proofs of the Brunn–Minkowski inequality.

The Knöthe map. We fix a coordinate system in $\mathbb{R}^n$. The properties of the Knöthe map [**78**] from K to T with respect to the given coordinate system are described in the following theorem.

Theorem 3.4. *Let K and T be open convex bodies in $\mathbb{R}^n$. There exists a map $\varphi : K \to T$ with the following properties (for a proof see* [**113**]*):*

(a) *φ is triangular: the ith coordinate function of φ depends only on $x_1, \dots, x_i$, i.e.,*

$$\varphi(x_1, \dots, x_n) = (\varphi_1(x_1), \varphi_2(x_1, x_2), \dots, \varphi_n(x_1, \dots, x_n)). \tag{3.3.1}$$

(b) *The partial derivatives $\dfrac{\partial \varphi_i}{\partial x_i}$ exist and they are positive on K, and the determinant of the Jacobian of φ is constant. More precisely, for every $x \in K$*

$$|\det J_\varphi(x)| = \prod_{i=1}^{n} \frac{\partial \varphi_i}{\partial x_i}(x) = \frac{|T|}{|K|}. \tag{3.3.2}$$

The Brenier map. For any two open convex bodies K and T there exists a volume preserving transformation from K onto T, called the Brenier map [**33**], which is the gradient of a C^2 convex function. The existence of this remarkable map is a consequence of a more general transportation of measure result which we briefly describe.

Consider the space $\mathcal{P}(\mathbb{R}^n)$ of Borel probability measures on $\mathbb{R}^n$ as a subset of the unit ball of $C_\infty(\mathbb{R}^n)^*$ (the dual of the space of continuous functions which vanish uniformly at infinity). Let $\mu, \nu \in \mathcal{P}(\mathbb{R}^n)$. If $T : \mathbb{R}^n \to \mathbb{R}^n$ is a measurable function which is defined μ-almost everywhere and satisfies $\nu(B) = \mu(T^{-1}(B))$ for every Borel subset B of $\mathbb{R}^n$, we say that T pushes forward μ to ν and write $T\mu = \nu$. It is easy to see that $T\mu = \nu$ if and only if for every bounded Borel measurable $g : \mathbb{R}^n \to \mathbb{R}$ we have

$$\int_{\mathbb{R}^n} g(y) d\nu(y) = \int_{\mathbb{R}^n} g(T(x)) d\mu(x). \tag{3.3.3}$$

Generalizing work of Brenier, McCann [**96**] proved the following.

Theorem 3.5. *Let $\mu, \nu \in \mathcal{P}(\mathbb{R}^n)$ and assume that μ is absolutely continuous with respect to Lebesgue measure. Then, there exists a convex function $f : \mathbb{R}^n \to \mathbb{R}$ such that $\nabla f : \mathbb{R}^n \to \mathbb{R}^n$ is defined μ-almost everywhere, and $(\nabla f)\mu = \nu$.*

The proof of Theorem 3.5 is based on the notion of cyclical monotonicity from convex analysis: A subset G of $\mathbb{R}^n \times \mathbb{R}^n$ is called cyclically monotone if for every $m \geqslant 2$ and $(x_i, y_i) \in G$, $i \leqslant m$, we have

$$\langle y_1, x_2 - x_1\rangle + \langle y_2, x_3 - x_2\rangle + \cdots + \langle y_m, x_1 - x_m\rangle \leqslant 0. \qquad (3.3.4)$$

Fact 1. *Let μ and ν be Borel probability measures on $\mathbb{R}^n$. There exists a joint probability measure γ on $\mathbb{R}^n \times \mathbb{R}^n$ which has cyclically monotone support and marginals μ, ν, i.e., for all bounded Borel measurable $f, g : \mathbb{R}^n \to \mathbb{R}$ we have*

$$\int_{\mathbb{R}^n} f(x)d\mu(x) = \int_{\mathbb{R}^n \times \mathbb{R}^n} f(x)d\gamma(x,y) \qquad (3.3.5)$$

and

$$\int_{\mathbb{R}^n} g(y)d\nu(y) = \int_{\mathbb{R}^n \times \mathbb{R}^n} g(y)d\gamma(x,y). \qquad (3.3.6)$$

The second ingredient is the connection of cyclically monotone sets with convex functions (see [**124**]). For every proper convex function $f : \mathbb{R}^n \to \mathbb{R}$ we consider the subdifferential of f

$$\partial(f) = \{(x,y) \in \mathbb{R}^n \times \mathbb{R}^n : f(z) \geqslant f(x) + \langle y, z - x\rangle,\ z \in \mathbb{R}^n\}. \quad (3.3.7)$$

The subdifferential parametrizes the supporting hyperplanes of f: the set $\partial(f)(x) = \{y : (x,y) \in \partial(f)\}$ is a closed and bounded convex set, and differentiability of f at x is equivalent to the existence of a unique $y \in \partial f(x)$, in which case $\nabla f(x) = y$.

Fact 2. *Let $G \subset \mathbb{R}^n \times \mathbb{R}^n$. Then, G is contained in the subdifferential of a proper convex function $f : \mathbb{R}^n \to \mathbb{R}$ if and only if G is cyclically monotone.*

We can now sketch the proof of Theorem 3.5. By Fact 1, there exists a probability measure γ on $\mathbb{R}^n \times \mathbb{R}^n$ which has cyclically monotone support and marginals μ, ν. Fact 2 shows that the support of γ is contained in the subdifferential of a proper convex function $f : \mathbb{R}^n \to \mathbb{R}$. Since f is convex and μ is absolutely continuous with respect to Lebesgue measure, f is differentiable μ-almost everywhere. Since $\text{supp}(\gamma) \subset \partial(f)$, by the definition of the subdifferential we have $y = \nabla f(x)$ for almost all pairs (x,y) with respect to γ. Then, for every bounded Borel measurable

$g : \mathbb{R}^n \to \mathbb{R}$ we see that

$$\int g(y)d\nu(y) = \int g(y)d\gamma(x, y)$$
$$= \int g(\nabla f(x))d\gamma(x, y) = \int g(\nabla f(x))d\mu(x), \tag{3.3.8}$$

which shows that $(\nabla f)\mu = \nu$.

Assume that μ and ν are the normalized Lebesgue measures on some convex bodies K and T. Regularity results of Caffarelli show that in this case f may be assumed twice continuously differentiable. This proves the following.

Theorem 3.6. *Let K and T be open convex bodies in $\mathbb{R}^n$. There is a convex function $f \in C^2(K)$ such that $\varphi = \nabla f : K \to T$ is one to one, onto and volume preserving.*

We can now show the Brunn–Minkowski inequality using either the Knöthe or the Brenier map. In each case we have $(I + \varphi)(K) \subseteq K + T$. If φ denotes the Knöthe map, $J_{I+\varphi}(x)$ is triangular and this implies

$$|\det J_{I+\varphi}(x)|^{1/n} = \prod_{i=1}^{n} \left(1 + \frac{\partial \varphi_i(x)}{\partial x_i}\right)^{1/n}$$
$$\geqslant 1 + |\det J_{\varphi}(x))^{1/n} = 1 + \left(\frac{|T|}{|K|}\right)^{1/n}. \tag{3.3.9}$$

If φ is the Brenier map, it is clear that the Jacobian $J_\varphi = \mathrm{Hess} f$ is a symmetric positive definite matrix for every $x \in K$. Therefore,

$$|\det J_{I+\varphi}(x)| = |\det(I + \mathrm{Hess} f)(x)| = \prod_{i=1}^{n}(1 + \lambda_i(x)) \tag{3.3.10}$$

where $\lambda_i(x)$ are the non negative eigenvalues of $\mathrm{Hess} f$. Moreover, by the volume preserving property of φ, we have $\prod_{i=1}^{n} \lambda_i(x) = |T|/|K|$ for every $x \in K$. Therefore, the arithmetic-geometric means inequality gives

$$|\det J_{I+\varphi}(x)|^{1/n} \geqslant 1 + \left(\frac{|T|}{|K|}\right)^{1/n}. \tag{3.3.11}$$

In both cases,

$$|K+T| \geqslant \int_{(I+\varphi)K} dx = \int_K |\det J_{I+\varphi}(x)| dx$$

$$\geqslant |K| \left(1 + (|T|/|K|)^{1/n}\right)^n , \qquad (3.3.12)$$

which is the Brunn–Minkowski inequality.

For an arbitrary pair of open convex bodies K_1 and K_2 it would be desirable to achieve a volume preserving transformation $\psi : K_1 \to K_2$ for which $(I+\psi)(K_1) = K_1 + K_2$. This was recently done in [**4**]. There are two ingredients in the construction: the first one is a regularity result of Caffarelli [**35**] (see also [**4**]):

Fact 3. *If T is an open convex body in $\mathbb{R}^n$, f is a probability density on $\mathbb{R}^n$, and g is a probability density on T such that f is locally bounded and bounded away from zero on compact sets, and there exist $c_1, c_2 > 0$ such that $c_1 \leqslant g(y) \leqslant c_2$ for every $y \in T$, then the Brenier map $\nabla f : (\mathbb{R}^n, f dx) \to (\mathbb{R}^n, g dx)$ is continuous and belongs locally to the Hölder class C^α for some $\alpha > 0$.*

The second is a theorem of Gromov [**67**] (see also [**4**]):

Fact 4. *Let $f : \mathbb{R}^n \to \mathbb{R}$ be a C^2-smooth convex function with strictly positive Hessian. Then, the image of the gradient map* Im(∇f) *is an open convex set. Also, if f_1, f_2 are two such functions, then*

$$\mathrm{Im}(\nabla f_1 + \nabla f_2) = \mathrm{Im}(\nabla f_1) + \mathrm{Im}(\nabla f_2). \qquad (3.3.13)$$

Having these tools in hand and given two open convex bodies K_1 and K_2 of volume 1 in $\mathbb{R}^n$, we choose a smooth strictly positive density ρ on $\mathbb{R}^n$ and consider the Brenier maps

$$\psi_i = \nabla f_i : (\mathbb{R}^n, \rho dx) \to (K_i, dx) \quad , \quad i = 1, 2. \qquad (3.3.14)$$

Fact 3 shows that ψ_1 and ψ_2 are C^1-smooth. Applying Fact 4, we see that, for every $\lambda > 0$,

$$K_1 + \lambda K_2 = \{\nabla f_1(x) + \lambda \nabla f_2(x) : x \in \mathbb{R}^n\}. \qquad (3.3.15)$$

Then, the map $\psi = \psi_2 \circ (\psi_1)^{-1} : K_1 \to K_2$ is a volume preserving C^1-diffeomorphism and satisfies $K_1 + \lambda K_2 = (I + \lambda\psi)(K_1)$ for all $\lambda > 0$.

This construction reveals the close relation between mixed volumes and mixed discriminants. Let $K_1, \dots, K_n$ be open convex bodies K_i

with normalized volume $|K_i| = 1$, and consider the Brenier maps

$$\varphi_i : (\mathbb{R}^n, \gamma_n) \to K_i, \tag{3.3.16}$$

where γ_n is the standard Gaussian probability density on $\mathbb{R}^n$. We have $\varphi_i = \nabla f_i$, where f_i are convex functions on $\mathbb{R}^n$. By Caffarelli's regularity result, all the φ_i's are smooth maps. Then, the image of $(\mathbb{R}^n, \gamma_n)$ by $\sum t_i \varphi_i$ is the interior of $\sum t_i K_i$. Since each φ_i is a measure preserving map, we have

$$\det\left(\frac{\partial^2 f_i}{\partial x_k \partial x_l}\right)(x) = \gamma_n(x) \quad , \quad i = 1, \dots, n. \tag{3.3.17}$$

It follows that

$$\left|\sum_{i=1}^n t_i K_i\right| = \int_{\mathbb{R}^n} \det\left(\sum_{i=1}^n t_i (\frac{\partial^2 f_i}{\partial x_k \partial x_l})\right) dx$$

$$= \sum_{i_1,\dots,i_n=1}^n t_{i_1} \dots t_{i_n} \int_{\mathbb{R}^n} D\left(\frac{\partial^2 f_{i_1}(x)}{\partial x_k \partial x_l}, \dots, \frac{\partial^2 f_{i_n}(x)}{\partial x_k \partial x_l}\right) dx. \tag{3.3.18}$$

In this way, we recover Minkowski's theorem on $|\sum t_i K_i|$, and see the connection between the mixed discriminants $D(\mathrm{Hess} f_{i_1}, \dots, \mathrm{Hess} f_{i_n})$ and the mixed volumes

$$V(K_{i_1}, \dots, K_{i_n}) = \int_{\mathbb{R}^n} D(\mathrm{Hess} f_{i_1}(x), \dots, \mathrm{Hess} f_{i_n}(x)) dx. \tag{3.3.19}$$

The Alexandrov–Fenchel inequalities do not follow from the corresponding mixed discriminant inequalities, but the deep connection between the two theories is obvious. Also, some particular cases are indeed simple consequences. For example (see **[4]**), as a consequence of a similar inequality for mixed discriminants one can prove that

$$V(K_1, \dots, K_n) \geqslant \prod_{i=1}^n |K_i|^{1/n}. \tag{3.3.20}$$

4. Extremal Problems and Isotropic Positions

4.1. Classical positions of convex bodies. The family of positions of a convex body K in $\mathbb{R}^n$ is the class $\{T(K) \mid T \in GL(n)\}$. The right choice of a position is often quite important for the study of geometric quantities. For example, let K be a symmetric convex body in $\mathbb{R}^n$ and consider the volume product $s(K) = \left(|K| \cdot |K^\circ|\right)^{1/n}$. The Blaschke–Santaló inequality asserts that $s(K)$ is maximized if and only if K is an ellipsoid (note that $s(K)$ is invariant under $GL(n)$). On the other hand, a simple application of Hölder's inequality shows that

$$\frac{|A|}{|B_2^n|} = \int_{S^{n-1}} \|\theta\|_A^{-n} \sigma(d\theta) \geqslant w(A^\circ)^{-n} \tag{4.1.1}$$

for every symmetric convex body A in $\mathbb{R}^n$. This implies that

$$\frac{s(B_2^n)}{s(K)} \leqslant \min_{T \in GL(n)} w(TK) w((TK)^\circ). \tag{4.1.2}$$

Therefore, in order to obtain a reverse Blaschke–Santaló inequality it is useful to study the quantity

$$\max_K \min_{T \in GL(n)} w(TK) w((TK)^\circ). \tag{4.1.3}$$

One way to estimate this minimum is using the ℓ-position of K, and Pisier's inequality shows that the above quantity is bounded by $C \log n$. Thus, the ℓ-position provides a first quite nontrivial reverse inequality for the volume product $s(K)$.

All classical positions of convex bodies arise as solutions of such extremal problems. We often normalize the volume of K to be 1 and ask for the maximum or minimum of $f(TK)$ over all $T \in SL(n)$, where f is some functional on convex bodies (in the example above, f is the product of the mean widths of a body and its polar). Another useful normalization is $|K| = |B_2^n|$: we then say that the volume radius of K is equal to 1. Below we describe some classical positions of a given convex body K which solve natural extremal problems. An interesting feature of this procedure is that a simple variational method leads to a geometric description of the extremal position, and that in many cases this position satisfies an isotropic condition for an appropriate measure on S^{n-1}. We say that a Borel measure μ on S^{n-1} is isotropic if

$$\int_{S^{n-1}} \langle x, \theta \rangle^2 \mu(d\theta) = \frac{\|\mu\|}{n} |x|^2 \quad \text{for all } x \in \mathbb{R}^n. \tag{4.1.4}$$

John's position. A symmetric convex body K is in John's position if the maximal volume ellipsoid of K is the Euclidean unit ball. John's theorem [**73**] asserts that, in this case, there exist contact points $u_1, \ldots, u_m$ of K and B_2^n (common points of their boundaries) and positive real numbers $c_1, \ldots, c_m$ such that

$$I = \sum_{j=1}^{m} c_j u_j \otimes u_j. \tag{4.1.5}$$

In particular, this decomposition of the identity implies that

$$|x|^2 = \sum_{j=1}^{m} c_j \langle x, u_j \rangle^2 \tag{4.1.6}$$

for every $x \in \mathbb{R}^n$. A direct consequence of (4.1.6) is the fact that $K \subset \sqrt{n} B_2^n$ (in other words, $d(X_K, \ell_2^n) \leqslant \sqrt{n}$). The condition in (4.1.6) may be viewed as an isotropic one: the measure μ supported by $\{u_1, \ldots, u_m\}$ which gives mass c_j to u_j is isotropic. Moreover, Ball observed that this condition is also sufficient in the following sense.

Theorem 4.1. *Let K be a symmetric convex body in $\mathbb{R}^n$ such that $B_2^n \subseteq K$. Then, K is in John's position if and only if there exists an isotropic measure μ on S^{n-1} which is supported by the set of contact points of K and B_2^n.*

There exists an analogue of this fact for the not necessarily symmetric case (see, for example, [**54**]). From John's decomposition of the identity one can recover all the available information about John's position: for example, the Dvoretzky–Rogers lemma is a simple consequence of (4.1.5).

John's decomposition of the identity holds in a much more general context: If K and L are (not necessarily symmetric) convex bodies in $\mathbb{R}^n$, we say that L *is of maximal volume in* K if $L \subseteq K$ and, for every $w \in \mathbb{R}^n$ and $T \in SL_n$, the affine image $w + T(L)$ of L is not contained in the interior of K. If L is of maximal volume in K then for every $z \in \text{int}(L)$, one can find contact points $v_1, \ldots, v_m$ of $K - z$ and $L - z$, contact points $u_1, \ldots, u_m$ of $(K - z)^\circ$ and $(L - z)^\circ$, and positive reals $c_1, \ldots, c_m$, such that $\sum c_j u_j = 0$, $\langle u_j, v_j \rangle = 1$, and

$$I = \sum_{j=1}^{m} c_j u_j \otimes v_j.$$

Moreover, there exists an optimal choice of the "center" z so that, setting $z = 0$, we simultaneously have $\sum c_j u_j = \sum c_j v_j = 0$. This fact was proved in **[57]** under some conditions on K and L (in the symmetric case it had been observed by Milman, see **[147]**). A second proof was recently given in **[66]**, where the decomposition is also used to establish that for *any* symmetric convex body K in $\mathbb{R}^n$ the Banach–Mazur distance (see § 6.1) $d(K,T)$ is less than or equal to n for every convex body T in $\mathbb{R}^n$ and the distance $d(K,S_n)$ to the simplex S_n is *equal* to n.

Minimal mean width position. Recall that the mean width of a convex body K in $\mathbb{R}^n$ is the quantity $w(K) = \int_{S^{n-1}} h_K(\theta)\sigma(d\theta)$, where h_K is the support function of K (the mean width is clearly invariant under translation). We fix the volume of K to be equal to 1 and say that K has minimal mean width if $w(K) \leqslant w(TK)$ for all $T \in SL(n)$.

Let ν_K be the Borel measure on S^{n-1} with density h_K with respect to σ. An isotropic characterization of the minimal mean width position is proved in **[53]**.

Theorem 4.2. *Let K be a convex body in $\mathbb{R}^n$. Then, K has minimal mean width if and only if the measure ν_K is isotropic, i.e., if*

$$w(K) = n \int_{S^{n-1}} h_K(\theta)\langle\theta, x\rangle^2 \sigma(d\theta) \tag{4.1.7}$$

for all $x \in S^{n-1}$. Moreover, this position is uniquely determined up to $O(n)$.

An interesting question is to determine the order of growth of the quantity

$$w(n) = \max_K \min_{T\in SL(n)} w(TK) \tag{4.1.8}$$

as n tends to infinity, where the maximum is over all convex bodies of volume 1 in $\mathbb{R}^n$. If $|K| = 1$, Urysohn's inequality implies that $w(K) \geqslant c\sqrt{n}$, where $c > 0$ is an absolute constant. Pisier's inequality shows that $w(n) \leqslant c_1\sqrt{n}\log n$, and the example of the ℓ_1^n ball shows that $w(n) \geqslant c_2\sqrt{n\log n}$.

Minimal surface area position. Recall that the area measure of a convex body K is the Borel measure σ_K on S^{n-1} with

$$\sigma_K(A) = \nu(\{x \in \mathrm{bd}(K) : \text{ the outer normal to } K \text{ at } x \text{ is in } A\}),$$

where ν is the $(n-1)$-dimensional surface measure on K. The surface area of K is $\partial(K) = \|\sigma_K\|$. Again, we fix the volume of K to be equal to 1 and say that K has minimal surface area if $\partial(K) \leqslant \partial(TK)$ for all $T \in SL(n)$.

An isotropic characterization of the minimal surface area position was proved by Petty [**117**] (see also [**56**]).

Theorem 4.3. *Let K be a convex body in $\mathbb{R}^n$. Then, K has minimal surface area if and only if the measure σ_K is isotropic, i.e., if*

$$\partial(K) = n \int_{S^{n-1}} \langle \theta, x\rangle^2 \sigma_K(d\theta) \tag{4.1.9}$$

for all $x \in S^{n-1}$. Moreover, this position is uniquely determined up to $O(n)$.

As in the case of the mean width, it is natural to study the quantity

$$\partial(n) = \max_K \min_{T \in SL(n)} \partial(TK) \tag{4.1.10}$$

and its behavior as n tends to infinity, where the maximum is over all convex bodies of volume 1 in $\mathbb{R}^n$. If $|K| = 1$, the isoperimetric inequality implies that $\partial(K) \geqslant c\sqrt{n}$, where $c > 0$ is an absolute constant. A sharp upper bound for $\partial(n)$ was given by Ball ([**12**], see § 4.4). The extremal bodies are: the cube in the symmetric case and the simplex in the general case.

4.2. Isotropic position and the slicing problem. The slicing problem asks if there exists an absolute constant $c > 0$ with the following property: for every convex body K of volume 1 in $\mathbb{R}^n$, with center of mass at the origin, there exists $\theta \in S^{n-1}$ such that $|K \cap \theta^{\perp}| \geqslant c$. This is an important question in modern convex geometry, which is deeply connected with the asymptotic versions of several classical geometric problems.

The question is in a sense equivalent to the study of linear functionals on convex bodies. Indeed, by Brunn's principle, for any $\theta \in S^{n-1}$ the function $f_{K,\theta}(t) = |K \cap (\theta^{\perp} + t\theta)|$ is log-concave, and this implies that

$$\frac{c_1}{|K \cap \theta^{\perp}|^2} \leqslant \int_K \langle x, \theta\rangle^2 dx \leqslant \frac{c_2}{|K \cap \theta^{\perp}|^2}, \tag{4.2.1}$$

where $c_1, c_2 > 0$ are absolute constants. In this way, the volume of sections is measured by the moments of inertia of the body.

This brings into play the Binet ellipsoid $E_B(K)$ of K, a notion coming from classical mechanics. The norm of the Binet ellipsoid is defined by

$$||y||^2_{E_B(K)} = \frac{1}{|K|}\int_K \langle x, y\rangle^2 dx \tag{4.2.2}$$

and a suitable homothet of its polar (the Legendre ellipsoid $E_L(K)$ of K) satisfies the equation

$$\int_{E_L(K)} \langle x, y\rangle^2 dx = \int_K \langle x, y\rangle^2 dx \tag{4.2.3}$$

for every $y \in \mathbb{R}^n$ (it has the same moments of inertia as K).

We say that a convex body K of volume 1 with center of mass at the origin is isotropic if the Legendre ellipsoid $E_L(K)$ is a multiple of B_2^n. Equivalently, if there exists a constant $L_K > 0$ such that

$$\int_K \langle y, \theta\rangle^2 dy = L_K^2 \tag{4.2.4}$$

for every $\theta \in S^{n-1}$. Every convex body (in fact, every compact set) has an isotropic position, which is unique up to orthogonal transformations. This position may again be described as the solution of an extremal problem of the type we discussed in the previous subsection (see **[111]** for an extensive survey of all these facts).

Theorem 4.4. *Let K be a convex body of volume 1 in $\mathbb{R}^n$, with center of mass at the origin. Then,*

$$\int_K |x|^2 dx \leqslant \int_{TK} |x|^2 dx \tag{4.2.5}$$

for every $T \in SL(n)$ if and only if there exists a constant $L_K > 0$ such that

$$\int_K \langle y, \theta\rangle^2 dy = L_K^2 \tag{4.2.6}$$

for every $\theta \in S^{n-1}$.

Uniqueness of the isotropic position up to $O(n)$ shows that this isotropic constant L_K is invariant for the class of K. It is easily proved that $L_K \geqslant L_{B_2^n} \geqslant c > 0$ for every convex body K in $\mathbb{R}^n$, where $c > 0$ is an absolute constant. For an isotropic convex body K, (4.2.1) shows that all $(n-1)$-dimensional sections through the origin are approximately equal to $1/L_K$. Therefore, the slicing problem becomes a question about the uniform boundedness of L_K. In fact, it is not hard to see that an affirmative answer to the slicing problem is in full generality equivalent to the following statement:

> There exists an absolute constant $C > 0$ such that $L_K \leqslant C$ for every convex body K of volume 1 with center of mass at the origin.

One can easily obtain the estimate $L_K = O(\sqrt{n})$ for every convex body K. In the symmetric case, this is an immediate consequence of John's theorem, while in the general case it can be deduced from Blaschke's identity which connects the matrix of inertia of K with the expected value of the volume of a random simplex inside K. Uniform boundedness of L_K is known for some classes of bodies: unit balls of spaces with a 1-unconditional basis, zonoids and their polars, etc. For partial answers to the question, see [**111, 9**]. The best known general upper estimate is due to Bourgain [**23**]: $L_K \leqslant c\sqrt[4]{n}\log n$ for every convex body K in $\mathbb{R}^n$. For a sketch of the proof, see [**54**] (the argument follows the presentation of [**38**], see also [**116**] for the not-necessarily symmetric case).

There is a renewed interest in the problem. We mention here a very recent result of Bourgain, Klartag and Milman [**24**] which reduces the question to convex bodies with bounded volume ratio. There exists a constant $A > 1$ with the following property: if for all n and all convex bodies K in $\mathbb{R}^n$ with $\mathrm{vr}(K) \leqslant A$ we have $L_K \leqslant \alpha$ for some constant α, then for all n and all convex bodies K in $\mathbb{R}^n$ we have $L_K \leqslant c(\alpha)$ for some constant $c(\alpha)$ depending only on α. Actually, the dependence of $c(\alpha)$ on α is almost linear. The proof of this fact uses two tools: Steiner symmetrization and the existence and properties of M-ellipsoids (see § 5.2).

4.3. Brascamp–Lieb inequality and its reverse form. The Brascamp–Lieb inequality concerns the multilinear operator $I : L^{p_1}(\mathbb{R}) \times \cdots \times L^{p_m}(\mathbb{R}) \to \mathbb{R}$ defined by

$$I(f_1, \ldots, f_m) = \int_{\mathbb{R}^n} \prod_{j=1}^{m} f_j(\langle u_j, x\rangle)\, dx, \tag{4.3.1}$$

where $m \geqslant n$, $p_1, \ldots, p_m \geqslant 1$ with $\frac{1}{p_1} + \cdots + \frac{1}{p_m} = n$, and $u_1, \ldots, u_m \in \mathbb{R}^n$.

Brascamp and Lieb [**31**] proved that the norm of I is the supremum D of

$$\frac{I(g_1,\dots,g_m)}{\prod_{j=1}^{m}\|g_j\|_{p_j}} \tag{4.3.2}$$

over all centered Gaussian functions $g_1,\dots,g_m$, i.e., over all functions of the form $g_j(t)=e^{-\lambda_j t^2}$, $\lambda_j>0$. This fact is a generalization of Young's convolution inequality $\|f*g\|_r \leqslant C_{p,q}\|f\|_p\|g\|_q$ for all $f\in L^p(\mathbb{R})$ and $g\in L^q(\mathbb{R})$, where $p,q,r\geqslant 1$ and $1/p+1/q=1+1/r$. The best constants $C_{p,q}=A_pA_qA_{r'}$ (where $A_s=(s^{1/s}/(s')^{1/s'})^{1/2}$ and s' is the conjugate exponent of s) had been also obtained by Beckner [**19**] who showed that Gaussian functions play the role of maximizers.

The original proof of the Brascamp–Lieb inequality was based on a general rearrangement inequality of Brascamp, Lieb, and Luttinger [**32**], who showed that if f^* is the symmetric decreasing rearrangement of a Borel measurable function f vanishing at infinity, then

$$I(f_1,\dots,f_m)\leqslant I(f_1^*,\dots,f_m^*). \tag{4.3.3}$$

A generalization of this fact to functions of several variables (based on Steiner symmetrization) and the fact that radial functions in high dimensions behave like Gaussian functions were the key ingredients of the original proof. Setting $c_j=1/p_j$ and replacing f_j by $f_j^{c_j}$ one can reformulate the Brascamp–Lieb inequality as follows.

Theorem 4.5. *If $m\geqslant n$, $u_1,\dots,u_m\in\mathbb{R}^n$ and $c_1,\dots,c_m>0$ with $c_1+\cdots+c_m=n$, then*

$$\int_{\mathbb{R}^n}\prod_{j=1}^{m} f_j^{c_j}(\langle x,u_j\rangle)dx \leqslant D\cdot\prod_{j=1}^{m}\Big(\int_{\mathbb{R}} f_j\Big)^{c_j} \tag{4.3.4}$$

for all integrable functions $f_j:\mathbb{R}\to\mathbb{R}^+$.

Testing on the Gaussians, one can see that $D=1/\sqrt{F}$, where

$$F=\inf\left\{\frac{\det\Big(\sum_{j=1}^{m}c_j\lambda_j u_j\otimes u_j\Big)}{\prod_{j=1}^{m}\lambda_j^{c_j}} \mid \lambda_j>0\right\}. \tag{4.3.5}$$

Barthe [**16**] proved the following reverse form of Theorem 4.5 which was conjectured by Ball.

Theorem 4.6. *Let $m \geqslant n$, $c_1, \dots, c_m > 0$ with $c_1 + \cdots + c_m = n$, and $u_1, \dots, u_m \in \mathbb{R}^n$. If $h_1, \dots, h_m : \mathbb{R} \to \mathbb{R}^+$ are measurable functions, we set*

$$K(h_1, \dots, h_m) = \int_{\mathbb{R}^n}^{*} \sup \left\{ \prod_{j=1}^{m} h_j^{c_j}(\theta_j) \mid \theta_j \in \mathbb{R} \,,\; x = \sum_{j=1}^{m} \theta_j c_j u_j \right\} dx, \tag{4.3.6}$$

where $\int^*$ denotes the outer integral. Then,

$$\inf \left\{ K(h_1, \dots, h_m) \mid \int_{\mathbb{R}} h_j = 1 \,,\; j = 1, \dots, m \right\} = \sqrt{F}. \tag{4.3.7}$$

The proof is remarkably elegant and, at the same time, it gives a new direct proof of the Brascamp–Lieb inequality. We will briefly discuss the argument. Again, first testing on centered Gaussian functions, one observes that

$$\inf \left\{ K(h_1, \dots, h_m) \mid \int_{\mathbb{R}} h_j = 1 \,,\; j = 1, \dots, m \right\} \leqslant \sqrt{F}. \tag{4.3.8}$$

The main step in Barthe's argument is the following proposition.

Proposition 4.1. *Let $f_1, \dots, f_m : \mathbb{R} \to \mathbb{R}^+$ and $h_1, \dots, h_m : \mathbb{R} \to \mathbb{R}^+$ be integrable functions with*

$$\int_{\mathbb{R}} f_j(t)dt = \int_{\mathbb{R}} h_j(t)dt = 1, \quad j = 1, \dots, m.$$

Then,

$$F \cdot I(f_1, \dots, f_m) \leqslant K(h_1, \dots, h_m). \tag{4.3.9}$$

Proof. We may assume that f_j, h_j are continuous and strictly positive. We may also assume that $0 < F < +\infty$ (F is not degenerated). We use the transportation of measure idea that was used for the proof of the Prékopa–Leindler inequality: For every $j = 1, \dots, m$ we define $T_j : \mathbb{R} \to \mathbb{R}$ by the equation

$$\int_{-\infty}^{T_j(t)} h_j(s)ds = \int_{-\infty}^{t} f_j(s)ds. \tag{4.3.10}$$

Then, each T_j is strictly increasing, 1-1 and onto, and

$$T_j'(t) h_j(T_j(t)) = f_j(t), \quad t \in \mathbb{R}. \tag{4.3.11}$$

We now define $W : \mathbb{R}^n \to \mathbb{R}^n$ by

$$W(y) = \sum_{j=1}^{m} c_j T_j(\langle y, u_j\rangle) u_j. \tag{4.3.12}$$

A simple computation shows that $J(W)(y) = \sum_{j=1}^{m} c_j T_j'(\langle y, u_j\rangle) u_j \otimes u_j$. This implies that $\langle [J(W)(y)](v), v\rangle > 0$ if $v \neq 0$ and hence, W is injective. Consider the function

$$m(x) = \sup\left\{ \prod_{j=1}^{m} h_j^{c_j}(\theta_j) \mid x = \sum_{j=1}^{m} \theta_j c_j u_j \right\}.$$

Then, (4.3.12) shows that

$$m(W(y)) \geqslant \prod_{j=1}^{m} h_j^{c_j}(T_j(\langle y, u_j\rangle)) \tag{4.3.13}$$

for every $y \in \mathbb{R}^n$. It follows that

$$\begin{aligned}\int_{\mathbb{R}^n} m(x)dx &\geqslant \int_{W(\mathbb{R}^n)} m(x)dx = \int_{\mathbb{R}^n} m(W(y)) \cdot |J(W)(y)|\, dy \\ &\geqslant \int_{\mathbb{R}^n} \prod_{j=1}^{m} h_j^{c_j}(T_j(\langle y, u_j\rangle)) \det\left(\sum_{j=1}^{m} c_j T_j'(\langle y, u_j\rangle) u_j \otimes u_j \right) dy.\end{aligned}$$

By the definition of F, we have

$$\det\left(\sum_{j=1}^{m} c_j T_j'(\langle y, u_j\rangle) u_j \otimes u_j \right) \geqslant F \cdot \prod_{j=1}^{m} \left(T_j'(\langle y, u_j\rangle)\right)^{c_j}. \tag{4.3.14}$$

Therefore, taking (4.3.11) into account we have

$$\begin{aligned}\int_{\mathbb{R}^n} m(x)dx &\geqslant F \cdot \int_{\mathbb{R}^n} \prod_{j=1}^{m} h_j^{c_j}(T_j(\langle y, u_j\rangle)) \cdot \prod_{j=1}^{m} \left(T_j'(\langle y, u_j\rangle)\right)^{c_j} dy \\ &= F \cdot \int_{\mathbb{R}^n} \prod_{j=1}^{m} f_j^{c_j}(\langle y, u_j\rangle) dy = F \cdot I(f_1, \dots, f_m).\end{aligned}$$

In other words, $F \cdot I(f_1, \dots, f_m) \leqslant K(h_1, \dots, h_m)$. □

One can now prove simultaneously Theorems 4.5 and 4.6. The computation leading to (4.3.5) shows that

$$\sup\left\{I(f_1,\dots,f_m) \mid \int_{\mathbb{R}} f_j = 1\,,\ j=1,\dots,m\right\} \geqslant \frac{1}{\sqrt{F}}. \qquad (4.3.15)$$

From Proposition 4.1, (4.3.8), and (4.3.15) we get

$$\begin{aligned}\frac{1}{\sqrt{F}} &\leqslant \sup\left\{I(f_1,\dots,f_m) \mid \int_{\mathbb{R}} f_j = 1\right\} \\ &\leqslant \frac{1}{F}\cdot \inf\left\{K(h_1,\dots,h_m) \mid \int_{\mathbb{R}} h_j = 1\right\} \leqslant \frac{1}{\sqrt{F}}.\end{aligned}$$

We must have equality everywhere, and this ends the proof(s).

There is a multidimensional generalization of both inequalities. Let $S^+(\mathbb{R}^k)$ be the set of $k\times k$ symmetric, positive definite matrices. If $A\in S^+(\mathbb{R}^k)$, we write G_A for the centered Gaussian function $G_A:\mathbb{R}^k\to\mathbb{R}$ defined by $G_A(x)=\exp(-\langle Ax,x\rangle)$, and $L_1^+(\mathbb{R}^k)$ for the class of integrable nonnegative functions $f:\mathbb{R}^k\to\mathbb{R}$. Let $m\geqslant n$, and assume we are given real numbers $c_1,\dots,c_m>0$ and integers $n_1,\dots,n_m$ less than or equal to n, such that $\sum_{j=1}^{m} c_j n_j = n$. We are also given linear maps $B_j:\mathbb{R}^n\to\mathbb{R}^{n_j}$ which are onto and satisfy $\bigcap_{j=1}^m \mathrm{Ker}(B_j)=\{0\}$.

Consider the operators $I,K : L_1^+(\mathbb{R}^{n_1})\times\cdots\times L_1^+(\mathbb{R}^{n_m})\to\mathbb{R}$ defined by

$$I(f_1,\dots,f_m)=\int_{\mathbb{R}^n}\prod_{j=1}^m f_j^{c_j}(B_jx)dx \qquad (4.3.16)$$

and

$$K(h_1,\dots,h_m)=\int_{\mathbb{R}^m}^{*} m(x)dx, \qquad (4.3.17)$$

where

$$m(x)=\sup\left\{\prod_{j=1}^m h_j^{c_j}(y_j) \mid y_j\in\mathbb{R}^{n_j} \text{ and } \sum_{j=1}^m c_jB_j^*y_j = x\right\}. \qquad (4.3.18)$$

Let E be the largest constant for which

$$K(h_1,\dots,h_m)\geqslant E\cdot\prod_{j=1}^m\left(\int_{\mathbb{R}^{n_j}} h_j\right)^{c_j} \qquad (4.3.19)$$

holds true for all $h_j \in L_1^+(\mathbb{R}^{n_j})$, and let F be the smallest constant for which

$$I(f_1,\dots,f_m) \leqslant F \cdot \prod_{j=1}^{m}\Big(\int_{\mathbb{R}^{n_j}} f_j\Big)^{c_j} \tag{4.3.20}$$

holds true for all $f_j \in L_1^+(\mathbb{R}^{n_j})$. Then, the following holds true.

Theorem 4.7. *The constants E and F can be computed using centered Gaussian functions. Moreover, if D is the largest real number for which*

$$\det\Big(\sum_{j=1}^{m} c_j B_j^* A_j B_j\Big) \geqslant D \cdot \prod_{j=1}^{m}(\det A_j)^{c_j}, \tag{4.3.21}$$

for all $A_j \in S^+(\mathbb{R}^{n_j})$, we have

$$E = \sqrt{D} \quad \text{and} \quad F = 1/\sqrt{D}. \tag{4.3.22}$$

The multidimensional version of the Brascamp–Lieb inequality was first established by Lieb in [**84**]. The simultaneous proof of both this inequality and its reverse form is due to Barthe [**16**] and follows the idea of the proof of the one-dimensional case. However, instead of the direct transportation of measure argument there, one now has to make essential use of the Brenier map.

4.4. Sharp geometric inequalities. As § 4.1 shows, isotropic positions of convex bodies and the corresponding decompositions of the identity are typical in the asymptotic theory: isotropicity may be viewed as the ultimate form of nondegeneracy. Ball made the very important observation that the constants in the Brascamp–Lieb inequality and its reverse form take a surprisingly simple form in the presence of such a decomposition of the identity.

Theorem 4.8. *Assume that the vectors $u_1,\dots,u_m \in S^{n-1}$ and the positive weights $c_1,\dots,c_m$ satisfy the isotropic condition*

$$I = \sum_{j=1}^{m} c_j u_j \otimes u_j. \tag{4.4.1}$$

Then, the constant $F = F(\{u_j\},\{c_j\})$ in Theorems 4.5 and 4.6 is equal to 1.

Ball applied the Brascamp–Lieb inequality in this context to solve purely geometric problems. A well-known example is his reverse isoperimetric inequality [**12**], which gives the exact value of the constant $\partial(n)$ in (4.1.4). We ask for the best constant $\partial(n)$ for which every symmetric convex body K in $\mathbb{R}^n$ has a position $\widetilde{K}$ satisfying

$$\partial(\widetilde{K}) \leqslant \partial(n)|\widetilde{K}|^{(n-1)/n}. \tag{4.4.2}$$

The natural position of K is the minimal surface area position. However, Ball's solution of the problem employs John's position. Assume that B_2^n is the maximal volume ellipsoid of K. Then,

$$\partial(K) = \lim_{t\to 0+} \frac{|K + tB_2^n| - |K|}{t} \leqslant \lim_{t\to 0+} \frac{|K + tK| - |K|}{t} = n|K|. \tag{4.4.3}$$

We claim that among all bodies in John's position the cube has maximal volume.

Theorem 4.9. *Let $Q_n = [-1,1]^n$ be the unit cube in $\mathbb{R}^n$. If K is a symmetric convex body in John's position in $\mathbb{R}^n$, then $|K| \leqslant 2^n = |Q_n|$.*

For the proof we use John's representation of the identity (4.4.1), where the u_j's are contact points of K and B_2^n. Observe that

$$K \subseteq M := \{x : |\langle x, u_j\rangle| \leqslant 1,\ j = 1, \ldots, m\}. \tag{4.4.4}$$

Therefore,

$$\begin{aligned} |K| &\leqslant |M| = \int_{\mathbb{R}^n} \prod_{j=1}^{m} \chi_{[-1,1]}^{c_j}(\langle x, u_j\rangle)dx \\ &\leqslant \prod_{j=1}^{m} \Big(\int_{\mathbb{R}} \chi_{[-1,1]}(t)dt \Big)^{c_j} = 2^{\sum_{j=1}^{m} c_j} = 2^n, \end{aligned}$$

where we used the Brascamp–Lieb inequality together with the observation of Theorem 4.8, and the fact that $\sum_{j=1}^{m} c_j = n$, which is a simple consequence of (4.4.1). □

Now, (4.4.3) shows that $\partial(K) \leqslant n|K| \leqslant 2n|K|^{(n-1)/n}$, and since K was arbitrary, $\partial(n) \leqslant 2n$. There is equality in the case of the cube, and this shows that $\partial(n) = 2n$.

Theorem 4.9 shows that the cube has maximal volume ratio among all symmetric convex bodies. In the general case, one can show that the simplex Δ_n is the extremal convex body. The reverse Brascamp–Lieb inequality can be used for the dual statements: consider the external volume ratio $\mathrm{evr}(K) = \inf\left(|E|/|K|\right)^{1/n}$, where the infimum is taken over all ellipsoids containing K. Then, $\mathrm{evr}(K) \leqslant \mathrm{evr}(\Delta_n)$ for every convex body K in $\mathbb{R}^n$. In the symmetric case the extremal body is the cross-polytope (the unit ball of ℓ_1^n).

The Brascam–Lieb inequality and its reverse form were also used for sharp estimates on the volume of sections and projections of the unit ball B_p^n of ℓ_p^n [**10**]. If $p > 0$ and H is a k-dimensional subspace of $\mathbb{R}^n$, then $|B_p^n \cap H| \leqslant |B_p^k|$ if $p \leqslant 2$, and

$$|B_p^n \cap H| \leqslant \left(\frac{n}{k}\right)^{k(1/2-1/p)} |B_p^k| \tag{4.4.5}$$

if $p \geqslant 2$. This last estimate is sharp if k divides n. On the other hand, $|P_H(B_p^n)| \geqslant |B_p^k|$ if $p \geqslant 2$, and

$$|P_H(B_p^n)| \geqslant \left(\frac{k}{n}\right)^{k(1/p-1/2)} |B_p^k| \tag{4.4.6}$$

if $p \leqslant 2$. This last estimate is sharp if $p > 1$ and k divides n. The proof of all these inequalities is based on the observation that if $\{e_1, \dots, e_n\}$ is the standard orthonormal basis in $\mathbb{R}^n$, then the obvious representation $I = \sum_{j=1}^{n} e_j \otimes e_j$ of the identity implies that

$$P_H = \sum_{j=1}^{n} a_j^2 u_j \otimes u_j, \tag{4.4.7}$$

where $a_j = |P_H(e_j)|$ and $u_j = P_H(e_j)/a_j$.

The multidimensional version of the reverse Brascamp–Lieb inequality is used in the proof of the following Brunn–Minkowski type inequality of Barthe [**16**]. Let m, n be integers. Let E_i, $i \leqslant m$ be linear subspaces of $\mathbb{R}^n$. Assume that there exist positive c_i's such that $I = \sum_{i \leqslant m} c_i P_i$, where P_i is the orthogonal projection onto E_i. Then, the inequality

$$\left|\sum c_i K_i\right| \geqslant \prod |K_i|^{c_i} \tag{4.4.8}$$

holds for any compact subsets K_i of E_i, where $|K_i|$ is the volume of K_i in E_i. In the case where each K_i is a line segment, this reduces to an inequality of Ball [**11**] which was proved by induction on the dimension.

Another extremal property of the simplex was proved by Barthe [**17**]. Assume that K is a convex body whose minimal volume ellipsoid is B_2^n. Then, $M(K) \leqslant M(\Delta_n)$, where Δ_n is the regular simplex inscribed in B_2^n. In the symmetric case one has $M(K) \leqslant M(B_1^n)$ (this is much simpler and was observed by Schechtman and Schmuckenschläger [**133**]). The proof of both inequalities makes use of the reverse Brascamp–Lieb inequality. In John's position, the simplex and the cube are the extremal bodies for $M(K)$.

For a different application, consider a polytope K with facets F_j and normals u_j, $j = 1, \ldots, m$. If K is in minimal surface area position, Petty's theorem 4.3 is equivalent to the statement

$$I = \sum_{j=1}^{m} \frac{n|F_j|}{\partial(K)} u_j \otimes u_j. \tag{4.4.9}$$

The projection body ΠK of K is defined by

$$h_{\Pi K}(x) = \frac{1}{2} \int_{S^{n-1}} |\langle x, z\rangle| \sigma_K(dz). \tag{4.4.10}$$

In our case,

$$\Pi K = \frac{\partial(K)}{2n} \sum_{j=1}^{m} c_j [-u_j, u_j],$$

and using (4.4.9) one can give a lower bound of its volume [**56**]. Namely,

$$|\Pi K| \geqslant 2^n \left(\frac{\partial(K)}{2n}\right)^n. \tag{4.4.11}$$

The example of the cube shows that this inequality is sharp for bodies with minimal surface area.

Combined with Theorem 4.2 this volume estimate leads to a sharp reverse Urysohn inequality for zonoids [**55**]. If Z be a zonoid in $\mathbb{R}^n$ with volume 1 and minimal mean width, then

$$w(Z) \leqslant w(Q_n) = \frac{2\omega_{n-1}}{\omega_n}. \tag{4.4.12}$$

For the proof, recall that Z is the projection body ΠK of some convex body K. Using (4.4.10) and the characterizations of Theorems 4.2 and

4.3 we check that K has minimal surface area. We have

$$w(Z) = 2 \int_{S^{n-1}} h_Z(x)\sigma(dx)$$
$$= \int_{S^{n-1}} \int_{S^{n-1}} |\langle x, z\rangle| \sigma_K(dz)\sigma(dx) = \frac{2\omega_{n-1}}{n\omega_n}\partial(K), \tag{4.4.13}$$

and (4.4.11) shows that $w(Z) \leqslant 2\omega_{n-1}/\omega_n$. We have equality when K is a cube, and this corresponds to the case $Z = (1/2)Q_n$.

4.5. Study of geometric probabilities. In this short subsection we describe some recent results from **[63]** on random properties of the uniform distribution over a convex body K in $\mathbb{R}^n$. To fix terminology, for any (measurable) set $A \subset \mathbb{R}^n$, the geometric probability of A is $P(A) := |A \cap K|/|K|$.

Theorem 4.10. *Let T_i be measurable sets in $\mathbb{R}^n$, $i = 1, \dots, m$, and K be a star-shaped body with $0 \in \mathrm{int}(K)$. Assume that $|K| = |T_1| = \cdots = |T_m|$. Consider the positively homogeneous function*

$$|||\widetilde{\lambda}||| = \frac{1}{\prod_{i=1}^{m} |T_i|} \int_{T_1} \cdots \int_{T_m} \left\| \sum_{i=1}^{m} \lambda_i x_i \right\|_K dx_m \dots dx_1 \tag{4.5.1}$$

on $\mathbb{R}^m$. Then,

$$|||\widetilde{\lambda}||| \geqslant c\sqrt{\sum \lambda_i^2} \tag{4.5.2}$$

for every $\widetilde{\lambda} \in \mathbb{R}^m$, where $c > 0$ is an absolute constant ($c \geqslant c_n/\sqrt{2}$, where $c_n \to 1$ as $n \to \infty$).

The proof of Theorem 4.10 is a direct consequence of the following fact: If K and T_i are as above and if $|K| = |T_i| = |B_2^n|$ for every i, then, for any scalars λ_i, $i = 1, \dots, m$ and for any $t > 0$, we have

$$P\left\{ (x_i \in T_i)_{i=1}^m : \left\| \sum_1^m \lambda_i x_i \right\|_K < t \right\}$$
$$\leqslant P\left\{ (x_i \in B_2^n)_{i=1}^m : \left\| \sum_1^m \lambda_i x_i \right\|_{B_2^n} < t \right\}. \tag{4.5.3}$$

One then knows that the extremal case is $K = T_1 = \cdots = T_m = B_2^n$ and a simple argument based on Kahane's inequality leads to the lower bound.

The proof of (4.5.3) uses the rearrangement inequality of Brascamp, Lieb and Luttinger [**32**] which was the starting point for the first proof of the Brascamp–Lieb inequality.

An interesting question is to give exact estimates for the probability in (4.5.3) in terms of $\{\lambda_i\}$ and t. This is done in [**63**] with a method which uses the sharp multivariable version of Young's inequality, proved by Brascamp and Lieb [**31**]. [This approach was first used by Arias-de-Reyna, Ball and Villa in [**6**] to establish the case $m = 2$, $\lambda_1 = -\lambda_2 = 1/\sqrt{2}$, $T_i = K$ (where K is a symmetric convex body)]:

Fact. *Assume that* $|K| = |T_1| = \cdots |T_m| = 1$. *Then, for any scalars* $\lambda_i \in \mathbb{R}$ *and any* $0 < t \leqslant 1$,

$$P\Big\{(x_i \in T_i)_{i=1}^m : \Big\|\sum_1^m \lambda_i x_i\Big\|_K < t\sqrt{\sum_1^m \lambda_i^2}\Big\} \leqslant t^n \exp\Big[\frac{(1-t^2)}{2}n\Big]. \tag{4.5.4}$$

A consequence of (4.5.4) is the fact that every n-dimensional normed space X has *random cotype 2* with constant bounded by an absolute constant $C > 0$ (see [**63**]).

We say that X has *random cotype* 2 *with constant* $A > 0$ if with probability greater than $1 - e^{-an}$ ($a > 0$ is a fixed universal number), n independent random vectors $\{x_i\}_1^n$ uniformly distributed over the unit ball K of X satisfy for every $\lambda_i \in \mathbb{R}$

$$\mathrm{Ave}_{\varepsilon_i=\pm 1}\Big\|\sum_1^n \varepsilon_i \lambda_i x_i\Big\| \geqslant \frac{1}{A}\sqrt{\sum_1^n |\lambda_i|^2}. \tag{4.5.5}$$

Note that the norms $\|x_i\|$ do not enter in the definition, since with probability exponentially close to 1 we have $1/2 \leqslant \|x_i\| \leqslant 1$ and hence the norms are absorbed in A.

5. Asymptotic Results with a Classical Convexity Flavor

5.1. Classical symetrizations. Symmetrization procedures play an important role in classical convexity. Until recently, the bounds on the number of successive symmetrizations of a certain type which are needed in order to obtain from a given body K a body $\widetilde{K}$ which is close to a ball were at least exponential in the dimension. The methods of asymptotic convex geometry show that a linear in the dimension number of steps is enough.

Minkowski symmetrization. Consider a convex body K in $\mathbb{R}^n$ and a direction $u \in S^{n-1}$. The Minkowski symmetrization of K with respect to u is the convex body $\frac{1}{2}(K + \pi_u K)$, where π_u denotes the reflection with respect to $u^\perp$. This operation is linear and preserves mean width. A random Minkowski symmetrization of K is a body $\pi_u K$, where u is chosen randomly on S^{n-1} with respect to the probability measure σ. Bourgain, Lindenstrauss, and Milman [**25**] proved that for every $\varepsilon > 0$ there exists $n_0(\varepsilon)$ such that for every $n \geqslant n_0$ and every convex body K, if we perform $N = Cn \log n + c(\varepsilon)n$ independent random Minkowski symmetrizations on K we receive a convex body $\widetilde{K}$ such that

$$(1-\varepsilon)w(K)B_2^n \subset \widetilde{K} \subset (1+\varepsilon)w(K)B_2^n \tag{5.1.1}$$

with probability greater than $1 - \exp(-c_1(\varepsilon)n)$. The method of proof is closely related to the concentration phenomenon for $SO(n)$.

Recently, Klartag [**75**] showed that if we perform a specific non-random choice of $5n$ Minkowski symmetrizations we may transform any convex body into an approximate Euclidean ball. We briefly describe the process. We may clearly start with the normalization $w(K) = 1$. We fix an orthonormal basis $\{e_1, \dots, e_n\}$ and first symmetrize K with respect to the e_j's. In this way we obtain a 1-unconditional convex body K_1 with the property $K_1 \subseteq c\sqrt{n}B_1^n$.

Let $Q = \sqrt{n}B_1^n$ and consider a "Walsh basis" of $\mathbb{R}^n$. This is an orthonormal basis $\{u_1, \dots, u_n\}$ satisfying $|\langle u_i, e_j\rangle| \leqslant 2/\sqrt{n}$ for every $i, j \leqslant n$. If we symmetrize Q with respect to $u_1, \dots, u_{n-1}$, we obtain a new body $\widetilde{Q}$ with $\mathrm{diam}(\widetilde{Q}) \leqslant c\sqrt{\log n}$. Applying the same sequence of symmetrizations to K_1 we arrive at a convex body K_2 with $w(K_2) = 1$ and

$$K_2 \subset c\sqrt{n}B_1^n \cap tB_2^n \tag{5.1.2}$$

with respect to a new orthonormal basis, where $t = \mathrm{diam}(K_2) \leqslant c\sqrt{\log n}$.

The next step shows that one can achieve a logarithmic decay of the diameter under $2n$ additional symmetrizations with respect to two independent random orthonormal bases.

Claim. *Let* $Q_t = \sqrt{n}B_1^n \cap tB_2^n$. *If* $\{v_j\}$, $\{w_j\}$ *are independent random orthonormal bases of* $\mathbb{R}^n$, *then symmetrization of* Q_t *with respect to* $v_1, \dots, v_{n-1}$ *and* $w_1, \dots, w_{n-1}$ *produces with high probability a convex body* $\widetilde{Q}_t$ *with* $\widetilde{Q}_t \subset (C \log t)B_2^n$.

It follows that the same sequence of symmetrizations applied to K_2 produces a convex body K_3 with $\mathrm{diam}(K_3) \leqslant c \log\log n$. One may then iterate this step and arrive at a body for which $\mathrm{diam}(K_s)$ is bounded by a universal constant. Then, the proof of **[25]** shows that $O(n)$ symmetrizations of K_s bring it close to a ball. Instead of this, one can show by concentration techniques that a second application of the claim's symmetrization process to the body K_3 is enough.

Even more recently, using spherical harmonics, Klartag **[76]** showed that for every convex body K and any $0 < \varepsilon < 1/2$ there exist $cn \log(1/\varepsilon)$ successive Minkowski symmetrizations which transform K to a convex body $\widetilde{K}$ satisfying $(1-\varepsilon)w(K)B_2^n \subseteq \widetilde{K} \subseteq (1+\varepsilon)w(K)B_2^n$.

Steiner symmetrization. It is well known that for any convex body K in $\mathbb{R}^n$ there exists a sequence of directions $\theta_j \in S^{n-1}$ such that $(S_{\theta_n} \circ \cdots \circ S_{\theta_1})(K)$ converges to a ball in the Hausdorff metric (S_θ is the Steiner symmetrization in the direction of θ). In fact, Mani **[90]** has proved that if we choose an infinite random sequence of directions $\theta_j \in S^{n-1}$ and apply successive Steiner symmetrizations S_{θ_j} of K in these directions, then we almost surely get a sequence of convex bodies converging to a ball.

Bourgain, Lindenstrauss, and Milman **[26]** proved an isomorphic version of this fact. There exist absolute constants $c, c_1, c_2 > 0$ with the following property: if K is a convex body in $\mathbb{R}^n$, there exist $k \leqslant cn \log n$ unit vectors θ_j such that successive Steiner symmetrizations in the directions of θ_j transform K into a convex body K_1 with

$$c_1 \rho B_2^n \subseteq K_1 \subseteq c_2 \rho B_2^n, \tag{5.1.3}$$

where B_2^n is the Euclidean unit ball and $|K| = |\rho B_2^n|$. This was a dramatic improvement with respect to the previously known estimate $(cn)^{n/2}$ of Hadwiger (1955). An essentially best possible result was recently obtained by Klartag and Milman **[77]**.

Theorem 5.1. *For every* $\varepsilon > 0$ *there exist constants* $c_1(\varepsilon), c_2(\varepsilon) > 0$ *such that: for every convex body* K *in* $\mathbb{R}^n$ *with* $|K| = |B_2^n|$, *there exist* $k \leqslant (2+\varepsilon)n$ *unit vectors* θ_j *such that successive Steiner symmetrizations* *in the directions of θ_j transform K into a convex body K' with*

$$c_1(\varepsilon)B_2^n \subseteq K' \subseteq c_2(\varepsilon)B_2^n. \tag{5.1.4}$$

The main steps of the argument are the following. Starting with a convex body of volume 1, we need $2n$ Steiner symmetrizations in order to obtain a convex body K_2 which is 1-unconditional (symmetric with respect to the coordinate subspaces) and "almost isotropic" in the following sense: for every $\theta \in S^{n-1}$,

$$\int_{K_2} \langle x, \theta\rangle^2 dx \leqslant 2. \tag{5.1.5}$$

The first n symmetrizations lead to a 1-unconditional body K_1. If the polar of the Binet ellipsoid of K_1 is transformed into a ball by n additional symmetrizations, it is proved that the same sequence of symmetrizations, applied to K_1, produces K_2. By recent results of Bobkov and Nazarov [**21**], it follows that

$$P_n \subseteq K_2 \subseteq cnB_1^n, \tag{5.1.6}$$

where P_n is a box with respect to the same coordinate system, having volume $|P_n|^{1/n} \simeq 1$ (equivalently, one may use a classical result of Losanovskii and a modification of this argument). This implies that it is enough to symmetrize P_n and the cross-polytope B_1^n. The same sequence of symmetrizations will transform K_2 into an isomorphic ball.

The analysis for these two particular bodies already proves that $(4+\varepsilon)n$ Steiner symmetrizations are enough. Employing this fact and using the quotient of subspace theorem (Theorem 2.13), one can build an iteration scheme which reduces the number of symmetrizations to $(2+\varepsilon)n$.

Floating bodies — centroid bodies. We close this subsection with some interesting observations on the connections of the Legendre ellipsoid with the centroid and floating bodies (for the proofs of these facts, see [**111**]). Let K be a symmetric convex body in $\mathbb{R}^n$ with $|K| = 1$. The centroid body of K is defined by $Z(K) = \int_K [0, x]dx$, where $[0, x]$ is the line segment from 0 to x. Equivalently, its dual norm is given by

$$\|y\|_{Z(K)^\circ} = \frac{1}{2}\int_K |\langle x, y\rangle| dx. \tag{5.1.7}$$

A consequence of the Brunn–Minkowski inequality is that $Z(K)$ is uniformly (i.e., up to an absolute constant) equivalent to the Legendre ellipsoid of K in the Hausdorff sense.

For every $0 < \delta < 1/2$, the floating body K_δ of K is defined to be the envelope of all hyperplanes that cut off a set of volume δ from K. It can be proved that K_δ is convex (this was observed by Meyer and Reisner, and independently by Ball). Moreover, K_δ is $C(\delta)$ equivalent to the Legendre ellipsoid of K, where $C(\delta)$ is a constant depending only on δ.

The process of forming the floating body may be viewed as a "one step symmetrization." One arrives at an "isomorphic ellipsoid" although one would expect that K_δ will stay close to K for small values of $\delta > 0$.

5.2. Isomorphic symmetrization. In this subsection we describe isomorphic geometric inequalities which are proved by the method of isomorphic symmetrization. This is our second main example of a body of results which answer deep questions of the Brunn–Minkowski theory, at least in their asymptotic version. Here, the main ideas and methods we described in § 2 find applications to classical convexity.

Our first example is the inverse Blaschke–Santaló inequality of Bourgain and Milman [**28**], which gives an "affirmative answer" to Mahler's conjecture (see § 4.1).

Theorem 5.2. *There exists an absolute constant $c > 0$ such that*

$$0 < c \leqslant \frac{s(K)}{s(B_2^n)} \leqslant 1 \tag{5.2.1}$$

for every symmetric convex body in $\mathbb{R}^n$.

The inequality on the right is the Blaschke–Santaló inequality. The left handside inequality answers the question of Mahler in the asymptotic sense: For every symmetric convex body K, the quantity $s(K)$ is of the order of $1/n$.

The original proof of Theorem 5.1 used a dimension descending procedure which was based on the quotient of subspace theorem. We will describe a proof using the method of isomorphic symmetrization [**105**]. This is closer to classical convexity and much more geometric in nature since it preserves dimension: however, it is a symmetrization scheme which is in many ways different from the classical symmetrizations. In each step, none of the natural parameters of the body is being preserved, but the ones which are of interest remain under control. After a finite number of steps, the body has come close to an ellipsoid, but there is no natural notion of convergence to an ellipsoid.

Since $s(K)$ is an affine invariant, we may start from a position of K which satisfies the inequality $M(K)M^*(K) \leqslant c\log[d(X_K,\ell_2^n)+1]$ (this is allowed by Theorems 2.9 and 2.10). We may also normalize so that $M(K)=1$. We define

$$\lambda_1 = M^*(K)a_1 \quad , \quad \lambda_1' = M(K)a_1, \tag{5.2.2}$$

for some $a_1 > 1$, and consider the new body

$$K_1 = \mathrm{co}\left((K \cap \lambda_1 B_2^n) \cup \frac{1}{\lambda_1'} B_2^n \right). \tag{5.2.3}$$

Sudakov's inequality (Theorem 2.14) and elementary properties of the covering numbers show that

$$|K_1| \geqslant |K \cap \lambda_1 B_2^n| \geqslant |K|/N(K,\lambda_1 B_2^n) \geqslant |K|\exp(-cn/a_1^2). \tag{5.2.4}$$

In an analogous way, using the dual Sudakov inequality (Theorem 2.15) one can show that

$$|K_1| \leqslant |\mathrm{co}(K \cup (1/\lambda_1')B_2^n)| \leqslant \exp(cn/a_1^2). \tag{5.2.5}$$

By the definition of K_1, one can apply the same reasoning to K_1°, and this shows that

$$\exp(-c/a_1^2) \leqslant \frac{s(K_1)}{s(K)} \leqslant \exp(c/a_1^2). \tag{5.2.6}$$

By construction, for the new body K_1 we have

$$d(X_{K_1},\ell_2^n) \leqslant M(K)M^*(K)a_1^2$$

and, since $s(K_1)$ is an affine invariant, we may assume that

$$M(K_1)M^*(K_1) \leqslant c\log[d(X_{K_1},\ell_2^n)+1] \quad \text{and} \quad M(K_1)=1.$$

If we set $\lambda_2 = M^*(K_1)a_2$, $\lambda_2' = M(K_1)a_2$ and define

$$K_2 = \mathrm{co}\left((K_1 \cap \lambda_2 B_2^n) \cup \frac{1}{\lambda_2'} B_2^n \right),$$

we obtain

$$\exp(-c/a_2^2) \leqslant \frac{s(K_2)}{s(K_1)} \leqslant \exp(c/a_2^2). \tag{5.2.7}$$

We now iterate this procedure, choosing $a_1 = \log n$, $a_2 = \log\log n, \dots,$ $a_t = \log^{(t)} n$ is the t-iterated logarithm of n, and stop the procedure at the first t for which $a_t < 2$. It is easy to check that $d(X_{K_t}, \ell_2^n) \leqslant C$. Therefore,

$$\frac{1}{C} \leqslant \frac{s(K_t)}{s(B_2^n)} \leqslant C. \tag{5.2.8}$$

On the other hand,

$$\begin{aligned} c_1 \leqslant \exp\left(-c\left(\frac{1}{a_1^2} + \cdots + \frac{1}{a_t^2}\right)\right) &\leqslant \frac{s(K_t)}{s(K)} \\ &\leqslant \exp\left(c\left(\frac{1}{a_1^2} + \cdots + \frac{1}{a_t^2}\right)\right), \end{aligned} \tag{5.2.9}$$

which proves the theorem (observe that the series $\frac{1}{a_1^2} + \cdots + \frac{1}{a_t^2} + \cdots$ remains bounded by an absolute constant). □

As a second important application of the method we prove the existence of "M-ellipsoids" associated to any convex body.

Theorem 5.3. *There exists an absolute constant $c > 0$ with the following property: For every symmetric convex body K in $\mathbb{R}^n$ there exists an ellipsoid M_K such that $|K| = |M_K|$ and for every body T in $\mathbb{R}^n$*

$$\frac{1}{c}\left|M_K + T\right|^{1/n} \leqslant \left|K + T\right|^{1/n} \leqslant c\left|M_K + T\right|^{1/n} \tag{5.2.10}$$

and

$$\frac{1}{c}|M_K^\circ + T|^{1/n} \leqslant |K^\circ + T|^{1/n} \leqslant c|M_K^\circ + T|^{1/n}. \tag{5.2.11}$$

For the proof of Theorem 5.3 we define the same sequence of bodies as in Theorem 5.1. For every s, we check that

$$\exp(-cn/a_s^2) \leqslant \frac{|K_s + T|}{|K_{s-1} + T|} \leqslant \exp(cn/a_s^2), \tag{5.2.12}$$

for every convex body T, and the same holds true for K_s°. After t steps, we arrive at a body K_t which is c-isomorphic to an ellipsoid M. Our volume estimates show that $|K_t|^{1/n} \simeq |K|^{1/n}$ up to an absolute constant. If we define $M_K = \rho M$, where $\rho > 0$ is such that $|M_K| = |K|$, then $\rho \simeq 1$ and the result follows. □

A consequence of Theorem 5.3 is that for every body K in $\mathbb{R}^n$ there exists a position $\widetilde{K} = u_K(K)$ of volume $|\widetilde{K}| = |K|$ such that for every pair of convex bodies K_1 and K_2 in $\mathbb{R}^n$,

$$|t_1\widetilde{K_1} + t_2\widetilde{K_2}|^{1/n} \leqslant c\left(t_1|\widetilde{K_1}|^{1/n} + t_2|\widetilde{K_2}|^{1/n}\right), \tag{5.2.13}$$

for all $t_1, t_2 > 0$, where $c > 0$ is an absolute constant. This statement is the "reverse Brunn–Minkowski inequality" (Milman, [**101**]).

The ellipsoid M_K in Theorem 5.3 is called an M-ellipsoid for K. The symmetry of K is not really needed (see, for example, [**112**]). It can be proved that the existence of an M-ellipsoid for K is equivalent to the following statement: There exists a constant $c > 0$ such that for every body K we can find an ellipsoid M_K with $|M_K| = |K|$ and $N(K, M_K) \leqslant \exp(cn)$.

Interchanging the roles of K and M_K, we say that a convex body K is in M-position (with constant c) if $|K| = |B_2^n|$ and $N(K, B_2^n) \leqslant \exp(cn)$. With this terminology, Theorem 5.3 is equivalent to the existence of a constant $c > 0$ such that in the affine class of any convex body there exists a representative which is in M-position with constant c. This condition on $N(K, B_2^n)$ implies that

$$\max\{N(B_2^n, K), N(K^\circ, B_2^n), N(B_2^n, K^\circ)\} \leqslant \exp(c_1 n)$$

for some constant c_1 which depends only on c. If K_1 and K_2 are in M-position with constant c, using these estimates one can easily check that

$$\begin{aligned} |K_1 + K_2|^{1/n} &\leqslant C(|K_1|^{1/n} + |K_2|^{1/n}), \\ |K_1^\circ + K_2^\circ|^{1/n} &\leqslant C(|K_1^\circ|^{1/n} + |K_2^\circ|^{1/n}), \end{aligned} \tag{5.2.14}$$

where C is a constant depending only on c (one just uses the volume estimate $|A + B| \leqslant N(A, B) \cdot |2B|$). If K is in M-position with constant c, setting $K_1 = K$, $K_2 = B_2^n$ and using the reverse Santaló inequality (Theorem 5.2), we get

$$c^n |K| \cdot |K^\circ| \leqslant |K \cap B_2^n| \cdot |\mathrm{co}(K^\circ \cup B_2^n)| \leqslant |K \cap B_2^n| \cdot |K^\circ + B_2^n|, \tag{5.2.15}$$

which, combined with (5.2.14), gives

$$|K \cap B_2^n| \geqslant c^n |K|. \tag{5.2.16}$$

The next fact about the M-position which is used in many applications is the following statement: If K is in M-position with constant c, then for any $\lambda \in (0, 1)$ a random orthogonal projection $P_E(K)$ onto a $[\lambda n]$-dimensional subspace E has volume ratio bounded by a constant $C(c, \lambda)$. To see this, note that $|\mathrm{co}(K^\circ \cup B_2^n)|^{1/n} \leqslant C|B_2^n|^{1/n}$, where C depends on c (this is a consequence of (5.2.14)). In other words, $W = \mathrm{co}(K^\circ \cup B_2^n)$ has bounded volume ratio, and Theorem 2.7 shows that for a random $E \in G_{n,[\lambda n]}$,

$$K^\circ \cap E \subseteq W \cap E \subseteq C(c, \lambda) B_E. \tag{5.2.17}$$

By duality, this means that $P_E(K)$ contains a ball rB_E of radius $r \geqslant 1/C(c,\lambda)$. Since

$$|P_E(K)| \leqslant N(P_E(K), B_E)|B_E| \leqslant N(K, B_2^n)|B_E| \leqslant \exp(cn)|B_E|, \tag{5.2.18}$$

this implies a bound on $(|P_E(K)|/|rB_E|)^{1/n}$.

Pisier (see [**121**, Chapter 7]) offers a different approach to these results, which provides a construction of special M-ellipsoids with regularity estimates on the covering numbers. The precise statement is as follows: for every $\alpha > 1/2$ and every body K there exists an affine image $\widetilde{K}$ of K which satisfies $|\widetilde{K}| = |B_2^n|$ and

$$\begin{aligned} &\max\{N(K, tB_2^n), N(B_2^n, tK), N(K^\circ, tB_2^n), N(B_2^n, tK^\circ)\} \\ &\leqslant \exp(c(\alpha)nt^{-1/\alpha}) \end{aligned} \tag{5.2.19}$$

for every $t \geqslant 1$, where $c(\alpha)$ is a constant depending only on α, with $c(\alpha) = O((\alpha - \frac{1}{2})^{-1/2})$ as $\alpha \to \frac{1}{2}$. We then say that K is in M-position of order α or α-regular M-position.

6. Additional Information in the Spirit of Geometric Functional Analysis

6.1. Banach–Mazur distance estimates. Recall the definition of the Banach–Mazur distance: if X and Y are two n-dimensional normed spaces, then

$$d(X,Y) = \min\{\|T\|\,\|T^{-1}\| \mid T : X \to Y \text{ is an isomorphism}\}. \tag{6.1.1}$$

Let $\mathcal{B}_n$ be the collection of all equivalence classes of n-dimensional normed spaces, where $X_1 \sim X_2$ if X_1 and X_2 are isometrically isomorphic. The Banach–Mazur compactum (of order n) is the compact metric space $(\mathcal{B}_n, \log d)$.

The quantitative study of the geometry of the Banach–Mazur compactum essentially starts with John's theorem [**73**]. For every $X \in \mathcal{B}_n$ one has $d(X, \ell_2^n) \leqslant \sqrt{n}$, and the multiplicative triangle inequality for d shows that $\mathrm{diam}(\mathcal{B}_n) \leqslant n$. The right order of growth of $\mathrm{diam}(\mathcal{B}_n)$ as $n \to \infty$ was established by Gluskin [**58**] who showed that the Banach–Mazur distance of a typical pair of n-dimensional projections of the unit ball of ℓ_1^{2n} is asymptotically equivalent to n. Gluskin's theorem was the starting point for a deep study of "random spaces" and of random sections and projections of general convex bodies, which is briefly described in the next subsection.

In many interesting cases, the Banach–Mazur distance $d(X,Y)$ is significantly smaller than n. A first example is given by the classical estimates of Gurarii, Kadec, and Macaev: $d(\ell_p^n, \ell_q^n) = n^{1/p-1/q}$ if $1 \leqslant p \leqslant q \leqslant 2$ or $2 \leqslant p \leqslant q \leqslant \infty$, and $c_1 n^\alpha \leqslant d(\ell_p^n, \ell_q^n) \leqslant c_2 n^\alpha$, where $c_1, c_2 > 0$ are absolute constants and $\alpha = \max\{1/p - 1/2, 1/2 - 1/q\}$, if $1 \leqslant p < 2 < q \leqslant \infty$. This suggests that the diameter $\mathrm{diam}(\mathcal{A}_n)$ of some important families $\mathcal{A}_n \subset \mathcal{B}_n$ may be of lower order. This has been proved to be true in two important cases: Let $\mathcal{S}_n$ be the family of all 1-symmetric spaces. Tomczak-Jaegermann [**145**] showed that $\mathrm{diam}(\mathcal{S}_n) \simeq \sqrt{n}$ (Gluskin [**60**] and Tomczak-Jaegermann had previously obtained the upper bound $c\sqrt{n}\log^b n$). The same question remains open for the family $\mathcal{U}_n$ of 1-unconditional spaces. It is conjectured that the right order of $\mathrm{diam}(\mathcal{U}_n)$ is close to $\sqrt{n}$. Lindenstrauss and Szankowski [**88**] have shown that this quantity is bounded by cn^α for some $\alpha \leqslant 2/3$.

In many cases, the diameter of a subclass of $\mathcal{B}_n$ is estimated by probabilistic methods. The general idea is to estimate the distance $d(X,Y)$ by a suitable average of norm-products. The method of random orthogonal factorizations (which has its origin in work of Tomczak-Jaegermann, and was later developed and used by Benyamini and Gordon [**20**]) uses the integral

$$\int_{O(n)} \|T\|_{X\to Y}\|T^{-1}\|_{Y\to X}\, d\nu(T) \tag{6.1.2}$$

with respect to the probability Haar measure ν on $O(n)$ as an upper bound for $d(X,Y)$. An inequality of Marcus and Pisier allows one to pass from $O(n)$ to matrices whose entries are independent standard Gaussian variables and then use Chevet's inequality from the theory of Gaussian processes in order to control this average (see [**147**]). Using this method one can prove a general inequality in terms of the type-2 constants of the spaces [**39**]:

$$d(X,Y) \leqslant c\sqrt{n}[T_2(X) + T_2(Y^*)] \tag{6.1.3}$$

for every $X, Y \in \mathcal{B}_n$. This was further improved by Bourgain and Milman [**27**] to

$$d(X,Y) \leqslant c\Big(d(Y,\ell_2^n)T_2(X) + d(X,\ell_2^n)T_2(Y^*)\Big). \tag{6.1.4}$$

A similar technique is used in [**27**], where it is shown that $d(X, X^*) \leqslant c(\log n)^\gamma n^{5/6}$ for every $X \in \mathcal{B}_n$. All these results indicate that the distance between spaces whose unit balls are "quite different" is not of the order of n.

The Banach–Mazur distance $d(K, L)$ between two not necessarily symmetric convex bodies K and L is the smallest $d > 0$ for which there exist $z_1, z_2 \in \mathbb{R}^n$ and $T \in GL(n)$ such that $K - z_1 \subseteq T(L - z_2) \subseteq d(K - z_1)$. The question of the maximal distance between nonsymmetric bodies is open. John's theorem implies that $d(K, L) \leqslant n^2$. Better estimates were obtained with the method of random orthogonal factorizations and recent progress on the nonsymmetric analogue of Theorem 2.11. In [**15**] it was proved that every convex body K has an affine image K_1 such that $w(K_1)w(K_1^\circ) \leqslant c\sqrt{n}$, a bound which was improved to $cn^{1/3}\log^9 n$ in [**127**]. Using this fact, Rudelson showed that $d(K, L) \leqslant cn^{4/3}\log^9 n$ for any $K, L \in \mathcal{K}_n$.

In another direction, for every $X \in \mathcal{B}_n$ let us consider the "radius" $R_n(X)$ of the Banach–Mazur compactum $\mathcal{B}_n$ with respect to X, defined by

$$R_n(X) = \max\{d(X, Y) : Y \in \mathcal{B}_n\}. \tag{6.1.5}$$

In this terminology, John's theorem states that $R_n(\ell_2^n) = n^{1/2}$. A natural question asked by Pelczynski is to determine the order of the radius $R_n(\ell_p^n)$ for other values of p. In the case of the cube, one has the estimates $n^{1/2} \leqslant R_n(\ell_\infty^n) \leqslant n$ as a consequence of John's theorem. Bourgain and Szarek [**29**] proved that $R_n(\ell_\infty^n) = o(n)$ and gave a proportional version of the Dvoretzky–Rogers lemma on the contact points of a body and its minimal volume ellipsoid: Assume that B_2^n is the ellipsoid of minimal volume containing K. For every $\delta \in (0, 1)$ there exist $m \geqslant (1 - \delta)n$ and contact points $x_1, \ldots, x_m$ of K and B_2^n, such that

$$f(\delta)\Big(\sum_{i=1}^m t_i^2\Big)^{1/2} \leqslant \Big|\sum_{i=1}^m t_i x_i\Big| \leqslant \Big\|\sum_{i=1}^m t_i x_i\Big\|_K \leqslant \sum_{i=1}^m |t_i|. \tag{6.1.6}$$

for every choice of scalars $t_1, \ldots, t_m$. This fact can be stated as a proportional factorization theorem [**29**].

Theorem 6.1. *Let X be an n-dimensional space. For every $\delta \in (0, 1)$ one can find $m \geqslant (1 - \delta)n$ and two operators $\alpha : \ell_2^m \to X$, $\beta : X \to \ell_\infty^m$, such that the identity $\mathrm{id}_{2,\infty} : \ell_2^m \to \ell_\infty^m$ is written as $\mathrm{id}_{2,\infty} = \beta \circ \alpha$ and $\|\alpha\|\,\|\beta\| \leqslant 1/f(\delta)$, where $f(\delta)$ is a function depending only on the proportion $\delta \in (0, 1)$.* □

Using this result Bourgain and Szarek gave a final answer to the problem of the uniqueness up to constant of the center of the Banach–Mazur compactum. This can be made a precise question as follows: Does there exist a function $f(\lambda)$, $\lambda \geqslant 1$, such that for every $X \in \mathcal{B}_n$ with $R_n(X) \leqslant \lambda\sqrt{n}$ we must have $d(X, \ell_2^n) \leqslant f(\lambda)$? In other words, are all the "asymptotic centers" of the Banach–Mazur compactum close to the Euclidean space? The answer is negative and the main tool in the proof is Theorem 6.1: Let $X = \ell_2^s \oplus \ell_1^{n-s}$, where $s = [n/2]$. Then $R_n(X) \leqslant c\sqrt{n}$ for some absolute constant but $d(X, \ell_2^n) \geqslant (n/2)^{1/2}$. Therefore, there exist asymptotic centers of the Banach–Mazur compactum with distance to ℓ_2^n of the order of $R_n(\ell_2^n)$.

The same inequality allowed Bourgain and Szarek to show that

$$R_n(\ell_\infty^n) = o(n).$$

It is now known (see [**141, 49**]) that (3) holds true with $f(\delta) = c\delta$, and this gives a better upper bound for $R_n(\ell_\infty^n)$, which however does not seem to give the right order of the quantity: There exists an absolute constant $c > 0$ such that $R_n(\ell_\infty^n) \leqslant cn^{5/6}$ (see [**48**]). On the other hand, Szarek [**138**] using random spaces (see the next subsection) proved that $R_n(\ell_\infty^n) \geqslant c\sqrt{n}\log n$.

6.2. Random spaces. The theory of random spaces started with Gluskin's theorem [**58**] on the diameter of the Banach–Mazur compactum. He considered a class $X_{n,m}$ of random n-dimensional normed spaces and showed that with high probability the Banach–Mazur distance of two spaces $X_1, X_2 \in X_{n,2n}$ exceeds cn, where $c > 0$ is an absolute constant.

The class $X_{n,m}$ is defined as follows: we consider a sequence $g_1, \dots, g_m$ of independent standard Gaussian random variables on some probability space $(\Omega, \mathcal{A}, P)$, and for each $\omega \in \Omega$ we define the space $X(\omega)$ whose unit ball is the symmetric convex body

$$B_m(\omega) = \text{absconv}\{e_1, \dots, e_n, g_1(\omega), \dots, g_m(\omega)\}. \tag{6.2.1}$$

Alternatively, one can consider the class $Y_{n,m}$ of spaces $Y(\omega)$ with unit ball

$$\widetilde{B}_m(\omega) = \text{absconv}\{g_1(\omega), \dots, g_m(\omega)\}. \tag{6.2.2}$$

If $m \geqslant n$, then $\widetilde{B}_m(\omega)$ has nonempty interior almost surely and defines a norm on $\mathbb{R}^n$. The random space $X(\omega)$ or $Y(\omega)$ can be identified with a quotient of ℓ_1^{n+m} or ℓ_1^m respectively.

Fix $m = 2n$. The basic geometric properties of $B_m(\omega)$ are the following:

1. $B_m(\omega) \supseteq (1/\sqrt{n})B_2^n$.
2. $|B_m(\omega)|^{1/n} \leqslant c_1|(1/\sqrt{n})B_2^n|^{1/n}$, where $c_1 > 0$ is an absolute constant.

Consider the class of pairs $(X(\omega_1), X(\omega_2)) \in X_{n,m} \times X_{n,m}$. If we fix ω_2 and $T \in SL(n)$, using the above properties of $B_m(\omega_2)$ we see that

$$\text{Prob}\left(\omega_1 : \|T : X(\omega_2) \to X(\omega_1)\| \leqslant c_2\rho\sqrt{n}\right) < \rho^{2n^2} \tag{6.2.3}$$

for every $0 < \rho < 1$, where $c_2 > 0$ is an absolute constant. Our aim is to show that the probability $P_1 := \text{Prob}(\omega_1 : X(\omega_1) \in \mathcal{L}(\omega_2))$ is small, where

$$\mathcal{L}(\omega_2) := \{X(\omega_1) : \exists T \in SL(n) : \|T : X(\omega_1) \to X(\omega_2)\| \leqslant \alpha\sqrt{n}\} \tag{6.2.4}$$

for some constant $0 < \alpha < 1$ to be determined. To this end, we define

$$\mathcal{M}(\omega_2) = \{T \in SL(n) : \|Te_j\|_{X(\omega_2)} \leqslant \sqrt{n}, j = 1, \ldots, n\}, \tag{6.2.5}$$

and consider a ε-net $\mathcal{N}(\omega_2)$ of $\mathcal{M}(\omega_2)$ in the norm $\|\cdot : \ell_2^n \to \ell_2^n\|$. If $X(\omega_1) \in \mathcal{L}(\omega_2)$, then there exists $T \in \mathcal{M}(\omega_2)$ such that $\|T : X(\omega_1) \to X(\omega_2)\| \leqslant \alpha\sqrt{n}$. It follows that $\|S : X(\omega_1) \to X(\omega_2)\| \leqslant (\alpha + \varepsilon)\sqrt{n}$ for some $S \in \mathcal{N}(\omega_2)$. If we set $\alpha = \varepsilon = c\rho/2$, combining with (6.2.3) we see that

$$P_1 < |\mathcal{N}(\omega_2)| \cdot \rho^{2n^2}. \tag{6.2.6}$$

The cardinality of the net is smaller than $(c_3/\varepsilon)^{n^2} = (c_4/\rho)^{n^2}$, and this shows that $P_1 < (1/2)^{n^2}$ if ρ is chosen small enough.

It is now clear that with probability greater than $1 - 2(1/2)^{n^2}$ in $X_{n,m} \times X_{n,m}$ we have

$$\|T : X(\omega_1) \to X(\omega_2)\| \cdot \|T^{-1} : X(\omega_2) \to X(\omega_1)\| \geqslant \rho^2 n \tag{6.2.7}$$

for all $T \in SL(n)$, which implies $d(X(\omega_1), X(\omega_2)) \geqslant \rho^2 n$. This proves Gluskin's theorem:

Theorem 6.2. *There exists a constant* $c > 0$ *such that* $\text{diam}(\mathcal{B}_n) \geqslant cn$ *for every* $n \in \mathbb{N}$. □

Let us mention the following recent result of Rudelson [**128**] which complements Gluskin's theorem. If K_1, K_2 are symmetric convex bodies in $\mathbb{R}^n$ and if $k < n$, write $d_k(K_1, K_2)$ for the smallest Banach–Mazur distance between k-dimensional subspaces of K_1 and K_2 respectively.

If $D(n,k)$ is the supremum of $d_k(K_1, K_2)$ over all pairs of symmetric convex bodies in $\mathbb{R}^n$, then $D(n,k) \simeq \sqrt{k}$ if $k \leqslant n^{2/3}$ and $D(n,k) \simeq k^2/n$ if $k \geqslant n^{2/3}$ (in this statement, $\simeq$ means "up to a fixed power of $\log n$").

Theorem 6.2 was the starting point for a systematic study of random spaces. Random quotients of ℓ_1^m provided examples of the worst possible order for several parameters of the local theory. It turns out that a random space $X \in X_{n,m}$ has a rather "poor" family of bounded operators. It was observed by Gluskin [**59**], that a random space X_{n,n^2} has the following property: any projection P in X of rank $k \leqslant n/2$ satisfies

$$\|T : X \to X\| \geqslant ck/\sqrt{n \log n}. \tag{6.2.8}$$

As a consequence such a space has basis constant $bc(X) \geqslant c'\sqrt{n/\log n}$. [Recall that the basis constant $bc(X)$ of an n-dimensional normed space X is the infimum of the basis constants $bc\{x_1, \dots, x_n\}$ over all bases of X.] This follows immediately from the fact that, by the definition of the basis constant, in any n-dimensional normed space X there exists a projection P of rank $k = [n/2]$ with $\|P : X \to X\| \leqslant bc(X)$.

Szarek [**137**] modified the random structure on $X_{n,m}$ and was able to construct an n-dimensional normed space X with $bc(X) \geqslant c\sqrt{n}$. Because of John's theorem this order is optimal. Mankiewicz [**91**] applied the random spaces method to construct finite dimensional spaces with the worst (in order) possible symmetric constant. In this work Mankiewicz used the "space mixing" property of the irreducible group of operators. Szarek [**139**] explicitly introduced the notion of the class $M(k,\alpha)$ of mixing operators which is the set of all linear operators T, satisfying

$$dist(Tx, E) = |P_{E^\perp} Tx| \geqslant \alpha |x| \tag{6.2.9}$$

for some k-dimensional subspace E and every $x \in E$. It is not difficult to show that any projection P of rank $k \leqslant n/2$ is $(k, 1/2)$ mixing. Then, Szarek showed that the mixing property is sufficient for proving the results of [**91**], but also [**59**] and [**137**]. In particular, he proved that for a random space $X \in X_{n,n^2}$ one has

$$\|T : X \to X\| \geqslant \alpha ck/\sqrt{n \log n} \tag{6.2.10}$$

for any $T \in Mix(k,\alpha)$ and that for some modified probability in $X_{n,m}$ the following result holds.

Theorem 6.3. *For every* $0 < \alpha \leqslant 1/2$ *and* $\delta > 0$, *a random space* $X(\omega) \in X_{n,m}$, *where* $m = [\delta n]$, *satisfies* $\|T : X(\omega) \to X(\omega)\| \geqslant c(\alpha,\delta)\sqrt{n}$ *for every* $T \in \mathrm{Mix}(\alpha n, 1)$. □

It should be mentioned that the random space method allows us to construct a sequence of finite dimensional normed spaces, which serve as blocks for the construction of examples of infinite dimensional spaces with some unexpected properties: real isomorphic complex Banach spaces which are not complex isomorphic (Bourgain [**22**]), a Banach space without a basis which has the bounded approximation property (Szarek [**140**]) etc.

The class $Y_{n,m}$, $m \simeq n^{1+\delta}$, provides examples of random spaces with large Banach–Mazur distance to ℓ_1^n. The distribution of $Y(\omega)$ is the same with the distribution of ℓ_1^m/H, where H is a random $(m-n)$-dimensional subspace of ℓ_1^m, and thus $Y_{n,m}$ reflects completely the geometry of quotients of ℓ_1^m. The following theorem of Szarek [**138**] gives the only known example of a pair of spaces with distance significantly larger than $\sqrt{n}$, in which one of the two spaces is concrete.

Theorem 6.4. *For every* $\delta > 0$, *a random space* $Y(\omega) \in Y_{n,m}$, *where* $m = [n^{1+\delta}]$, *satisfies* $d(Y(\omega), \ell_1^n) \geqslant c(\delta)\sqrt{n}\log n$.

The proof involves a precise distributional inequality on the singular numbers s_i of random Gaussian matrices, which is a quantitative finite version of Wigner's semicircle law: if $G(\omega)$ is an $n \times n$ matrix with independent $N(0, 1/n)$ Gaussian entries, then

$$\text{Prob}(\omega : c_1 k/n \leqslant s_{n-k}(G(\omega)) \leqslant c_2 k/n) > 1 - c_3 \exp(-c_4 k^2), \quad (6.2.11)$$

for all $k \leqslant n/2$, where the c_i's are absolute positive constants.

In the last years it was understood that the ideas and arguments used in the study of random quotients of ℓ_1^{n+m} could be transferred to a much more general setting. The idea of studying random projections of arbitrary high-dimensional convex bodies comes from Bourgain, and it was developed in a whole theory by Mankiewicz and Tomczak-Jaegermann (see the survey article [**93**]). The starting observation is that the main geometric properties of a random space in $X_{n,m}$ can be satisfied by projections of an arbitrary convex body if they are put in a suitable position. More precisely, for fixed $0 < \rho < 1$ and for every n-dimensional convex body K, there exist a $[\rho n]$-dimensional projection $T = P_E(K)$ and a Euclidean norm on E satisfying the following properties:

1. $\text{vr}(T) \leqslant C_1(\rho)$.
2. $d(X_K, \ell_2^n)^{-1} B_E \subseteq T \subseteq 2B_E$.
3. There is an orthonormal basis $\{x_j\}$ in X_T with $\max_j \|x_j\|_T \leqslant C_2(\rho)$.

The proof of this fact makes use of the M-ellipsoids. Properties 1 and 2 correspond to the two geometric properties of $X(\omega) \in X_{n,m}$. The third one, which was also clear by construction in our previous discussion, is allowed in the general setting because of the proportional Dvoretzky–Rogers factorization (Theorem 6.1).

An example of this line of thought is the following recent result from [**92**]: If K_1 and K_2 are two symmetric convex bodies in $\mathbb{R}^n$ whose minimal volume ellipsoid is the Euclidean unit ball, then for every proportional dimension $k = \lambda n$ the average distance between k-dimensional projections $P_{H_1}(K_1)$ and $P_{H_2}(K_2)$ of K_1 and K_2 is bounded from below by the product of the average distances

$$\int_{G_{n,s}} d(P_{L_i}(K_i), \ell_2^s)\, d\mu_{n,s}(L_i),$$

where s can be taken equal to $s = (1/2 - \varepsilon)k$ for any small $\varepsilon > 0$.

Random spaces were used very recently by Szarek and Tomczak-Jaegermann [**144**] to provide a strong negative answer to a series of questions raised in the mid-eighties (see [**102**]), which roughly speaking asked if the cotype properties of every n-dimensional normed space improve by passing to quotients of proportional dimension. A typical example is the following: Is it true that there is an absolute constant $C > 0$ such that every n-dimensional space X has a quotient X_1 of dimension $\dim(X_1) \geqslant n/2$ such that the cotype-2 constant of X_1 is bounded by C? Recall that this is true if we replace bounded cotype-2 constant by bounded volume ratio (and, by a result of Bourgain and Milman [**28**], the first property implies the second). A positive answer would be of obvious importance, since all the theory of type and cotype would enter decisively in the study of general convex bodies.

For any given finite dimensional space W, Szarek and Tomczak-Jaegermann construct a space X of an appropriately larger dimension, which is well saturated with W. The precise statement is the following: Let n and m_0 be positive integers with $\sqrt{n \log n} \leqslant m_0 \leqslant n$. If W is a normed space with $\dim(W) \leqslant c \min\{m_0/\sqrt{n}, m_0^2/(n \log n)\}$, there exists an n-dimensional normed space X such that: if $m_0 \leqslant m \leqslant n$, every m-dimensional quotient X_1 of X contains a 1-complemented subspace isometric to W.

Let us give a direct application of this fact: If we choose $W = \ell_\infty^k$ with $k \simeq \sqrt{n}$ and consider an n-dimensional space X as above, taking

m_0 proportional to n we see that the cotype-2 constant of every m_0-dimensional quotient X_1 of X is at least of the order of $\sqrt[4]{n}$ (and the cotype-q constant of every such X_1 is at least of the order of $n^{1/(2q)}$).

References

1. A. D. Alexandrov, *On the theory of mixed volumes of convex bodies II: New inequalities between mixed volumes and their applications* [in Russian], Mat. Sb. N. S. **2** (1937), 1205–1238.
2. A. D. Alexandrov, *On the theory of mixed volumes of convex bodies IV: Mixed discriminants and mixed volumes* [in Russian], Mat. Sb. N. S. **3** (1938), 227–251.
3. S. Alesker, *Remark on Szarek–Talagrand theorem*, Combin. Probab. Comput. **6** (1997), 139–144.
4. S. Alesker, S. Dar and V. D. Milman, *A remarkable measure preserving diffeomorphism between two convex bodies in* $\mathbb{R}^n$, Geom. Dedicata **74** (1999), 201–212.
5. D. Amir and V. D. Milman, *Unconditional and symmetric sets in n-dimensional normed spaces*, Israel J. Math. **37** (1980), 3–20.
6. J. Arias-de-Reyna, K. Ball and R. Villa, *Concentration of the distance in finite dimensional normed spaces*, Mathematika **45** (1998), 245–252.
7. S. Artstein, V. D. Milman and S. J. Szarek, *More on the duality conjecture for entropy numbers*, C. R. Math. Acad. Sci. Paris **337** (2003), no. 11, 711–714.
8. S. Artstein, V. D. Milman and S. J. Szarek, *Duality of the entropy numbers when one of the two spaces is a Hilbert space*, Ann. Math. [to appear].
9. K. M. Ball, *Normed spaces with a weak Gordon–Lewis property*, Lect. Notes Math. **1470**, (1991), 36–47.
10. K. M. Ball, *Volumes of sections of cubes and related problems*, Lect. Notes Math. **1376**, (1989), 251–260.
11. K. M. Ball, *Shadows of convex bodies*, Trans. Am. Math. Soc. **327** (1991), 891–901.
12. K. M. Ball, *Volume ratios and a reverse isoperimetric inequality*, J. Lond. Math. Soc. (2) **44** (1991), 351–359.
13. K. M. Ball, *Convex geometry and Functional analysis*, Handbook of the Geometry of Banach spaces (Lindenstrauss-Johnson, Eds.), Elsevier (2001), 161–194.
14. K. M. Ball, *An elementary introduction to modern convex geometry*, Flavors of Geometry, Math. Sci. Res. Inst. Publ. **31**, Cambridge Univ. Press (1997), 1–58.
15. W. Banaszczyk, A. Litvak, A. Pajor and S. J. Szarek, *The flatness theorem for nonsymmetric convex bodies via the local theory of Banach spaces*, Math. Oper. Res. **24** (1999), 728–750.
16. F. Barthe, *On a reverse form of the Brascamp–Lieb inequality*, Invent. Math. **134** (1998), 335–361.
17. F. Barthe, *An extremal property of the mean width of the simplex*, Math. Ann. **310** (1998), 685–693.
18. E. F. Beckenbach and R. Bellman, *Inequalities*, Springer-Verlag, 1971.
19. W. Beckner, *Inequalities in Fourier analysis*, Annals of Math. **102** (1975), 159–182.

20. Y. Benyamini and Y. Gordon, *Random factorization of operators between Banach spaces*, J. d'Analyse Math. **39** (1981), 45–74.
21. S. G. Bobkov and F. L. Nazarov, *On convex bodies and log-concave probability measures with unconditional basis*, Geom. Aspects of Funct. Analysis (Milman–Schechtman, Eds.), Lect. Notes Math. **1807** (2003), 53–69.
22. J. Bourgain, *Real isomorphic complex spaces need not to be complex isomorphic*, Proc. Am. Math. Soc. **96** (1986), 221–226.
23. J. Bourgain, *On the distribution of polynomials on high dimensional convex sets*, Lect. Notes Math. **1469**, Springer, Berlin (1991), 127–137.
24. J. Bourgain, B. Klartag and V. D. Milman, *A reduction of the slicing problem to finite volume ratio bodies*, C. R. Acad. Sci. Paris, Ser. I **336** (2003), 331–334.
25. J. Bourgain, J. Lindenstrauss and V. D. Milman, *Minkowski sums and symmetrizations*, Lect. Notes Math. **1317** (1988), 44–66.
26. J. Bourgain, J. Lindenstrauss and V. D. Milman, *Estimates related to Steiner symmetrizations*, Lect. Notes Math. **1376** (1989), 264–273.
27. J. Bourgain and V. D. Milman, *Distances between normed spaces, their subspaces and quotient spaces*, Integral Eq. Operator Th. **9** (1986), 31–46.
28. J. Bourgain and V. D. Milman, *New volume ratio properties for convex symmetric bodies in* $\mathbb{R}^n$, Invent. Math. **88** (1987), 319–340.
29. J. Bourgain and S. J. Szarek, *The Banach–Mazur distance to the cube and the Dvoretzky–Rogers factorization*, Israel J. Math. **62** (1988), 169–180.
30. J. Bourgain and L. Tzafriri, *Invertibility of "large" submatrices with applications to the geometry of Banach spaces and harmonic analysis*, Israel J. of Math. **57** (1987), 137–224.
31. H. J. Brascamp and E. H. Lieb, *Best constants in Young's inequality, its converse and its generalization to more than three functions*, Adv. in Math. **20** (1976), 151–173.
32. H. J. Brascamp, E. H. Lieb and J. M. Luttinger, *A general rearrangement inequality for multiple integrals*, J. Funct. Anal. **17** (1974), 227–237.
33. Y. Brenier, *Polar factorization and monotone rearrangement of vector-valued functions*, Comm. Pure Appl. Math. **44** (1991), 375–417.
34. Y. D. Burago and V. A. Zalgaller, *Geometric Inequalities*, Springer Series in Soviet Mathematics, Springer-Verlag, Berlin-New York (1988).
35. L. A. Caffarelli, *A-priori estimates and the geometry of the Monge-Ampère equation*, Park City/IAS Mathematics Series **2** (1992).
36. D. Cordero-Erausquin, *Inégalité de Prékopa-Leindler sur la sphère*, C. R. Acad. Sci. Paris Sér. I Math. **329** (1999), no. 9, 789–792.
37. D. Cordero-Erausquin, R. J. McCann, M. Schmuckenschläger, *A Riemannian interpolation inequality à la Borell, Brascamp and Lieb*, Invent. Math. **146** (2001), no. 2, 219–257.
38. S. Dar, *Remarks on Bourgain's problem on slicing of convex bodies*, in Geometric Aspects of Functional Analysis, Operator Theory: Advances and Applications **77** (1995), 61–66.
39. W. J. Davis, V. D. Milman and N. Tomczak-Jaegermann, *The distance between certain n-dimensional spaces*, Israel J. Math. **39** (1981), 1–15.

40. A. Dvoretzky, *A theorem on convex bodies and applications to Banach spaces*, Proc. Nat. Acad. Sci. U.S.A **45** (1959), 223–226.
41. A. Dvoretzky, *Some results on convex bodies and Banach spaces*, in Proc. Sympos. Linear Spaces, Jerusalem (1961), 123–161.
42. A. Dvoretzky and C. A. Rogers, *Absolute and unconditional convergence in normed linear spaces*, Proc. Nat. Acad. Sci., U.S.A **36** (1950), 192–197.
43. J. Elton, *Sign-embeddings of ℓ_1^n*, Trans. Am. Math. Soc. **279** (1983), 113–124.
44. T. Figiel, J. Lindenstrauss and V. D. Milman, *The dimension of almost spherical sections of convex bodies*, Acta Math. **139** (1977), 53–94.
45. T. Figiel and N. Tomczak-Jaegermann, *Projections onto Hilbertian subspaces of Banach spaces*, Israel J. Math. **33** (1979), 155–171.
46. M. Fradelizi, A. Giannopoulos, and M. Meyer, *Some inequalities about mixed volumes*, Israel J. Math. **135** (2003), 157–180.
47. J. Galambos, *The Asymptotic Theory of Extreme Order Statistics*, Wiley: New York, 1978.
48. A. A. Giannopoulos, *A note on the Banach–Mazur distance to the cube*, in Geometric Aspects of Functional Analysis, Operator Theory: Advances and Applications **77** (1995), 67–73.
49. A. A. Giannopoulos, *A proportional Dvoretzky–Rogers factorization result*, Proc. Am. Math. Soc. **124** (1996), 233–241.
50. A. A. Giannopoulos, M. Hartzoulaki, and G. Paouris, *On a local version of the Aleksandrov–Fenchel inequalities for the quermassintegrals of a convex body*, Proc. Am. Math. Soc. **130** (2002), 2403–2412.
51. A. A. Giannopoulos and V. D. Milman, *Low M^*-estimates on coordinate subspaces*, Journal of Funct. Analysis **147** (1997), 457–484.
52. A. A. Giannopoulos and V. D. Milman, *On the diameter of proportional sections of a symmetric convex body*, International Mathematics Research Notices (1997) **1**, 5–19.
53. A. A. Giannopoulos and V. D. Milman, *Extremal problems and isotropic positions of convex bodies*, Israel J. Math. **117** (2000), 29–60.
54. A. A. Giannopoulos and V. D. Milman, *Euclidean structure in finite dimensional normed spaces*, Handbook of the Geometry of Banach spaces (Lindenstrauss-Johnson, Eds.), Elsevier (2001), 707–779.
55. A. Giannopoulos, V. D. Milman, and M. Rudelson, *Convex bodies with minimal mean width*, Geometric Aspects of Functional Analysis (Milman–Schechtman, Eds.), Lect. Notes Math. **1745** (2000), 81–93.
56. A. A. Giannopoulos and M. Papadimitrakis, *Isotropic surface area measures*, Mathematika **46** (1999), 1–13.
57. A. A. Giannopoulos, I. Perissinaki, and A. Tsolomitis, *John's theorem for an arbitrary pair of convex bodies*, Geom. Dedicata **84** (2001), 63–79.
58. E. D. Gluskin, *The diameter of the Minkowski compactum is approximately equal to n*, Funct. Anal. Appl. **15** (1981), 72–73.
59. E. D. Gluskin, *Finite dimensional analogues of spaces without basis*, Dokl. Akad. Nauk USSR **216** (1981), 1046–1050.
60. E. D. Gluskin, *On distances between some symmetric spaces*, J. Soviet Math. **22** (1983), 1841–1846.

61. E. D. Gluskin, *Probability in the geometry of Banach spaces*, Proc. Int. Congr. Berkeley, Vol. 2 (1986), 924–938.
62. E. D. Gluskin, *On the sum of intervals*, Geometric Aspects of Functional Analysis (Milman–Schechtman, Eds.), Lect. Notes Math. **1807** (2003), 122–130.
63. E. D. Gluskin and V. D. Milman, *Randomizing properties of convex high-dimensional bodies and some geometric inequalities*, C.R.Acad. Sci. Paris, Ser. I **334** (2002), no. 10, 875–879.
64. Y. Gordon, *On Milman's inequality and random subspaces which escape through a mesh in* $\mathbb{R}^n$, Lect. Notes Math. **1317** (1988), 84–106.
65. Y. Gordon, H. König and C. Schütt, *Geometric and probabilistic estimates for entropy and approximation numbers of operators*, J. Approx. Theory **49** (1987), 219–239.
66. Y. Gordon, A. Litvak, M. Meyer, and A. Pajor, *John's Decomposition in the General Case and Applications*, Preprint.
67. M. Gromov, *Convex sets and Kähler manifolds*, in: Advances in Differential Geometry and Topology, World Scientific Publishing, Teaneck NJ (1990), 1-38.
68. M. Gromov, *Metric Structures for Riemannian and Non-Riemannian Spaces*, based on "Structures métriques des variétés Riemanniennes" (L. LaFontaine, P. Pansu, Eds.), English translation by Sean M. Bates, Birkhäuser, Boston–Basel–Berlin, 1999 (with Appendices by M. Katz, P. Pansu and S. Semmes).
69. M. Gromov, *Spaces and question*, Proceedings of Visions in Mathematics Conference, Israel 1999, GAFA, Special Volume issue 1, GAFA 2000.
70. M. Gromov and V. D. Milman, *A topological application of the isoperimetric inequality*, Am. J. Math. **105** (1983), 843–854.
71. M. Gromov and V. D. Milman, *Generalization of the spherical isoperimetric inequality to uniformly convex Banach spaces*, Compositio Math. **62** (1987), 263–282.
72. L. H. Harper, *Optimal numberings and isoperimetric problems on graphs*, J. Combin. Theory **1** (1966), 385–393.
73. F. John, *Extremum problems with inequalities as subsidiary conditions*, Courant Anniversary Volume, Interscience, New York (1948), 187–204.
74. B. S. Kashin, *Sections of some finite-dimensional sets and classes of smooth functions*, Izv. Akad. Nauk. SSSR Ser. Mat. **41** (1977), 334–351.
75. B. Klartag, $5n$ *Minkowski symmetrizations suffice to arrive at an approximate Euclidean ball*, Annals of Math. **156** (2002), 947–960.
76. B. Klartag, *Rate of convergence of Minkowski symmetrizations*, Geom. Funct. Anal. **14** (2004), no. 5 [to appear.]
77. B. Klartag and V. D. Milman, *Isomorphic Steiner symmetrization*, Invent. Math. **153** (2003), no. 3, 463–485.
78. H. Knöthe, *Contributions to the theory of convex bodies*, Michigan Math. J. **4** (1957), 39–52.
79. H. König and V. D. Milman, *On the covering numbers of convex bodies*, Geometric Aspects of Functional Analysis (Lindenstrauss-Milman, Eds.), Lect. Notes Math. **1267**, Springer (1987), 82–95.
80. M. Ledoux, *The concentration of measure phenomenon*, Math. Surveys and Monographs **89**, Am. Math. Soc. (2001).

81. M. Ledoux and M. Talagrand, *Probability in Banach spaces*, Ergeb. Math. Grenzgeb., 3. Folge, Vol. 23 Springer, Berlin (1991).
82. P. Lévy, *Problèmes Concrets d'Analyse Fonctionelle*, Gauthier-Villars, Paris (1951).
83. D. R. Lewis, *Ellipsoids defined by Banach ideal norms*, Mathematika **26** (1979), 18–29.
84. E. H. Lieb, *Gaussian kernels have only Gaussian maximizers*, Inventiones Mathematicae **102** (1990), 179–208.
85. M. A. Lifshits, *Gaussian Random Functions*, Mathematics and its Applications **322**, Kluwer Academic Publishers, Dordrecht, 1995.
86. J. Lindenstrauss, *Almost spherical sections, their existence and their applications*, Jber. Deutsch. Math.-Vereinig., Jubiläumstagung 1990 (Teubner, Stuttgart), 39–61.
87. J. Lindenstrauss and V. D. Milman, *The Local Theory of Normed Spaces and its Applications to Convexity*, Handbook of Convex Geometry (P. M. Gruber and J. M. Wills, Eds.), Elsevier 1993, 1149–1220.
88. J. Lindenstrauss and A. Szankowski, *On the Banach–Mazur distance between spaces having an unconditional basis*, Vol. 122 of Math. Studies, North-Holland, 1986.
89. A. Litvak, V. D. Milman and G. Schechtman, *Averages of norms and quasi-norms*, Math. Ann. **312** (1998), 95–124.
90. P. Mani, *Random Steiner symmetrizations*, Studia Sci. Math. Hungar. **21** (1986), 373–378.
91. P. Mankiewicz, *Finite dimensional spaces with symmetry constant of order* $\sqrt{n}$, Studia Math. **79** (1984), 193–200.
92. P. Mankiewicz and N. Tomczak-Jaegermann, *Families of random projections of symmetric convex bodies*, Geom. Funct. Anal. **11** (2001), 1282–1326.
93. P. Mankiewicz and N. Tomczak-Jaegermann, *Quotients of finite-dimensional Banach spaces; random phenomena*, Handbook of the Geometry of Banach spaces (Lindenstrauss-Johnson, Eds.), Elsevier (2003), 1201–1246.
94. B. Maurey, *Constructions de suites symétriques*, C. R. Acad. Sci. Paris **288** (1979), 679–681.
95. B. Maurey, *Some deviation inequalities*, Geom. Funct. Anal. **1** (1991), no.2, 188–197.
96. R. J. McCann, *Existence and uniqueness of monotone measure preserving maps*, Duke Math. J. **80** (1995), 309–323.
97. V. D. Milman, *New proof of the theorem of Dvoretzky on sections of convex bodies*, Funct. Anal. Appl. **5** (1971), 28–37.
98. V. D. Milman, *Geometrical inequalities and mixed volumes in the Local Theory of Banach spaces*, Astérisque **131** (1985), 373–400.
99. V. D. Milman, *Random subspaces of proportional dimension of finite dimensional normed spaces: approach through the isoperimetric inequality*, Lect. Notes Math. **1166** (1985), 106–115.
100. V. D. Milman, *Almost Euclidean quotient spaces of subspaces of finite dimensional normed spaces*, Proc. Am. Math. Soc. **94** (1985), 445–449.

101. V. D. Milman, *Inegalité de Brunn–Minkowski inverse et applications à la théorie locale des espaces normés*, C. R. Acad. Sci. Paris **302** (1986), 25–28.
102. V. D. Milman, *The concentration phenomenon and linear structure of finite-dimensional normed spaces*, Proceedings of the ICM, Berkeley (1986), 961–975.
103. V. D. Milman, *A few observations on the connection between local theory and some other fields*, Lect. Notes Math. **1317** (1988), 283–289.
104. V. D. Milman, *The heritage of P. Lévy in geometrical functional analysis*, Astérisque **157–158** (1988), 273–302.
105. V. D. Milman, *Isomorphic symmetrization and geometric inequalities*, Lect. Notes Math. **1317** (1988), 107–131.
106. V. D. Milman, *A note on a low M^*-estimate*, in "Geometry of Banach spaces, Proceedings of a conference held in Strobl, Austria, 1989" (P. F. Muller and W. Schachermayer, Eds.), LMS Lecture Note Series, Vol. 158, Cambridge University Press (1990), 219–229.
107. V. D. Milman, *Spectrum of a position of a convex body and linear duality relations*, in Israel Math. Conf. Proceedings 3, Festschrift in Honor of Professor I. Piatetski–Shapiro, Weizmann Science Press of Israel (1990), 151–162.
108. V. D. Milman, *Some applications of duality relations*, Lect. Notes Math. **1469** (1991), 13–40.
109. V. D. Milman, *Dvoretzky's theorem - Thirty years later*, Geom. Funct. Anal. **2** (1992), 455–479.
110. V. D. Milman, *Topics in asymptotic geometric analysis*, Proceedings of "Visions in Mathematics - Towards 2000," GAFA 2000, Special Volume (2000), 792–815.
111. V. D. Milman and A. Pajor, *Isotropic position and inertia ellipsoids and zonoids of the unit ball of a normed n-dimensional space*, Lect. Notes Math. **1376** (1989), 64–104.
112. V. D. Milman and A. Pajor, *Entropy and asymptotic geometry of nonsymmetric convex bodies*, Adv. Math. **152** (2000), 314–335.
113. V. D. Milman and G. Schechtman, *Asymptotic Theory of Finite Dimensional Normed Spaces*, Lect. Notes Math. **1200** (1986).
114. V. D. Milman and G. Schechtman, *An "isomorphic" version of Dvoretzky's theorem*, C. R. Acad. Sci. Paris **321** (1995), 541–544.
115. V. D. Milman and G. Schechtman, *Global versus Local asymptotic theories of finite-dimensional normed spaces*, Duke Math. Journal **90** (1997), 73–93.
116. G. Paouris, *On the isotropic constant of nonsymmetric convex bodies*, Lect. Notes Math. **1745** (2000), 239–243.
117. C. M. Petty, *Surface area of a convex body under affine transformations*, Proc. Am. Math. Soc. **12** (1961), 824–828.
118. A. Pietsch, *Theorie der Operatorenideale* (Zusammenfassung), Friedrich–Schiller-Universität Jena (1972).
119. G. Pisier, *Holomorphic semi-groups and the geometry of Banach spaces*, Ann. of Math. **115** (1982), 375–392.
120. G. Pisier, *Probabilistic methods in the geometry of Banach spaces*, Lect. Notes Math. **1206** (1986), 167–241.
121. G. Pisier, *The Volume of Convex Bodies and Banach Space Geometry*, Cambridge Tracts in Mathematics **94** (1989).

122. A. Pajor and N. Tomczak-Jaegermann, *Remarques sur les nombres d'entropie d'un opérateur et de son transposé*, C. R. Acad. Sci. Paris **301** (1985), 743–746.
123. A. Pajor and N. Tomczak-Jaegermann, *Subspaces of small codimension of finite dimensional Banach spaces*, Proc. Am. Math. Soc. **97** (1986), 637–642.
124. T. R. Rockafellar, *Convex Analysis*, Princeton University Press (1970).
125. C. A. Rogers and G. Shephard, *The difference body of a convex body*, Arch. Math. **8** (1957), 220–233.
126. M. Rudelson, *Contact points of convex bodies*, Israel J. Math. **101** (1997), 93–124.
127. M. Rudelson, *Distances between nonsymmetric convex bodies and the MM^*-estimate*, Positivity **4** (2000), 161–178.
128. M. Rudelson, *Distances between sections of convex bodies*, Geom. Funct. Anal. **14** (2004), no. 5 [to appear].
129. M. Rudelson and R. Vershynin, *Embedding of Lévy families into Banach spaces*, Geom. Finct. Anal. **12** (2002), 183–198.
130. M. Rudelson and R. Vershynin, *Combinatorics of Random Processes and Sections of Convex Bodies*, Preprint.
131. G. Schechtman, *Lévy type inequality for a class of metric spaces*, Martingale theory in harmonic analysis and Banach spaces, Springer-Verlag, Berlin-New York (1981), 211–215.
132. G. Schechtman, *Concentration results and applications*, Handbook of the Geometry of Banach spaces (Lindenstrauss-Johnson, Eds.), Elsevier (2003), 1603–1634.
133. G. Schechtman and M. Schmuckenschläger, *A concentration inequality for harmonic measures on the sphere*, in Geometric Aspects of Functional Analysis, Operator Theory: Advances and Applications **77** (1995), 255–273.
134. R. Schneider, *Convex Bodies: The Brunn–Minkowski Theory*, Encyclopedia of Mathematics and its Applications **44**, Cambridge University Press, Cambridge (1993).
135. V. N. Sudakov, *Gaussian random processes and measures of solid angles in Hilbert spaces*, Soviet Math. Dokl. **12** (1971), 412–415.
136. S. J. Szarek, *On Kashin's almost Euclidean orthogonal decomposition of ℓ_1^n*, Bull. Acad. Polon. Sci. **26** (1978), 691–694.
137. S. J. Szarek, *The finite dimensional basis problem, with an appendix on nets of Grassman manifold*, Acta Math. **159** (1983), 153–179.
138. S. J. Szarek, *Spaces with large distance to ℓ_∞^n and random matrices*, Am. J. Math. **112** (1990), 899–942.
139. S. J. Szarek, *On the Existence and Uniqueness of Complex Structure and Spaces with "Few" Operators*, Trans. Am. Math. Soc. **293** (1986), 339–353.
140. S. J. Szarek, *A Banach space without a basis which has the bounded approximation property*, Acta Math. **159** (1987), 81–98.
141. S. J. Szarek and M. Talagrand, *An isomorphic version of the Sauer–Shelah lemma and the Banach–Mazur distance to the cube*, Lect. Notes Math. **1376** (1989), 105–112.
142. S. J. Szarek and M. Talagrand, *On the convexified Sauer–Shelah theorem*, J. Combinat. Theory, Ser. B **69** (1997), 183–192.

143. S. J. Szarek and N. Tomczak-Jaegermann, *On nearly Euclidean decompositions of some classes of Banach spaces*, Compositio Math. **40** (1980), 367–385.
144. S. J. Szarek and N. Tomczak-Jaegermann, *Saturating constructions for normed spaces*, Geom. Funct. Anal. **14** (2004), no. 6 [to appear].
145. N. Tomczak-Jaegermann, *The Banach–Mazur distance between symmetric spaces*, Israel J. Math. **46** (1983), 40–66.
146. N. Tomczak-Jaegermann, *Dualité des nombres d'entropie pour des opérateurs à valeurs dans un espace de Hilbert*, C. R. Acad. Sci. Paris **305** (1987), 299–301.
147. N. Tomczak-Jaegermann, *Banach–Mazur Distances and Finite Dimensional Operator Ideals*, Pitman Monographs **38** (1989), Pitman, London.

Singularities of Special Lagrangian Submanifolds

Dominic Joyce[†]
Lincoln College
Oxford, UK

We survey what is known about singularities of special Lagrangian submanifolds (SL m-folds) in (almost) Calabi–Yau manifolds. The bulk of the paper summarizes the author's work [**18, 19, 20, 21, 22**] on SL m-folds X with *isolated conical singularities*. That is, near each singular point x, X is modelled on an SL cone C in $\mathbb{C}^m$ with isolated singularity at 0. We also discuss directions for future research, and give a list of open problems.

1. Introduction

Special Lagrangian m-folds (SL *m-folds*) are a distinguished class of real m-dimensional minimal submanifolds which may be defined in $\mathbb{C}^m$, or in *Calabi–Yau m-folds*, or more generally in *almost Calabi–Yau m-folds* (compact Kähler m-folds with trivial canonical bundle). They are of

[†]Supported by an EPSRC Advanced Research Fellowship.

interest to Differential Geometers, to String Theorists (a species of theoretical physicist), and perhaps in the future to Algebraic Geometers.

This article will discuss the *singularities* of SL m-folds, a field which has received little attention until quite recently. We begin in §2 with a brief introduction to special Lagrangian geometry and (almost) Calabi–Yau m-folds. Sections 3–7 survey the author's series of papers [**18, 19, 20, 21, 22**] on SL m-folds with *isolated conical singularities*, a large class of singularities which are simple enough to study in detail. The last and longest section, §8, suggests directions for future research and gives some open problems.

We say that a compact SL m-fold X in an almost Calabi–Yau m-fold M for $m > 2$ has *isolated conical singularities* if it has only finitely many singular points $x_1, \ldots, x_n$ in M, such that for some *special Lagrangian cones* C_i in $T_{x_i}M \cong \mathbb{C}^m$ with $C_i \setminus \{0\}$ nonsingular, X approaches C_i near x_i, in an asymptotic C^1 sense. The exact definition is given in §3.3.

Section 4 discusses the *regularity* of SL m-folds X with conical singularities $x_1, \ldots, x_n$, i.e., how quickly X converges to the cone C_i near x_i, with all derivatives. In §5 we consider the *deformation theory* of compact SL m-folds X with conical singularities. We find that the *moduli space* $\mathcal{M}_X$ of deformations of X in M is locally homeomorphic to the zeroes of a smooth map $\varphi : \mathcal{I}_{X'} \to \mathcal{O}_{X'}$ between finite-dimensional vector spaces, and if the *obstruction space* $\mathcal{O}_{X'}$ is zero then $\mathcal{M}_X$ is a smooth manifold.

Section 6 is an aside on *Asymptotically Conical SL m-folds* (*AC SL m-folds*) in $\mathbb{C}^m$, i.e., nonsingular, noncompact SL m-folds L in $\mathbb{C}^m$ which are asymptotic at infinity to an SL cone C at a prescribed *rate* λ. In §7 we explain how to *desingularize* of a compact SL m-fold X with conical singularities x_i with cones C_i for $i = 1, \ldots, n$ in an almost Calabi–Yau m-fold M. We take AC SL m-folds L_i in $\mathbb{C}^m$ asymptotic to C_i at infinity, and glue tL_i into X at x_i for small $t > 0$ to get a smooth family of compact, *nonsingular* SL m-folds $\tilde{N}^t$ in M, with $\tilde{N}^t \to X$ as $t \to 0$.

For brevity I generally give only statements of results, with at most brief sketches of proofs. For the same reason I have left out several subjects I would like to discuss. Some particular omissions are:

- We give very few *examples* of SL m-folds. But many examples are known in $\mathbb{C}^m$, in [**2, 4, 3, 7, 8, 9, 11, 12, 13, 14, 15, 16, 17**] and other papers.
- We give no *applications* of the results of §3–§7. See [**22**, §8–§10].

- We do not discuss *smooth families* of almost Calabi–Yau m-folds. However, all the main results of §2.4, §5 and §7 have extensions to families, which can be found in [**18, 19, 20, 21, 22**]. The discussion of *index* of singularities in §8.1, and its applications in §8.3 and §8.4, would also be improved by extending it to families.

2. Special Lagrangian Geometry

We begin with some background from symplectic geometry. Then special Lagrangian submanifolds (SL m-folds) are introduced both in $\mathbb{C}^m$ and in *almost Calabi–Yau m-folds*. We also describe the *deformation theory* of compact SL m-folds. Some references for this section are McDuff and Salamon [**26**], Harvey and Lawson [**3**], McLean [**28**], and the author [**13**].

2.1. Background from symplectic geometry. We start by recalling some elementary symplectic geometry, which can be found in McDuff and Salamon [**26**]. Here are the basic definitions.

Definition 2.1. Let M be a smooth manifold of even dimension $2m$. A closed 2-form ω on M is called a *symplectic form* if the $2m$-form ω^m is nonzero at every point of M. Then (M, ω) is called a *symplectic manifold.* A submanifold N in M is called *Lagrangian* if $\dim N = m = \frac{1}{2}\dim M$ and $\omega|_N \equiv 0$.

The simplest example of a symplectic manifold is $\mathbb{R}^{2m}$.

Definition 2.2. Let $\mathbb{R}^{2m}$ have coordinates $(x_1, \ldots, x_m, y_1, \ldots, y_m)$, and define the standard metric g' and symplectic form ω' on $\mathbb{R}^{2m}$ by

$$g' = \textstyle\sum_{j=1}^m (\mathrm{d}x_j^2 + \mathrm{d}y_j^2) \quad \text{and} \quad \omega' = \textstyle\sum_{j=1}^m \mathrm{d}x_j \wedge \mathrm{d}y_j.$$

Then $(\mathbb{R}^{2m}, \omega')$ is a symplectic manifold. When we wish to identify $\mathbb{R}^{2m}$ with $\mathbb{C}^m$, we take the complex coordinates $(z_1, \ldots, z_m)$ on $\mathbb{C}^m$ to be $z_j = x_j + iy_j$. For $R > 0$, define B_R to be the open ball of radius R about 0 in $\mathbb{R}^{2m}$.

Darboux's Theorem [**26**, Theorem 3.15] says that every symplectic manifold is locally isomorphic to $(\mathbb{R}^{2m}, \omega')$. Our version easily follows.

Theorem 2.3. *Let (M, ω) be a symplectic $2m$-manifold and $x \in M$. Then there exists $R > 0$ and an embedding $\Upsilon : B_R \to M$ with $\Upsilon(0) = x$ such that $\Upsilon^*(\omega) = \omega'$, where ω' is the standard symplectic* *form on* $\mathbb{R}^{2m} \supset B_R$. *Given an isomorphism* $\upsilon : \mathbb{R}^{2m} \to T_xM$ *with* $\upsilon^*(\omega|_x)\omega'$, *we can choose* Υ *with* $\mathrm{d}\Upsilon|_0 = \upsilon$.

Let N be a real m-manifold. Then its tangent bundle T^*N has a *canonical symplectic form* $\widehat{\omega}$, defined as follows. Let $(x_1, \ldots, x_m)$ be local coordinates on N. Extend them to local coordinates $(x_1, \ldots, x_m, y_1, \ldots, y_m)$ on T^*N such that $(x_1, \ldots, y_m)$ represents the 1-form $y_1 \mathrm{d}x_1 + \cdots + y_m \mathrm{d}x_m$ in $T^*_{(x_1,\ldots,x_m)}N$. Then $\widehat{\omega} = \mathrm{d}x_1 \wedge \mathrm{d}y_1 + \cdots + \mathrm{d}x_m \wedge \mathrm{d}y_m$.

Identify N with the zero section in T^*N. Then N is a *Lagrangian submanifold* of T^*N. The *Lagrangian Neighborhood Theorem* [**26**, Theorem 3.33] shows that any compact Lagrangian submanifold N in a symplectic manifold looks locally like the zero section in T^*N.

Theorem 2.4. *Let* (M, ω) *be a symplectic manifold and* $N \subset M$ *a compact Lagrangian submanifold. Then there exists an open tubular neighborhood* U *of the zero section* N *in* T^*N, *and an embedding* $\varphi : U \to M$ *with* $\varphi|_N = \mathrm{id} : N \to N$ *and* $\varphi^*(\omega) = \widehat{\omega}$, *where* $\widehat{\omega}$ *is the canonical symplectic structure on* T^*N.

We shall call U, φ a *Lagrangian neighborhood* of N. Such neighborhoods are useful for parametrizing nearby Lagrangian submanifolds of M. Suppose that $\widetilde{N}$ is a Lagrangian submanifold of M which is C^1-close to N. Then $\widetilde{N}$ lies in $\varphi(U)$, and is the image $\varphi(\Gamma(\alpha))$ of the graph $\Gamma(\alpha)$ of a unique C^1-small 1-form α on N.

As $\widetilde{N}$ is Lagrangian and $\varphi^*(\omega) = \widehat{\omega}$ we see that $\widehat{\omega}|_{\Gamma(\alpha)} \equiv 0$. But one can easily show that $\widehat{\omega}|_{\Gamma(\alpha)} = -\pi^*(\mathrm{d}\alpha)$, where $\pi : \Gamma(\alpha) \to N$ is the natural projection. Hence $\mathrm{d}\alpha = 0$, and α is a *closed 1-form*. This establishes a 1-1 correspondence between small closed 1-forms on N and Lagrangian submanifolds $\widetilde{N}$ close to N in M, which is an essential tool in proving later results.

2.2. Special Lagrangian submanifolds in $\mathbb{C}^m$. We define *calibrations* and *calibrated submanifolds*, following [**3**].

Definition 2.5. Let (M, g) be a Riemannian manifold. An *oriented tangent k-plane* V on M is a vector subspace V of some tangent space T_xM to M with $\dim V = k$, equipped with an orientation. If V is an oriented tangent k-plane on M then $g|_V$ is a Euclidean metric on V, so combining $g|_V$ with the orientation on V gives a natural *volume form* vol_V on V, which is a k-form on V.

Now let φ be a closed k-form on M. We say that φ is a *calibration* on M if for every oriented k-plane V on M we have $\varphi|_V \leqslant \mathrm{vol}_V$. Here $\varphi|_V = \alpha \cdot \mathrm{vol}_V$ for some $\alpha \in \mathbb{R}$, and $\varphi|_V \leqslant \mathrm{vol}_V$ if $\alpha \leqslant 1$. Let N be an oriented submanifold of M with dimension k. Then each tangent space T_xN for $x \in N$ is an oriented tangent k-plane. We say that N is a *calibrated submanifold* if $\varphi|_{T_xN} = \mathrm{vol}_{T_xN}$ for all $x \in N$.

It is easy to show that calibrated submanifolds are automatically *minimal submanifolds* [**3**, Theorem II.4.2]. Here is the definition of special Lagrangian submanifolds in $\mathbb{C}^m$, taken from [**3**, §III].

Definition 2.6. Let $\mathbb{C}^m$ have complex coordinates $(z_1, \ldots, z_m)$, and define a metric g', a real 2-form ω' and a complex m-form Ω' on $\mathbb{C}^m$ by

$$\begin{aligned} g' &= |\mathrm{d}z_1|^2 + \cdots + |\mathrm{d}z_m|^2, \\ \omega' &= \tfrac{i}{2}(\mathrm{d}z_1 \wedge \mathrm{d}\bar{z}_1 + \cdots + \mathrm{d}z_m \wedge \mathrm{d}\bar{z}_m), \\ \Omega' &= \mathrm{d}z_1 \wedge \cdots \wedge \mathrm{d}z_m. \end{aligned} \tag{1}$$

Then g', ω' are as in Definition 2.2, and $\mathrm{Re}\,\Omega'$ and $\mathrm{Im}\,\Omega'$ are real m-forms on $\mathbb{C}^m$. Let L be an oriented real submanifold of $\mathbb{C}^m$ of real dimension m. We say that L is a *special Lagrangian submanifold* of $\mathbb{C}^m$, or *SL m-fold* for short, if L is calibrated with respect to $\mathrm{Re}\,\Omega'$, in the sense of Definition 2.5.

Harvey and Lawson [**3**, Corollary III.1.11] give the following alternative characterization of special Lagrangian submanifolds:

Proposition 2.7. *Let L be a real m-dimensional submanifold of $\mathbb{C}^m$. Then L admits an orientation making it into an SL submanifold of $\mathbb{C}^m$ if and only if $\omega'|_L \equiv 0$ and* $\mathrm{Im}\,\Omega'|_L \equiv 0$.

Thus special Lagrangian submanifolds are *Lagrangian* submanifolds satisfying the extra condition that $\mathrm{Im}\,\Omega'|_L \equiv 0$, which is how they get their name.

2.3. Almost Calabi–Yau m-folds and SL m-folds. We shall define special Lagrangian submanifolds not just in Calabi–Yau manifolds, as usual, but in the much larger class of *almost Calabi–Yau manifolds.*

Definition 2.8. Let $m \geqslant 2$. An *almost Calabi–Yau m-fold* is a quadruple (M, J, ω, Ω) such that (M, J) is a compact m-dimensional complex manifold, ω is the Kähler form of a Kähler metric g on M, and Ω is a non-vanishing holomorphic $(m, 0)$-form on M.

We call (M, J, ω, Ω) a *Calabi–Yau m-fold* if in addition ω and Ω satisfy

$$\omega^m/m! = (-1)^{m(m-1)/2}(i/2)^m \Omega \wedge \bar{\Omega}. \tag{2}$$

Then for each $x \in M$ there exists an isomorphism $T_x M \cong \mathbb{C}^m$ that identifies g_x, ω_x and Ω_x with the flat versions g', ω', Ω' on $\mathbb{C}^m$ in (1). Furthermore, g is Ricci-flat and its holonomy group is a subgroup of $\mathrm{SU}(m)$.

This is not the usual definition of a Calabi–Yau manifold, but is essentially equivalent to it.

Definition 2.9. Let (M, J, ω, Ω) be an almost Calabi–Yau m-fold, and N a real m-dimensional submanifold of M. We call N a *special Lagrangian submanifold*, or *SL m-fold* for short, if $\omega|_N \equiv \mathrm{Im}\,\Omega|_N \equiv 0$. It easily follows that $\mathrm{Re}\,\Omega|_N$ is a nonvanishing m-form on N. Thus N is orientable, with a unique orientation in which $\mathrm{Re}\,\Omega|_N$ is positive.

Again, this is not the usual definition of SL m-fold, but is essentially equivalent to it. Suppose (M, J, ω, Ω) is an almost Calabi–Yau m-fold, with metric g. Let $\psi : M \to (0, \infty)$ be the unique smooth function such that

$$\psi^{2m}\omega^m/m! = (-1)^{m(m-1)/2}(i/2)^m \Omega \wedge \bar{\Omega}, \tag{3}$$

and define $\tilde{g}$ to be the conformally equivalent metric $\psi^2 g$ on M. Then $\mathrm{Re}\,\Omega$ is a *calibration* on the Riemannian manifold $(M, \tilde{g})$, and SL m-folds N in (M, J, ω, Ω) are calibrated with respect to it, so that they are minimal with respect to $\tilde{g}$.

If M is a Calabi–Yau m-fold then $\psi \equiv 1$ by (2), so $\tilde{g} = g$, and an m-submanifold N in M is special Lagrangian if and only if it is calibrated w.r.t. $\mathrm{Re}\,\Omega$ on (M, g), as in Definition 2.6. This recovers the usual definition of special Lagrangian m-folds in Calabi–Yau m-folds.

2.4. Deformations of compact SL m-folds. The *deformation theory* of special Lagrangian submanifolds was studied by McLean [**28**, §3], who proved the following result in the Calabi–Yau case. The extension to the almost Calabi–Yau case is described in [**13**, §9.5].

Theorem 2.10. *Let N be a compact SL m-fold in an almost Calabi–Yau m-fold (M, J, ω, Ω). Then the moduli space $\mathcal{M}_N$ of special Lagrangian deformations of N is a smooth manifold of dimension $b^1(N)$.*

SKETCH OF PROOF. There is a natural orthogonal decomposition $TM|_N = TN \oplus \nu$, where $\nu \to N$ is the *normal bundle* of N in M. As N is Lagrangian, the complex structure $J : TM \to TM$ gives an isomorphism $J : \nu \to TN$. But the metric g gives an isomorphism $TN \cong T^*N$. Composing these two gives an isomorphism $\nu \cong T^*N$.

Let T be a small *tubular neighborhood* of N in M. Then we can identify T with a neighborhood of the zero section in ν. Using the isomorphism $\nu \cong T^*N$, we have an identification between T and a neighborhood of the zero section in T^*N. This can be chosen to identify the Kähler form ω on T with the natural symplectic structure on T^*N. Let $\pi : T \to N$ be the obvious projection.

Under this identification, submanifolds N' in $T \subset M$ which are C^1 close to N are identified with the graphs of small smooth sections α of T^*N. That is, submanifolds N' of M close to N are identified with 1-*forms* α on N We need to know: which 1-forms α are identified with *special Lagrangian* submanifolds N'?

Well, N' is special Lagrangian if $\omega|_{N'} \equiv \operatorname{Im}\Omega|_{N'} \equiv 0$. Now $\pi|_{N'} : N' \to N$ is a diffeomorphism, so we can push $\omega|_{N'}$ and $\operatorname{Im}\Omega|_{N'}$ down to N, and regard them as functions of α. Calculation shows that $\pi_*\big(\omega|_{N'}\big) = \mathrm{d}\alpha$ and $\pi_*\big(\operatorname{Im}\Omega|_{N'}\big) = F(\alpha, \nabla\alpha)$, where F is a nonlinear function of its arguments. Thus, the moduli space $\mathcal{M}_N$ is locally isomorphic to the set of small 1-forms α on N.such that $\mathrm{d}\alpha \equiv 0$ and $F(\alpha, \nabla\alpha) \equiv 0$.

Now it turns out that F satisfies $F(\alpha, \nabla\alpha) \approx \mathrm{d}(*\alpha)$ when α is small. Therefore $\mathcal{M}_N$ is locally approximately isomorphic to the vector space of 1-forms α with $\mathrm{d}\alpha = \mathrm{d}(*\alpha) = 0$. But by Hodge theory, this is isomorphic to the de Rham cohomology group $H^1(N, \mathbb{R})$, and is a manifold with dimension $b^1(N)$.

To carry out this last step rigorously requires some technical machinery: one must work with certain *Banach spaces* of sections of T^*N, $\Lambda^2 T^*N$ and $\Lambda^m T^*N$, use *elliptic regularity results* to show the map $\alpha \mapsto \big(\mathrm{d}\alpha, F(\alpha, \nabla\alpha)\big)$ has *closed image* in these Banach spaces, and then use the *Implicit Function Theorem for Banach spaces* to show that the kernel of the map is what we expect. □

3. SL Cones and Conical Singularities

We begin in §3.1 with some definitions on *special Lagrangian cones*. Section 3.2 gives *examples* of SL cones, and §3.3 defines *SL m-folds with conical singularities*, the subject of the paper. Section 3.4 discusses *homology* and *cohomology* of SL m-folds with conical singularities.

3.1. Preliminaries on special Lagrangian cones. We define *special Lagrangian cones*, and some notation.

Definition 3.1. A (singular) SL m-fold C in $\mathbb{C}^m$ is called a *cone* if $C = tC$ for all $t > 0$, where $tC = \{t\mathbf{x} : \mathbf{x} \in C\}$. Let C be a closed SL cone in $\mathbb{C}^m$ with an isolated singularity at 0. Then $\Sigma = C \cap \mathcal{S}^{2m-1}$ is a compact, nonsingular $(m-1)$-submanifold of $\mathcal{S}^{2m-1}$, not necessarily connected. Let g_Σ be the restriction of g' to Σ, where g' is as in (1).

Set $C' = C \setminus \{0\}$. Define $\iota : \Sigma \times (0,\infty) \to \mathbb{C}^m$ by $\iota(\sigma, r) = r\sigma$. Then ι has image C'. By an abuse of notation, *identify* C' with $\Sigma \times (0,\infty)$ using ι. The *cone metric* on $C' \cong \Sigma \times (0,\infty)$ is $g' = \iota^*(g') = \mathrm{d}r^2 + r^2 g_\Sigma$.

For $\alpha \in \mathbb{R}$, we say that a function $u : C' \to \mathbb{R}$ is *homogeneous of order* α if $u \circ t \equiv t^\alpha u$ for all $t > 0$. Equivalently, u is homogeneous of order α if $u(\sigma, r) \equiv r^\alpha v(\sigma)$ for some function $v : \Sigma \to \mathbb{R}$.

In [**18**, Lemma 2.3] we study *homogeneous harmonic functions* on C'.

Lemma 3.2. *In the situation of Definition* 3.1, *let* $u(\sigma, r) \equiv r^\alpha v(\sigma)$ *be a homogeneous function of order* α *on* $C' = \Sigma \times (0,\infty)$, *for* $v \in C^2(\Sigma)$. *Then*

$$\Delta u(\sigma, r) = r^{\alpha-2}\big(\Delta_\Sigma v - \alpha(\alpha + m - 2)v\big),$$

where Δ, Δ_Σ *are the Laplacians on* (C', g') *and* (Σ, g_Σ). *Hence, u is harmonic on C' if and only if v is an eigenfunction of Δ_Σ with eigenvalue* $\alpha(\alpha + m - 2)$.

Following [**18**, Definition 2.5], we define:

Definition 3.3. In Definition 3.1, suppose $m > 2$ and define

$$\mathcal{D}_\Sigma = \big\{\alpha \in \mathbb{R} : \alpha(\alpha + m - 2) \text{ is an eigenvalue of } \Delta_\Sigma\big\}. \tag{4}$$

Then $\mathcal{D}_\Sigma$ is a countable, discrete subset of $\mathbb{R}$. By Lemma 3.2, an equivalent definition is that $\mathcal{D}_\Sigma$ is the set of $\alpha \in \mathbb{R}$ for which there exists a nonzero homogeneous harmonic function u of order α on C'.

Define $m_\Sigma : \mathcal{D}_\Sigma \to \mathbb{N}$ by taking $m_\Sigma(\alpha)$ to be the multiplicity of the eigenvalue $\alpha(\alpha + m - 2)$ of Δ_Σ, or equivalently the dimension of the vector space of homogeneous harmonic functions u of order α on C'.

Define $N_\Sigma : \mathbb{R} \to \mathbb{Z}$ by

$$N_\Sigma(\delta) \begin{cases} -\sum_{\alpha\in\mathcal{D}_\Sigma\cap(\delta,0)} m_\Sigma(\alpha), & \delta < 0, \\ \sum_{\alpha\in\mathcal{D}_\Sigma\cap[0,\delta]} m_\Sigma(\alpha), & \delta \geqslant 0. \end{cases} \tag{5}$$

Then N_Σ is monotone increasing and upper semicontinuous, and is discontinuous exactly on $\mathcal{D}_\Sigma$, increasing by $m_\Sigma(\alpha)$ at each $\alpha \in \mathcal{D}_\Sigma$. As the eigenvalues of Δ_Σ are nonnegative, we see that $\mathcal{D}_\Sigma \cap (2-m, 0) = \varnothing$ and $N_\Sigma \equiv 0$ on $(2-m, 0)$.

We define the *stability index* of C, and *stable* and *rigid* cones [**19**, Definition 3.6].

Definition 3.4. Let C be an SL cone in $\mathbb{C}^m$ for $m > 2$ with an isolated singularity at 0, let G be the Lie subgroup of $\mathrm{SU}(m)$ preserving C, and use the notation of Definitions 3.1 and 3.3. Then [**19**, Equation (8)] shows that

$$m_\Sigma(0) = b^0(\Sigma), \quad m_\Sigma(1) \geqslant 2m \quad \text{and} \quad m_\Sigma(2) \geqslant m^2 - 1 - \dim G. \tag{6}$$

Define the *stability index* s-ind(C) to be

$$\text{s-ind}(C) = N_\Sigma(2) - b^0(\Sigma) - m^2 - 2m + 1 + \dim G.$$

Then s-ind$(C) \geqslant 0$ by (6), as $N_\Sigma(2) \geqslant m_\Sigma(0) + m_\Sigma(1) + m_\Sigma(2)$ by (5). We call C *stable* if s-ind$(C) = 0$.

Following [**18**, Definition 6.7], we call C *rigid* if $m_\Sigma(2) = m^2 - 1 - \dim G$. As

$$\text{s-ind}(C) \geqslant m_\Sigma(2) - (m^2 - 1 - \dim G) \geqslant 0,$$

we see that *if C is stable, then C is rigid.*

We shall see in §5 that s-ind(C) is the dimension of an obstruction space to deforming an SL m-fold X with a conical singularity with cone C, and that if C is *stable* then the deformation theory of X simplifies. An SL cone C is *rigid* if all infinitesimal deformations of C as an SL cone come from $\mathrm{SU}(m)$ rotations of C. This will be useful in the Geometric Measure Theory material of §4.

3.2. Examples of special Lagrangian cones. In our first example we can compute the data of §3.1 very explicitly.

Example 3.5. Here is a family of special Lagrangian cones constructed by Harvey and Lawson [**3**, §III.3.A]. For $m \geqslant 3$, define

$$\begin{aligned} C^m_{\text{HL}} = \big\{(z_1, \dots, z_m) \in \mathbb{C}^m : i^{m+1} z_1 \cdots z_m \in [0, \infty), \\ |z_1| = \cdots = |z_m|\big\}. \end{aligned} \tag{7}$$

Then C^m_{HL} is a *special Lagrangian cone* in $\mathbb{C}^m$ with an isolated singularity at 0, and $\Sigma^m_{\text{HL}} = C^m_{\text{HL}} \cap \mathcal{S}^{2m-1}$ is an $(m-1)$-torus T^{m-1}. Both C^m_{HL} and Σ^m_{HL} are invariant under the $\mathrm{U}(1)^{m-1}$ subgroup of $\mathrm{SU}(m)$ acting by

$$\begin{aligned} (z_1, \dots, z_m) \longmapsto (e^{i\theta_1} z_1, \dots, e^{i\theta_m} z_m) \\ \text{for } \theta_j \in \mathbb{R} \text{ with } \theta_1 + \cdots + \theta_m = 0. \end{aligned} \tag{8}$$

In fact $\pm C^m_{\text{HL}}$ for m odd, and $C^m_{\text{HL}}, iC^m_{\text{HL}}$ for m even, are the unique SL cones in $\mathbb{C}^m$ invariant under (8), which is how Harvey and Lawson constructed them.

The metric on $\Sigma^m_{\text{HL}} \cong T^{m-1}$ is flat, so it is easy to compute the eigenvalues of $\Delta_{\Sigma^m_{\text{HL}}}$. This was done by Marshall [**25**, §6.3.4]. There is a 1-1 correspondence between $(n_1, \dots, n_{m-1}) \in \mathbb{Z}^{m-1}$ and eigenvectors of $\Delta_{\Sigma^m_{\text{HL}}}$ with eigenvalue

$$m \sum_{i=1}^{m-1} n_i^2 - \sum_{i,j=1}^{m-1} n_i n_j. \tag{9}$$

Using (9) and a computer we can find the eigenvalues of $\Delta_{\Sigma^m_{\text{HL}}}$ and their multiplicities. The Lie subgroup G^m_{HL} of $\mathrm{SU}(m)$ preserving C^m_{HL} has identity component the $\mathrm{U}(1)^{m-1}$ of (8), so that $\dim G^m_{\text{HL}} = m-1$. Thus we can calculate $\mathcal{D}_{\Sigma^m_{\text{HL}}}$, $m_{\Sigma^m_{\text{HL}}}$, $N_{\Sigma^m_{\text{HL}}}$, and s-ind(C^m_{HL}). This was done by Marshall [**25**, Table 6.1] and the author [**19**, §3.2]. Table 1 gives the data $m, N_{\Sigma^m_{\text{HL}}}(2), m_{\Sigma^m_{\text{HL}}}(2)$ and s-ind(C^m_{HL}) for $3 \leqslant m \leqslant 12$.

m	3	4	5	6	7	8	9	10	11	12
$N_{\Sigma^m_{\text{HL}}}(2)$	13	27	51	93	169	311	331	201	243	289
$m_{\Sigma^m_{\text{HL}}}(2)$	6	12	20	30	42	126	240	90	110	132
s-ind(C^m_{HL})	0	6	20	50	112	238	240	90	110	132

TABLE 1. Data for $\mathrm{U}(1)^{m-1}$-invariant SL cones C^m_{HL} in $\mathbb{C}^m$

One can also prove that

$$\begin{aligned} N_{\Sigma^m_{\rm HL}}(2) &= 2m^2+1, \\ m_{\Sigma^m_{\rm HL}}(2) &= \text{s-ind}(C^m_{\rm HL}) = m^2-m \quad \text{for } m \geqslant 10. \end{aligned} \tag{10}$$

As $C^m_{\rm HL}$ is *stable* when s-ind$(C^m_{\rm HL})=0$ we see from Table 1 and (10) that $C^3_{\rm HL}$ is a *stable* cone in $\mathbb{C}^3$, but $C^m_{\rm HL}$ is *unstable* for $m \geqslant 4$. Also $C^m_{\rm HL}$ is *rigid* when $m_{\Sigma^m_{\rm HL}}(2) = m^2-m$, as $\dim G^m_{\rm HL} = m-1$. Thus $C^m_{\rm HL}$ is *rigid* if and only if $m \neq 8,9$, by Table 1 and (10).

Here is an example chosen from [**7**, Example 9.4] as it is easy to write down.

Example 3.6. Let $a_1,\ldots,a_m \in \mathbb{Z}$ with $a_1+\cdots+a_m = 0$ and highest common factor 1, such that $a_1,\ldots,a_k > 0$ and $a_{k+1},\ldots,a_m < 0$ for $0<k<m$. Define

$$L_0^{a_1,\ldots,a_m} = \big\{ \big(i e^{ia_1\theta}x_1, e^{ia_2\theta}x_2,\ldots,e^{ia_m\theta}x_m\big) : \theta \in [0,2\pi), \\ x_1,\ldots,x_m \in \mathbb{R}, \qquad a_1x_1^2+\cdots+a_mx_m^2 = 0\big\}.$$

Then $L_0^{a_1,\ldots,a_m}$ is an *immersed SL cone* in $\mathbb{C}^m$, with an isolated singularity at 0.

Define $C^{a_1,\ldots,a_m} = \big\{(x_1,\ldots,x_m)\in\mathbb{R}^m : a_1x_1^2+\cdot\cdot+a_mx_m^2=0\big\}$. Then $C^{a_1,\ldots,a_m}$ is a quadric cone on $\mathcal{S}^{k-1}\times\mathcal{S}^{m-k-1}$ in $\mathbb{R}^m$, and $L_0^{a_1,\ldots,a_m}$ is the image of an immersion $\varphi : C^{a_1,\ldots,a_m}\times\mathcal{S}^1 \to \mathbb{C}^m$, which is generically 2:1. Therefore $L_0^{a_1,\ldots,a_m}$ is an immersed SL cone on $(\mathcal{S}^{k-1}\times\mathcal{S}^{m-k-1}\times\mathcal{S}^1)/\mathbb{Z}_2$.

Further examples of SL cones are constructed by Harvey and Lawson [**3**, §III.3], Haskins [**4**], the author [**7**, **8**], and others. Special Lagrangian cones in $\mathbb{C}^3$ are a special case, which may be treated using the theory of *integrable systems.* In principle this should yield a *classification* of all SL cones on T^2 in $\mathbb{C}^3$. For more information see McIntosh [**27**] or the author [**12**].

3.3. Special Lagrangian m-folds with conical singularities.

Now we can define *conical singularities* of SL m-folds, following [**18**, Definition 3.6].

Definition 3.7. Let (M,J,ω,Ω) be an almost Calabi–Yau m-fold for $m>2$, and define $\psi : M \to (0,\infty)$ as in (3). Suppose X is a compact singular SL m-fold in M with singularities at distinct points $x_1,\ldots,x_n \in X$, and no other singularities.

Fix isomorphisms $v_i : \mathbb{C}^m \to T_{x_i}M$ for $i = 1, \ldots, n$ such that $v_i^*(\omega) = \omega'$ and $v_i^*(\Omega) = \psi(x_i)^m \Omega'$, where ω', Ω' are as in (1). Let $C_1, \ldots, C_n$ be SL cones in $\mathbb{C}^m$ with isolated singularities at 0. For $i = 1, \ldots, n$ let $\Sigma_i = C_i \cap \mathcal{S}^{2m-1}$, and let $\mu_i \in (2, 3)$ with

$$(2, \mu_i] \cap \mathcal{D}_{\Sigma_i} = \varnothing, \quad \text{where } \mathcal{D}_{\Sigma_i} \text{ is defined in (4).} \tag{11}$$

Then we say that X has a *conical singularity* or *conical singular point* at x_i, with *rate* μ_i and *cone* C_i for $i = 1, \ldots, n$, if the following holds.

By Theorem 2.3 there exist embeddings $\Upsilon_i : B_R \to M$ for $i = 1, \ldots, n$ satisfying $\Upsilon_i(0) = x_i$, $\mathrm{d}\Upsilon_i|_0 = v_i$ and $\Upsilon_i^*(\omega) = \omega'$, where B_R is the open ball of radius R about 0 in $\mathbb{C}^m$ for some small $R > 0$. Define $\iota_i : \Sigma_i \times (0, R) \to B_R$ by $\iota_i(\sigma, r) = r\sigma$ for $i = 1, \ldots, n$.

Define $X' = X \setminus \{x_1, \ldots, x_n\}$. Then there should exist a compact subset $K \subset X'$ such that $X' \setminus K$ is a union of open sets $S_1, \ldots, S_n$ with $S_i \subset \Upsilon_i(B_R)$, whose closures $\overline{S}_1, \ldots, \overline{S}_n$ are disjoint in X. For $i = 1, \ldots, n$ and some $R' \in (0, R]$ there should exist a smooth $\varphi_i : \Sigma_i \times (0, R') \to B_R$ such that $\Upsilon_i \circ \varphi_i : \Sigma_i \times (0, R') \to M$ is a diffeomorphism $\Sigma_i \times (0, R') \to S_i$, and

$$\left|\nabla^k(\varphi_i - \iota_i)\right| = O(r^{\mu_i - 1 - k}) \quad \text{as } r \to 0 \text{ for } k = 0, 1. \tag{12}$$

Here ∇ is the Levi–Civita connection of the cone metric $\iota_i^*(g')$ on $\Sigma_i \times (0, R')$, $|\,.\,|$ is computed using $\iota_i^*(g')$. If the cones $C_1, \ldots, C_n$ are *stable* in the sense of Definition 3.4, then we say that X has *stable conical singularities.*

We will see in Theorems 4.1 and 4.2 that if (12) holds for $k = 0, 1$ and some μ_i satisfying (11), then we can choose a natural φ_i for which (12) holds for *all* $k \geqslant 0$, and for *all* rates μ_i satisfying (11). Thus the number of derivatives required in (12) and the choice of μ_i both make little difference. We choose $k = 0, 1$ in (12), and some μ_i in (11), to make the definition as weak as possible.

We suppose $m > 2$ for two reasons. Firstly, the only SL cones in $\mathbb{C}^2$ are finite unions of SL planes $\mathbb{R}^2$ in $\mathbb{C}^2$ intersecting only at 0. Thus any SL 2-fold with conical singularities is actually *nonsingular* as an immersed 2-fold, so there is really no point in studying them. Secondly, $m = 2$ is a special case in the analysis of [**18**, §2], and it is simpler to exclude it. Therefore we will suppose $m > 2$ throughout the paper.

Here are the reasons for the conditions on μ_i in Definition 3.7:

- We need $\mu_i > 2$, or else (12) does not force X to approach C_i near x_i.

- The definition involves a choice of $\Upsilon_i : B_R \rightarrow M$. If we replace Υ_i by a different choice $\widetilde{\Upsilon}_i$ then we should replace φ_i by $\widetilde{\varphi}_i = (\widetilde{\Upsilon}_i^{-1} \circ \Upsilon_i) \circ \varphi_i$ near 0 in B_R. Calculation shows that as $\Upsilon_i, \widetilde{\Upsilon}_i$ agree up to second order, we have $\left|\nabla^k(\widetilde{\varphi}_i - \varphi_i)\right| = O(r^{2-k})$.

  Therefore we choose $\mu_i < 3$ so that these $O(r^{2-k})$ terms are absorbed into the $O(r^{\mu_i - 1-k})$ in (12). This makes the definition independent of the choice of Υ_i, which it would not be if $\mu_i > 3$.
- Condition (11) is needed to prove the regularity result Theorem 4.2, and also to reduce to a minimum the obstructions to deforming compact SL m-folds with conical singularities studied in §5.

3.4. Homology and cohomology. Next we discuss *homology* and *cohomology* of SL m-folds with conical singularities, following [**18**, §2.4]. For a general reference, see for instance Bredon [**1**]. When Y is a manifold, write $H^k(Y, \mathbb{R})$ for the kth *de Rham cohomology group* and $H^k_{\text{cs}}(Y, \mathbb{R})$ for the kth *compactly-supported de Rham cohomology group* of Y. If Y is compact then $H^k(Y, \mathbb{R}) = H^k_{\text{cs}}(Y, \mathbb{R})$. The *Betti numbers* of Y are $b^k(Y) = \dim H^k(Y, \mathbb{R})$ and $b^k_{\text{cs}}(Y) = \dim H^k_{\text{cs}}(Y, \mathbb{R})$.

Let Y be a topological space, and $Z \subset Y$ a subspace. Write $H_k(Y, \mathbb{R})$ for the kth *real singular homology group* of Y, and $H_k(Y; Z, \mathbb{R})$ for the kth *real singular relative homology group* of $(Y; Z)$. When Y is a manifold and Z a submanifold we define $H_k(Y, \mathbb{R})$ and $H_k(Y; Z, \mathbb{R})$ using *smooth* simplices, as in [**1**, §V.5]. Then the pairing between (singular) homology and (de Rham) cohomology is defined at the chain level by integrating k-forms over k-simplices.

Let X be a compact SL m-fold in M with conical singularities $x_1, \ldots, x_n$ and cones $C_1, \ldots, C_n$, and set $X' = X \setminus \{x_1, \ldots, x_n\}$ and $\Sigma_i = C_i \cap \mathcal{S}^{2m-1}$, as in §3.3. Then X' is the interior of a compact manifold $\overline{X}'$ with boundary $\coprod_{i=1}^n \Sigma_i$. Using this we show in [**18**, §2.4] that there is a natural long exact sequence

$$\cdots \rightarrow H^k_{\text{cs}}(X', \mathbb{R}) \rightarrow H^k(X', \mathbb{R}) \rightarrow \bigoplus_{i=1}^n H^k(\Sigma_i, \mathbb{R}) \rightarrow H^{k+1}_{\text{cs}}(X', \mathbb{R}) \rightarrow \cdots, \tag{13}$$

and natural isomorphisms

$$H_k(X; \{x_1, \ldots, x_n\}, \mathbb{R})^* \cong H^k_{\text{cs}}(X', \mathbb{R}) \cong H_{m-k}(X', \mathbb{R}) \cong H^{m-k}(X', \mathbb{R})^*$$
$$\text{and} \quad H^k_{\text{cs}}(X', \mathbb{R}) \cong H_k(X, \mathbb{R})^* \quad \text{for all } k > 1.$$

The inclusion $\iota : X \rightarrow M$ induces homomorphisms $\iota_* : H_k(X, \mathbb{R}) \rightarrow H_k(M, \mathbb{R})$.

4. The Asymptotic Behavior of X near x_i

We now review the work of [**18**] on the *regularity* of SL m-folds with conical singularities. Let M be an almost Calabi–Yau m-fold and X an SL m-fold in M with conical singularities at $x_1, \ldots, x_n$, with identifications υ_i and cones C_i. We study how quickly X converges to the cone $\upsilon(C_i)$ in $T_{x_i}M$ near x_i.

Roughly speaking, we work by arranging for φ_i in Definition 3.7 to satisfy an *elliptic equation*, and then use *elliptic regularity results* to deduce asymptotic bounds for $\varphi_i - \iota_i$ and all its derivatives. Now φ_i is not uniquely defined, but is a more-or-less arbitrary parametrization of $\Upsilon_i^*(X')$ near 0 in $\mathbb{C}^m$. To make φ_i satisfy an elliptic equation we impose an *extra condition*, that $(\varphi_i - \iota_i)(\sigma, r)$ is orthogonal to $T_{\iota_i(\sigma,r)}C_i$ w.r.t. the metric g' on $\mathbb{C}^m$, for all $(\sigma, r) \in \Sigma_i \times (0, R')$. By [**18**, Theorem 4.4] this also fixes φ_i uniquely, given $\upsilon_i, R, \Upsilon_i$ and R'.

Theorem 4.1. *Let (M, J, ω, Ω) be an almost Calabi–Yau m-fold, and X a compact SL m-fold in M with conical singularities at $x_1, \ldots, x_n$ with identifications $\upsilon_i : \mathbb{C}^m \to T_{x_i}M$ and cones $C_1, \ldots, C_n$. Choose $R > 0$ and $\Upsilon_i : B_R \to M$ as in Definition 3.7. Then for sufficiently small $R' \in (0, R]$ there exist unique φ_i, S_i for $i = 1, \ldots, n$ satisfying the conditions of Definition 3.7 and*

$$(\varphi_i - \iota_i)(\sigma, r) \perp T_{\iota_i(\sigma,r)}C_i \quad \text{in } \mathbb{C}^m \text{ for all } (\sigma, r) \in \Sigma_i \times (0, R'). \tag{14}$$

In fact [**18**, Theorem 4.4] characterizes φ_i in terms of a *Lagrangian neighborhood* U_{C_i}, φ_{C_i} of C_i in $\mathbb{C}^m$, but examining the proof of [**18**, Theorem 4.2] shows this is equivalent to (14). In [**18**, §5] we study the asymptotic behavior of the maps φ_i of Theorem 4.1. Combining [**18**, Theorems 5.1 and 5.5, Lemma 4.5] proves:

Theorem 4.2. *In the situation of Theorem 4.1, suppose $\mu_i' \in (2, 3)$ with $(2, \mu_i'] \cap \mathcal{D}_{\Sigma_i} = \emptyset$ for $i = 1, \ldots, n$. Then*

$$\left|\nabla^k(\varphi_i - \iota_i)\right| = O(r^{\mu_i'-1-k}) \qquad \text{for all } k \geqslant 0 \text{ and } i = 1, \ldots, n. \tag{15}$$

Hence X has conical singularities at x_i with cone C_i and rate μ_i', for all possible rates μ_i' allowed by Definition 3.7. Therefore, the definition of conical singularities is essentially independent of the choice of rate μ_i.

Theorem 4.2 in effect *strengthens* the definition of SL m-folds with conical singularities, Definition 3.7, as it shows that (12) actually implies the much stronger condition (15) on all derivatives.

To prove Theorem 4.2, we show using an analogue of Theorem 2.4 for C_i in $\mathbb{C}^m$ that as $\Upsilon_i^*(X')$ is *Lagrangian* in B_R, we may regard φ_i as the graph of a *closed* 1-*form* η_i on $\Sigma_i \times (0, R')$. The asymptotic condition (12) implies that η_i is *exact*, so we may write $\eta_i = \mathrm{d}A_i$ for smooth $A_i : \Sigma_i \times (0, R') \to \mathbb{R}$. As $\operatorname{Im}\Omega|_{X'} \equiv 0$, we find that A_i satisfies the second-order nonlinear p.d.e.

$$\mathrm{d}^*\big(\psi^m \mathrm{d}A_i\big)(\sigma, r)Q\big(\sigma, r, \mathrm{d}A_i(\sigma, r), \nabla^2 A_i(\sigma, r)\big) \qquad (16)$$

for $(\sigma, r) \in \Sigma_i \times (0, R')$, where Q is a smooth nonlinear function.

When r is small the Q term in (16) is also small and (16) approximates $\Delta_i A_i = 0$, where Δ_i is the Laplacian on the cone C_i. Therefore (16) is *elliptic* for small r. Using known results on the regularity of solutions of nonlinear second-order elliptic p.d.e.s, and a theory of analysis on weighted Sobolev spaces on manifolds with cylindrical ends developed by Lockhart and McOwen [**24**], we can then prove (15).

Our next result [**18**, Theorem 6.8] is an application of *Geometric Measure Theory*. For an introduction to the subject, see Morgan [**29**]. Geometric Measure Theory studies measure-theoretic generalizations of submanifolds called *integral currents*, which may be very singular, and is particularly powerful for *minimal* submanifolds. As from §2 SL m-folds are minimal submanifolds w.r.t. an appropriate metric, many major results of Geometric Measure Theory immediately apply to *special Lagrangian integral currents*, a very general class of singular SL m-folds with strong compactness properties.

Theorem 4.3. *Let (M, J, ω, Ω) be an almost Calabi–Yau m-fold and define $\psi : M \to (0, \infty)$ as in* (3). *Let $x \in M$ and fix an isomorphism $\upsilon : \mathbb{C}^m \to T_x M$ with $\upsilon^*(\omega) = \omega'$ and $\upsilon^*(\Omega) = \psi(x)^m \Omega'$, where ω', Ω' are as in* (1).

Suppose that T is a special Lagrangian integral current in M with $x \in T^\circ$, where $T^\circ = \operatorname{supp} T \setminus \operatorname{supp} \partial T$, and that $\upsilon_*(C)$ is a multiplicity 1 tangent cone to T at x, where C is a rigid special Lagrangian cone in $\mathbb{C}^m$, in the sense of Definition 3.4. *Then T has a conical singularity at x, in the sense of Definition* 3.7.

This is a *weakening* of Definition 3.7 for *rigid* cones C. Theorem 4.3 also holds for the larger class of *Jacobi integrable* SL cones C, defined in [**18**, Definition 6.7]. Basically, Theorem 4.3 shows that if a singular SL m-fold T in M is locally modelled on a rigid SL cone C in only a very weak sense, then it necessarily satisfies Definition 3.7. One moral of Theorems 4.2 and 4.3 is that, at least for rigid SL cones C, more-or-less *any* sensible definition of SL m-folds with conical singularities is equivalent to Definition 3.7.

Theorem 4.3 is proved by applying regularity results of Allard and Almgren, and Adams and Simon, mildly adapted to the special Lagrangian situation, which roughly say that if a tangent cone C_i to X at x_i has an isolated singularity at 0, is multiplicity 1, and rigid, then X has a parametrization φ_i near x_i which satisfies (12) for some $\mu_i > 2$. It then quickly follows that X has a conical singularity at x_i, in the sense of Definition 3.7.

As discussed in [**18**, §6.3], one can use other results from Geometric Measure Theory to argue that for tangent cones C which are not Jacobi integrable, Definition 3.7 may be *too strong*, in that there could exist examples of singular SL m-folds with tangent cone C which are not covered by Definition 3.7, as the decay conditions (12) are too strict.

5. Moduli of SL m-Folds with Conical Singularities

Next we review the work of [**19**] on *deformation theory* for compact SL m-folds with conical singularities. Following [**19**, Definition 5.4], we define the space $\mathcal{M}_X$ of compact SL m-folds $\hat{X}$ in M with conical singularities deforming a fixed SL m-fold X with conical singularities.

Definition 5.1. Let (M, J, ω, Ω) be an almost Calabi–Yau m-fold and X a compact SL m-fold in M with conical singularities at $x_1, \ldots, x_n$ with identifications $\upsilon_i : \mathbb{C}^m \to T_{x_i}M$ and cones $C_1, \ldots, C_n$. Define the *moduli space $\mathcal{M}_X$ of deformations of X* to be the set of $\hat{X}$ such that

(i) $\hat{X}$ is a compact SL m-fold in M with conical singularities at $\hat{x}_i$, with cones C_i and identifications $\hat{\upsilon}_i : \mathbb{C}^m \to T_{\hat{x}_i}M$, for $i = 1, \ldots, n$.

(ii) There exists a homeomorphism $\hat{\iota} : X \to \hat{X}$ with $\hat{\iota}(x_i) = \hat{x}_i$ for $i = 1, \ldots, n$ such that $\hat{\iota}|_{X'} : X' \to \hat{X}'$ is a diffeomorphism and $\hat{\iota}$ and ι are isotopic as continuous maps $X \to M$, where $\iota : X \to M$ is the inclusion.

In [**19**, Definition 5.6] we define a *topology* on $\mathcal{M}_X$, and explain why it is the natural choice. We will not repeat the complicated definition here; readers are referred to [**19**, §5] for details. In [**19**, Theorem 6.10] we describe $\mathcal{M}_X$ near X, in terms of a smooth map φ between the *infinitesimal deformation space* $\mathcal{I}_{X'}$ and the *obstruction space* $\mathcal{O}_{X'}$.

Theorem 5.2. *Suppose* (M, J, ω, Ω) *is an almost Calabi–Yau m-fold and* X *a compact SL m-fold in* M *with conical singularities at* $x_1, \ldots, x_n$ *and cones* $C_1, \ldots, C_n$. *Let* $\mathcal{M}_X$ *be the moduli space of deformations of* X *as an SL m-fold with conical singularities in* M, *as in Definition* 5.1. *Set* $X' = X \setminus \{x_1, \ldots, x_n\}$.

Then there exist natural finite-dimensional vector spaces $\mathcal{I}_{X'}$, $\mathcal{O}_{X'}$ *such that* $\mathcal{I}_{X'}$ *is isomorphic to the image of* $H^1_{\mathrm{cs}}(X', \mathbb{R})$ *in* $H^1(X', \mathbb{R})$ *and* $\dim \mathcal{O}_{X'} = \sum_{i=1}^n \text{s-ind}(C_i)$, *where* s-ind$(C_i)$ *is the stability index of Definition* 3.4. *There exists an open neighborhood* U *of* 0 *in* $\mathcal{I}_{X'}$, *a smooth map* $\varphi : U \to \mathcal{O}_{X'}$ *with* $\varphi(0) = 0$, *and a map* $\Xi : \{u \in U : \varphi(u) = 0\} \to \mathcal{M}_X$ *with* $\Xi(0) = X$ *which is a homeomorphism with an open neighborhood of* X *in* $\mathcal{M}_X$.

Here is a sketch of the proof. For simplicity, consider first the subset of $\widehat{X} \in \mathcal{M}_X$ which have the same singular points $x_1, \ldots, x_n$ and identifications $v_1, \ldots, v_n$ as X. Generalizing Theorem 2.10, in [**18**, Theorem 4.3] we define a *Lagrangian neighborhood* $U_{X'}, \varphi_{X'}$ for X', with certain compatibilities with Υ_i, φ_i near x_i. If $\widehat{X}$ is C^1 close to X in an appropriate sense then $\widehat{X}' = \varphi_{X'}(\Gamma(\alpha))$, where $\Gamma(\alpha) \subset U_{X'}$ is the graph of a small 1-form α on X'.

Since $\widehat{X}'$ is Lagrangian, α is *closed*, as in §2.1. Also, applying Theorem 4.2 to $X, \widehat{X}$ and noting that α on S_i corresponds to $\widehat{\varphi}_i - \varphi_i$ on $\Sigma_i \times (0, R')$, we find that if $i = 1, \ldots, n$ and $\mu'_i \in (2,3)$ with $(2, \mu'_i] \cap \mathcal{D}_{\Sigma_i} = \varnothing$ then

$$\left|\nabla^k \alpha(x)\right| = O\big(d(x, x_i)^{\mu'_i - 1 - k}\big) \quad \text{near } x_i \text{ for all } k \geqslant 0. \tag{17}$$

As α is closed it has a cohomology class $[\alpha] \in H^1(X', \mathbb{R})$. Since (17) implies that α decays quickly near x_i, it turns out that α must be *exact* near x_i. Therefore $[\alpha]$ can be represented by a compactly-supported form on X', and lies in the image of $H^1_{\mathrm{cs}}(X', \mathbb{R})$ in $H^1(X', \mathbb{R})$.

Choose a vector space $\mathcal{I}_{X'}$ of compactly-supported 1-forms on X' representing the image of $H^1_{\mathrm{cs}}(X', \mathbb{R})$ in $H^1(X', \mathbb{R})$. Then we can write $\alpha = \beta + \mathrm{d}f$, where $\beta \in \mathcal{I}_{X'}$ with $[\alpha] = [\beta]$ is unique, and $f \in C^\infty(X')$ is unique up to addition of constants. As $\widehat{X}'$ is special Lagrangian we find that f satisfies a *second-order nonlinear elliptic p.d.e.* similar to (16):

$$\mathrm{d}^*\bigl(\psi^m(\beta+\mathrm{d}f)\bigr)(x)Q\bigl(x,(\beta+\mathrm{d}f)(x),(\nabla\beta+\nabla^2 f)(x)\bigr) \qquad (18)$$

for $x \in X'$. The *linearization* of (18) at $\beta = f = 0$ is $\mathrm{d}^*\bigl(\psi^m(\beta+\mathrm{d}f)\bigr) = 0$.

We study the family of small solutions β, f of (18) for which f has the decay near x_i required by (17). There is a ready-made theory of analysis on manifolds with cylindrical ends developed by Lockhart and McOwen [**24**], which is well-suited to this task. We work on certain *weighted Sobolev spaces* $L^p_{k,\mu}(X')$ of functions on X'.

By results from [**24**] it turns out that the operator $f \mapsto \mathrm{d}^*(\psi^m \mathrm{d}f)$ is a *Fredholm* map $L^p_{k,\mu}(X') \to L^p_{k-2,\mu-2}(X')$, with cokernel of dimension $\sum_{i=1}^n N_{\Sigma_i}(2)$. This cokernel is in effect the *obstruction space* to deforming X with x_i, v_i fixed, as it is the obstruction space to solving the linearization of (18) in f at $\beta = f = 0$.

By varying the x_i and v_i, and allowing f to converge to different constant values on the ends of X' rather than zero, we can overcome many of these obstructions. This reduces the dimension of the obstruction space $\mathcal{O}_{X'}$ from $\sum_{i=1}^n N_{\Sigma_i}(2)$ to $\sum_{i=1}^n \text{s-ind}(C_i)$. The obstruction map φ is constructed using the Implicit Mapping Theorem for Banach spaces. This concludes our sketch.

If the C_i are *stable* then $\mathcal{O}_{X'} = \{0\}$ and we deduce [**19**, Corollary 6.11]:

Corollary 5.3. *Suppose (M, J, ω, Ω) is an almost Calabi–Yau m-fold and X a compact* SL *m-fold in M with stable conical singularities, and let $\mathcal{M}_X$ and $\mathcal{I}_{X'}$ be as in Theorem* 5.2. *Then $\mathcal{M}_X$ is a smooth manifold of dimension* $\dim \mathcal{I}_{X'}$.

This has clear similarities with Theorem 2.10. Here is another simple condition for $\mathcal{M}_X$ to be a manifold near X, [**19**, Definition 6.12].

Definition 5.4. Let (M, J, ω, Ω) be an almost Calabi–Yau m-fold and X a compact SL m-fold in M with conical singularities, and let $\mathcal{I}_{X'}, \mathcal{O}_{X'}, U$ and φ be as in Theorem 5.2. We call X *transverse* if the linear map $\mathrm{d}\varphi|_0 : \mathcal{I}_{X'} \to \mathcal{O}_{X'}$ is surjective.

If X is transverse then $\{u \in U : \varphi(u) = 0\}$ is a manifold near 0, so Theorem 5.2 yields [**19**, Corollary 6.13]:

Corollary 5.5. *Suppose* (M, J, ω, Ω) *is an almost Calabi–Yau m-fold and X a transverse compact* SL *m-fold in M with conical singularities, and let* $\mathcal{M}_X, \mathcal{I}_{X'}$ *and* $\mathcal{O}_{X'}$ *be as in Theorem* 5.2. *Then* $\mathcal{M}_X$ *is near X a smooth manifold of dimension* $\dim \mathcal{I}_{X'} - \dim \mathcal{O}_{X'}$.

Now there are a number of well-known moduli space problems in geometry where in general moduli spaces are obstructed and singular, but after a generic perturbation they become smooth manifolds. For instance, moduli spaces of instantons on 4-manifolds can be made smooth by choosing a generic metric, and similar things hold for Seiberg–Witten equations, and moduli spaces of pseudo-holomorphic curves in symplectic manifolds.

In [**19**, §9] we try (but do not quite succeed) to replicate this for moduli spaces of SL m-folds with conical singularities, by choosing a *generic Kähler metric* in a fixed Kähler class. This is the idea behind [**19**, Conjecture 9.5]:

Conjecture 5.6. *Let* (M, J, ω, Ω) *be an almost Calabi–Yau m-fold, X a compact* SL *m-fold in M with conical singularities, and* $\mathcal{I}_{X'}, \mathcal{O}_{X'}$ *be as in Theorem* 5.2. *Then for a second category subset of Kähler forms* $\breve{\omega}$ *in the Kähler class of* ω*, the moduli space* $\breve{\mathcal{M}}_X$ *of compact SL m-folds* $\widehat{X}$ *with conical singularities in* $(M, J, \breve{\omega}, \Omega)$ *isotopic to X consists solely of transverse* $\widehat{X}$*, and so is a manifold of dimensi on* $\dim \mathcal{I}_{X'} - \dim \mathcal{O}_{X'}$.

A partial proof of this is given in [**19**, §9]. If we could treat the moduli spaces $\mathcal{M}_X$ as compact, the conjecture would follow from [**19**, Theorem 9.3]. However, without knowing $\mathcal{M}_X$ is compact, the condition that $\breve{\mathcal{M}}_X$ is smooth everywhere is in effect the intersection of an infinite number of genericity conditions on $\breve{\omega}$, and we do not know that this intersection is dense (or even nonempty) in the Kähler class.

Notice that Conjecture 5.6 constrains the topology and cones of SL m-folds X with conical singularities that can occur in a generic almost Calabi–Yau m-fold, as we must have $\dim \mathcal{I}_{X'} \geqslant \dim \mathcal{O}_{X'}$.

6. Asymptotically Conical SL m-Folds

We now discuss *Asymptotically Conical* SL m-folds L in $\mathbb{C}^m$, [**18**, Definition 7.1].

Definition 6.1. Let C be a closed SL cone in $\mathbb{C}^m$ with isolated singularity at 0 for $m > 2$, and let $\Sigma = C \cap \mathcal{S}^{2m-1}$, so that Σ is a compact, nonsingular $(m-1)$-manifold, not necessarily connected. Let g_Σ be the metric on Σ induced by the metric g' on $\mathbb{C}^m$ in (1), and r the radius function on $\mathbb{C}^m$. Define $\iota : \Sigma \times (0, \infty) \to \mathbb{C}^m$ by $\iota(\sigma, r) = r\sigma$. Then the image of ι is $C \setminus \{0\}$, and $\iota^*(g') = r^2 g_\Sigma + dr^2$ is the cone metric on $C \setminus \{0\}$.

Let L be a closed, nonsingular SL m-fold in $\mathbb{C}^m$. We call L *Asymptotically Conical (AC)* with *rate* $\lambda < 2$ and *cone* C if there exists a compact subset $K \subset L$ and a diffeomorphism $\varphi : \Sigma \times (T, \infty) \to L \setminus K$ for $T > 0$, such that

$$\left|\nabla^k(\varphi - \iota)\right| = O(r^{\lambda-1-k}) \quad \text{as } r \to \infty \text{ for } k = 0, 1.$$

Here $\nabla, |\,.\,|$ are computed using the cone metric $\iota^*(g')$.

This is very similar to Definition 3.7, and in fact there are strong parallels between the theories of SL m-folds with conical singularities and of Asymptotically Conical SL m-folds. We continue to assume $m > 2$ throughout.

6.1. Regularity and deformation theory of AC SL m-folds. Here are the analogues of Theorems 4.1 and 4.2, proved in [**18**, Theorems 7.4 and 7.11].

Theorem 6.2. *Suppose L is an AC SL m-fold in $\mathbb{C}^m$ with cone C, and let Σ, ι be as in Definition* 6.1. *Then for sufficiently large $T > 0$ there exist unique K, φ satisfying the conditions of Definition* 6.1 *and $(\varphi - \iota)(\sigma, r) \perp T_{\iota(\sigma,r)}C$ in $\mathbb{C}^m$ for all $(\sigma, r) \in \Sigma \times (T, \infty)$.*

Theorem 6.3. *In Theorem* 6.2, *if either $\lambda = \lambda'$, or λ, λ' lie in the same connected component of $\mathbb{R} \setminus \mathcal{D}_\Sigma$, then L is an AC SL m-fold with rate λ' and $\left|\nabla^k(\varphi - \iota)\right| = O(r^{\lambda'-1-k})$ for all $k \geqslant 0$. Here $\nabla, |\,.\,|$ are computed using the cone metric $\iota^*(g')$ on $\Sigma \times (T, \infty)$.*

The *deformation theory* of Asymptotically Conical SL m-folds in $\mathbb{C}^m$ has been studied independently by Pacini [**30**] and Marshall [**25**]. Pacini's results are earlier, but Marshall's are more complete.

Definition 6.4. Suppose L is an Asymptotically Conical SL m-fold in $\mathbb{C}^m$ with cone C and rate $\lambda < 2$, as in Definition 6.1. Define the *moduli space $\mathcal{M}_L^\lambda$ of deformations of L with rate λ* to be the set of AC SL m-folds $\hat{L}$ in $\mathbb{C}^m$ with cone C and rate λ, such that $\hat{L}$ is diffeomorphic to

L and isotopic to L as an Asymptotically Conical submanifold of $\mathbb{C}^m$. One can define a natural *topology* on $\mathcal{M}_L^\lambda$, in a similar way to the conical singularities case of [**19**, Definition 5.6].

Note that if L is an AC SL m-fold with rate λ, then it is *also* an AC SL m-fold with rate λ' for any $\lambda' \in [\lambda, 2)$. Thus we have defined a 1-*parameter family* of moduli spaces $\mathcal{M}_L^{\lambda'}$ for L, and not just one. Since we did not impose any condition on λ in Definition 6.1 analogous to (11) in the conical singularities case, it turns out that $\mathcal{M}_L^\lambda$ depends nontrivially on λ.

The following result can be deduced from Marshall [**25**, Theorem 6.2.15] and [**25**, Table 5.1]. (See also Pacini [**30**, Theorems 2 and 3].) It implies conjectures by the author in [**6**, Conjecture 2.12] and [**13**, §10.2].

Theorem 6.5. *Let L be an Asymptotically Conical* SL *m-fold in $\mathbb{C}^m$ with cone C and rate $\lambda < 2$, and let $\mathcal{M}_L^\lambda$ be as in Definition* 6.4. *Set $\Sigma = C \cap \mathcal{S}^{2m-1}$, and let $\mathcal{D}_\Sigma, N_\Sigma$ be as in* §3.1 *and $b^k(L), b^k_{\mathrm{cs}}(L)$ as in* §3.4. *Then*

(a) *If $\lambda \in (0,2) \setminus \mathcal{D}_\Sigma$ then $\mathcal{M}_L^\lambda$ is a manifold with*

$$\dim \mathcal{M}_L^\lambda = b^1(L) - b^0(L) + N_\Sigma(\lambda).$$

Note that if $0 < \lambda < \min(\mathcal{D}_\Sigma \cap (0,\infty))$ then $N_\Sigma(\lambda) = b^0(\Sigma)$.

(b) *If $\lambda \in (2-m, 0)$ then $\mathcal{M}_L^\lambda$ is a manifold of dimension $b^1_{\mathrm{cs}}(L) = b^{m-1}(L)$.*

This is the analogue of Theorems 2.10 and 5.2 for AC SL m-folds. If $\lambda \in (2-m, 2) \setminus \mathcal{D}_\Sigma$ then the deformation theory for L with rate λ is *unobstructed* and $\mathcal{M}_L^\lambda$ is a *smooth manifold* with a given dimension. This is similar to the case of nonsingular compact SL m-folds in Theorem 2.10, but different to the conical singularities case in Theorem 5.2.

6.2. Cohomological invariants of AC SL m-folds. Let L be an AC SL m-fold in $\mathbb{C}^m$ with cone C, and set $\Sigma = C \cap \mathcal{S}^{2m-1}$. Using the notation of §3.4, as in (13) there is a long exact sequence

$$\cdots \to H^k_{\mathrm{cs}}(L,\mathbb{R}) \to H^k(L,\mathbb{R}) \to H^k(\Sigma,\mathbb{R}) \to H^{k+1}_{\mathrm{cs}}(L,\mathbb{R}) \to \cdots. \quad (19)$$

Following [**18**, Definition 7.2] we define *cohomological invariants* $Y(L), Z(L)$ of L.

Definition 6.6. Let L be an AC SL m-fold in $\mathbb{C}^m$ with cone C, and let $\Sigma = C \cap \mathcal{S}^{2m-1}$. As $\omega', \operatorname{Im}\Omega'$ in (1) are closed forms with $\omega'|_L \equiv$

$\operatorname{Im}\Omega'|_L \equiv 0$, they define classes in the relative de Rham cohomology groups $H^k(\mathbb{C}^m; L, \mathbb{R})$ for $k = 2, m$. But for $k > 1$ we have the exact sequence

$$0 = H^{k-1}(\mathbb{C}^m, \mathbb{R}) \to H^{k-1}(L, \mathbb{R}) \xrightarrow{\cong} H^k(\mathbb{C}^m; L, \mathbb{R}) \to H^k(\mathbb{C}^m, \mathbb{R}) = 0.$$

Let $Y(L) \in H^1(\Sigma, \mathbb{R})$ be the image of $[\omega']$ in $H^2(\mathbb{C}^m; L, \mathbb{R}) \cong H^1(L, \mathbb{R})$ under $H^1(L, \mathbb{R}) \to H^1(\Sigma, R)$ in (19), and $Z(L) \in H^{m-1}(\Sigma, \mathbb{R})$ be the image of $[\operatorname{Im}\Omega']$ in $H^m(\mathbb{C}^m; L, \mathbb{R}) \cong H^{m-1}(L, \mathbb{R})$ under $H^{m-1}(L, \mathbb{R}) \to H^{m-1}(\Sigma, R)$ in (19).

Here are some conditions for $Y(L)$ or $Z(L)$ to be zero, [**18**, Proposition 7.3].

Proposition 6.7. *Let L be an AC SL m-fold in $\mathbb{C}^m$ with cone C and rate λ, and let $\Sigma = C \cap \mathcal{S}^{2m-1}$. If $\lambda < 0$ or $b^1(L) = 0$ then $Y(L) = 0$. If $\lambda < 2 - m$ or $b^0(\Sigma) = 1$ then $Z(L) = 0$.*

6.3. Examples. Examples of AC SL m-folds L are constructed by Harvey and Lawson [**3**, §III.3], the author [**7, 8, 9, 11**], and others. Nearly all the known examples (up to translations) have minimum rate λ either 0 or $2 - m$, which are topologically significant values by Proposition 6.7. For instance, all examples in [**8**] have $\lambda = 0$, and [**7**, Theorem 6.4] constructs AC SL m-folds with $\lambda = 2 - m$ in $\mathbb{C}^m$ from any SL cone C in $\mathbb{C}^m$. The only explicit, nontrivial examples known to the author with $\lambda \neq 0, 2 - m$ are in [**9**, Theorem 11.6], and have $\lambda = \frac{3}{2}$.

We shall give three families of examples of AC SL m-folds L in $\mathbb{C}^m$ explicitly. The first family is adapted from Harvey and Lawson [**3**, §III.3.A].

Example 6.8. Let C^m_{HL} be the SL cone in $\mathbb{C}^m$ of Example 3.5. We shall define a family of AC SL m-folds in $\mathbb{C}^m$ with cone C^m_{HL}. Let $a_1, \ldots, a_m \geqslant 0$ with exactly two of the a_j zero and the rest positive. Write $\mathbf{a} = (a_1, \ldots, a_m)$, and define

$$\begin{aligned} L^{\mathbf{a}}_{\mathrm{HL}} = \bigl\{&(z_1, \ldots, z_m) \in \mathbb{C}^m : i^{m+1} z_1 \cdots z_m \in [0, \infty), \\ &|z_1|^2 - a_1 = \cdots = |z_m|^2 - a_m\bigr\}. \end{aligned} \tag{20}$$

Then $L^{\mathbf{a}}_{\mathrm{HL}}$ is an AC SL m-fold in $\mathbb{C}^m$ diffeomorphic to $T^{m-2} \times \mathbb{R}^2$, with cone C^m_{HL} and rate 0. It is invariant under the $\mathrm{U}(1)^{m-1}$ group (8). It is surprising that equations of the form (20) should define a nonsingular submanifold of $\mathbb{C}^m$ *without boundary*, but in fact they do.

Now suppose for simplicity that $a_1, \dots, a_{m-2} > 0$ and $a_{m-1} = a_m = 0$. As $\Sigma^m_{\mathrm{HL}} \cong T^{m-1}$ we have $H^1(\Sigma^m_{\mathrm{HL}}, \mathbb{R}) \cong \mathbb{R}^{m-1}$, and calculation shows that $Y(L^{\mathbf{a}}_{\mathrm{HL}}) = (\pi a_1, \dots, \pi a_{m-2}, 0) \in \mathbb{R}^{m-1}$ in the natural coordinates. Since $L^{\mathbf{a}}_{\mathrm{HL}} \cong T^{m-2} \times \mathbb{R}^2$ we have $H^1(L^{\mathbf{a}}_{\mathrm{HL}}, \mathbb{R}) = \mathbb{R}^{m-2}$, and $Y(L^{\mathbf{a}}_{\mathrm{HL}})$ lies in the image $\mathbb{R}^{m-2} \subset \mathbb{R}^{m-1}$ of $H^1(L^{\mathbf{a}}_{\mathrm{HL}}, \mathbb{R})$ in $H^1(\Sigma^m_{\mathrm{HL}}, \mathbb{R})$, as in Definition 6.6. As $b^0(\Sigma^m_{\mathrm{HL}}) = 1$, Proposition 6.7 shows that $Z(L^{\mathbf{a}}_{\mathrm{HL}}) = 0$.

Take $C = C^m_{\mathrm{HL}}$, $\Sigma = \Sigma^m_{\mathrm{HL}}$ and $L = L^{\mathbf{a}}_{\mathrm{HL}}$ in Theorem 6.5, and let $0 < \lambda < \min\bigl(\mathcal{D}_\Sigma \cap (0, \infty)\bigr)$. Then $b^1(L) = m - 2$, $b^0(L) = 1$ and $N_\Sigma(\lambda) = b^0(\Sigma) = 1$, so part (a) of Theorem 6.5 shows that $\dim \mathcal{M}^\lambda_L = m - 2$. This is consistent with the fact that L depends on $m - 2$ real parameters $a_1, \dots, a_{m-2} > 0$.

The family of all $L^{\mathbf{a}}_{\mathrm{HL}}$ has $\frac{1}{2}m(m-1)$ connected components, indexed by which two of $a_1, \dots, a_m$ are zero. Using the theory of §7, these can give many *topologically distinct* ways to desingularize SL m-folds with conical singularities with these cones.

Our second family, from [**7**, Example 9.4], was chosen as it is easy to write down.

Example 6.9. Let $m, a_1, \dots, a_m, k$ and $L^{a_1, \dots, a_m}_0$ be as in Example 3.6. For $0 \neq c \in \mathbb{R}$ define

$$L^{a_1, \dots, a_m}_c = \bigl\{ \bigl(i e^{i a_1 \theta} x_1, e^{i a_2 \theta} x_2, \dots, e^{i a_m \theta} x_m \bigr) : \theta \in [0, 2\pi), \\ x_1, \dots, x_m \in \mathbb{R}, \qquad a_1 x_1^2 + \dots + a_m x_m^2 = c \bigr\}.$$

Then $L^{a_1, \dots, a_m}_c$ is an AC SL m-fold in $\mathbb{C}^m$ with rate 0 and cone $L^{a_1, \dots, a_m}_0$. It is diffeomorphic as an immersed SL m-fold to $(\mathcal{S}^{k-1} \times \mathbb{R}^{m-k} \times \mathcal{S}^1)/\mathbb{Z}_2$ if $c > 0$, and to $(\mathbb{R}^k \times \mathcal{S}^{m-k-1} \times \mathcal{S}^1)/\mathbb{Z}_2$ if $c < 0$.

Our third family was first found by Lawlor [**23**], made more explicit by Harvey [**2**, p. 139–140], and discussed from a different point of view by the author in [**8**, §5.4(b)]. Our treatment is based on that of Harvey.

Example 6.10. Let $m > 2$ and $a_1, \dots, a_m > 0$, and define polynomials p, P by

$$p(x) = (1 + a_1 x^2) \cdots (1 + a_m x^2) - 1 \quad \text{and} \quad P(x) = \frac{p(x)}{x^2}.$$

Define real numbers $\varphi_1, \ldots, \varphi_m$ and A by

$$\varphi_k = a_k \int_{-\infty}^{\infty} \frac{\mathrm{d}x}{(1 + a_k x^2)\sqrt{P(x)}} \quad \text{and} \quad A = \omega_m (a_1 \cdots a_m)^{-1/2}, \tag{21}$$

where ω_m is the volume of the unit sphere in $\mathbb{R}^m$. Clearly $\varphi_k, A > 0$. But writing $\varphi_1 + \cdots + \varphi_m$ as one integral gives

$$\varphi_1 + \cdots + \varphi_m = \int_0^{\infty} \frac{p'(x)\mathrm{d}x}{(p(x)+1)\sqrt{p(x)}} = 2\int_0^{\infty} \frac{\mathrm{d}w}{w^2+1} = \pi,$$

making the substitution $w = \sqrt{p(x)}$. So $\varphi_k \in (0, \pi)$ and $\varphi_1 + \cdots + \varphi_m = \pi$. This yields a 1-1 correspondence between m-tuples $(a_1, \ldots, a_m)$ with $a_k > 0$, and $(m+1)$-tuples $(\varphi_1, \ldots, \varphi_m, A)$ with $\varphi_k \in (0, \pi)$, $\varphi_1 + \cdots + \varphi_m = \pi$ and $A > 0$.

For $k = 1, \ldots, m$ and $y \in \mathbb{R}$, define a function $z_k : \mathbb{R} \to \mathbb{C}$ by

$$z_k(y) = \mathrm{e}^{i\psi_k(y)} \sqrt{a_k^{-1} + y^2}, \quad \text{where} \quad \psi_k(y) = a_k \int_{-\infty}^{y} \frac{\mathrm{d}x}{(1 + a_k x^2)\sqrt{P(x)}}.$$

Now write $\boldsymbol{\varphi} = (\varphi_1, \ldots, \varphi_n)$, and define a submanifold $L^{\boldsymbol{\varphi}, A}$ in $\mathbb{C}^m$ by

$$L^{\boldsymbol{\varphi}, A} = \big\{ (z_1(y)x_1, \ldots, z_m(y)x_m) : y \in \mathbb{R},\ x_k \in \mathbb{R},\ x_1^2 + \cdots + x_m^2 = 1 \big\}.$$

Then $L^{\boldsymbol{\varphi}, A}$ is closed, embedded, and diffeomorphic to $\mathcal{S}^{m-1} \times \mathbb{R}$, and Harvey [**2**, Theorem 7.78] shows that $L^{\boldsymbol{\varphi}, A}$ is *special Lagrangian*. One can also show that $L^{\boldsymbol{\varphi}, A}$ is *Asymptotically Conical*, with rate $2 - m$ and cone the union $\Pi^0 \cup \Pi^{\boldsymbol{\varphi}}$ of two special Lagrangian m-planes $\Pi^0, \Pi^{\boldsymbol{\varphi}}$ in $\mathbb{C}^m$ given by

$$\Pi^0 = \big\{ (x_1, \ldots, x_m) : x_j \in \mathbb{R} \big\},$$
$$\Pi^{\boldsymbol{\varphi}} = \big\{ (\mathrm{e}^{i\varphi_1} x_1, \ldots, \mathrm{e}^{i\varphi_m} x_m) : x_j \in \mathbb{R} \big\}.$$

As $\lambda = 2 - m < 0$ we have $Y(L^{\boldsymbol{\varphi}, A}) = 0$ by Proposition 6.7. Now $L^{\boldsymbol{\varphi}, A} \cong \mathcal{S}^{m-1} \times \mathbb{R}$ so that $H^{m-1}(L^{\boldsymbol{\varphi}, A}, \mathbb{R}) \cong \mathbb{R}$, and $\Sigma = (\Pi^0 \cup \Pi^{\boldsymbol{\varphi}}) \cap \mathcal{S}^{2m-1}$ is the disjoint union of two unit $(m-1)$-spheres $\mathcal{S}^{m-1}$, so $H^{m-1}(\Sigma, \mathbb{R}) \cong \mathbb{R}^2$. The image of $H^{m-1}(L^{\boldsymbol{\varphi}, A}, \mathbb{R})$ in $H^{m-1}(\Sigma, \mathbb{R})$ is $\big\{ (x, -x) : x \in \mathbb{R} \big\}$ in the natural coordinates. Calculation shows that

$Z(L^{\varphi,A})(A,-A) \in H^{m-1}(\Sigma,\mathbb{R})$, which is why we defined A this way in (21).

Apply Theorem 6.5 with $L = L^{\varphi,A}$ and $\lambda \in (2-m,0)$. As $L \cong \mathcal{S}^{m-1} \times \mathbb{R}$ we have $b^1_{\rm cs}(L) = 1$, so part (b) of Theorem 6.5 shows that $\dim \mathcal{M}^\lambda_L = 1$. This is consistent with the fact that when φ is fixed, $L^{\varphi,A}$ depends on one real parameter $A > 0$. Here φ is fixed in $\mathcal{M}^\lambda_L$ as the cone $C = \Pi^0 \cup \Pi^\varphi$ of L depends on φ, and all $\widehat{L} \in \mathcal{M}^\lambda_L$ have the same cone C, by definition.

7. Desingularizing Singular SL m-Folds

We now discuss the work of [**20, 21**] on *desingularizing* compact SL m-folds with conical singularities. Here is the basic idea. Let (M,J,ω,Ω) be an almost Calabi–Yau m-fold, and X a compact SL m-fold in M with conical singularities $x_1,\ldots,x_n$ and cones $C_1,\ldots,C_n$. Suppose $L_1,\ldots,L_n$ are AC SL m-folds in $\mathbb{C}^m$ with the same cones $C_1,\ldots,C_n$ as X.

If $t > 0$ then $tL_i = \{t\mathbf{x} : \mathbf{x} \in L_i\}$ is also an AC SL m-fold with cone C_i. We construct a 1-parameter family of compact, nonsingular *Lagrangian* m-folds N^t in (M,ω) for $t \in (0,\delta)$ by gluing tL_i into X at x_i, using a partition of unity.

When t is small, N^t is *close to special Lagrangian* (its phase is nearly constant), but also *close to singular* (it has large curvature and small injectivity radius). We prove using analysis that for small $t \in (0,\delta)$ we can deform N^t to a *special* Lagrangian m-fold $\widetilde{N}^t$ in M, using a small Hamiltonian deformation.

The proof involves a delicate balancing act, showing that the advantage of being close to special Lagrangian outweighs the disadvantage ofbeing nearly singular. Doing this in full generality is rather complex. Here is our simplest desingularization result, [**20**, Theorem 6.13].

Theorem 7.1. *Suppose (M,J,ω,Ω) is an almost Calabi–Yau m-fold and X a compact SL m-fold in M with conical singularities at $x_1,\ldots,x_n$ and cones $C_1,\ldots,C_n$. Let $L_1,\ldots,L_n$ be Asymptotically Conical* SL *m-folds in $\mathbb{C}^m$ with cones $C_1,\ldots,C_n$ and rates $\lambda_1,\ldots,\lambda_n$. Suppose $\lambda_i < 0$ for $i = 1,\ldots,n$, and $X' = X \setminus \{x_1,\ldots,x_n\}$ is connected.*

Then there exists $\varepsilon > 0$ *and a smooth family* $\{\tilde{N}^t : t \in (0, \varepsilon]\}$ *of compact, nonsingular SL m-folds in* (M, J, ω, Ω), *such that* $\tilde{N}^t$ *is constructed by gluing* tL_i *into* X *at* x_i *for* $i = 1, \ldots, n$. *In the sense of currents,* $\tilde{N}^t \to X$ *as* $t \to 0$.

The theorem contains two *simplifying assumptions*:

(a) that X' is connected, and

(b) that $\lambda_i < 0$ for all i.

These avoid two kinds of *obstructions* to desingularizing X using the L_i.

In [**20**, Theorem 7.10] we remove assumption (a), allowing X' not connected.

Theorem 7.2. *Suppose* (M, J, ω, Ω) *is an almost Calabi–Yau m-fold and* X *a compact SL m-fold in* M *with conical singularities at* $x_1, \ldots, x_n$ *and cones* $C_1, \ldots, C_n$. *Define* $\psi : M \to (0, \infty)$ *as in* (3). *Let* $L_1, \ldots, L_n$ *be Asymptotically Conical* SL *m-folds in* $\mathbb{C}^m$ *with cones* $C_1, \ldots, C_n$ *and rqtes* $\lambda_1, \ldots, \lambda_n$. *Suppose* $\lambda_i < 0$ *for* $i = 1, \ldots, n$. *Write* $X' = X \setminus \{x_1, \ldots, x_n\}$ *and* $\Sigma_i = C_i \cap \mathcal{S}^{2m-1}$.

Set $q = b^0(X')$, *and let* $X'_1, \ldots, X'_q$ *be the connected components of* X'. *For* $i = 1, \ldots, n$ *let* $l_i = b^0(\Sigma_i)$, *and let* $\Sigma_i^1, \ldots, \Sigma_i^{l_i}$ *be the connected components of* Σ_i. *Define* $k(i,j) = 1, \ldots, q$ *by* $\Upsilon_i \circ \varphi_i(\Sigma_i^j \times (0, R')) \subset X'_{k(i,j)}$ *for* $i = 1, \ldots, n$ *and* $j = 1, \ldots, l_i$. *Suppose that*

$$\sum_{\substack{1 \leqslant i \leqslant n,\ 1 \leqslant j \leqslant l_i: \\ k(i,j) = k}} \psi(x_i)^m Z(L_i) \cdot [\Sigma_i^j] = 0 \quad \text{for all } k = 1, \ldots, q. \tag{22}$$

Suppose also that the compact m-manifold N *obtained by gluing* L_i *into* X' *at* x_i *for* $i = 1, \ldots, n$ *is connected. A sufficient condition for this to hold is that* X *and* L_i *for* $i = 1, \ldots, n$ *are connected.*

Then there exists $\varepsilon > 0$ *and a smooth family* $\{\tilde{N}^t : t \in (0, \varepsilon]\}$ *of compact, nonsingular* SL *m-folds in* (M, J, ω, Ω) *diffeomorphic to* N, *such that* $\tilde{N}^t$ *is constructed by gluing* tL_i *into* X *at* x_i *for* $i = 1, \ldots, n$. *In the sense of currents in Geometric Measure Theory,* $\tilde{N}^t \to X$ *as* $t \to 0$.

The new issue here is that if X' is not connected then there is an *analytic obstruction* to deforming N^t to $\tilde{N}^t$, because the Laplacian Δ^t on functions on N^t has *small eigenvalues* of size $O(t^{m-2})$. As in §6.2 the L_i have *cohomological invariants* $Z(L_i)$ in $H^{m-1}(\Sigma_i, \mathbb{R})$ derived from the relative cohomology class of $\operatorname{Im}\Omega'$. It turns out that we can only deform N^t to $\tilde{N}^t$ if the $Z(L_i)$ satisfy (22). This equation arises by requiring the projection of an error term to the eigenspaces of Δ^t with small eigenvalues to be zero.

In [**21**, Theorem 6.13] we remove assumption (b), extending Theorem 7.1 to the case $\lambda_i \leqslant 0$, and allowing $Y(L_i) \neq 0$.

Theorem 7.3. *Let (M, J, ω, Ω) be an almost Calabi–Yau m-fold for $2 < m < 6$, and X a compact SL m-fold in M with conical singularities at $x_1, \ldots, x_n$ and cones $C_1, \ldots, C_n$. Let $L_1, \ldots, L_n$ be Asymptotically Conical* SL *m-folds in $\mathbb{C}^m$ with cones $C_1, \ldots, C_n$ and rates $\lambda_1, \ldots, \lambda_n$. Suppose that $\lambda_i \leqslant 0$ for $i = 1, \ldots, n$, that $X' = X \setminus \{x_1, \ldots, x_n\}$ is connected, and that there exists $\varrho \in H^1(X', \mathbb{R})$ such that $\big(Y(L_1), \ldots, Y(L_n)\big)$ is the image of ϱ under the map $H^1(X', \mathbb{R}) \to \bigoplus_{i=1}^n H^1(\Sigma_i, \mathbb{R})$ in (13), where $\Sigma_i = C_i \cap \mathcal{S}^{2m-1}$.*

Then there exists $\varepsilon > 0$ and a smoot h family $\{\tilde{N}^t : t \in (0, \varepsilon]\}$ of compact, nonsingular SL m-folds in (M, J, ω, Ω), such that $\tilde{N}^t$ is constructed by gluing tL_i into X at x_i for $i = 1, \ldots, n$. In the sense of currents, $\tilde{N}^t \to X$ as $t \to 0$.

From §6.3, the L_i have *cohomological invariants* $Y(L_i)$ in $H^1(\Sigma_i, \mathbb{R})$ derived from the relative cohomology class of ω'. The new issue in Theorem 7.3 is that if $Y(L_i) \neq 0$ then there are obstructions to the existence of N^t as a *Lagrangian* m-fold. That is, we can only define N^t if the $Y(L_i)$ satisfy an equation. This did not appear in Theorem 7.1, as $\lambda_i < 0$ implies that $Y(L_i) = 0$.

To define the N^t when $Y(L_i) \neq 0$ we must also use a more complicated construction. This introduces new errors. To overcome these errors when we deform N^t to $\tilde{N}^t$ we must assume that $m < 6$. There is also [**21**, Theorem 6.12] a result combining the modifications of Theorems 7.2 and 7.3, but for brevity we will not give it.

8. Directions for Future Research

Finally we discuss directions the field of special Lagrangian singularities might develop in the future, giving a number of problems the author believes are worth attention. Some of these problems may be too difficult to solve completely, but can still serve as a guide.

8.1. The index of singularities of SL m-folds. We now consider the *boundary* $\partial\mathcal{M}_N$ of a moduli space $\mathcal{M}_N$ of SL m-folds.

Definition 8.1. Let (M, J, ω, Ω) be an almost Calabi–Yau m-fold, N a compact, nonsingular SL m-fold in M, and $\mathcal{M}_N$ the moduli space of deformations of N in M. Then $\mathcal{M}_N$ is a smooth manifold of dimension $b^1(N)$, in general noncompact. We can construct a natural *compactification* $\overline{\mathcal{M}}_N$ as follows.

Regard $\mathcal{M}_N$ as a moduli space of special Lagrangian *integral currents* in the sense of Geometric Measure Theory, as discussed in [**18**, §6]. Let $\overline{\mathcal{M}}_N$ be the closure of $\mathcal{M}_N$ in the space of integral currents. As elements of $\mathcal{M}_N$ have uniformly bounded volume, $\overline{\mathcal{M}}_N$ is *compact*. Define the *boundary* $\partial\mathcal{M}_N$ to be $\overline{\mathcal{M}}_N \setminus \mathcal{M}_N$. Then elements of $\partial\mathcal{M}_N$ are *singular SL integral currents*.

In good cases, say if (M, J, ω, Ω) is suitably generic, it seems reasonable that $\partial\mathcal{M}_N$ should be divided into a number of *strata*, each of which is a moduli space of singular SL m-folds with singularities of a particular type, and is itself a manifold with singularities. In particular, some or all of these strata could be moduli spaces $\mathcal{M}_X$ of SL m-folds with isolated conical singularities, as in §5.

Let $\mathcal{M}_N$ be a moduli space of compact, nonsingular SL m-folds N in (M, J, ω, Ω), and $\mathcal{M}_X$ a moduli space of singular SL m-folds in $\partial\mathcal{M}_N$ with singularities of a particular type, and $X \in \mathcal{M}_X$. Following [**22**, §8.3], we (loosely) define the *index* of the singularities of X to be $\operatorname{ind}(X) = \dim\mathcal{M}_N - \dim\mathcal{M}_X$, provided $\mathcal{M}_X$ is smooth near X. Note that $\operatorname{ind}(X)$ depends on N as well as X.

In [**22**, Theorem 8.10] we use the results of [**19, 20, 21**] to compute $\operatorname{ind}(X)$ when X is *transverse* with conical singularities, in the sense of Definition 5.4. Here is a simplified version of the result, where we assume that $H^1_{\rm cs}(L_i, \mathbb{R}) \to H^1(L_i, \mathbb{R})$ is surjective to avoid a complicated correction term to $\operatorname{ind}(X)$ related to the obstructions to defining N^t as a Lagrangian m-fold.

Theorem 8.2. *Let X be a compact, transverse SL m-fold in (M, J, ω, Ω) with conical singularities at $x_1, \ldots, x_n$ and cones $C_1, \ldots, C_n$. Let $L_1, \ldots, L_n$ be AC SL m-folds in $\mathbb{C}^m$ with cones $C_1, \ldots, C_n$, such that the natural projection $H^1_{\rm cs}(L_i, \mathbb{R}) \to H^1(L_i, \mathbb{R})$ is surjective. Construct desingularizations N of X by gluing AC SL m-folds $L_1, \ldots, L_n$ in at $x_1, \ldots, x_n$, as in § 7. Then*

$$\operatorname{ind}(X) = 1 - b^0(X') + \textstyle\sum_{i=1}^n b^1_{\rm cs}(L_i) + \sum_{i=1}^n \text{s-ind}(C_i). \qquad (23)$$

If the cones C_i are not *rigid*, for instance if $C_i \setminus \{0\}$ is not connected, then (23) should be corrected, as in [**22**, §8.3]. If Conjecture 5.6 is true then for a *generic* Kähler form ω, *all* compact SL m-folds X with conical singularities are transverse, and so Theorem 8.2 and [**22**, Theorem 8.10] allow us to calculate $\text{ind}(X)$.

Now singularities with *small index* are the most commonly occurring, and so arguably the most interesting kinds of singularity. Also, as $\text{ind}(X) \leqslant \dim \mathcal{M}_N$, for various problems, such as those in §8.3 and §8.4, it will only be necessary to know about singularities with index up to a certain value. This motivates the following:

Problem 8.3. Classify types of singularities of SL 3-folds with *small index* in suitably generic almost Calabi–Yau 3-folds, say with index 1, 2 or 3.

Here we restrict to $m = 3$ to make the problem more feasible, though still difficult. Note, however, that we do *not* restrict to isolated conical singularities, so a complete, rigorous answer would require a theory of more general kinds of singularities of SL 3-folds.

One can make some progress on this problem simply by studying the many examples of singular SL 3-folds in [**3, 4**] and [**8, 9, 10, 11, 12, 13, 14, 15, 16, 17**], calculating or guessing the index of each, and ruling out other kinds of singularities by plausible-sounding arguments. Using these techniques I have a conjectural classification of index 1 singularities of SL 3-folds, which involves the SL T^2-cone C^3_{HL} of (7), and several different kinds of singularity whose tangent cone is two copies of $\mathbb{R}^3$, intersecting in 0, $\mathbb{R}$ or $\mathbb{R}^3$.

Coming from another direction, *integrable systems* techniques may yield rigorous classification results for SL T^2-cones by index. Haskins [**5**, Theorem A] has used them to prove that the SL T^2-cone C^3_{HL} in $\mathbb{C}^3$ of (7) is up to SU(3) equivalence the *unique* SL T^2-cone C with $\text{s-ind}(C) = 0$. Now the index of a singularity modelled on C is at least $\text{s-ind}(C) + 1$, so this implies that C^3_{HL} is the unique SL T^2-cone with index 1 in Problem 8.3.

8.2. Singularities which are not isolated conical. Singularities of SL m-folds which are not 'isolated conical singularities' in the sense of Definition 3.7 are an important, but virtually unexplored, subject. Here are some known classes of nontrivial examples when $m = 3$.

(i) In [**11**] we study *ruled* SL 3-folds in $\mathbb{C}^3$, i.e., SL 3-folds N fibred by a 2-dimensional family Σ of real straight lines in $\mathbb{C}^3$. When Σ

is nonsingular N can still have singularities, and examples may be written down very explicitly, as in [**11**, Theorem 7.1].

The *tangent cones* of such singularities, in the sense of Geometric Measure Theory, are generally $\mathbb{R}^3$ with multiplicity $k > 1$. Near the singular point, the SL 3-fold resembles a k-fold branched cover of $\mathbb{R}^3$, branched along $\mathbb{R}$. A similar class of singularities of SL 3-folds, with tangent cone $\mathbb{R}^3$ with multiplicity 2, is studied in [**9**, § 6].

(ii) In [**14, 15, 16**] we study SL 3-folds in $\mathbb{C}^3$ invariant under the U(1)-action

$$e^{i\theta} : (z_1, z_2, z_3) \mapsto (e^{i\theta} z_1, e^{-i\theta} z_2, z_3) \quad \text{for } e^{i\theta} \in \mathrm{U}(1).$$

The three papers are surveyed in [**17**]. A U(1)-invariant SL 3-fold N may locally be written in the form

$$N = \big\{(z_1, z_2, z_3) \in \mathbb{C}^3 : z_1 z_2 = v(x, y) + iy, \quad z_3 = x + iu(x, y), \\ |z_1|^2 - |z_2|^2 = 2a, \quad (x, y) \in S\big\},$$

where S is a domain in $\mathbb{R}^2$, $a \in \mathbb{R}$ and $u, v : S \to \mathbb{R}$ satisfy (in a weak sense if $a = 0$) the *nonlinear Cauchy–Riemann equations*

$$\frac{\partial u}{\partial x} = \frac{\partial v}{\partial y} \quad \text{and} \quad \frac{\partial v}{\partial x} = -2\big(v^2 + y^2 + a^2\big)^{1/2} \frac{\partial u}{\partial y}. \tag{24}$$

Using analytic techniques, we construct and study solutions u, v of (24) satisfying boundary conditions on a strictly convex domain S. These include many *singular solutions*, and we show in [**16**, §9–§10] that we can construct countably many distinct geometrical-topological types of isolated SL 3-fold singularities, whose tangent cone is the union of two $\mathbb{R}^3$'s in $\mathbb{C}^3$, intersecting in $\mathbb{R}$.

There appear to the author to be two ways of studying special Lagrangian singularities which are not isolated conical. The first is to try and study *all* singularities of special Lagrangian integral currents, using Geometric Measure Theory. As far as the author understands (which is not very far), it will be difficult to use the special Lagrangian condition in GMT, or to say anything nontrivial about special Lagrangian singularities in this generality.

The second way is to define some restricted class of singularities and then study them, just as we did in §3–§7. The problem here is to decide upon a suitable kind of *local model* for the singularities, and appropriate *asymptotic conditions* for how the SL m-fold approaches the local model near the singularity. Now not just any local model and asymptotic conditions will do.

For a class of singularities to be worth studying, they should occur fairly often in "real life," so that, for instance, examples of such singularities might occur in compact SL m-folds in fairly generic almost Calabi–Yau m-folds. A good test of this is whether the *deformation theory* of compact SL m-folds with this kind of singularity is well-behaved. That is, the analogue of Theorem 5.2 should hold, with *finite-dimensional* obstruction space $\mathcal{O}_{X'}$.

One very obvious way to make examples of SL m-folds with nonisolated singularities is to consider $C \times \mathbb{R}^{m-k}$ in $\mathbb{C}^k \times \mathbb{C}^{m-k} = \mathbb{C}^m$, where C is an SL cone in $\mathbb{C}^k$ with isolated singularity at 0, and $3 \leqslant k < m$. So we could study SL m-folds with singularities locally modelled on $C \times \mathbb{R}^{m-k}$. Calculations by the author indicate that the deformation theory of such singular SL m-folds will be well-behaved if and only if C is *stable*. Therefore we propose:

Problem 8.4. Let $3 \leqslant k < m$, and suppose C is an SL cone in $\mathbb{C}^k$ with an isolated singularity at 0 which is *stable*, in the sense of Definition 3.4. Study compact SL m-folds N in almost Calabi–Yau m-folds (M, J, ω, Ω), where the singular set S of N is a compact $(m-k)$-submanifold of M, and N is modelled on $C \times \mathbb{R}^{m-k}$ in $\mathbb{C}^k \times \mathbb{C}^{m-k} = \mathbb{C}^m$ at each singular point $s \in S$.

Here we have not defined what we mean by 'modelled on'. There should be some fairly natural asymptotic condition, along the lines of (12). Perhaps, as in Theorem 4.3, it will be equivalent to N having tangent cone $C \times \mathbb{R}^{m-k}$ with multiplicity 1 at each $s \in S$.

A related problem is to classify the possible stable C:

Problem 8.5. Classify special Lagrangian cones C in $\mathbb{C}^m$ for $m \geqslant 3$ with an isolated singularity at 0 which are *stable*, in the sense of Definition 3.4.

As above, by Haskins [**5**, Theorem A] the SL T^2-cone $C^3_{\rm HL}$ in $\mathbb{C}^3$ of (7) is up to SU(3) equivalence the *unique* stable SL T^2-cone C in $\mathbb{C}^3$. In fact $C^3_{\rm HL}$ is the *only* example of a stable SL cone in $\mathbb{C}^m$ for $m \geqslant 3$ known to the author.

It is conceivable that it really is the only example, so that the answer to Problem 8.5 is $C^3_{\rm HL}$ and no others.

We can also look for other interesting classes of singularities with well-behaved deformation theory. The key is to find suitable asymptotic conditions.

Problem 8.6. Let C be an SL cone in $\mathbb{C}^m$ with nonisolated singularity at 0, or with multiplicity $k > 1$. Can you find a good, natural set of asymptotic conditions for SL m-folds with isolated singularities with tangent cone C?

One way to approach this is through *examples*: we find some class of examples of singular SL m-folds, calculate their asymptotic behavior near their singularities, and try and abstract the important features. For the examples in (i) above this may be easy, as they are very explicit. But for those in (ii) above the author failed miserably to understand the asymptotic behavior.

8.3. The SYZ Conjecture. *Mirror Symmetry* is a mysterious relationship between pairs of Calabi–Yau 3-folds $M, \widehat{M}$, arising from a branch of physics known as *String Theory*, and leading to some very strange and exciting conjectures about Calabi–Yau 3-folds.

Roughly speaking, String Theorists believe that each Calabi–Yau 3-fold M has a quantization, a *Super Conformal Field Theory* (SCFT). If $M, \widehat{M}$ have SCFT's isomorphic under a certain simple involution of SCFT structure, we say that $M, \widehat{M}$ are *mirror* Calabi–Yau 3-folds. One can argue using String Theory that $H^{1,1}(M) \cong H^{2,1}(\widehat{M})$ and $H^{2,1}(M) \cong H^{1,1}(\widehat{M})$. The mirror transform also exchanges things related to the complex structure of M with things related to the symplectic structure of $\widehat{M}$, and vice versa.

The *SYZ Conjecture*, due to Strominger, Yau and Zaslow [**31**] in 1996, gives a geometric explanation of Mirror Symmetry. Here is an attempt to state it.

Conjecture 8.7 (Strominger–Yau–Zaslow). *Suppose M and $\widehat{M}$ are mirror Calabi–Yau 3-folds. Then (under some additional conditions) there should exist a compact topological 3-manifold B and surjective, continuous maps $f : M \to B$ and $\widehat{f} : \widehat{M} \to B$ with fibres $X_b = f^{-1}(b)$ and $\widehat{X}_b = \widehat{f}^{-1}(b)$ for $b \in B$, such that*

(i) *There exists a dense open set $B_0 \subset B$, such that for each $b \in B_0$, the fibres $X_b, \widehat{X}_b$ are nonsingular special Lagrangian 3-tori T^3 in M and $\widehat{M}$, which are in some sense dual to one another.*

(ii) *For each* $b \in \Delta = B \setminus B_0$, *the fibres* X_b, $\widehat{X}_b$ *are expected to be singular special Lagrangian 3-folds in* M *and* $\widehat{M}$.

We call $f, \widehat{f}$ *special Lagrangian fibrations*, and the set of singular fibres Δ is called the *discriminant*. It is not yet clear what the final form of the SYZ Conjecture should be. Much work has been done on it, working primarily with *Lagrangian* fibrations, by authors such as Mark Gross and Wei-Dong Ruan. For references see [**10**].

The author's approach to the SYZ Conjecture, focussing primarily on *special Lagrangian singularities*, is set out in [**10**], and we do not have space to discuss it here. Very briefly, we argue that for generic (almost) Calabi–Yau 3-folds (ii) will not hold, as the discriminants $\Delta, \widehat{\Delta}$ of $f, \widehat{f}$ cannot be homeomorphic near certain kinds of singular fibre. We also suggest that the final form of the SYZ Conjecture should be an *asymptotic statement* about 1-*parameter families* of Calabi-Yau 3-folds approaching the *large complex structure limit.*

Problem 8.8. Study *special Lagrangian fibrations* $f : M \to B$ of almost Calabi–Yau 3-folds (M, J, ω, Ω), particularly when ω is *generic* in its Kähler class. Clarify/prove/disprove the SYZ Conjecture.

Note that the ideas of §8.1 will be helpful here. As B has dimension 3, we see that $\operatorname{ind}(X_b) \leqslant 3$ for all $b \in \Delta$. If Conjecture 5.6 holds, ω is generic, and $f^{-1}(b)$ has isolated conical singularities, then X_b is *transverse.* We can then use Theorem 8.2 or [**22**, Theorem 8.10] to calculate $\operatorname{ind}(X_b)$, and $\operatorname{ind}(X_b) \leqslant 3$ will severely restrict the possible singular behavior.

8.4. Invariants from counting special Lagrangian homology spheres. In [**6**] the author proposed to define an invariant of almost Calabi–Yau 3-folds (M, J, ω, Ω) by counting special Lagrangian rational homology 3-spheres N (which occur in 0-dimensional moduli spaces) in a given homology class, with a certain topological weight. This invariant will only be interesting if it is conserved under deformations of the underlying almost Calabi–Yau 3-fold, or at least transforms in a rigid way as the cohomology classes $[\omega], [\Omega]$ change.

During such a deformation, nonsingular SL 3-folds can develop singularities and disappear, or new ones appear, which might change the invariant. In [**6**] the author showed that if we count rational SL homology spheres N with weight $|H_1(N, \mathbb{Z})|$, then under two kinds of singular behavior of SL 3-folds, the resulting invariant is independent of $[\omega]$, and transforms according to certain rules as $[\Omega]$ crosses real hypersurfaces in complex structure moduli space where phases of $\alpha, \beta \in H_3(M, \mathbb{Z})$ become equal.

Again, the ideas of §8.1 will be helpful here. It is enough for us to study how the invariant changes along *generic* 1-*parameter families* of almost Calabi–Yau 3-folds. The only kinds of singularities of SL homology 3-spheres that arise in such families will have index 1. So if we can complete the index 1 classification in Problem 8.3, we should be able to resolve the conjectures of [**6**].

In fact, I now believe that interesting invariants of almost Calabi–Yau m-folds by 'counting' SL m-folds can be defined for all $m \geqslant 3$. The definition, properties and transformation laws of these invariants are formidably complex and difficult, even to state. The best approach I have to them is to use Homological Mirror Symmetry to translate the problem to the derived category $\mathcal{T} = D^b(\mathrm{Fuk}(M, \omega))$ of the Fukaya category of (M, ω).

Then SL m-folds conjecturally correspond to *stable objects* of the triangulated category $\mathcal{T}$, under a stability condition à la Tom Bridgeland. The invariants are Euler characteristics of moduli spaces of *configurations* in $\mathcal{T}$, which are finite collections of (stable or semistable) objects and morphisms in $\mathcal{T}$ satisfying some axioms. In this set-up, using algebra and category theory, I can rigorously develop the definition and properties of the invariants, and their transformation rules under change of stability condition (effectively, deformation of J, Ω). I am writing (yet) another series of papers about this.

Problem 8.9. Try to use moduli spaces of compact SL m-folds (possibly immersed, or singular) to define systems of invariants of an almost Calabi–Yau m-fold (M, J, ω, Ω) for $m \geqslant 3$. These invariants should be defined for ω generic in its Kähler class, and the key property we want is that they should be *independent of* ω. Compute the invariants for the quintic. Calculate the transformation rules for the invariants under deformation of J, Ω. Relate them to Homological Mirror Symmetry, and to 'branes' in String Theory.

Acknowledgements. Many people have helped me develop my ideas on special Lagrangian geometry. Amongst them I would like to thank Tom Bridgeland, Adrian Butscher, Mark Haskins, Nigel Hitchin, Stephen Marshall, Ian McIntosh, Sema Salur and Richard Thomas.

References

1. G.E. Bredon, *Topology and Geometry*, Graduate Texts in Mathematics **139**, Springer-Verlag, Berlin, 1993.
2. R. Harvey, *Spinors and Calibrations*, Academic Press, San Diego, 1990.
3. R. Harvey and H. B. Lawson, *Calibrated geometries*, Acta Math. **148** (1982), 47–157.
4. M. Haskins, *Special Lagrangian Cones*, math.DG/0005164, 2000. [To appear in Am. J. Math.]
5. M. Haskins, *The geometric complexity of special Lagrangian T^2-cones*, math.DG/0307129, 2003. [To appear in Invent. Math.]
6. D. D. Joyce, *On counting special Lagrangian homology 3-spheres*, pages 125–151 in *Topology and Geometry: Commemorating SISTAG*, editors A.J. Berrick, M.C. Leung and X.W. Xu, Contemporary Mathematics **314**, Am. Math. Soc., Providence, RI, 2002. hep-th/9907013, 1999.
7. D. D. Joyce, *Special Lagrangian m-folds in $\mathbb{C}^m$ with symmetries*, Duke Math. J. **115** (2002), 1-51. math.DG/0008021.
8. D. D. Joyce, *Constructing special Lagrangian m-folds in $\mathbb{C}^m$ by evolving quadrics*, Math. Ann. **320** (2001), 757–797. math.DG/0008155.
9. D. D. Joyce, *Evolution equations for special Lagrangian 3-folds in $\mathbb{C}^3$*, Ann. Global Anal. Geom. **20** (2001), 345–403. math.DG/0010036.
10. D. D. Joyce, *Singularities of special Lagrangian fibrations and the SYZ Conjecture*, Comm. Anal. Geom. **11** (2003), 859–907. math.DG/0011179.
11. D. D. Joyce, *Ruled special Lagrangian 3-folds in $\mathbb{C}^3$*, Proc. London Math. Soc. **85** (2002), 233–256. math.DG/0012060.
12. D. D. Joyce, *Special Lagrangian 3-folds and integrable systems*, math.DG/0101249, 2001. [To appear in volume 1 of the Proceedings of the Mathematical Society of Japan's 9th International Research Institute on *Integrable Systems in Differential Geometry*, Tokyo, 2000.]
13. D. D. Joyce, *Lectures on Calabi–Yau and special Lagrangian geometry*, math.DG/0108088, 2001. Published, with extra material, as Part I of M. Gross, D. Huybrechts and D. Joyce, *Calabi–Yau Manifolds and Related Geometries*, Universitext series, Springer, Berlin, 2003.
14. D. D. Joyce, U(1)-*invariant special Lagrangian 3-folds. I. Nonsingular solutions*, math.DG/0111324, 2001. [To appear in Adv. Math.]
15. D. D. Joyce, U(1)-*invariant special Lagrangian 3-folds. II. Existence of singular solutions*, math.DG/0111326, 2001. [To appear in Adv. Math.]
16. D. D. Joyce, U(1)-*invariant special Lagrangian 3-folds. III. Properties of singular solutions*, math.DG/0204343, 2002. [To appear in Adv. Math.]
17. D. D. Joyce, U(1)-*invariant special Lagrangian 3-folds in $\mathbb{C}^3$ and special Lagrangian fibrations*, Turkish Math. J. **27** (2003), 99–114. math.DG/0206016.
18. D. D. Joyce, *Special Lagrangian submanifolds with isolated conical singularities. I. Regularity*, Ann. Global Anal. Geom. **25** (2004), 201–251. math.DG/0211294.
19. D. D. Joyce, *Special Lagrangian submanifolds with isolated conical singularities. II. Moduli spaces*, Ann. Global Anal. Geom. **25** (2004), 301–352. math.DG/0211295.

20. D. D. Joyce, *Special Lagrangian submanifolds with isolated conical singularities. III. Desingularization, the unobstructed case*, Ann. Global Anal. Geom. **26** (2004), 1–58. math.DG/0302355.
21. D. D. Joyce, *Special Lagrangian submanifolds with isolated conical singularities. IV. Desingularization, obstructions and families*, math.DG/0302356, 2003. [To appear in Ann. Global Anal. Geom.]
22. D.D. Joyce, *Special Lagrangian submanifolds with isolated conical singularities. V. Survey and applications*, J. Differ. Geom. **63** (2003), 279–347. math.DG/0303272.
23. G. Lawlor, *The angle criterion*, Invent. Math. **95** (1989), 437–446.
24. R. Lockhart, *Fredholm, Hodge and Liouville Theorems on noncompact manifolds*, Trans. Am. Math. Soc. **301** (1987), 1–35.
25. S. P. Marshall, *Deformations of special Lagrangian submanifolds*, Oxford D. Phil. Thesis, 2002.
26. D. McDuff and D. Salamon, *Introduction to symplectic topology*, second ed., Oxford University Press, Oxford, 1998.
27. I. McIntosh, *Special Lagrangian cones in $\mathbb{C}^3$ and primitive harmonic maps*, J. London Math. Soc. **67** (2003), 769–789. math.DG/0201157.
28. R. C. McLean, *Deformations of calibrated submanifolds*, Comm. Anal. Geom. **6** (1998), 705–747.
29. F. Morgan, *Geometric Measure Theory, A Beginner's Guide*, Academic Press, San Diego, 1995.
30. T. Pacini, *Deformations of Asymptotically Conical Special Lagrangian Submanifolds*, math.DG/0207144, 2002.
31. A. Strominger, S.-T. Yau, and E. Zaslow, *Mirror symmetry is T-duality*, Nucl. Phys. **B479** (1996), 243–259. hep-th/9606040.

Curvature Functionals, Optimal Metrics, and the Differential Topology of 4–Manifolds

Claude LeBrun†
SUNY at Stony Brook
New York, USA

This paper investigates the question of which smooth compact 4-manifolds admit Riemannian metrics that minimize the L^2-norm of the curvature tensor. Metrics with this property are called *optimal*; Einstein metrics and *scalar-flat anti-self-dual* metrics provide us with two interesting classes of examples. Using twistor methods, optimal metrics of the second type are constructed on the connected sums $k\overline{\mathbb{CP}}_2$ for $k > 5$. However, related constructions also show that large classes of simply connected 4-manifolds *do not* admit any optimal metrics at all. Interestingly, the difference between existence and nonexistence turns out to delicately depend on one's choice of smooth structure; there are smooth 4-manifolds which carry optimal metrics, but which are homeomorphic to infinitely many distinct smooth 4-manifolds on which no optimal metric exists.

---

†Supported in part by NSF grant DMS-0305865.

1. Introduction

"Does every smooth compact manifold admit a best metric?" René Thom allegedly first posed this naïve but fundamental question to Marcel Berger [**10**] at some point in the early 1960s. By the early 1990s, it had emerged from the world of informal discussion to find itself in print, as the leading entry on one of S.-T. Yau's celebrated problem lists [**61**].

Of course, Thom's question, as formulated above, seems to be less a problem than a *meta-problem*; after all, we are being asked to find some interpretation of the word "best" which will lead to an interesting conclusion. Nevertheless, the question always had some unambiguous content, because Thom had clarified his question by means of a paradigmatic example. What he of course had in mind was the classical uniformization theorem, which tells us that every compact 2-manifold carries metrics of constant curvature. This paradigm moreover gives us some vital clues concerning what we ought to look for. First, the definition of "best metric" should somehow involve the Riemannian curvature, and should be invariant under the action of the diffeomorphism group. Second, one might hope that metrics of constant sectional curvature, when they exist, would turn out to be the "best metrics" on the manifold in question. And third, we should not expect our "best" metric to necessarily be absolutely unique; a finite-dimensional moduli space of "best metrics" would certainly be quite acceptable.

If we agree that the flat metrics are the best metrics on the n-dimensional torus $T^n = S^1 \times \cdots \times S^1$, then it seems rather natural to look for metrics on other manifolds which are "as flat as possible," in the sense that they minimize some norm of curvature. For example, one might try to minimize the L^p-norm of the Riemann curvature tensor for some fixed $p > 1$. However, this is simply not a sensible problem for most choices of p; one can typically find a sequence of metrics for which the L^p norm of curvature tends to zero by just multiplying a fixed metric by a suitable sequence of constants. Indeed, there is only one value of p for which this trick does *not* work: namely, $p = n/2$, where n is the dimension of the manifold.

Given a smooth compact n-manifold M, and letting

$$\mathscr{G}_M = \{ \text{ smooth Riemannian metrics } g \text{ on } M\},$$

we are thus led to consider the functional

$$\mathcal{K} : \mathscr{G}_M \longrightarrow \mathbb{R}$$

given by

$$\mathcal{K}(g) = \int_M |\mathcal{R}_g|_g^{n/2} d\mu_g,$$

where $\mathcal{R}$ denotes the Riemann curvature tensor, $|\mathcal{R}|$ is its point-wise norm with respect to the metric, and $d\mu$ is the n-dimensional volume measure determined by the metric. Berger **[10]** has suggested the minima of $\mathcal{K}$ as natural candidates for Thom's "best metrics." Let us codify this proposal:

Definition 1.1. Let M be a smooth compact n-dimensional manifold, $n \geqslant 3$. A smooth Riemannian metric g on M will be called an *optimal metric* if it is an absolute minimizer of the functional $\mathcal{K}$, in the sense that

$$\mathcal{K}(g') \geqslant \mathcal{K}(g)$$

for every smooth Riemannian metric g' on M.

Notice that we have defined an optimal metric to be a *minimum*, not just a critical point, of the functional $\mathcal{K}$. This brings into play a natural diffeomorphism invariant which is defined even in the absence of an optimal metric:

Definition 1.2. For any smooth compact n-dimensional manifold M, we define $\mathcal{I}_{\mathcal{R}}(M)$ to be the nonnegative real number given by

$$\mathcal{I}_{\mathcal{R}}(M) = \inf_{g \in \mathcal{G}_M} \mathcal{K}(g) = \inf_g \int_M |\mathcal{R}_g|_g^{n/2} d\mu_g.$$

Thus $\mathcal{I}_{\mathcal{R}}$ coincides with the number Berger **[10]** calls $\min ||\mathcal{R}||^{n/2}$. Of course, our definitions have precisely been chosen so that any metric on g on M automatically satisfies

$$\mathcal{K}(g) \geqslant \mathcal{I}_{\mathcal{R}}(M),$$

with equality iff g is an optimal metric.

While it generally remains unclear to what extent optimal metrics really represent an appropriate response to Thom's question, the situation in dimension 4 is rather encouraging. In particular **[10, 11]**, an Einstein metric on any compact 4-manifold is optimal. However, the converse is by no means true, and the primary purpose of this article is to explore this aspect of the problem. Here is what will emerge:

- We will construct infinitely many new examples of non-Einstein optimal metrics on simply connected compact 4-manifolds.

- We will show that there are many simply connected compact 4-manifolds which do not admit optimal metrics.
- We will see that the existence or nonexistence of optimal metrics depends strictly on the diffeotype of a simply connected 4-manifold; it is not determined by the homeotype alone.
- We will calculate the invariant $\mathcal{I}_{\mathcal{R}}$ for many simply connected 4-manifolds (some common-garden, others a bit more exotic).
- We will show that the value of $\mathcal{I}_{\mathcal{R}}$ depends strictly on the diffeotype of a 4-manifold. Different differentiable structures on an underlying topological 4-manifold can often be distinguished by the fact that the corresponding values of $\mathcal{I}_{\mathcal{R}}$ are different. However, we will also see many examples of distinct differentiable structures which *cannot* be distinguished in this way.

2. Four-Dimensional Geometry

Our investigation of optimal metrics on 4-manifolds will necessarily presuppose a certain familiarity with the rudiments of 4-dimensional geometry and topology. The present section will attempt to offer a quick introduction to some of this essential background material.

The notion of self-duality plays a fundamental role in four-dimensional Riemannian geometry. If (M, g) is an oriented Riemannian 4-manifold, the Hodge star operator

$$\star : \Lambda^2 \to \Lambda^2$$

satisfies $\star^2 = 1$, and so yields a decomposition

$$\Lambda^2 = \Lambda^+ \oplus \Lambda^-, \tag{1}$$

where Λ^+ is the $(+1)$-eigenspace of $\star$, and Λ^- is the (-1)-eigenspace. Both Λ^+ and Λ^- are rank-3 vector bundles over M. Reversing the orientation of M interchanges these two bundles.

Definition 2.1. On any smooth oriented 4-manifold, sections of Λ^+ are called *self-dual 2-forms*, whereas sections of Λ^- are called *anti-self-dual 2-forms*.

Because the curvature of any connection is a bundle-valued 2-form, the decomposition (1) allows one to break any curvature tensor up into more primitive pieces. This idea has particularly important ramifications when applied to the Riemannian curvature of the metric itself. Indeed, first notice that, by raising an index, the Riemann curvature tensor may be reinterpreted as a linear map $\Lambda^2 \to \Lambda^2$, called the *curvature operator*. But decomposing the 2-forms according to (1) then allows us to view this linear map as consisting of four blocks:

$$\mathcal{R} = \left(\begin{array}{c|c} W_+ + \frac{s}{12} & \overset{\circ}{r} \\ \hline \overset{\circ}{r} & W_- + \frac{s}{12} \end{array} \right). \tag{2}$$

Here $W_\pm$ are the trace-free pieces of the appropriate blocks, and are called the self-dual and anti-self-dual Weyl curvatures respectively. The scalar curvature s is understood to act by scalar multiplication, whereas the trace-free Ricci curvature $\overset{\circ}{r} = r - \frac{s}{4}g$ acts on 2-forms by

$$\varphi_{ab} \mapsto \overset{\circ}{r}_{ac}\varphi^c{}_b - \overset{\circ}{r}_{bc}\varphi^c{}_a.$$

An important feature of the decomposition (1) is that it is *conformally invariant*, in the sense that it is unchanged if g is replaced by $u^2 g$, where u is an arbitrary smooth positive function. Similarly, the self-dual and anti-self-dual Weyl curvatures are also conformally invariant (when considered as sections of $\Lambda^2 \otimes \mathrm{End}(TM)$).

Since our objective is to better understand metrics on 4-manifolds which minimize the quadratic curvature integral

$$\mathcal{K}(g) = \int |\mathcal{R}|^2 d\mu,$$

it is highly relevant that there are two other quadratic curvature integrals which actually compute topological invariants. Indeed, no matter which metric g we choose on a smooth compact oriented 4-manifold M, the generalized Gauss–Bonnet theorem [**1**] tells us that the Euler characteristic is given by

$$\chi(M) = \frac{1}{8\pi^2} \int_M \left(\frac{s^2}{24} + |W_+|^2 + |W_-|^2 - \frac{|\overset{\circ}{r}|^2}{2} \right) d\mu, \tag{3}$$

while the Hirzebruch signature theorem [**31**] tells us that the *signature* is given by

$$\tau(M) = \frac{1}{12\pi^2}\int_M \left(|W_+|^2 - |W_-|^2\right) d\mu \ . \tag{4}$$

Let us recall that the signature of a smooth compact 4-manifold may be defined in terms of the *intersection pairing*

$$\begin{array}{rcl} \smile : H^2(M,\mathbb{R}) \times H^2(M,\mathbb{R}) & \longrightarrow & \mathbb{R} \\ ([\varphi]\,,\,[\psi]) & \mapsto & \int_M \varphi \wedge \psi \end{array}$$

on de Rham cohomology. By Poincaré duality, this is a non-degenerate pairing; and it is symmetric, since 2-forms commute with respect to the wedge product. We may therefore find a basis for $H^2(M,\mathbb{R})$ in which the intersection pairing is represented by the diagonal matrix

$$\left[\begin{array}{cc} \underbrace{\begin{matrix} 1 & & \\ & \ddots & \\ & & 1 \end{matrix}}_{b_+(M)} & \\ & b_-(M)\left\{\begin{matrix} -1 & & \\ & \ddots & \\ & & -1 \end{matrix}\right. \end{array}\right],$$

and the numbers $b_\pm(M)$ are then topological invariants of M. Their difference

$$\tau(M) = b_+(M) - b_-(M)$$

is the signature of M, whereas their sum

$$b_2(M) = b_+(M) + b_-(M)$$

is just the second Betti number.

A more concrete interpretation of the numbers $b_\pm(M)$ can be given by using a bit of Hodge theory. Since every de Rham class on M has a unique harmonic representative with respect to g, we have a canonical identification

$$H^2(M,\mathbb{R}) = \{\varphi \in \Gamma(\Lambda^2) \mid d\varphi = 0,\ d\star\varphi = 0\}.$$

But the Hodge star operator $\star$ defines an involution of the right-hand side. We thus obtain a direct sum decomposition

$$H^2(M, \mathbb{R}) = \mathcal{H}_g^+ \oplus \mathcal{H}_g^-, \tag{5}$$

where

$$\mathcal{H}_g^\pm = \{\varphi \in \Gamma(\Lambda^\pm) \mid d\varphi = 0\}$$

are the spaces of self-dual and anti-self-dual harmonic forms. The intersection form is then positive-definite on $\mathcal{H}_g^+$, and negative-definite on $\mathcal{H}_g^-$, so we have

$$b_\pm(M) = \dim \mathcal{H}_g^\pm.$$

Notice that the spaces $\mathcal{H}_g^\pm$ only depend on the conformal class of the metric.

One can easily construct 4-manifolds with any desired values of b_+ and b_- by means of the following construction:

Definition 2.2. Let M_1 and M_2 be smooth connected compact oriented n-manifolds.

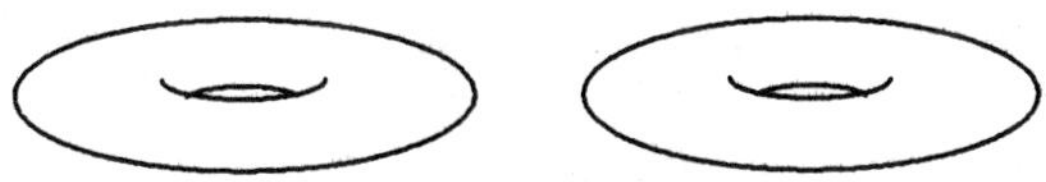

Their connected sum $M_1 \# M_2$ is then the smooth connected oriented n-manifold obtained by deleting a small ball from each manifold

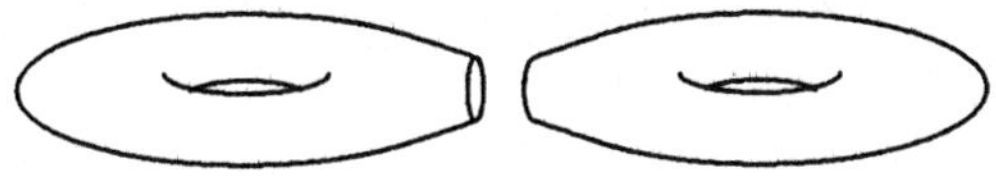

and identifying the resulting S^{n-1} boundaries

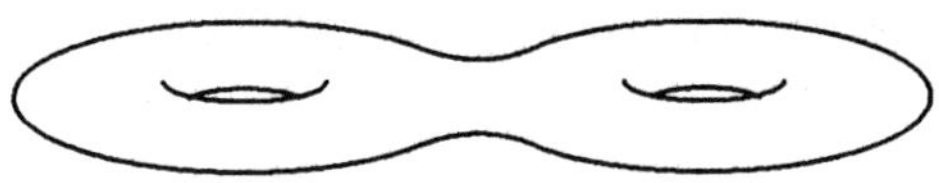

via a reflection.

If M_1 and M_2 are simply connected 4-manifolds, then $M = M_1 \# M_2$ is also simply connected, and has $b_\pm(M) = b_\pm(M_1) + b_\pm(M_2)$. Now let us use $\mathbb{CP}_2$ denote the complex projective plane with its *standard* orientation, and $\overline{\mathbb{CP}}_2$ denote the same smooth 4-manifold with the *opposite* orientation. Then the iterated connected sum

$$j\mathbb{CP}_2 \# k\overline{\mathbb{CP}}_2 = \underbrace{\mathbb{CP}_2 \# \cdots \# \mathbb{CP}_2}_{j} \# \underbrace{\overline{\mathbb{CP}}_2 \# \cdots \# \overline{\mathbb{CP}}_2}_{k}$$

is a simply connected 4-manifold with $b_+ = j$ and $b_- = k$. Notice that $\chi(j\mathbb{CP}_2 \# k\overline{\mathbb{CP}}_2) = 2 + j + k$ and that $\tau(j\mathbb{CP}_2 \# k\overline{\mathbb{CP}}_2) = j - k$.

These 4-manifolds are *non-spin*, meaning that their tangent bundles have $w_2 \neq 0$. For a simply connected compact 4-manifold M, this is equivalent to saying that M contains a compact oriented surface of odd self-intersection.

Is this a complete list of the simply connected non-spin 4-manifolds? Well, yes and no. In the affirmative direction, the remarkable work of Michael Freedman **[20]** tells us the following:

Theorem 2.3 (Freedman). *Two smooth compact simply connected oriented 4-manifolds are orientedly homeomorphic if and only if*

- *they have the same value of* b_+;
- *they have the same value of* b_-; *and*
- *both are spin, or both are non-spin.*

Thus, up to *homeomorphism*, the connected sums $j\mathbb{CP}_2 \# k\overline{\mathbb{CP}}_2$ provide us with a complete list of the simply connected *non-spin* 4-manifolds. However, many of these topological 4-manifolds turn out to have infinitely many distinct *known* smooth structures, and it is generally thought that many of these manifolds will turn out to have exotic smooth structures that no one has even yet imagined.

Freedman's surgical techniques also allow one to classify simply connected *topological* 4-manifolds, but the classification is much more involved. One of the key ingredients that makes it possible to give Theorem 2.3 the simple phrasing used above is the main theorem of Donaldson's thesis **[16]**, which showed that the anti-self-dual Yang–Mills equations implied previously unsuspected constraints on the homotopy types of smooth 4-manifolds:

Theorem 2.4 (Donaldson). *Let M be any smooth compact simply connected 4-manifold with $b_+ = 0$. Then M is homotopy equivalent to S^4 or to a connected sum $k\overline{\mathbb{CP}}_2$.*

In particular, if M is a simply connected differentiable 4-manifold with $b_+(M) = 0$ and $b_-(M) \neq 0$, this result tells us that M cannot be spin.

We have just seen that Theorem 2.3 allows us to compile a complete list of simply connected non-spin homeotypes. In the spin case, the situation remains more unsettled, but a conjectural complete list of smoothable simply connected spin homeotypes consists of S^4, $S^2 \times S^2$, $K3$, their connect sums, and orientation reverses of these. Here $K3$ means the unique simply connected smooth compact 4-manifold admitting a complex structure with $c_1 = 0$. This 4-manifold has $b_+ = 3$ and $b_- = 19$. An interesting model of $K3$ was discovered by Kummer, who considered the involution of T^4 with 16 fixed points which arises as the product of two copies of the Weierstrass involution of an elliptic curve:

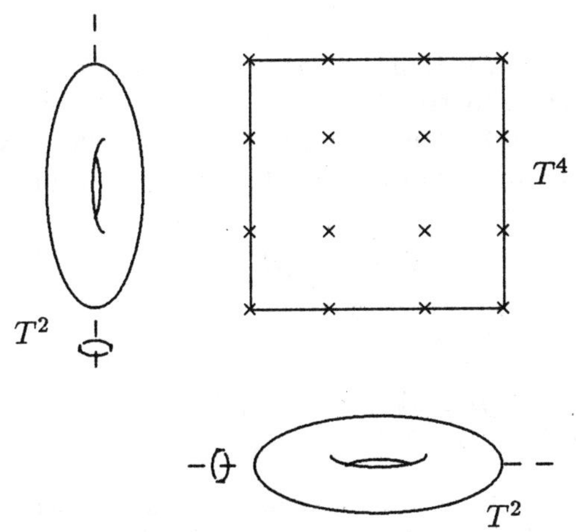

Kummer's model of $K3$ is then obtained from the orbifold $T^4/\mathbb{Z}_2$ by replacing each singular point with a $\mathbb{CP}_1$ of self-intersection -2. Analogous constructions will turn out to play a central role in this paper.

3. Optimal Geometries in Dimension Four

A Riemannian metric is said to be *Einstein* if it has constant Ricci curvature. Since the Ricci curvature of g is by definition the function $v \mapsto r(v,v)$ on the unit tangent bundle $g(v,v) = 1$, where r denotes the Ricci tensor, this is clearly equivalent to the requirement that

$$r = \lambda g$$

for some constant λ. This in turn can be rewritten as the pair of conditions

$$\mathring{r} = 0, \quad s = \text{constant},$$

where $s = \text{trace}_g(r)$ is the scalar curvature, and where $\mathring{r} = r - \frac{s}{n} g$ is the trace-free part of the Ricci tensor. However, double contraction of the second Bianchi identity tells us that

$$\nabla \cdot \mathring{r} = \frac{n-2}{n}\, ds,$$

so, in any dimension $n \neq 2$, a metric is Einstein iff it satisfies the equation

$$\mathring{r} = 0.$$

An important motivation for the study of optimal metrics is that **[10, 11]** any Einstein metric is optimal in dimension 4. Indeed, let M be a smooth compact 4-manifold, and let g be an arbitrary Riemannian metric on M. Then the 4-dimensional Gauss–Bonnet formula (3) allows us to rewrite

$$\mathcal{K}(g) = \int_M |\mathcal{R}|^2 d\mu_g$$

as

$$\mathcal{K}(g) = 8\pi^2 \chi(M) + \int_M |\mathring{r}_g|^2 d\mu_g \,. \qquad (6)$$

Thus any metric g with $\mathring{r} = 0$ minimizes $\mathcal{K}$; and when such a metric exists, the Einstein metrics are the *only* optimal metrics on M.

However, similar arguments also show that smooth compact 4-manifolds often do not admit Einstein metrics, even in the simply connected case. Indeed, a judicious combination of (3) with (4) reads

$$(2\chi + 3\tau)(M) = \frac{1}{4\pi^2} \int_M \left(\frac{s^2}{24} + 2|W_+|^2 - \frac{|\mathring{r}|^2}{2} \right) d\mu_g, \qquad (7)$$

so that a compact oriented 4-manifold M can only admit an Einstein metric if it satisfies the *Hitchin–Thorpe inequality* [**32, 57**]

$$(2\chi + 3\tau)(M) \geqslant 0.$$

Thus, for example, the simply connected 4-manifolds $k\overline{\mathbb{CP}}_2$ do not admit an Einstein metric when $k > 4$, since these spaces have $2\chi + 3\tau = 4 - k$.

In the boundary case of the Hitchin–Thorpe inequality, we also get a striking amount of additional information. Indeed, if $(2\chi + 3\tau)(M) = 0$ and M admits an Einstein metric g, then the Riemannian connection on $\Lambda^+ \to M$ must be flat, since its curvature tensor is algebraically determined by r and W_+; such a metric is said to be *locally hyper-Kähler*. Any locally hyper-Kähler 4-manifold is finitely covered [**11, 32**] by a flat 4-torus or a Calabi–Yau $K3$. In particular, a simply connected 4-manifold with $2\chi + 3\tau = 0$ can admit an Einstein metric only if it is diffeomorphic to $K3$. In particular, $4\overline{\mathbb{CP}}_2$ does not admit an Einstein metric, since it has $2\chi + 3\tau = 0$, but is not even homotopy equivalent to $K3$.

It should therefore seem a bit reassuring that there are many simply connected 4-manifolds which carry optimal metrics which are not Einstein [**45**]. To see this, we may begin by observing that (3) and (4) allow one to re-express $\mathcal{K}$ as

$$\mathcal{K}(g) = -8\pi^2(\chi + 3\tau)(M) + 2\int_M \left(\frac{s^2}{24} + 2|W_+|^2\right) d\mu_g\,. \tag{8}$$

This brings another important class of metrics to the fore:

Definition 3.1. If M is a smooth oriented 4-manifold, a Riemannian metric g on M will be said to be *anti-self-dual* (or, for brevity, *ASD*) if its self-dual Weyl curvature is identically zero:

$$W_+ \equiv 0.$$

A metric g will be called *scalar-flat* (or, more briefly, *SF*) if it satisfies

$$s = 0.$$

Finally, we will say that g is *scalar-flat anti-self-dual* (or *SFASD*) if it satisfies both of these conditions.

Equation (8) then immediately yields the following:

Proposition 3.2. *Suppose that M is a smooth compact oriented 4-manifold. If M carries a scalar-flat anti-self-dual metric g, then g is* *optimal. When this happens, moreover, every other optimal metric g' on M is SFASD, too.*

This fact seems to have first been noticed by Lafontaine [**42**], who simultaneously discovered an important obstruction to the existence of SFASD metrics. Indeed, equation (7) and our previous discussion of the locally hyper-Kähler manifolds immediately gives us the following upside-down version of the Hitchin–Thorpe inequality:

Proposition 3.3 (Lafontaine). *Let (M, g) be a compact scalar-flat anti-self-dual 4-manifold. Then*

$$(2\chi + 3\tau)(M) \leqslant 0,$$

with equality if and only if (M, g) is finitely covered by a flat 4-torus or a Calabi–Yau $K3$.

However, a completely different set of topological constraints is imposed by the following result [**43**]:

Proposition 3.4. *Let (M, g) be a compact scalar-flat anti-self-dual 4-manifold. Then either*

- $b_+(M) = 0$; *or*
- $b_+(M) = 1$, *and g is a scalar-flat Kähler metric; or else*
- $b_+(M) = 3$, *and g is a hyper-Kähler metric.*

PROOF. Recall that $b_+(M)$ is exactly the dimension of the space of harmonic self-dual 2-forms on M. However, any self-dual 2-form φ on any Riemannian 4-manifold satisfies the Weitzenböck formula [**13**]

$$(d + d^*)^2\varphi = \nabla^*\nabla\varphi - 2W_+(\varphi, \cdot) + \frac{s}{3}\varphi,$$

so if φ is harmonic and if g is SFASD we obtain

$$0 = \int_M \langle\varphi, \nabla^*\nabla\varphi\rangle d\mu = \int_M |\nabla\varphi|^2 d\mu.$$

Thus, when (M, g) is a compact SFASD manifold, $b_+(M)$ is exactly the dimension of the space of the parallel self-dual 2-forms.

Now $SO(4)$ is a double cover of $SO(3) \times SO(3)$, where the factor projections to $SO(3)$ are given by its action on $\Lambda^\pm$. The subgroup of $SO(4)$ stabilizing a nonzero element of Λ^+ is thus the double cover $U(2)$ of $SO(2) \times SO(3)$, whereas the subgroup acting trivially on Λ^+, or even on a 2-dimensional subspace of it, is the universal cover $SU(2) = Sp(1)$

of $SO(3)$. Thus an oriented Riemannian 4-manifold with a non-trivial parallel self-dual 2-form has holonomy $\subset U(2)$, and is Kähler, whereas the existence of 2 independent parallel self-dual 2-forms would force the manifold to have holonomy $\subset Sp(1)$, and so to be hyper-Kähler. □

Let us now assume that (M, g) is a *simply connected* SFASD manifold, and see what the above results now tell us. By Proposition 3.4, the only possibilities for $b_+(M)$ are $0, 1$ and 3. If $b_+ = 3$, we would have a simply connected hyper-Kähler manifold; such an object is necessarily a $K3$ surface. If $b_+ = 1$, we have a simply connected complex surface with nontrivial canonical line bundle and a Kähler metric of zero scalar curvature; by a plurigenus vanishing theorem of Yau [**59**] and the Enriques–Kodaira classification [**7**], such a complex surface must be obtained from $\mathbb{CP}_2$ by blowing up and down, and hence diffeomorphic to either $S^2 \times S^2$ or a connected sum $\mathbb{CP}_2 \# k\overline{\mathbb{CP}}_2$; and since Proposition 3.3 tells us that $2\chi + 3\tau = 4 - 4b_1 + 5b_+ - b_- = 9 - b_-$ is negative, we conclude in this case that M is diffeomorphic to $\mathbb{CP}_2 \# k\overline{\mathbb{CP}}_2$ for some $k > 9$. Finally, if $b_+ = 0$, Theorems 2.4 and 2.3 tell us that M is at least *homeomorphic* to either S^4 or a connected sum $k\overline{\mathbb{CP}}_2$; and since Proposition 3.3 tells us that $2\chi + 3\tau = 4 - b_-$ is negative, we conclude in this case that M is homeomorphic to $k\overline{\mathbb{CP}}_2$ for some $k > 4$. Summarizing, we have [**43**]

Proposition 3.5. *Let M be a smooth compact simply connected 4-manifold. If M admits a scalar-flat anti-self-dual metric g, then*

- *M is homeomorphic to $k\overline{\mathbb{CP}}_2$ for some $k \geqslant 5$; or*
- *M is diffeomorphic to $\mathbb{CP}_2 \# k\overline{\mathbb{CP}}_2$ for some $k \geqslant 10$; or else*
- *M is diffeomorphic to $K3$.*

A major objective of this paper is to prove the following partial converse:

Theorem A. *A simply connected 4-manifold M admits scalar-flat anti-self-dual metrics if*

- *M is diffeomorphic to $k\overline{\mathbb{CP}}_2$ for some $k \geqslant 6$; or*
- *M is diffeomorphic to $\mathbb{CP}_2 \# k\overline{\mathbb{CP}}_2$ for some $k \geqslant 14$; or*
- *M is diffeomorphic to $K3$.*

In particular, each of these manifolds carries optimal metrics.

It is worth emphasizing that, except in the $K3$ case, the optimal metrics of Theorem A are necessarily non-Einstein.

On the other hand, Corollary 3.5 and a computation of $\mathcal{I}_{\mathcal{R}}$ will allow us to show that the existence of optimal metrics is highly sensitive to the choice of smooth structure:

Theorem B. *For each $k \geqslant 9$, the topological manifold $\mathbb{CP}_2\#k\overline{\mathbb{CP}}_2$ admits infinitely many "exotic" smooth structures for which the corresponding smooth compact 4-manifold does not admit optimal metrics. Similarly, the topological manifold $K3$ admits infinitely many exotic smooth structures for which the corresponding smooth 4-manifold does not admit optimal metrics.*

Similar ideas will also allow us to prove the nonexistence of optimal metrics for smooth manifolds representing many more homeotypes:

Theorem C. *If $j \geqslant 2$ and $k \geqslant 9j$, the smooth simply connected 4-manifold $j\mathbb{CP}_2\#k\overline{\mathbb{CP}}_2$ does not admit optimal metrics. Moreover, if $j \geqslant 5$ and $j \not\equiv 0 \bmod 8$, the underlying topological manifold of this space admits infinitely many distinct differentiable structures for which no optimal metric exists.*

4. Constructing Anti-Self-Dual Metrics

The condition of anti-self-duality is *conformally invariant*; if g is an ASD metric, so is u^2g, for any $u > 0$. The strategy of our proof of Theorem A will be to first construct a family of anti-self-dual conformal classes of metrics on $k\overline{\mathbb{CP}}_2$, $k \geqslant 6$, and then show that some of the constructed conformal classes contain scalar-flat metrics. Our approach to both aspects of this problem will be carried out using methods of complex analysis via the Penrose *twistor correspondence*, to which we now provide a brief introduction.

Given any oriented Riemannian 4-manifold (M, g), one can construct an associated almost-complex 6-manifold (Z, J), where $\pi : Z \to M$ is the S^2-bundle $S(\Lambda^+)$ of unit self-dual 2-forms. The almost-complex structure $J : TZ \to TZ$ preserves the decomposition of TZ into horizontal and vertical components with respect to the Levi–Civita connection. On the tangent spaces of each fiber S^2, J simply acts by rotation by $-90°$. Meanwhile, in the horizontal sub-bundle, which we identify with π^*TM, J acts at $\phi \in S(\Lambda^+)$ by $v \mapsto \sqrt{2}(v \lrcorner \phi)^\sharp$. Each fiber S^2 of $S(\Lambda^+) \to M$ is thus a J-holomorphic curve, and the fiber-wise antipodal map $\sigma : S(\Lambda^+) \to S(\Lambda^+)$ is J-anti-holomorphic, in the sense that

$\sigma_* \circ J = -J \circ \sigma_*$. A remarkable and non-obvious feature of this construction is that the almost-complex structure J is actually *conformally invariant*, despite the fact that replacing g with $u^2 g$ alters the horizontal subspaces on $Z = (\Lambda^+ - 0)/\mathbb{R}^+$. Now recall that an almost-complex manifold is a complex manifold iff it admits sufficiently many local holomorphic functions. In general, the obstruction [**52**] to the existence of such functions is the *Nijenhuis tensor*, but in the present case the Nijenhuis tensor of (Z, J) just amounts to the self-dual Weyl curvature W_+ of (M, g). When (M, g) is anti-self-dual, (Z, J) thus acquires the structure of a complex manifold [**4, 53**]:

Theorem 4.1 (Penrose/Atiyah–Hitchin–Singer). *The almost-complex manifold (Z, J) is a complex 3-manifold iff $W_+ = 0$. Moreover, a complex 3-manifold arises by this construction iff it admits a fixed-point-free anti-holomorphic involution $\sigma : Z \to Z$ and a foliation by σ-invariant rational curves $\mathbb{CP}_1$, each of which has normal bundle $\mathcal{O}(1) \oplus \mathcal{O}(1)$. Finally, the complex manifold (Z, J) and the real structure σ suffice to determine the metric g on M up to conformal rescaling.*

Definition 4.2. The complex 3-manifold (Z, J) associated with an anti-self-dual 4-manifold (M, g) by Theorem 4.1 is called the *twistor space* of $(M, [g])$.

Definition 4.3. Let (M, g) be an anti-self-dual 4-manifold, and let (Z, J) be its twistor space, and let $x \in M$. Then the holomorphic curve $P_x \subset Z$ given by $\pi^{-1}(x) = S(\Lambda^+_x)$ will be called the *real twistor line* corresponding to x.

The moduli space of holomorphic curves $\mathbb{CP}_1 \subset Z$ near the real twistor lines is a complex 4-manifold $\mathcal{M}$, and is a complexification of the original real 4-manifold M. The term *complex twistor line* (or just *twistor line*) is used to refer to any $\mathbb{CP}_1$ in this larger family.

One of the cornerstones of the theory of anti-self-dual manifolds is the connected sum construction of Donaldson and Friedman [**15**]. If M_1 and M_2 admit anti-self-dual metrics, this allows one to construct anti-self-dual metrics on the connected sum $M_1 \# M_2$, provided the twistor spaces Z_1 and Z_2 of the given manifolds satisfy $H^2(Z_j, \mathcal{O}(TZ_j)) = 0$, $j = 1, 2$. An orbifold generalization of this construction was later developed by the present author in collaboration with Michael Singer [**48**], and allows one to build up nonsingular anti-self-dual manifolds by gluing special orbifolds across $\mathbb{RP}^3 \times \mathbb{R}$ necks. We will now review those features of this generalized construction which will be needed in what follows.

Let (Y,g) be a compact anti-self-dual manifold with twistor space Z, and let us assume from the outset that $H^2(Z,\mathcal{O}(TZ)) = 0$. Let $\phi : Y \to Y$ be an isometry of (Y,g) with $\phi^2 = \mathrm{id}_Y$, and assume that ϕ has exactly k fixed points, for some positive integer k. Each fixed point is therefore isolated, and in geodesic normal coordinates around any fixed point, ϕ therefore just becomes the involution $\vec{v} \to -\vec{v}$ of $\mathbb{R}^4$; in particular, ϕ is orientation-preserving. We thus have an induced map $\phi^* : S(\Lambda^+) \to S(\Lambda^+)$, and this map may be viewed as a holomorphic involution $\widehat{\phi} : Z \to Z$. The fixed point set of $\widehat{\phi}$ then consists of a disjoint union of k real twistor lines, one for each fixed point of ϕ. Let $\widetilde{Z}$ be obtained by blowing up Z along these k twistor lines, and notice that $\widehat{\phi}$ induces a holomorphic involution $\widetilde{\phi}$ of $\widetilde{Z}$. The fixed point set of $\widetilde{\phi}$ then consists of k quadrics $\mathbb{CP}_1 \times \mathbb{CP}_1$ with normal bundle $\mathcal{O}(1,-1)$, and the $\widetilde{\phi}$ acts on their normal bundles by multiplication by -1. The quotient $\widetilde{Z}/\mathbb{Z}_2$ is therefore a nonsingular compact complex 3-fold containing k hypersurfaces $Q_1, \ldots Q_k$, each biholomorphic to $\mathbb{CP}_1 \times \mathbb{CP}_1$, and each with normal bundle $\mathcal{O}(2,-2)$. The complement of these hypersurfaces is just the twistor space of $[Y - \{\text{fixed points}\}]/\mathbb{Z}_2$. In this complement, choose $\ell \geqslant 0$ twistor lines, and blow them up to obtain ℓ hypersurfaces $Q_{k+1}, \ldots, Q_{k+\ell}$, each biholomorphic to $\mathbb{CP}_1 \times \mathbb{CP}_1$, and each with normal bundle $\mathcal{O}(1,-1)$. Let Z_+ denote this blow-up of $\widetilde{Z}/\mathbb{Z}_2$, and notice that our original anti-holomorphic involution σ of Z induces an anti-holomorphic involution $\sigma_+ : Z_+ \to Z_+$.

The next ingredient we will need is a compactification of the twistor space of the Eguchi–Hanson metric. The usual Eguchi–Hanson metric [**19**] is a locally asymptotically flat hyper-Kähler metric on T^*S^2 which, up to homothety, is the metric-space completion of the Riemannian metric

$$g_{EH,\epsilon} = \frac{d\varrho^2}{1-\varrho^{-4}} + \varrho^2 \left(\sigma_1^2 + \sigma_2^2 + \left[1-\varrho^{-4}\right]\sigma_3^2\right)$$

on $(1,\infty) \times S^3/\mathbb{Z}_2$, where $\{\sigma_j\}$ is the standard left-invariant co-frame on $S^3/\mathbb{Z}_2 = SO(3)$. However, because this metric is asymptotic to the flat metric on $(\mathbb{R}^4 - \{0\})/\mathbb{Z}_2$ as $\varrho \to \infty$, its conformal class naturally extends to an orbifold ASD conformal metric on $T^*S^2 \cup \{\infty\}$, where the added point is singular, with a neighborhood modeled on $\mathbb{R}^4/\mathbb{Z}_2$. Blowing up the twistor line of this added "point at infinity" then yields a nonsingular complex 3-fold $\widetilde{Z}_{EH}$ which contains a hypersurface $Q \cong \mathbb{CP}_1 \times \mathbb{CP}_1$ with normal bundle $\mathcal{O}(2,-2)$ arising as the exceptional divisor of the blow-up.

The complex 3-fold $\widetilde{Z}_{EH}$ is **[48]** a small resolution of the hypersurface

$$xy = z^2 - t^2\zeta_1^2\zeta_2^2$$

in the $\mathbb{CP}_3$-bundle $\mathbb{P}(\mathcal{O}(2)^{\oplus 3} \oplus \mathcal{O})$ over $\mathbb{CP}_1$, where $x, y, z \in \mathcal{O}(2)$, $t \in \mathcal{O}$, and $[\zeta_1 : \zeta_2]$ are the homogeneous coordinates on $\mathbb{CP}_1$. The small resolutions replace the two singular points $x = y = z = 0$ with rational curves.

Finally, consider the Fubini–Study metric on $\mathbb{CP}_2$, which, up to homothety, may be characterized as the unique $SU(3)$-invariant metric on the complex projective plane. Because the isotropy subgroup $U(2) \subset SU(3)$ is so large, the Fubini–Study metric has $W_- = 0$ for representation-theoretic reasons; reversing the orientation, the Fubini–Study metric thus becomes an ASD metric on $\overline{\mathbb{CP}}_2$. Let Z_{FS} denote the twistor space of this metric, and let $\widetilde{Z}_{FS}$ denote its blow-up along a real twistor line. Explicitly, Z_{FS} may be realized **[4]** as

$$\{([x_1 : x_2 : x_3], [y_1 : y_2 : y_3]) \in \mathbb{CP}_2 \times \mathbb{CP}_2 \mid \sum_j x_j y_j = 0\},$$

and we may take the relevant twistor line to be given by $x_1 = y_1 = 0$. Blowing up of this twistor line provides us with a preferred hypersurface $\mathbb{CP}_1 \times \mathbb{CP}_1$ with normal bundle $\mathcal{O}(1, -1)$.

Now let Z_- be the disjoint union of k copies of $\widetilde{Z}_{EH}$ and ℓ copies of $\widetilde{Z}_{FS}$, and let σ_- be the real structure it inherits from the twistor spaces of the Eguchi–Hanson and Fubini–Study metrics. Let $Q_- \subset Z_-$ be the disjoint union of $k + \ell$ copies of $\mathbb{CP}_1 \times \mathbb{CP}_1$, each being a copy of the constructed exceptional divisor in a copy of $\widetilde{Z}_{EH}$ or $\widetilde{Z}_{FS}$. Remembering that Z_+ also contains a disjoint union $Q_+ = Q_1 \sqcup \cdots \sqcup Q_{k+\ell}$ of the same number of copies of $\mathbb{CP}_1 \times \mathbb{CP}_1$, we may thus form a 3-dimensional complex-analytic space Z_0 with normal crossing singularities by identifying Q_+ with Q_-. However, we carry out this identification according to a few simple rules. First of all, the quadrics $Q_1, \ldots, Q_k$ are each to be identified with a quadric in a copy of $\widetilde{Z}_{EH}$, while the remaining quadrics $Q_{k+1}, \ldots Q_{k+\ell}$ are each to be identified with a quadric in a copy of $\widetilde{Z}_{FS}$. Secondly, we always interchange the factors of $\mathbb{CP}_1 \times \mathbb{CP}_1$ when gluing Q_+ to Q_-, thereby making the normal bundles of each quadric relative to Z_+ an Z_- dual to each other. Finally, we always make our identifications in such a way that the real structures σ_+ and σ_- agree on the locus Q obtained by identifying Q_+ with Q_-, so that Z_0 comes equipped with an anti-holomorphic involution $\sigma_0 : Z_0 \to Z_0$.

Because we have assumed that $H^2(Z, \mathcal{O}(TZ)) = 0$, one can show [**48**] that $\mathbf{Ext}^2_{Z_0}(\Omega^1, \mathcal{O}) = 0$, and a generalization of Kodaira–Spencer theory [**21**] then yields a versal deformation of Z_0, parameterized by a neighborhood of the origin in $\mathbf{Ext}^1_{Z_0}(\Omega^1, \mathcal{O})$. The generic fiber of this family is nonsingular

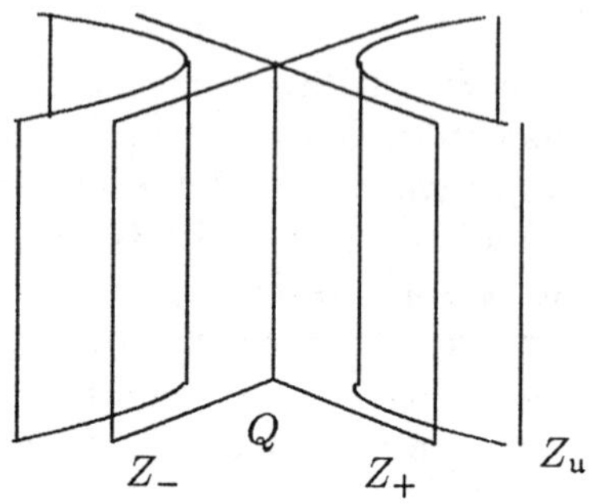

and the real structure σ_0 extends to act on the total space of this family.

Rather than working with the entire versal family, it is convenient to restrict ones attention to certain subfamilies, called *standard deformations*. A 1-*parameter standard deformation* of Z_0 is by definition a flat proper holomorphic map $\varpi : \mathcal{Z} \to \mathcal{U}$ together with an anti-holomorphic involution $\sigma : \mathcal{Z} \to \mathcal{Z}$, such that

- $\mathcal{Z}$ is a complex 4-manifold;
- $\mathcal{U} \subset \mathbb{C}$ is an open neighborhood of 0;
- $\varpi^{-1}(0) = Z_0$;
- $\sigma|_{Z_0} = \sigma_0$;
- σ covers complex conjugation in $\mathcal{U}$;
- ϖ is a submersion away from $Q \subset Z_0$; and
- near any point of Q, ϖ is modeled on $(x, y, z, w) \mapsto xy$.

When $\mathrm{u} \in \mathcal{U} \subset \mathbb{C}$ is real, non-zero, and sufficiently small, the corresponding fiber $Z_{\mathrm{u}} = \varpi^{-1}(\mathrm{u})$ is a twistor space, and we obtain the following result [**15, 48**]:

Theorem 4.4. *Let (Y, g) be a compact anti-self-dual 4-manifold equipped with an isometric $\mathbb{Z}_2$-action with exactly ℓ fixed points, for some positive integer ℓ. Let Z denote the twistor space of (Y, g), and suppose that $H^2(Z, \mathcal{O}(TZ)) = 0$. Let $X = Y/\mathbb{Z}_2$, and let $\widetilde{X}$ be the oriented manifold obtained by replacing each singularity of X with a 2-sphere of self-intersection -2. Then, for any integer $k \geqslant 0$, there are anti-self-dual* *conformal classes on* $\widetilde{X}\#k\overline{\mathbb{CP}}_2$ *whose twistor spaces arise as fibers in a* 1*-parameter standard deformation of* $Z_0 = (\widetilde{Z}/\mathbb{Z}_2) \cup \ell\widetilde{Z}_{EH} \cup k\widetilde{Z}_{FS}$.

Similarly, one can define standard deformations depending on several parameters. For example, a 2*-parameter standard deformation* of Z_0 is by definition a flat proper holomorphic map $\varpi : \mathcal{Z} \to \mathcal{U}$ together with an anti-holomorphic involution $\sigma : \mathcal{Z} \to \mathcal{Z}$, such that

- $\mathcal{Z}$ is a complex 5-manifold;
- $\mathcal{U} \subset \mathbb{C}^2$ is an open neighborhood of $(0,0)$;
- $\varpi^{-1}(0,0) = Z_0$;
- $\sigma|_{Z_0} = \sigma_0$;
- σ covers complex conjugation in $\mathcal{U}$;
- ϖ is a submersion away from $Q \subset Z_0$; and
- near any point of Q, ϖ is modeled on $(x,y,z,v,w) \mapsto (xy,z)$.

When $(\mathfrak{u}_1,\mathfrak{u}_2) \in \mathcal{U} \subset \mathbb{C}^2$ is real, and sufficiently close to $(0,0)$, with $\mathfrak{u}_1 \neq 0$, the corresponding fiber $Z_{\mathfrak{u}} = \varpi^{-1}(\mathfrak{u}_1,\mathfrak{u}_2)$ is a twistor space. By extracting such standard deformations from the versal deformation of Z_0, the same proof tells one the following:

Theorem 4.5. *Let* Y *be a compact real-analytic oriented* 4*-manifold equipped with a* $\mathbb{Z}_2$*-action with exactly* ℓ *fixed points, for some positive integer* ℓ*. Let* $[g_t]$, $t \in (-\varepsilon,\varepsilon)$, *be a real-analytic* 1*-parameter family of* $\mathbb{Z}_2$*-invariant anti-self-dual conformal metrics on* Y*. Let* $X = Y/\mathbb{Z}_2$*, and let* $\widetilde{X}$ *be the oriented manifold obtained by replacing each singularity of* X *with a* 2*-sphere of self-intersection* -2*. Let* Z *denote the twistor space of* $(Y,[g_0])$*, and suppose that* $H^2(Z,\mathcal{O}(TZ)) = 0$*. Then, for any* $k \geqslant 0$*, there is a* 2*-parameter standard deformation of* $Z_0 = (\widetilde{Z}/\mathbb{Z}_2) \cup \ell\widetilde{Z}_{EH} \cup k\widetilde{Z}_{FS}$ *such that, for all small real numbers* $\mathfrak{u}_1 > 0$ *and* $\mathfrak{u}_2$*, the fiber* $Z_{(\mathfrak{u}_1,\mathfrak{u}_2)}$ *is the twistor spaces of a family of an anti-self-dual metric on* $\widetilde{X}\#k\overline{\mathbb{CP}}_2$*, and such that, for all sufficiently small real numbers* t*, the fiber over* $Z_{(0,t)}$ *is the complex-analytic space with normal crossings built from the twistor space of* $(Y,[g_t])$ *by analogy to the construction of* Z_0.

5. Conformal Green's Functions

Let (M,g) be a compact Riemannian 4-manifold, and assume that its Yamabe Laplacian $\Delta + s/6$ has trivial kernel; the latter is automatic if the conformal class $[g]$ is contains a metric with $s > 0$, never happens if

$[g]$ is contains a metric with $s \equiv 0$, and may or may not happen if $[g]$ contains a metric with $s < 0$. Since the operator $\Delta + s/6$ is self-adjoint, it also has trivial cokernel, and the equation

$$(\Delta + s/6)u = f$$

therefore has a unique smooth solution u for any smooth function f; it follows that it also has a unique distributional solution u for any distribution f. If $y \in M$ is any point, and if δ_y is the Dirac delta distribution centered at y, we thus have a unique distributional solution G_y of the equation

$$(\Delta + s/6)G_y = \delta_y.$$

Since δ_y is identically zero on $M - \{y\}$, elliptic regularity tells us that G_y is actually smooth away from y. In general, one has an expansion

$$G_y = \frac{1}{4\pi^2}\frac{1}{\varrho^2} + O(\log \varrho)$$

near ϱ denotes the distance from y, but when (M, g) is anti-self-dual it in fact turns out [**3**] that

$$G_y = \frac{1}{4\pi^2}\frac{1}{\varrho^2} + \text{ bounded terms.}$$

In this article, the function G_y will be called the *conformal Green's function* of (M, g, y).

The motivation for this terminology is that the Yamabe Laplacian is a *conformally invariant* differential operator when viewed as a map between sections of suitable real line bundles; the geometric reason for this is that for any smooth function $u \not\equiv 0$, the expression $6u^{-3}(\Delta + s/6)u$ computes the scalar curvature of the conformally related metric $u^2 g$ on the open set $u \neq 0$. One useful consequence of this is that, for any smooth function $u > 0$ on M, a constant times $u^{-1}G_y$ is the conformal Green's function of $(M, u^2 g, y)$.

Now the celebrated proof of the Yamabe conjecture [**49**] t ells us that any conformal class on any compact manifold contains metrics of constant scalar curvature. In particular, any conformal class contains metrics whose scalar curvature has the same sign at every point. But actually, this last assertion is much more elementary. Indeed, if $u \not\equiv 0$ is an eigenfunction corresponding to the lowest eigenvalue λ of the Yamabe Laplacian, then u has empty nodal set, and $u^2 g$ is therefore a conformally related metric whose scalar curvature has the same sign as λ everywhere on M. Similar considerations also show that if two metrics with scalar curvatures of fixed signs are conformally related, then their scalar curvatures have the same sign. The *sign of Yamabe constant* of a conformal class, meaning the sign of the constant scalar curvature of the metric produced by the proof of the Yamabe conjecture, therefore coincides with the sign of the smallest Yamabe eigenvalue λ for any metric in the conformal class.

Here is another way of determining the sign of the Yamabe constant:

Lemma 5.1. *Let (M, g) be a compact Riemannian 4-manifold whose Yamabe Laplacian $\Delta + s/6$ has trivial kernel. Let $y \in M$ be any point. Then the conformal class $[g]$ contains a metric of positive scalar curvature if and only if $G_y(x) \neq 0$ for all $x \in M - \{y\}$. Moreover, if $[g]$ contains a metric of negative scalar curvature, then $G_y(x) < 0$ for some $x \in M$.*

These assertions also hold for any finite sum $G_{y_1} + \cdots + G_{y_m}$, or for any other finite linear combinations of conformal Green's functions with positive coefficients.

PROOF. Since the Yamabe Laplacian is conformally invariant when viewed as acting on functions of the appropriate conformal weight, we may assume from the outset that either $s > 0$ everywhere, or else $s < 0$ everywhere.

Now notice that

$$\frac{1}{6}\int sG_y \, d\mu = \int (\Delta + s/6)G_y \, d\mu = \int \delta_y \, d\mu = 1 > 0.$$

Thus, if $s < 0$, G_y must be negative somewhere, and since $G_y \to +\infty$ at y, the Green's function must also have a zero by continuity. Notice that the same argument works for any finite linear combinations of Green's functions with positive coefficients.

On the other hand, $G_y^{-1}((-\infty, a])$ is compact for any $a \in \mathbb{R}$, and it follows that G_y has a minimum. But if $s > 0$, then $G_y = \frac{1}{s}\nabla \cdot \nabla G_y$ on $M - \{y\}$, and hence $G_y \geqslant 0$ at its minimum. Moreover, if the minimum were actually zero, we could apply Hopf's strong maximum principle [**26, 54**] to $-G_y$, and conclude that $G_y \equiv 0$, contradicting the fact that $G_y \to \infty$ at y. Thus $G_y > 0$ everywhere, as claimed. As a consequence, any finite linear combinations of Green's functions with positive coefficients is a sum of positive functions, and so is positive at all points where its value is defined. □

Now suppose that (M, g) is a compact *anti-self-dual* Riemannian 4-manifold, and let Z be its twistor space. If $U \subset M$ is any open subset, and if $Z_U \subset Z$ is its inverse image in the twistor space, the Penrose transform [**5, 33**] gives a natural one-to-one correspondence between $H^1(Z_U, \mathcal{O}(K^{1/2}))$ and the smooth complex-valued functions on U which solve $(\Delta + s/6)u = 0$. Given a cohomology class $\psi \in H^1(Z_U, \mathcal{O}(K^{1/2}))$, the value of the corresponding function u_ψ at $x \in U$ is obtain by restricting u_ψ to the real twistor line $P_x \subset Z$ to obtain an element of $H^1(P_x, \mathcal{O}(K_Z^{1/2})) \cong H^1(\mathbb{CP}_1, \mathcal{O}(-2)) \cong \mathbb{C}$. Note that u_ψ is ostensibly only a section of a line bundle, but the choice of a metric g in the conformal class turns out to determine a canonical trivialization of this line bundle [**33**], and u_ψ then becomes a function in the ordinary sense.

In particular, our compact anti-self-dual 4-manifold (M, g) satisfies $\ker(\Delta + s/6) = 0$ iff its twistor space Z satisfies $H^1(Z, \mathcal{O}(K^{1/2})) = 0$, and Serre duality tells us that the latter happens iff $H^2(Z, \mathcal{O}(K^{1/2})) = 0$. When any of these thee equivalent conditions is met, we then have a conformal Green's function G_y for any chosen $y \in M$, and G_y then corresponds to a particular element of $H^1(Z - P_y, \mathcal{O}(K^{1/2}))$, where P_y is the twistor line corresponding to y. What is this mysterious cohomology class? The answer was discovered by Atiyah [**3**], and involves a construction largely due to Serre [**56**] and Horrocks [**34**]:

Lemma 5.2. *Let W be a (possibly non-compact) complex manifold, and let $V \subset W$ be a closed complex submanifold of complex codimension 2. Let $N \to V$ denote the normal bundle TW/TV of V, and suppose that there is a holomorphic line bundle $L \to W$ such that*

- $L|_V \cong \wedge^2 N$;
- $H^1(W, \mathcal{O}(L)) = 0$; *and*
- $H^2(W, \mathcal{O}(L)) = 0$.

Then there is a rank-2 holomorphic vector bundle $E \to W$, together with a holomorphic section $\zeta \in \Gamma(W, \mathcal{O}(E))$ such that

- $\wedge^2 E \cong L$;
- $\zeta = 0$ *exactly at* W; *and*
- $d\zeta : N \to E$ *is an isomorphism.*

This (E, ζ) is unique up to isomorphism if we also demand that the isomorphism $\det d\zeta : \wedge^2 N \to \wedge^2 E|_V$ *should agree with a given isomorphism* $\wedge^2 N \to L|_V$. *The pair (E, ζ) gives rise to an extension*

$$0 \to \mathcal{O}(L^*) \to \mathcal{O}(E^*) \xrightarrow{\cdot\zeta} \mathscr{I}_V \to 0,$$

where $\mathscr{I}_V$ is the ideal sheaf of V; and by restriction to $W - V$, this extension determines an element of $H^1(W - V, \mathcal{O}(L^*))$.

PROOF. Because [**2**, **28**] $V \subset W$ is smooth and of codimension 2,

$$\mathcal{E}xt^q(\mathcal{O}_V, \mathcal{O}(L^*)) \begin{cases} 0, & q = 0, 1, \\ \mathcal{O}_V(L^* \otimes \wedge^2 N), & q = 2, \end{cases}$$

and the spectral sequence

$$E_2^{p,q} = H^p(W, \mathcal{E}xt^q(\mathcal{O}_V, \mathcal{O}(L^*))) \Longrightarrow \mathbf{Ext}_W^{p+q}(\mathcal{O}_V, \mathcal{O}(L^*))$$

therefore tells us that

$$\mathbf{Ext}_W^2(\mathcal{O}_V, \mathcal{O}(L^*)) = \Gamma(V, \mathcal{O}_V(L^* \otimes \wedge^2 N)).$$

On the other hand, the tautological short exact sequence of sheaves

$$0 \to \mathscr{I}_V \to \mathcal{O} \to \mathcal{O}_V \to 0$$

induces a long exact sequence

$$\begin{array}{rll} \cdots \to & \mathbf{Ext}_W^1(\mathcal{O}, \mathcal{O}(L^*)) & \to \mathbf{Ext}_W^1(\mathscr{I}_V, \mathcal{O}(L^*)) \\ \to \mathbf{Ext}_W^2(\mathcal{O}_V, \mathcal{O}(L^*)) \to & \mathbf{Ext}_W^2(\mathcal{O}, \mathcal{O}(L^*)) & \to \cdots \end{array}$$

and since

$$\mathbf{Ext}_W^j(\mathcal{O}, \mathcal{O}(L^*)) = H^j(W, \mathcal{O}(L^*))$$

is assumed to vanish when $j = 1, 2$, the Bockstein map of this long exact sequence therefore gives us an isomorphism

$$\mathbf{Ext}_W^1(\mathscr{I}_V, \mathcal{O}(L^*)) \cong \Gamma(V, \mathcal{O}_V(L^* \otimes \wedge^2 N)).$$

In particular, any choice of isomorphism $\wedge^2 N \to L|_V$ gives us an extension

$$0 \to \mathcal{O}(L^*) \to \mathcal{O}(E^*) \to \mathscr{I}_V \to 0 \tag{9}$$

of sheaves on W. The class of this extension is called the *Serre class* $\lambda(V) \in \mathbf{Ext}_W^1(\mathscr{I}_V, \mathcal{O}(L^*))$, and its restriction to $W - V$ is an element of $H^1(W - V, \mathcal{O}(L^*))$. Strictly speaking, the Serre class depends on a choice of isomorphism $\wedge^2 N \to L|_V$, but any two such extensions are intertwined by an automorphism of $L|_V$. Since any isomorphism $\wedge^2 N \to L|_V$ corresponds to a section of $\mathcal{E}xt^2(\mathcal{O}_V, \mathcal{O}(L^*))$ which is non-zero at each point of V, the corresponding extension (9) is locally free, with

$\Lambda^2 E^* = L^*$. Tensoring the inclusion $\mathcal{O}(L^*) \hookrightarrow \mathcal{O}(E^*)$ by L, we thus obtain an inclusion

$$\mathcal{O} \hookrightarrow \mathcal{O}(E^* \otimes \Lambda^2 E) = \mathcal{O}(E),$$

and the image of 1 under this map is then a section $\zeta \in \Gamma(W, \mathcal{O}(E))$ with all the advertised properties. □

Proposition 5.3 (Atiyah). *Let (M, g) be a compact anti-self-dual 4-manifold with twistor space Z, and assume that (M, g) has*

$$\ker(\Delta + s/6) = 0.$$

Let $y \in M$ be any point, and let $P_y \subset Z$ be the corresponding twistor line. Then the image of the Serre class $\lambda(P_y) \in \mathbf{Ext}^1_Z(\mathscr{I}_{P_y}, \mathcal{O}(K^{1/2}))$ in $H^1(Z - P_y, \mathcal{O}(K^{1/2}))$ is the Penrose transform of the Green's function G_y times a non-zero constant.

Indeed, if one identifies $K_Z^{1/2}|_{P_y}$ with K_{P_y} according the isomorphism determined by g and the conventions of [**33**], the relevant constant turns out to be exactly 4π.

Combining this remarkable result with Lemma 5.1 now gives us a twistorial criterion for determining whether an anti-self-dual conformal class has positive Yamabe constant:

Proposition 5.4. *Let Z be the twistor space of a compact anti-self-dual 4-manifold $(M, [g])$, and let $P_y \subset Z$ be a real twistor line. Then the conformal class $[g]$ contains a metric g of positive scalar curvature if and only if*

- *$H^1(Z, \mathcal{O}(K^{1/2})) = 0$, and*
- *the holomorphic vector bundle $E \to Z$ with $\wedge^2 E \cong K^{-1/2}$ associated to P_y by Lemma 5.2 satisfies $E|_{P_x} \cong \mathcal{O}(1) \oplus \mathcal{O}(1)$ for every real twistor line P_x.*

Proof. Let us first recall that

$$\ker(\Delta + s/6) = H^1(Z, \mathcal{O}(K^{1/2})) = [H^2(Z, \mathcal{O}(K^{1/2}))]^*,$$

so that a necessary condition for the positivity of the Yamabe constant is certainly the vanishing of $H^1(Z, \mathcal{O}(K^{1/2}))$. When this happens, Lemma 5.2 then allows us to construct $E \to Z$. On $Z - P_y$, E is then given by an extension

$$0 \to \mathcal{O} \to \mathcal{O}(E) \to \mathcal{O}(K^{-1/2}) \to 0,$$

and this extension is represented by an element of $H^1(Z - P_y, \mathcal{O}(K^{1/2}))$. The value of the Penrose transform of this class at $x \neq y$ is obtained via the restriction map

$$H^1(Z - P_y, \mathcal{O}(K^{1/2})) \to H^1(P_x, \mathcal{O}_{P_x}(K^{1/2})) \cong \mathbb{C}$$

and its value at x is therefore non-zero iff the induced extension

$$0 \to \mathcal{O} \to E|_{P_x} \to \mathcal{O}(2) \to 0$$

does not split; and this happens iff $E|_{P_x} \cong \mathcal{O}(1) \oplus \mathcal{O}(1)$. Since we also have $E|_{P_y} \cong \mathcal{O}(1) \oplus \mathcal{O}(1)$ by construction, the result now follows from lemma 5.1 and Proposition 5.3. □

6. The Sign of the Scalar Curvature

We are now ready to approach the problem of determining the sign of the Yamabe constant for the anti-self-dual conformal classes constructed in Theorem 4.4. The results obtained in this section are loosely inspired by the work of Dominic Joyce [**37**] on the Yamabe constants of certain conformal classes on connect sums, although the techniques employed here are completely different from Joyce's.

Lemma 6.1. *Let $\varpi : \mathcal{Z} \to \mathcal{U}$ be a 1-parameter standard deformation of Z_0, where Z_0 is as in Theorem 4.4, and $\mathcal{U} \subset \mathbb{C}$ is an open disk about the origin. Let $k\widetilde{Z}_{EH}$ be the union of the Eguchi–Hanson components of $Z_- \subset Z_0$, which is a nonsingular complex hypersurface in $\mathcal{Z}$, and let $\mathscr{I}_{k\widetilde{Z}_{EH}} \subset \mathcal{O}$ denotes its ideal sheaf. Then the invertible sheaf $\mathscr{I}_{k\widetilde{Z}_{EH}}(K_{\mathcal{Z}}) \subset \mathcal{O}_{\mathcal{Z}}(K_{\mathcal{Z}})$ has a square-root as a holomorphic line bundle.*

Proof. A holomorphic line bundle has a holomorphic square-root iff its second Stieffel–Whitney class $w_2 \in H^2(\mathbb{Z}_2)$ vanishes. Write $\mathcal{Z}$ as $U \cup V$, where U is a tubular neighborhood of Z_+, V is a tubular neighborhood of Z_-, and $U \cap V$ is a tubular neighborhood of Q, so that these open sets deform retract to Z_+, Z_-, and Q respectively. Since each component of Q is simply connected, the Mayer–Vietoris sequence

$$\cdots \to H^1(U \cap V, \mathbb{Z}_2) \to H^2(U \cup V, \mathbb{Z}_2) \to H^2(U, \mathbb{Z}_2) \oplus H^2(V, \mathbb{Z}_2) \to \cdots$$

therefore tells us that it is enough to check that the restrictions of our line bundle to Z_+ and Z_- both have square-roots.

It thus suffices to produce an explicit square-root of the restrictions of $\mathscr{I}_{k\widetilde{Z}_{EH}} \otimes K_{\mathcal{Z}}$ to each copy of $\widetilde{Z}_{FS}$, each copy of $\widetilde{Z}_{EH}$, and to Z_+. On each copy of $\widetilde{Z}_{FS}$, such a square-root is given by $[Q] \otimes K^{-1/2}_{Z_{FS}}$, where $[Q]$ is the divisor of the exceptional quadric, and where $Z^{1/2}_{FS}$ is the pull-back of $K^{1/2}$ from the twistor space Z_{FS} via the blowing-down map. On each copy of $\widetilde{Z}_{EH}$, such a square-root is given by the pull-back of $\mathcal{O}(-2)$ via the projection $\widetilde{Z}_{EH} \to \mathbb{CP}_1$. And on Z_+, there is a natural choice of square-root whose sections are the $\mathbb{Z}_2$ invariant sections of $K^{-1/2}_Z$, pulled-back to the blow-up $\widetilde{Z}$, twisted by the divisors $Q_{k+1}, \dots, Q_{k+\ell}$. That each of these bundles really has the correct square can be verified directly using the adjunction formula; the details are left as an exercise for the interested reader. □

Lemma 6.2. *Let $\varpi : \mathcal{Z} \to \mathcal{U}$ be a 1-parameter standard deformation of Z_0, where Z_0 is as in Theorem 4.4, and $\mathcal{U} \subset \mathbb{C}$ is a neighborhood of the origin. Let $L \to \mathcal{Z}$ be the holomorphic line bundle defined by*

$$\mathcal{O}(L^*) = [\mathscr{I}_{k\widetilde{Z}_{EH}}(K_{\mathcal{Z}})]^{1/2} \otimes \mathscr{I}_{\ell\widetilde{Z}_{FS}},$$

where the hypersurface $\ell\widetilde{Z}_{FS} \subset \mathcal{Z}$ is the union of the Fubini–Study components of Z_-. If the twistor space Z of $(Y,[g])$ satisfies $H^1(Z,\mathcal{O}(K^{1/2})) = 0$, then by possibly replacing $\mathcal{U}$ with a smaller neighborhood of $0 \in \mathbb{C}$ and simultaneously replacing $\mathcal{Z}$ with its inverse image, we can arrange for our complex 4-fold $\mathcal{Z}$ to satisfy

$$H^1(\mathcal{Z},\mathcal{O}(L^*)) = H^2(\mathcal{Z},\mathcal{O}(L^*)) = 0.$$

Proof. Since any open set in $\mathbb{C}$ is Stein, the Leray spectral sequence tells us that it would suffice to show that the direct image sheaves $\varpi^j_*\mathcal{O}(L^*)$ vanish for $j = 1,2$. But since ϖ is flat and we are allowed to shrink $\mathcal{U}$ if necessary, semi-continuity [6] asserts that it is enough to show that $H^j(Z_0,\mathcal{O}(L^*)) = 0$ for $j = 1,2$.

The normalization of Z_0 is the disjoint union $Z_+ \sqcup Z_-$, and we have an exact sequence

$$0 \to \mathcal{O}_{Z_0}(L^*) \to \nu_*\mathcal{O}_{Z_+}(L^*) \oplus \nu_*\mathcal{O}_{Z_-}(L^*) \to \mathcal{O}_Q(L^*) \to 0$$

where $\nu : Z_+ \sqcup Z_- \to Z_0$ is the identification map. However, $\mathcal{O}_{Z_0}(L^*)$ exactly consists of $\mathbb{Z}_2$-invariant sections of the pull-back of $K^{1/2}_Z$, and the Leray spectral sequence therefore tells us that

$$H^j(Z,\mathcal{O}(K^{1/2})) = 0 \Longrightarrow H^j(Z_+,\mathcal{O}(L^*)) = 0,$$

so our vanishing hypothesis guarantees that these groups vanish for all j. On the other hand,

$$H^j(\widetilde{Z}_{EH}, \mathcal{O}(L^*)) = H^j(\widetilde{Z}_{FS}, \mathcal{O}(L^*)) = H^j(Q_j, \mathcal{O}(L^*)) = \begin{cases} \mathbb{C}, & j = 1, \\ 0, & j \neq 1. \end{cases}$$

and each of the relevant restriction maps, from $H^1(\widetilde{Z}_{EH}, \mathcal{O}(L^*))$ or $H^1(\widetilde{Z}_{FS}, \mathcal{O}(L^*))$ to the cohomology group $H^1(\mathbb{CP}_1 \times \mathbb{CP}_1, \mathcal{O}(-2, 0))$ of the appropriate quadric Q_j, is an isomorphism. Hence $H^j(Z_0, \mathcal{O}(L^*)) = 0$ for all j, and the result follows. □

Now choose a real twistor line $P_x \in Z_+$, and extend this as a 1-parameter family of twistor lines in $P_{x_{\mathfrak{u}}} \in Z_{\mathfrak{u}}$ for $\mathfrak{u}$ near $0 \in \mathbb{C}$ and such that $P_{x_{\mathfrak{u}}}$ is a real twistor line for $\mathfrak{u}$ real. By possibly shrinking $\mathcal{U}$, we may then arrange that $\mathcal{P} = \cup_{\mathfrak{u}} P_{x_{\mathfrak{u}}}$ is a closed submanifold of $\mathcal{Z}$ and that $H^1(\mathcal{Z}, \mathcal{O}(L^*)) = H^2(\mathcal{Z}, \mathcal{O}(L^*)) = 0$. The hypotheses of Lemma 5.2 are then satisfied, and we thus obtain a holomorphic vector bundle $E \to \mathcal{Z}$ and a holomorphic section ζ vanishing exactly along $\mathcal{P}$; moreover, the corresponding extension

$$0 \to \mathcal{O}(L^*) \to \mathcal{O}(E^*) \to \mathcal{I}_{\mathcal{P}} \to 0$$

gives us an element of $\lambda(\mathcal{P}) \in H^1(\mathcal{Z} - \mathcal{P}, \mathcal{O}(L^*))$. Since the restriction of L^* to any smooth fiber $Z_{\mathfrak{u}}$, $\mathfrak{u} \neq 0$, is just $K^{1/2}$, Proposition 5.3 tells us that the restriction of $\lambda(\mathcal{P})$ to $Z_{\mathfrak{u}}$, $\mathfrak{u} > 0$, has Penrose transform equal to a positive constant times the conformal Green's function of $(\widetilde{X} \# \ell \overline{\mathbb{CP}}_2, g_{\mathfrak{u}}, x_{\mathfrak{u}})$ for any $\mathfrak{u} > 0$. However, we may also restrict (E, ζ) to Z_+, and, by pulling-back and pushing down, convert this into a $\mathbb{Z}_2$-invariant holomorphic vector bundle on the twistor space Z of (Y, g). This bundle on Z then has determinant line bundle $K^{-1/2}$, and comes equipped with a section vanishing exactly at the twistor lines of the two pre-images of y_1, y_2 of $x \in X$; by Proposition 5.3, the Penrose transform of this object corresponds, according to your taste, either to the Green's function G_x on(X, g) or to the sum $G_{y_1} + G_{y_2}$ on (Y, g). If g has negative scalar curvature, Lemma 5.1 thus tells us that is a region of X where $G_x < 0$, and deforming the twistor lines of this into $Z_{\mathfrak{u}}$ for small $\mathfrak{u} > 0$ then shows that the conformal Green's function of $[g_{\mathfrak{u}}]$ is negative somewhere for any small $\mathfrak{u}$. By Lemma 5.1, we thus obtain the following:

Theorem 6.3. *In Theorem 4.4, suppose that (Y, g) is an anti-self-dual manifold with $s < 0$ and $\ker(\Delta + s/6) = 0$. Then for all sufficiently* *small* $\mathfrak{u} > 0$*, the conformal class* $[g_{\mathfrak{u}}]$ *contains a metric of negative scalar curvature.*

The positive case is similar, but is slightly more delicate. Instead of just restricting $\lambda(\mathcal{P})$ on rational curves in Z_+, we must also consider what happens when we restrict this class to twistor lines in Z_-. However, we already saw in the proof of Lemma 6.2, an element of $H^1(\widetilde{Z}_{EH}, \mathcal{O}(L^*)) \cong \mathbb{C}$ or $H^1(\widetilde{Z}_{FS}, \mathcal{O}(L^*)) \cong \mathbb{C}$ is non-zero iff its restriction to the corresponding exceptional quadric is non-zero, and this has the effect that the restriction of the cohomology class to every twistor line in either of these spaces is non-zero if there is a rational curve in the quadric on which the class is non-zero. Thus, when the conformal Green's function G_x of (X, g) is positive, the vector bundle E determined by $\lambda(\mathcal{P})$ has splitting type $\mathcal{O}(1) \oplus \mathcal{O}(1)$ on all the σ_0-invariant rational curves in Z_0 which are limits of real twistor lines in $\mathcal{Z}_{\mathfrak{u}}$ as $\mathfrak{u} \to 0$. It therefore has the same splitting type on all the real twistor lines of $\mathcal{Z}_{\mathfrak{u}}$ for $\mathfrak{u}$ small, and Proposition 5.4 therefore tells us:

Theorem 6.4. *In Theorem* 4.4*, suppose that* (Y, g) *is an anti-self-dual manifold with* $s > 0$*. Then for all sufficiently small* $\mathfrak{u} > 0$*, the conformal class* $[g_{\mathfrak{u}}]$ *contains a metric of positive scalar curvature.*

In this positive case, it is interesting to re-examine the above construction in purely Riemannian terms. In this setting, the positivity of the Green's functions allows us to define a family of asymptotically flat, scalar-flat, anti-self-dual metrics $\widetilde{g}_{\mathfrak{u}} = G^2_{x_{\mathfrak{u}}} g_{\mathfrak{u}}$ on $(\widetilde{X} \# \ell \overline{\mathbb{CP}}_2) - \{pt\}$. What the above construction tells us is that these metrics converge, in the pointed Gromov–Hausdorff sense [**29**], to the orbifold metric $\widetilde{g} = G^2_x g$ on $X - x$. However, there is something else going on in certain regions, where viewing these metrics under higher and higher magnification results in a family that converges to the Eguchi–Hanson metric or to the *Burns metric*, meaning the Green's function rescaling of the Fubini–Study metric on $\overline{\mathbb{CP}}_2 - \{pt\}$. It is the appearance of the ideal sheaves in the definition of L^* which accounts for the fact that these seemingly incompatible pictures simultaneously apply at wildly different length-scales.

7. Conformally Flat Orbifolds

Consider the involution of $S^1 \times S^3 \subset \mathbb{C} \times \mathbb{H}$ given by $(z, q) \mapsto (\overline{z}, \overline{q})$. This involution has only four fixed points, namely $(z, q) = (\pm 1, \pm 1)$;

and near each of these isolated fixed points, the involution necessarily looks exactly like reflection through the origin in $\mathbb{R}^4$. We can therefore construct an involution of the connected sum $(S^1 \times S^3)\#(S^1 \times S^3)$ by cutting out a ball centered at a fixed point of the involution of each copy of $S^1 \times S^3$, and then being careful to carry out the usual gluing procedure in a $\mathbb{Z}_2$-equivariant manner. The resulting involution

$$\phi : (S^1 \times S^3)\#(S^1 \times S^3) \to (S^1 \times S^3)\#(S^1 \times S^3)$$

then has exactly 6 fixed points, and may usefully be thought of as a 4-dimensional analog of the hyperelliptic involution

of a Riemann surface of genus 2.

In this section, we will be interested in conformally flat orbifold metrics on $X = [(S^1 \times S^3)\#(S^1 \times S^3)]/\mathbb{Z}_2$, or in other words, in ϕ-invariant, conformally flat metrics on $(S^1 \times S^3)\#(S^1 \times S^3)$. The key result we will need is the following:

Proposition 7.1. *There is a real-analytic family g_t, $t \in (0,1)$, of Riemannian metrics on $(S^1 \times S^3)\#(S^1 \times S^3)$ with the following properties:*

- *for each t, the metric g_t is locally conformally flat;*
- *for each t, the involution ϕ is an isometry of g_t;*
- *for each t, the scalar curvature s of g_t has a fixed sign;*
- *when t is sufficiently close to 0, g_t has $s > 0$;*
- *when t is sufficiently close to 1, g_t has $s < 0$;*
- *the set of t for which* $\ker(\Delta + s/6) \neq \{0\}$ *is discrete;* and
- *there are only finitely many values of t for which g_t has $s \equiv 0$.*

Consequently, there is a $t_0 \in (0,1)$ and an $\varepsilon > 0$ such that, for all $t \in (t_0 - \varepsilon, t_0 + \varepsilon)$, the scalar curvature of g_t has the same sign as $t_0 - t$, and such that the Yamabe Laplacian $\Delta + s/6$ has trivial kernel for all $t \in (t_0 - \varepsilon, t_0) \cup (t_0, t_0 + \varepsilon)$.

Our proof of the existence of such a family hinges on a result of Schoen and Yau [**55**], and the construction used here is analogous to related constructions of Kim [**39**] and Nayatani [**51**]. Let us begin by observing that $(S^1 \times S^3)\#(S^1 \times S^3)$ can be obtained from S^4 by deleting four balls, and identifying the resulting boundary spheres in pairs via reflections. Now think of S^4 as $\mathbb{HP}_1 = \mathbb{H} \cup \{\infty\}$, and let $D_t \subset \mathbb{H} \cup \{\infty\}$ be the complement of the four open balls

$$B_t(\sqrt{3}), B_t(-\sqrt{3}), B_t(i), B_t(-i) \subset \mathbb{H}$$

of radius t; and we henceforth stipulate that $t < 1$, so as to guarantee that the closures of these four balls are pairwise disjoint. We may then think of $(S^1 \times S^3)\#(S^1 \times S^3)$ as obtained from D_t by identifying $\partial B_t(i)$ with $\partial B_t(-i)$ via the reflection $q \mapsto \overline{q}$, and identifying $\partial B_t(\sqrt{3})$ with $\partial B_t(-\sqrt{3})$ via the reflection $q \mapsto -\overline{q}$:

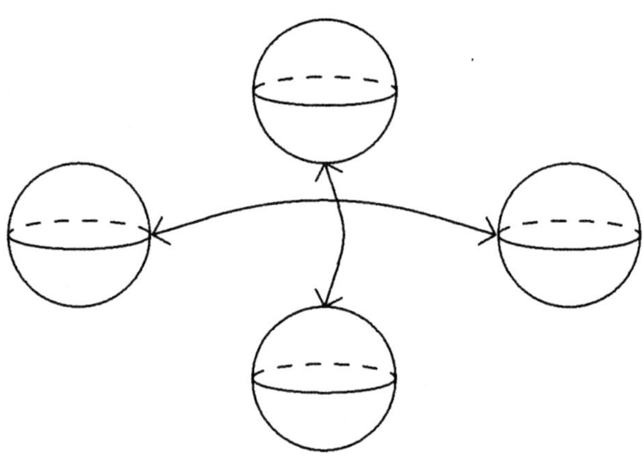

Since the reflections that we have used to identify the boundary components in pairs are just the restrictions to the relevant spheres of the global conformal transformations

$$q \mapsto t^2(q+i)^{-1} + i \text{ and } q \mapsto -t^2(q-\sqrt{3})^{-1} - \sqrt{3}$$

of $S^4 = \mathbb{H} \cup \{\infty\}$, we may therefore define a unique flat conformal metric $[g_t]$ on $(S^1 \times S^3)\#(S^1 \times S^3)$ by restricting the standard conformal metric on S^4 to D_t, and then pushing this structure down to $D_t/\sim$.

To complete the picture, we let ϕ act on $D_t/\sim$ by $q \mapsto -q$. Note that we obviously have $\phi^*[g_t] = [g_t]$. The six fixed points of ϕ are just

$i \pm t, \sqrt{3} \pm t, 0$ and ∞. That this ϕ coincides with the previously-described involution of $(S^1 \times S^3)\#(S^1 \times S^3)$ may be seen, when $t \in (0, 1/3)$, by first decomposing $\mathbb{H} \cup \{\infty\}$ into the hemispheres $||q|| \leqslant 4/3$ and $||q|| \geqslant 4/3$. Thus ϕ can be constructed by first letting $q \mapsto -q$ act on two separate copies of S^4 minus *two* balls, each with its boundary components identified via reflections, and then forming the $\mathbb{Z}_2$-equivariant connected sum of these manifolds; but each of these two building blocks looks like $S^3 \times ([-1, 1]/\{-1, 1\})$ equipped with the involution $(q, t) \mapsto (\overline{q}, -t)$, so the claim follows.

So far, we have only constructed a family $[g_t]$ of ϕ-invariant flat conformal classes, but we next need to worry about how nicely these conformal structures vary with t. However, it is not hard to see that they are real-analytic in t, since on a given open neighborhood U of a given $D_{t'} \subset S^4$, we are simply gluing together neighborhoods of the boundary spheres via the Möbius transformations

$$q \mapsto t^2(q + i)^{-1} + i, \text{ and } q \mapsto -t^2(q - \sqrt{3})^{-1} - \sqrt{3}$$

for t near t', and these transformations depend real-analytically on t. Since the sheaf of real-analytic functions is acyclic [**30**], we can now choose a real-analytic family of ϕ-invariant metrics h_t which represents the family of conformal classes $[g_t]$. Let λ_t be the smallest eigenvalue of the Yamabe Laplacian $\Delta_{h_t} + s_{h_t}/6$ of h_t, and let f_t be an eigenfunction of eigenvalue λ_t and integral 1. By the minimum principle [**26**], f_t is everywhere positive, and it follows that it must be unique; in particular, f_t must be ϕ-invariant. Moreover, this uniqueness tells us that λ_t has multiplicity 1. Hence λ_t never meets another eigenvalue as t varies, so perturbation theory [**38**] tells us that λ_t and f_t depend real-analytically on t. Now set

$$g_t = f_t^2 h_t,$$

and notice that the scalar curvature

$$s_{g_t} = f_t^{-3}(6\Delta_{h_t} + s_{h_t})f_t = 6\lambda_t f_t^{-2}$$

of this metric has the same sign as λ_t at every point. Thus g_t is a real-analytic family of ϕ-invariant metrics representing the constructed conformal classes $[g_t]$, with the desirable property that the scalar curvature is of a fixed sign for each t.

But what *is* the sign of the scalar curvature? To answer this, first observe that, for each $t \in (0, 1)$, the universal cover of Y can naturally be realized as an open set of S^4, namely the union Ω_t of all translates of

D_t via elements of the group generated by the Möbius transformations $q \mapsto t^2(q+i)^{-1}+i$ and $q \mapsto -t^2(q-\sqrt{3})^{-1}-\sqrt{3}$ of $S^4 = \mathbb{HP}_1$. In other words,

$$(S^1 \times S^3)\#(S^1 \times S^3) = \Omega_t/\mathbb{Z} * \mathbb{Z},$$

where Ω_t is the region of discontinuity of the Kleinian group

$$\mathbb{Z} * \mathbb{Z} \subset PSL(2,\mathbb{C}) \subset PGL(2,\mathbb{H})$$

generated by

$$\frac{1}{t}\begin{bmatrix} 1 & i-it^2 \\ -i & 1 \end{bmatrix} \text{ and } \frac{1}{t}\begin{bmatrix} \sqrt{3} & t^2-3 \\ -1 & \sqrt{3} \end{bmatrix} \in SL(2,\mathbb{C}).$$

These Kleinian groups are of the special type known as *Schottky groups* [**50**]. Henceforth, D_t is to be understood as a fundamental domain for the corresponding group action.

The complement $\Lambda_t = S^4 - \Omega_t$ of the region of discontinuity is called the *limit set* of the group action. If we think of S^4 as the boundary of the 5-disk, on whose interior $PGL(2,\mathbb{H}) = SO^\uparrow(5,1)$ acts by isometries of the hyperbolic metric, then the limit set may also be characterized as the accumulation points of the orbit of any point in the open 5-ball. Since we have arranged for each of our subgroups of $PGL(2,\mathbb{H}) = SO^\uparrow(5,1)$ to actually lie in $PSL(2,\mathbb{C}) = SO^\uparrow(3,1)$, it follows that we have $\Lambda_t \subset \mathbb{CP}_1 \subset \mathbb{HP}_1$. This will later allow us to use planar diagrams to understand the structure of the limit set.

For our purposes, the ultimate utility of the Kleinian point of view stems from a remarkable result of Schoen and Yau [**55**] that relates the scalar curvature of a uniformized conformally flat manifold to the size of the corresponding limit set. The form of this result we will use is actually a slight refinement due to Nayatani [**51**]:

Lemma 7.2 (Schoen–Yau, Nayatani). *Let (M,[g]) be a compact, locally conformally flat n-manifold, $n \geqslant 3$, which can be uniformized as*

$$M = \Omega/G,$$

where $G \subset SO^\uparrow(n+1,1)$ is a Kleinian group and where $\Omega \subset S^n$ is the region of discontinuity of G. Let $g \in [g]$ be a metric on M in the fixed conformal class for which the scalar curvature s does not change sign. Assume that the limit set Λ of G is infinite, and let $\dim(\Lambda) > 0$ *denotes* *its Hausdorff dimension. Then*

$$s > 0 \iff \dim(\Lambda) < \frac{n}{2} - 1$$
$$s = 0 \iff \dim(\Lambda) = \frac{n}{2} - 1$$
$$s < 0 \iff \dim(\Lambda) > \frac{n}{2} - 1.$$

The original argument given by Schoen and Yau is rather indirect, but Nayatani's proof actually constructs a particular metric for which the scalar curvature does not change sign; his conformal factor is obtained by convolving an appropriate power of the Euclidean distance with the Patterson–Sullivan measure of the limit set. Because this construction is so natural and canonical, it might seem tempting to simply use Nayatani's algorithm to define our family of metrics g_t. We have avoided doing so here, however, in order avoid the technical problem of proving that these metrics depend analytically on the parameter t.

To prove Proposition 7.1, we now proceed by showing that $\dim(\Lambda_t) < 1$ for t close to 0, and that $\dim(\Lambda_t) > 1$ for t close to 1.

Since the region of discontinuity Ω_t is the union of all translates of D_t, the limit set Λ_t may be thought of as the intersection of a nested sequence of balls in $\mathbb{R}^4 = \mathbb{H}$, where each of our original four balls contains the reflections of the other three, each of these in turn contains another three, and so forth. However, we have also observed that $\Lambda_t = \Lambda_t \cap \mathbb{C}$, so the limit set may instead be thought of as the generalized Cantor set in $\mathbb{C}$ given by the intersection of a nested sequence of disks, where, in passing from one level to the next, each disk is replaced by three smaller ones. Now the two generators of our Schottky group both have derivatives satisfying

$$\left|\frac{d}{dz}\left(\pm\frac{t^2}{z-c} \pm c\right)\right| \leqslant t^2 \text{ whenever } |z-c| \geqslant 1.$$

Since the disks at the kth level of the nesting are obtained by applying compositions of k generators to one of the original 4 disks, this implies that the the disks at the kth level have Euclidean radius $< t^{2k}$. There are $4 \cdot 3^k$ of these,

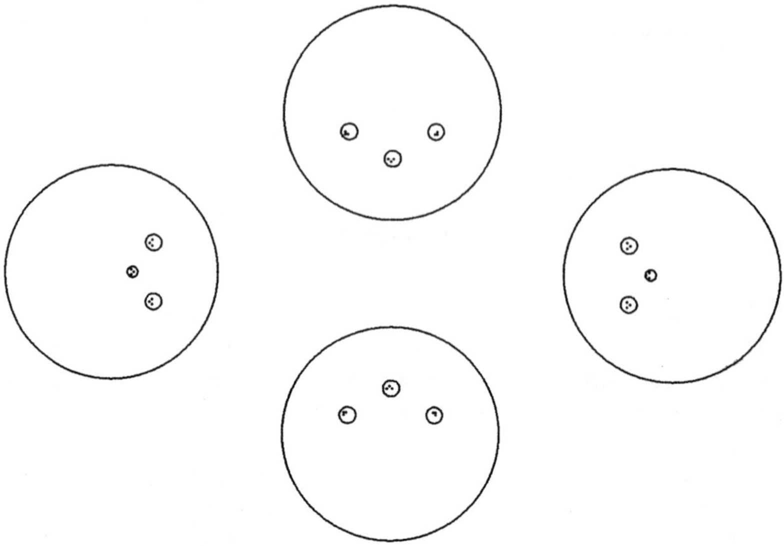

so the d-dimensional Hausdorff measure of Λ_t is less than a constant times $(3t^{2d})^k$ for all k, and vanishes if $\log 3 + 2d \log t < 0$. It therefore follows that

$$\dim(\Lambda_t) \leqslant -\frac{\log 3}{2 \log t}.$$

In particular, $\dim(\Lambda_t) < 1$ for all $t \in (0, 1/2]$, and for these values of t our ϕ-invariant conformally flat metrics g_t will have $s > 0$. The interested reader may enjoy the exercise of constructing explicit choices of g_t with $s > 0$ when t is extremely small, and comparing the results obtainable in this way with the predictions of the above limit-set argument.

Next, we need to show that $\dim(\Lambda_t) > 1$ when t is sufficiently close to 1. To see this, first consider the Kleinian group $G \subset PSL(2, \mathbb{C})$ generated by

$$\begin{bmatrix} 1 & 0 \\ -i & 1 \end{bmatrix} \text{ and } \begin{bmatrix} \sqrt{3} & -2 \\ -1 & \sqrt{3} \end{bmatrix} \in SL(2, \mathbb{C}).$$

which is the limiting case of our construction that arises by naïvely setting $t = 1$. We can still construct a fundamental domain D domain for this action as the complement of four balls, but certain pairs our four balls in $\mathbb{H}$ will now have a boundary point in common. The corresponding system of nested disks in $\mathbb{C}$ then contains two 'bracelets' of disks arranged around a pair of circles in the plane:

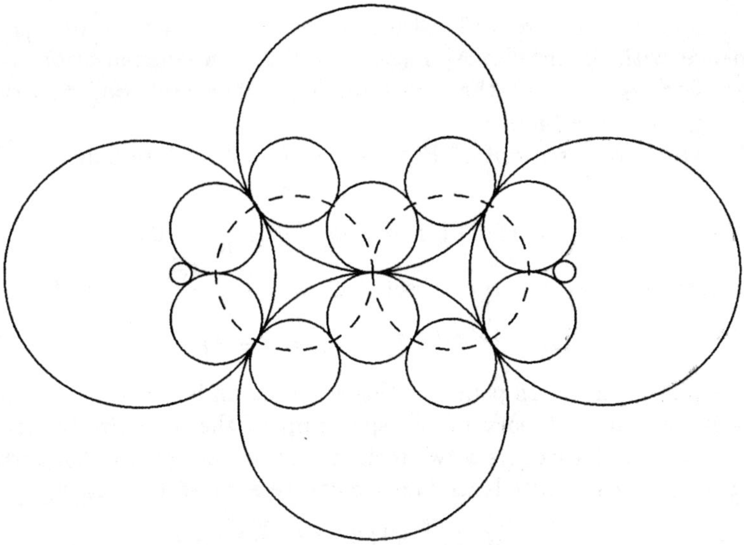

As one passes from one level of the nesting to the next, both of these circles continue to be completely covered by bracelets of smaller and smaller disks. Hence these two circles are both contained in the limit set of G. But, by a result of Bishop and Jones [**12**], this implies that the Hausdorff dimension of the corresponding limit set must be strictly greater than one:

Lemma 7.3 (Bishop–Jones). *Let $G \subset PSL(2, \mathbb{C})$ be a finitely generated Kleinian group with infinite limit set $\Lambda \subset \mathbb{CP}_1$. If Λ is not totally disconnected, then either Λ is a single geometric circle, or else* $\dim(\Lambda) > 1$.

Now, to clinch the argument, we would like to somehow use this to estimate the Hausdorff dimension of Λ_t as $t \to 1$. A second general result of Bishop and Jones [**12**] provides the machinery needed to do this:

Lemma 7.4 (Bishop–Jones). *Let $G \subset PSL(2, \mathbb{C})$ be a finitely generated Kleinian group, and suppose that G is the limit of subgroups $G_j \subset PSL(2, \mathbb{C})$, in the sense of convergence of a set of generators. If Λ is the limit set of G, and if Λ_j is the limit sets of G_j, then* $\liminf \dim(\Lambda_j) \geqslant \dim(\Lambda)$.

In our case, it follows that there is an $\epsilon > 0$ such that

$$\dim(\Lambda_t) > 1 \text{ whenever } t > 1 - \epsilon,$$

since otherwise there would exist a sequence $t_j \nearrow 1$ with $\dim(\Lambda_{t_j}) \leqslant 1$, and hence with $\liminf \dim(\Lambda_{t_j}) \leqslant 1 < \dim(\Lambda)$, in contradiction to the lemma. For $t \in (1-\epsilon, 1)$, the corresponding ϕ-invariant conformally flat metrics g_t therefore have $s < 0$.

To wrap up our proof of Proposition 7.1, it just remains to show that

$$B = \left\{t \in (0,1) \;\middle|\; g_t \text{ has } \ker\left(\Delta + \frac{s}{6}\right) \neq 0\right\}, \tag{10}$$

is a discrete set , so that the subset $A \subset B \cap [\frac{1}{2}, 1-\epsilon]$ defined by

$$A = \left\{t \in (0,1) \;\middle|\; g_t \text{ has } s \equiv 0\right\} \tag{11}$$

is consequently finite. In principle this could again be done by appealing to the perturbation theory of the spectrum of the Yamabe Laplacian, but, just for fun, let us give a twistorial proof, in the spirit of §5. Indeed, the Penrose transform tells us that we can re-express (10) as

$$B = \left\{t \in (0,1) \;\middle|\; H^1(Z_t, \mathcal{O}(K^{1/2})) \neq 0\right\},$$

where Z_t is the twistor space of (Y, g_t). However, Z_t is constructed by taking an open set in $\mathbb{CP}_3$ (namely the inverse image of some open neighborhood of $D_t \subset \mathbb{HP}_1$ via the twistor projection) and making identifications using the two biholomorphisms given by the $PSL(4,\mathbb{C})$-transformations arising from the generators of our Schottky group $\mathbb{Z} * \mathbb{Z}$ via the inclusions

$$PSL(2,\mathbb{C}) \hookrightarrow PGL(2,\mathbb{H}) \hookrightarrow PSL(4,\mathbb{C}).$$

But our generators depend algebraically on t, so we may extend our construction of the Z_t to t in an open neighborhood $\mathcal{U}$ of $(0,1) \subset \mathbb{C}$, giving us an analytic family of complex 3-folds. The semi-continuity principle **[6]** then implies that the set of $t \in \mathcal{U}$ for which $H^1(Z_t, \mathcal{O}(K^{1/2})) \neq 0$ is closed in the analytic Zariski topology; in other words, it is either discrete, or else is all of $\mathcal{U}$. Since $\Delta + \frac{s}{6}$ is a positive operator for t small, this shows that the set B defined by (10) is discrete. The compact set $A \subset B$ defined by (11) is thus finite, as claimed.

It follows that the element of A defined by

$$t_0 = \sup\{t \in (0,1) \mid g_t \text{ has } s > 0\}$$

has a neighborhood $(t_0 - \varepsilon, t_0 + \varepsilon) \subset (0,1)$ which does not meet $A - \{t_0\}$. For t in this neighborhood, the scalar curva ture s of g_t then has the same sign as $t_0 - t$, and our proof of Proposition 7.1 is therefore done.

8. A Vanishing Theorem

At this point, we have constructed an interesting family of locally conformally flat metrics on the orbifold $X = Y/\mathbb{Z}_2$, where $Y = 2(S^1 \times S^3)$. However, our aim is to eventually smooth the orbifold singularities of these metrics in order to produce a similar family of anti-self-dual metrics on a simply connected manifold. To carry this out, we will need to know that the Kodaira–Spencer deformation theory is unobstructed for the corresponding family of twistor spaces. In fact, the relevant vanishing theorem easily follows from a decade-old unpublished paper of Eastwood and Singer [**18**], whose beautiful ideas will be given a self-contained exposition in this section.

Let (Y, g) be an oriented, locally conformally flat Riemannian 4-manifold, and let $\mathscr{C}$ denote the complete presheaf of *conformal Killing fields* on (Y, g), defined by setting

$$\mathscr{C}_U = \{v \in \mathcal{E}_U(TY) \mid \pounds_v g \propto g\}$$

for any open set $U \subset Y$; here, as throughout, $\mathcal{E}$ is used to indicate the C^∞ sections of a given vector bundle. Now observe that there is a rank-15 vector bundle $F \to Y$, equipped with a flat connection ∇, such that $\mathscr{C}$ is the sheaf of parallel sections of (F, ∇). Indeed, if $U \subset Y$ is any *simply connected* open set, then we can conformally immerse U onto an open subset of S^4 by means of the *developing map* [**41**], and $\mathscr{C}_U$ is thereby identified with the 15-dimensional space $\mathfrak{so}(5, 1)$ of global conformal Killing fields on the round 4-sphere S^4. As we pass from one such choice of U to another, these identifications will be related to one another by elements of $SO^{\uparrow}(5, 1)$, acting on $\mathfrak{so}(5, 1)$ via the adjoint representation. These elements of $SO^{\uparrow}(5, 1)$ are exactly the transition of functions of F, relative to a collection of local trivializations of F in which the flat connection ∇ has vanishing connection 1-forms.

Now we could certainly construct a fine resolution of $\mathscr{C}$ by just considering the F-valued differential forms on Y, but this would involve using vector bundles of rather high rank. A more efficient resolution was first discovered by Gasqui and Goldschmidt [**24**] using Spencer cohomology, and later rediscovered by Eastwood and Rice [**17**] in the setting of Bernstein–Gelfand–Gelfand resolutions. This resolution takes the form

$$0 \to \mathscr{C} \to \mathcal{E}(TY) \xrightarrow{L_0} \mathcal{E}(\odot_0^2\Lambda^1) \xrightarrow{L_1} \mathcal{E}\begin{pmatrix} \odot_0^2\Lambda^+ \\ \oplus \\ \odot_0^2\Lambda^- \end{pmatrix} \xrightarrow{L_2} \mathcal{E}(\odot_0^2\Lambda^1) \xrightarrow{L_3} \mathcal{E}(TY) \to 0$$

where $\odot_0^2$ indicates the trace-free symmetric-tensor-product of a vector bundle with itself. Here L_0 is the first-order operator

$$L_0(v) = \text{trace-free part of } \mathcal{L}_v g$$

which measures the way the conformal class $[g]$ is distorted by the flow of a given vector field. The next step in the sequence is the linearization $L_1 = DW$ of the Weyl curvature tensor; and for our purposes, it will be important to recognize this second-order differential operator can be decomposed as $DW = DW_+ + DW_-$, where the operators

$$\begin{aligned} DW_+ : \ & \mathcal{E}(\odot_0^2\Lambda^1) \to \ \mathcal{E}(\odot_0^2\Lambda^+), \\ DW_- : \ & \mathcal{E}(\odot_0^2\Lambda^1) \to \ \mathcal{E}(\odot_0^2\Lambda^-) \end{aligned}$$

are the linearizations

$$DW_\pm(h) = \frac{d}{dt} W_\pm(g + th)\Big|_{t=0}$$

of the self-dual and anti-self-dual Weyl curvatures. The next step is again a second-order operator, and is given by

$$L_2 = (DW_+)^* - (DW_-)^*.$$

The sequence then culminates with the first-order operator

$$L_3 = L_0^*.$$

Clearly, all of these operators are conformally invariant, provided that each bundle in the complex is given the correct conformal weight.

Since each of the sheaves in the Gasqui–Goldschmidt resolution is fine, and hence acyclic, the abstract de Rham theorem [**58**] immediately tells us that the sheaf cohomology of Y with coefficients in $\mathscr{C}$ is exactly the cohomology of the corresponding complex of global sections:

$$H^p(Y, \mathscr{C}) = \frac{\ker L_p}{\operatorname{im} L_{p-1}}.$$

However, the Gasqui–Goldschmidt resolution is also an elliptic complex; thus, provided Y is *compact*, we have [**25**]

$$H^p(Y, \mathscr{C}) = \ker L_p \cap \ker L_{p-1}^*,$$

by a generalized form of the Hodge theorem. Since

$$DW_\pm = \frac{1}{2}(L_1^* \pm L_2),$$

this immediately gives us the following result:

Proposition 8.1. *Let (Y, g) be any compact, oriented, locally conformally flat 4-manifold. Then*

$$\begin{aligned} H^2(Y, \mathscr{C}) &= \ker(DW_+)^* \oplus \ker(DW_-)^* \\ &\subset \mathcal{E}_Y(\odot_0^2 \Lambda^+) \oplus \mathcal{E}_Y(\odot_0^2 \Lambda^+). \end{aligned}$$

Using this key observation, it is now easy to deduce the desired vanishing result:

Theorem 8.2 (Eastwood–Singer). *Let g be any conformally flat metric on the oriented 4-manifold*

$$Y = k(S^1 \times S^3) = \underbrace{(S^1 \times S^3)\# \cdots \#(S^1 \times S^3)}_{k},$$

$k \geqslant 1$, and let Z be the twistor space of (Y, g). Then

$$DW_+ : \mathcal{E}(\odot_0^2 \Lambda^1) \to \mathcal{E}(\odot_0^2 \Lambda^+)$$

is surjective on (Y, g), and

$$H^2(Z, \mathcal{O}(TZ)) = 0.$$

Proof. By Serre duality, $H^2(Z, \mathcal{O}(TZ))$ is the dual of $H^1(Z, \Omega^1(K))$. However, the latter sheaf cohomology group corresponds, via the Penrose transform [5], to $\mathbb{C} \otimes \ker(DW_+)^*$. By Proposition 8.1, it therefore suffices to show that $H^2(Y, \mathscr{C}) = 0$, where $\mathscr{C}$ is once again the sheaf of local conformal Killing fields of $[g]$.

Now Y can be obtained from S^4 by replacing k pairs of balls with k tubes modeled on $S^3 \times \mathbb{R}$. This allows us to express Y as the union

$$Y = U \cup V$$

of open sets

$$U = S^4 - \{p_1, \dots, p_{2k}\}$$

and

$$V \approx \underbrace{(S^3 \times \mathbb{R}) \sqcup \cdots \sqcup (S^3 \times \mathbb{R})}_{k}$$

such that

$$U \cap V \approx \underbrace{(S^3 \times \mathbb{R}) \sqcup \cdots \sqcup (S^3 \times \mathbb{R})}_{2k}.$$

We may thus proceed by examining the Mayer–Vietoris sequence

$$\to H^1(U \cap V, \mathscr{C}) \to H^2(U \cup V, \mathscr{C}) \to H^2(U, \mathscr{C}) \oplus H^2(V, \mathscr{C}) \to \cdots \tag{12}$$

Indeed, notice that V and $U \cap V$ are homotopy equivalent to disjoint unions of 3-spheres, while U is homotopy equivalent to a bouquet of 3-spheres. In particular, each of these sets is a disjoint union of simply connected spaces. Since $\mathscr{C}$ is the sheaf of covariantly constant sections of a flat rank-15 vector bundle (F, ∇), the restriction of $\mathscr{C}$ to any of these open sets may be identified with the constant sheaf $\mathbb{R}^{15}$, and the relevant sheaf cohomology therefore amounts to singular cohomology with coefficients in the Abelian group $\mathbb{R}^{15}$. By the homotopy invariance of singular cohomology, we thus have

$$\begin{aligned} H^1(U \cap V, \mathscr{C}) &\cong H^1(\underbrace{S^3 \sqcup \cdots \sqcup S^3}_{2k}, \mathbb{R}^{15}) = 0, \\ H^2(U, \mathscr{C}) &\cong H^2(\underbrace{S^3 \vee \cdots \vee S^3}_{2k-1}, \mathbb{R}^{15}) = 0, \\ H^2(V, \mathscr{C}) &\cong H^2(\underbrace{S^3 \sqcup \cdots \sqcup S^3}_{k}, \mathbb{R}^{15}) = 0, \end{aligned}$$

and (12) therefore tells us that

$$H^2(Y, \mathscr{C}) = H^2(U \cup V, \mathscr{C}) = 0,$$

as claimed. □

It is perhaps worth remarking that, for *any* compact oriented locally conformally flat 4-manifold (Y, g), one may use the index theorem to show that $\ker(DW_+)^*$ and $\ker(DW_-)^*$ have the same dimension. Proposition 8.1 and the Penrose transform therefore imply that

$$\dim_{\mathbb{R}} H^2(Y, \mathscr{C}) = 2 \dim_{\mathbb{C}} H^2(Z, \mathcal{O}(TZ)).$$

Thus the vanishing of $H^2(Y, \mathscr{C})$ is actually necessary, as well as sufficient, for the deformation theory of Z to be unobstructed.

9. Existence Results

We will now assemble the results of the last several sections into a proof of Theorem A.

Proposition 9.1. *For any integer $k \geqslant 6$, the connected sum*

$$k\overline{\mathbb{CP}}_2 = \underbrace{\overline{\mathbb{CP}}_2 \# \cdots \# \overline{\mathbb{CP}}_2}_{k}$$

admits a real-analytic 1*-parameter family of anti-self-dual conformal metrics* $[g_t]$, $t \in [a,b]$, *such that* $[g_a]$ *contains a metric with* $s > 0$ *everywhere, while* $[g_b]$ *contains a metric with* $s < 0$ *everywhere.*

PROOF. We once again let $Y = (S^1 \times S^3)\#(S^1 \times S^3)$, and let $X = Y/\mathbb{Z}_2$. Equip X with a conformally-flat scalar-flat orbifold metric g_0 which belongs to a real-analytic 1-parameter family of conformally flat metrics g_t, $t \in (-\varepsilon, \varepsilon)$ such that each g_t has $s > 0$ when $t < 0$, $s < 0$ when $t > 0$, and $\ker(\Delta + s/6) = 0$ for all $t \neq 0$; for example, Proposition 7.1 constructs such a family, after replacing t with $t - t_0$ for some t_0. Since Theorem 8.2 guarantees that the twistor space Z of (Y, g_0) has $H^2(Z, \mathcal{O}(TZ)) = 0$, Theorem 4.5 therefore tells us that, for any integer $\ell \geqslant 0$, there exists a connected 2-parameter family of anti-self-dual conformal metrics $[g_{(\mathfrak{u}_1,\mathfrak{u}_2)}]$, $(\mathfrak{u}_1, \mathfrak{u}_2) \in (0,\epsilon) \times (-\epsilon, \epsilon)$ on $\widetilde{X}\#\ell\overline{\mathbb{CP}}_2$ which arises from a 2-parameter standard deformation of the singular space $Z_0 = [\widetilde{Z}/\mathbb{Z}_2] \cup \widetilde{Z}_{EH} \cup \widetilde{Z}_{FS}$; here $\widetilde{X}$ is once again the oriented 4-manifold obtained from the orbifold X by replacing each singular point of X with a 2-sphere of self-intersection -2. Moreover this 2-parameter standard deformation can be chosen so that its restriction to $\mathfrak{u}_2 = t$ is a 1-parameter standard deformation of the complex 3-fold with normal crossings arising from $(X, [g_t])$, for every real number t in a neighborhood of 0. Let $a < 0$ and $b > 0$ be choices of t in this neighborhood. Theorems 6.3 and 6.4 then tell us that for any sufficiently small $w > 0$, the conformal class $[g_{(w,a)}]$ contains a metric with $s > 0$, while the conformal class $[g_{(w,b)}]$ contains a metric with $s < 0$. Thus $[g_t] := [g_{(w,t)}]$, $a \leqslant t \leqslant b$, is a family of ASD conformal metrics on $\widetilde{X}\#\ell\overline{\mathbb{CP}}_2$ with the desired scalar-curvature behavior .

It remains only to unmask the identity of the manifold $\widetilde{X}\#\ell\overline{\mathbb{CP}}_2$. To do this, first notice we may cut up Y into three punctured 4-spheres

in a manner which is compatible with the involution. Thus X can be expressed as a connected sum

$$X = V\#V\#V$$

of three copies of the orbifold $S^4/\mathbb{Z}_2$, where the connect sum is carried out in the vicinity of *nonsingular* points of V, and where the $\mathbb{Z}_2$ acts on $S^4 \subset \mathbb{R}^5$ by reflection through an axis. Hence

$$\widetilde{X} = \widetilde{V} \# \widetilde{V} \# \widetilde{V},$$

where $\widetilde{V}$ is obtained from V by replacing the two isolated singularities of $V = S^4/\mathbb{Z}_2$ by 2-spheres of self-intersection -2. However, $\overline{\mathbb{CP}}_2 \# \overline{\mathbb{CP}}_2$ is diffeomorphic to $\widetilde{V}$. Indeed, if E_1 and E_2 are the standard 2-spheres of self-intersection -1 in $\overline{\mathbb{CP}}_2 \# \overline{\mathbb{CP}}_2$, then $E_1 + E_2$ and $E_1 - E_2$ are represented by disjoint embedded 2-spheres of self-intersection -2, and $\overline{\mathbb{CP}}_2 \# \overline{\mathbb{CP}}_2$ is obtained by gluing tubular neighborhoods of these two 2-spheres along their boundaries. We therefore have $X \approx 6\overline{\mathbb{CP}}_2$, and hence $M \approx k\overline{\mathbb{CP}}_2$, where $k = 6 + \ell$. □

As a corollary, we now obtain one of the central results of this paper:

Theorem 9.2. *For any integer $k \geqslant 6$, the connected sum $k\overline{\mathbb{CP}}_2$ admits scalar-flat anti-self-dual metrics.*

PROOF. Consider the smooth family of conformal classes $[g_t]$ constructed in Proposition 9.1, and let $h_t \in [g_t]$ be any smooth family of metrics representing these conformal classes. Let λ_t denote the smallest eigenvalue of the Yamabe Laplacian $(\Delta + s/6)$ for the metric h_t. Then λ_t is a continuous function of t. But Proposition 9.1 tells us that $\lambda_a > 0$, whereas $\lambda_b < 0$. By continuity, there is thus some $c \in [a, b]$ for which $\lambda_c = 0$. Let u be a unit-integral eigenfunction of the Yamabe Laplacian $(\Delta + s/6)$ of h_c with eigenvalue $\lambda_c = 0$. By the minimum principle, u is a positive function. Thus $g = u^2 h_c$ is a scalar-flat anti-self-dual metric on $k\overline{\mathbb{CP}}_2$ for the given value of $k \geqslant 6$. □

Now this by no means represents the first construction ever of SFASD metrics on simply connected compact 4-manifolds. However, all the previous results depended on an essentially different idea: namely, that any Kähler metric on a complex surface with $s \equiv 0$ is SFASD. Through this observation, Yau's existence theorem for Ricci-flat Kähler metrics on $K3$ surfaces **[60]** provided a crucial early family of examples which largely drove the subsequent development of the entire subject. Much later, the present author and his collaborators showed **[40]** that $\mathbb{CP}_2 \# k\overline{\mathbb{CP}}_2$ admits scalar-flat anti-self-dual metrics if $k \geqslant 14$. The proof of this last result depends on a refinement of Theorem 4.4, set up so that the constructed twistor spaces carries a special divisor whose existence implies that the

ASD conformal class contains a Kähler metric. For a related re-proof of the existence of Calabi–Yau metrics on $K3$, see [**48**].

Putting these previous results together with Corollary 9.2, we have thus proved Theorem A:

Theorem 9.3. *The following smooth* 4*-manifolds admit scalar-flat anti-self-dual metrics*:

(*i*) $k\overline{\mathbb{CP}}_2$, *for every* $k \geqslant 6$;
(*ii*) $\mathbb{CP}_2 \# k\overline{\mathbb{CP}}_2$, *for every* $k \geqslant 14$; *and*
(*iii*) $K3$.

In particular, each of these simply connected compact 4*-manifolds admits optimal metrics; and any optimal metric on any one of them is SFASD.*

10. Nonexistence Results

We have now seen that many simply connected 4-manifolds admit non-Einstein optimal metrics. However, related ideas will now allow us to show there are also many simply connected 4-manifolds which do *not* admit optimal metrics. To see this, we begin by introducing a new concept:

Definition 10.1. Let M be a smooth compact oriented 4-dimensional manifold (respectively, orbifold). We will say that M admits an *anorexic sequence* if there is a sequence g_j of smooth Riemannian metrics (respectively, orbifold metrics) on M for which $\int s^2 d\mu \to 0$ and $\int |W_+|^2 d\mu \to 0$.

When a manifold admits such a sequence, we then know the value of $\mathcal{I}_{\mathcal{R}}(M)$, and stand a very good chance of determining whether it admits an optimal metric:

Lemma 10.2. *Let M be a smooth compact oriented* 4*-manifold which admits an anorexic sequence. Then any optimal metric on M is SFASD. Moreover,*

$$\mathcal{I}_{\mathcal{R}}(M) = -8\pi^2(\chi + 3\tau)(M).$$

PROOF. Recall that equation (8) tells us that

$$\mathcal{K}(g) = -8\pi^2(\chi + 3\tau)(M) + 2\int_M \left(\frac{s^2}{24} + 2|W_+|^2\right) d\mu_g \, .$$

If there is an anorexic sequence, the infimum of the right-hand side is thus obtained by dropping the curvature integral. Moreover, a metric minimizing $\mathcal{K}$ would necessarily have $s \equiv 0$ and $W_+ \equiv 0$. □

Now imagine a curvaceous young 4-manifold who, bedazzled by the glamorous starlets with optimal metrics she has been reading about in the tabloids, suddenly decides to go on a starvation diet to get rid of all that unwanted curvature. If she has the wrong body type, this misguided procedure will be dangerous to her health, and she will merely succeed in putting herself in the hospital:

Proposition 10.3. *Let M be a smooth compact oriented 4-manifold which admits an anorexic sequence. Then M does not admit an optimal metric if*

- $b_+(M) \geqslant 4$; *or*
- $b_+(M) = 2$; *or*
- $b_+(M) = 3$, $\pi_1(M) = 0$, *and M is not diffeomorphic to $K3$; or*
- $b_+(M) = 1$, $\pi_1(M) = 0$, *and M is not diffeomorphic to $\mathbb{CP}_2 \# k\overline{\mathbb{CP}}_2$ for some $k \geqslant 10$.*

PROOF. By Lemma 10.2, an optimal metric on such an M would necessarily be SFASD. However, Propositions 3.4 and 3.5 show that we would then obtain a contradiction in any of the scenarios considered above. □

Now we come to our main method of construction [**46**]:

Lemma 10.4. *Let X be an oriented compact 4-dimensional orbifold with only isolated singularities modeled on $\mathbb{R}^4/\mathbb{Z}_2$. Let $\widetilde{X}$ be the smooth oriented 4-manifold obtained by replacing each singular point by a 2-sphere of self-intersection -2. If X admits an anorexic sequence, then so does $\widetilde{X}$. Moreover, if there is an anorexic sequence on X with the property that $\int |r|^2 d\mu \to 0$, then $\widetilde{X}$ also admits an anorexic sequence with this property.*

PROOF. If h is an orbifold metric on X with

$$\int s^2 d\mu < \varepsilon$$

and

$$\int |W_+|^2 d\mu < \varepsilon,$$

we will show that $\widetilde{X}$ has a metric g with $\int s^2 d\mu < 2\varepsilon$ and $\int |W_+|^2 d\mu < 2\varepsilon$. Moreover, if h also has

$$\int |\mathring{r}|^2 d\mu < \varepsilon,$$

then we will be able to arrange for g to also satisfy

$$\int |\mathring{r}|^2 d\mu < 2\varepsilon.$$

To do this, we choose geodesic normal coordinates around a given orbifold point, so that we have

$$h = \left[\sum_j (dx^j)^2\right] + \alpha$$

in these coordinates, where α is a ($\mathbb{Z}_2$-invariant) smooth symmetric tensor field on a neighborhood of the origin in $\mathbb{R}^4$ with $|\alpha| < C|\vec{x}|^2$. We would like to delete a ball of small radius ρ around the origin, and glue in a copy of the Eguchi–Hanson metric on T^*S^2, with very small length scale. Recall that the restriction of the Eguchi–Hanson metric to the complement of the zero section in T^*S^2 is isometric to the metric

$$g_{EH,\epsilon} = \frac{d\varrho^2}{1-(\frac{\epsilon}{\varrho})^4} + \varrho^2\left(\sigma_1^2 + \sigma_2^2 + \left[1 - \left(\frac{\epsilon}{\varrho}\right)^4\right]\sigma_3^2\right)$$

on $(\epsilon, \infty) \times S^3/\mathbb{Z}_2$, where $\{\sigma_j\}$ is the standard left-invariant co-frame on $S^3/\mathbb{Z}_2 = SO(3)$; the constant $\epsilon > 0$ is herein referred to as the *length scale*. Now, for any fixed $\rho > 0$, this family of metrics converges uniformly in the C^2 topology to the Euclidean metric on the annulus $\varrho \in [\rho/2, \rho]$ as $\epsilon \to 0$. If $\varphi : (0, \infty) \to [0, 1]$ is a bump function which is $\equiv 0$ on $(0, 1/2]$ and $\equiv 1$ on $[1, \infty)$, then, for any fixed ρ, the metrics

$$g_{\epsilon,\rho} = \varphi\left(\tfrac{\varrho}{\rho}\right) h + \left[1 - \varphi\left(\tfrac{\varrho}{\rho}\right)\right] g_{EH,\epsilon}$$

therefore converge in the C^2 norm to

$$g_{0,\rho} = \left[\sum_j (dx^j)^2\right] + \varphi\left(\tfrac{\varrho}{\rho}\right)\alpha$$

on the annulus $\varrho \in [\rho/2, \rho]$, and in particular the curvature tensors of these metrics converge uniformly in the annulus to the curvature of $g_{0,\rho}$. On the other hand, since α is of magnitude $O(\varrho^2)$, the first and second coordinate partial derivatives of $g_{0,\rho}$ are uniformly bounded as $\rho \to 0$.

Thus we can choose a sequence of $(\epsilon_j, \rho_j) \to (0,0)$ such that the sectional curvatures of the metrics $g_j = g_{\epsilon_j,\rho_j}$ are uniformly bounded on the transition annuli $\varrho \in [\rho_j/2, \rho_j]$, while the volumes of these annuli simultaneously tend to zero. For j far out in the sequence, the transition annulus therefore makes a contribution to $\int s^2 d\mu$, $\int |\mathring{r}|^2 d\mu$, or $\int |W_+|^2 d\mu$ which is as small as we like — for example, smaller than the given ε divided by the number of orbifold singularities of X. We now take g to be given by such a choice of g_j in each glued region, h on the complement X minus a collection of balls or radius ρ_j about its orbifold singularities, and equal to the Eguchi–Hanson metric with length scale ϵ_j near the added 2-spheres. Since the Eguchi–Hanson metric has $r \equiv 0$ and $W_+ \equiv 0$, the only possible source of increase of $\int s^2 d\mu$, $\int |\mathring{r}|^2 d\mu$, or $\int |W_+|^2 d\mu$ comes from the transition annuli, which already have under control, and so we have succeeded in producing a metric g on $\widetilde{X}$ with all the claimed properties. □

Here is a simple application of this Lemma:

Lemma 10.5. *The 4-manifold $4\overline{\mathbb{CP}}_2$ admits an anorexic sequence.*

PROOF. Consider the involution of $S^3 \times S^1 \subset \mathbb{H} \times \mathbb{C}$ given by $(q, z) \mapsto (\overline{q}, \overline{z})$. Now equip $S^3 \times S^1$ with the product of the unit-sphere metric on S^3 and the radius-ϵ metric on S^1. These metrics descend to orbifold metrics on $(S^3 \times S^1)/\mathbb{Z}_2$ with bounded sectional curvature, but with arbitrarily small volume; thus we obtain an anorexic sequence of such metrics by taking any sequence $\epsilon_j \to 0$. The 4-manifold $\widetilde{X}$ obtained from X by replacing its orbifold singularities with 2-spheres of self-intersection -2 therefore also admits anorexic sequences by Lemma 10.4. But $X = V \# V$, where $V = S^4/\mathbb{Z}_2$, and so $\widetilde{X} = \widetilde{V} \# \widetilde{V}$. However, as we already noted in the proof of Proposition 9.1, $\widetilde{V} = 2\overline{\mathbb{CP}}_2$, and so $\widetilde{X} = 4\overline{\mathbb{CP}}_2$. □

We thus obtain our first nonexistence result:

Proposition 10.6. *The 4-manifold $4\overline{\mathbb{CP}}_2$ does not admit optimal metrics.*

PROOF. By Lemmas 10.2 and 10.5, an optimal metric on $4\overline{\mathbb{CP}}_2$ would have to be SFASD. However, $(2\chi + 3\tau)(4\overline{\mathbb{CP}}_2) = 0$, so Proposition 3.3 would imply that any SFASD metric on the simply connected 4-manifold $4\overline{\mathbb{CP}}_2$ would be hyper-Kähler.

But such a metric would entail the existence of non-trivial self-dual harmonic 2-forms, which is excluded here, since $b_+(4\overline{\mathbb{CP}}_2) = 0$. □

A rather more important application of 10.4 is the following:

Lemma 10.7. *The 4-manifold $\mathbb{CP}_2\#9\overline{\mathbb{CP}}_2$ admits an anorexic sequence.*

PROOF. Consider the involution of $S^2 \times T^2$ which is obtained as the product of a 180° rotation of S^2 around an axis and the Weierstrass involution of an elliptic curve:

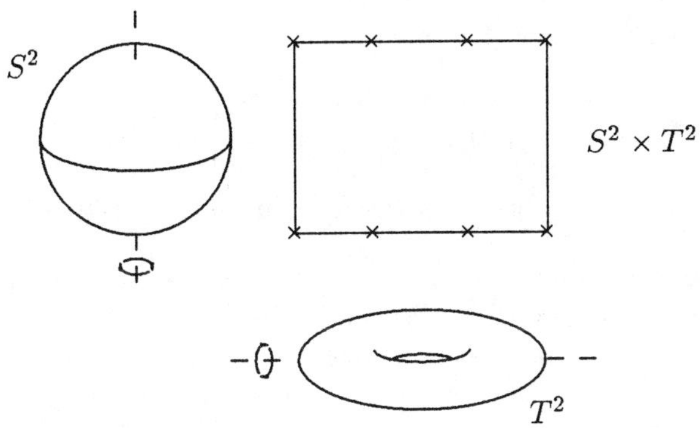

This involution has exactly 8 fixed points. Let $\widetilde{X}$ be the manifold which desingularizes the orbifold $X = [S^2 \times T^2]/\mathbb{Z}_2$ by replacing each of the resulting eight singular points with an S^2 of self-intersection -2.

Now it is easy to see that $S^2 \times T^2$ admits sequences of metrics with bounded sectional curvature, but with volume tending to zero: namely, equip T^2 with a sequence of flat metrics of smaller and smaller area, and take the Riemannian product of these metrics with the standard round metric on S^2. Moreover, the metrics given by this explicit recipe are all $\mathbb{Z}_2$-invariant, and so give rise to a sequence of orbifold metrics on $[S^2 \times T^2]/\mathbb{Z}_2$ with bounded sectional curvature for which the total volume tends to zero. Such a sequence is anorexic, and also has the special property that $\int |r|^2 d\mu \to 0$. By Lemma 10.4, $\widetilde{X}$ therefore also admits such a special anorexic sequence.

It only remains to show that $\widetilde{X}$ is diffeomorphic to $\mathbb{CP}_2\#9\overline{\mathbb{CP}}_2$. To se this, think of $S^2 \times T^2$ as $\mathbb{CP}_1 \times E$, where E is an elliptic curve.

Then $\widetilde{X}$ becomes a complex surface which has a branched double cover biholomorphic to $\mathbb{CP}_1 \times E$ blown up at eight points. This complex surface is simply connected, and it has Kodaira dimension $-\infty$ because it contains a $\mathbb{CP}_1$ with trivial normal bundle. By the Enriques–Kodaira classification [7], any such complex surface is rational, and hence must be diffeomorphic to either $S^2 \times S^2$ or a connected sum $\mathbb{CP}_2 \# k\overline{\mathbb{CP}}_2$. However,

$$(2\chi + 3\tau)(\widetilde{X}) = \frac{1}{4\pi^2} \int \left(\frac{s^2}{24} + 2|W_+|^2 - \frac{|\mathring{r}|^2}{2} \right) d\mu$$

must vanish, since the right-hand side will certainly tend to zero for our special anorexic sequence. Since

$$\begin{aligned} (2\chi + 3\tau)(S^2 \times S^2) &= 8, \\ (2\chi + 3\tau)(\mathbb{CP}_2 \# k\overline{\mathbb{CP}}_2) &= 9 - k, \end{aligned}$$

it thus follows that $\widetilde{X}$ must be diffeomorphic to $\mathbb{CP}_2 \# 9\overline{\mathbb{CP}}_2$. □

In fact, one does not need to appeal to any classification machinery to check that $\widetilde{X} \approx \mathbb{CP}_2 \# 9\overline{\mathbb{CP}}_2$. Working with one's bare hands [**46**], it is not difficult to show that, with the fixed complex structure used above, $\widetilde{X}$ is precisely the complex surface obtained by iteratively blowing up $\mathbb{CP}_2$ at a configuration of points arranged as in the following diagram, in which a pair of adjacent points on a line means that one is to blow up the first point, and then blow up the resulting exceptional divisor at the point corresponding to the direction of the line, while the adjacent triple of points on a line has an analogous interpretation:

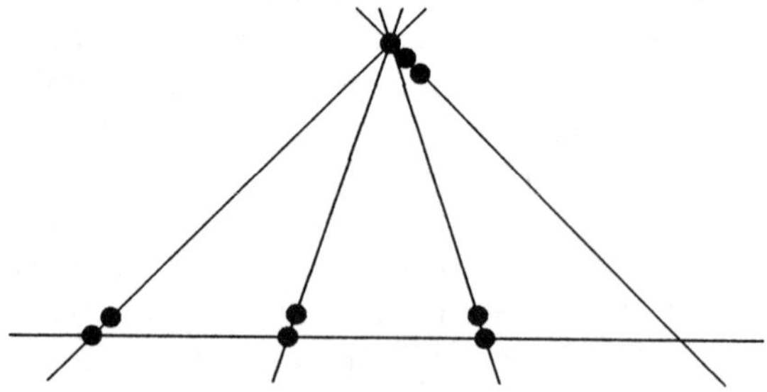

If we think of the right-hand line as the line at infinity, the horizontal line as the x-axis, and the three other lines as $x = 0$, $x = 1$ and $x = t$, then the elliptic curves $y^2 = Ax(x-1)(x-t)$ foliate an open dense set of the blow-up, and arise from the E factor of $[\mathbb{CP}_1 \times E]/\mathbb{Z}_2$.

Corollary 10.8. *The simply connected* 4*-manifold* $\mathbb{CP}_2\#9\overline{\mathbb{CP}}_2$ *does not admit optimal metrics.*

PROOF. Lemma 10.7 allows us to apply the last clause of Proposition 10.3. □

To build more complicated examples, first consider the *wormhole space* obtained by equipping $\mathbb{R}^4 - \{0\}$ with the metric

$$g_{wh} = \left(1 + \frac{\epsilon}{|\vec{x}|^2}\right)^2 \sum_j (dx^j)^2.$$

Because $1 + \epsilon/|\vec{x}|^2$ is a harmonic function, this conformally flat metric is scalar-flat. However, rewriting this metric in polar coordinates as

$$g_{wh} = \left(\varrho + \frac{\epsilon}{\varrho}\right)^2 \left[\frac{d\varrho^2}{\varrho^2} + h_{S^3}\right],$$

where h_{S^3} is the standard metric on the unit 3-sphere S^3, we immediately see that there is an isometry $\varrho \mapsto \epsilon/\varrho$ of the wormhole which interchanges the two ends of $\mathbb{R}^4 - \{0\} \approx \mathbb{R} \times S^3$. Thus our wormhole connects two asymptotically Euclidean ends, but has $s \equiv 0$ and $W_+ \equiv 0$. Now, on any fixed annulus $\varrho \in [\rho/2, \rho]$, the wormhole metric uniformly converges in C^2 to the Euclidean metric, and so exactly the same argument used to glue in Eguchi–Hanson metrics allows us to join two manifolds with $\int s^2 d\mu < \varepsilon$ and $\int |W_+|^2 d\mu < \varepsilon$ by a wormhole neck so as to obtain a new manifold with $\int s^2 d\mu < 3\varepsilon$ and $\int |W_+|^2 d\mu < 3\varepsilon$. Thus:

Lemma 10.9. *Suppose that M_1 and M_2 are two smooth compact oriented* 4*-manifolds which admit anorexic sequences. Then their connected sum $M_1\#M_2$ admits anorexic sequences, too.*

The final basic building block we will need is the Burns metric [**44**]. This is an asymptotically flat metric on $\overline{\mathbb{CP}}_2 - \{p\}$ with $s \equiv 0$ and $W_+ \equiv 0$. Rescaled versions of this metric, restricted to the complement of a $\mathbb{CP}_1$, are explicitly given by

$$g_{B,\epsilon} = \frac{d\varrho^2}{1 - (\frac{\epsilon}{\varrho})^2} + \varrho^2 \left(\sigma_1^2 + \sigma_2^2 + \left[1 - \left(\frac{\epsilon}{\varrho}\right)^2\right]\sigma_3^2\right)$$

as a metric on $(\epsilon, \infty) \times S^3$. As the length-scale parameter ϵ tends to zero, we once again get uniform C^2 convergence to the Euclidean metric on any fixed annulus $\varrho \in [\rho/2, \rho]$, and the same gluing argument as before therefore gives us the following result:

Lemma 10.10. *Suppose that M is a smooth compact oriented 4-manifold which admits an anorexic sequence. Then $M\#\overline{\mathbb{CP}}_2$ admits anorexic sequences, too.*

We now prove the first part of Theorem C:

Theorem 10.11. *Let j and k be integers such that $j \geqslant 2$ and $k \geqslant 9j$. Then the simply connected 4-manifold $j\mathbb{CP}_2\#k\overline{\mathbb{CP}}_2$ does not admit optimal metrics.*

PROOF. By induction on j, Lemmas 10.7 and 10.9 imply that the connected sum $j\mathbb{CP}_2\#9j\overline{\mathbb{CP}}_2$ of j copies of $\mathbb{CP}_2\#9\overline{\mathbb{CP}}_2$ admits an anorexic sequence. Lemma 10.10 and induction on $k - 9j$ then gives us the existence of an anorexic sequence on $j\mathbb{CP}_2\#k\overline{\mathbb{CP}}_2$ for any $k \geqslant 9j$, $j \geqslant 2$. Proposition 10.3 therefore tells us that there is no optimal metric on any of these non-spin simply connected 4-manifolds with $b_+ = j \geqslant 2$. □

Now let us consider some "exotic" smooth structures on $\mathbb{CP}_2\#9\overline{\mathbb{CP}}_2$. Let X denote the complex orbifold $[\mathbb{CP}_1 \times E]/\mathbb{Z}_2$ used in the proof of Lemma 10.7, and let us observe that we have a holomorphic orbifold submersion $X \to \mathbb{CP}_1/\mathbb{Z}_2$ given by projection to the first factor. Near any nonsingular point of $\mathbb{CP}_1/\mathbb{Z}_2$, this is a locally trivial holomorphic E-bundle. Let p and q be two relatively prime integers $\geqslant 2$, and choose two nonsingular points $a, b \in \mathbb{CP}_1/\mathbb{Z}_2$. Note that $\mathbb{CP}_1/\mathbb{Z}_2$ is really just a copy of $\mathbb{CP}_1$ with two marked points which are considered to be orbifold points of order 2. Let us now also mark the points a and b, and consider them to be orbifold points of order p and q. At the same time, we modify V to obtain a new orbifold $V^0_{p,q}$ by replacing the fiber over a with $E/\mathbb{Z}_p$ and the fiber over b with $E/\mathbb{Z}_q$, where the two actions are generated by translation of E of order p and order q respectively. This can be done via a *logarithmic transformation* in the sense of Kodaira; for example, in a neighborhood of a biholomorphic to the open unit disk $D \subset \mathbb{C}$, we $E \times D$ with $[E \times D]/\mathbb{Z}_p$, where $\mathbb{Z}_p$ acts on E as before, and simultaneously acts on D via the action generated by $z \mapsto e^{2\pi i/p}z$. We then have a holomorphic orbifold submersion from $V^0_{p,q}$ to our orbifold $\mathbb{CP}_1$ with four orbifold points. Now choose the compatible flat metric on E with unit area and use a partition of unity to patch the a product metric on V with local product metrics on $[E \times D]/\mathbb{Z}_p$ and $[E \times D]/\mathbb{Z}_q$. The result is a Riemannian submersion orbifold metric on V. If we now scale down the fiber E with keeping the metric on our orbifold $\mathbb{CP}_1$ fixed, the result is therefore a family of metrics on $V^0_{p,q}$ with volume tending to zero while the curvature remains uniformly bounded [**14, 46**]. In particular, this is an anorexic sequence on $V^0_{p,q}$, and Lemma 10.4 tells us that the complex surface $M^0_{p,q}$ obtain by replacing each of the 8 orbifold singularities of $V^0_{p,q}$ by (-2)-curves also admits anorexic sequences; moreover, the manifolds $M^0_{p,q}\#\ell\overline{\mathbb{CP}}_2$ all admit anorexic sequences, too, as a consequence of Lemma 10.10. However, the manifolds $M_{p,q}$ are the so-called *Dolgachev surfaces.* These Dolgachev surfaces are all homeomorphic to $\mathbb{CP}_2\#9\overline{\mathbb{CP}}_2$, but Donaldson or Seiberg–Witten invariants can be used [**22**] to show that no two of them are diffeomorphic. Moreover, the corresponding smooth structures on the blow-ups remain distinct, no matter how many times we blow up, and these smooth structures are moreover all distinct from the standard one on $\mathbb{CP}_2\#k\overline{\mathbb{CP}}_2$. By Proposition 10.3, it follows that none of these smooth manifolds admits an optimal metric, even though $\mathbb{CP}_2\#k\overline{\mathbb{CP}}_2$ *does* admit optimal metrics for $k \geqslant 14$.

The story is similar for homotopy $K3$ surfaces. Namely, we can view $T^4/\mathbb{Z}_2$ as an orbifold elliptic fibration over $\mathbb{CP}_1/\mathbb{Z}_2$, and so modify it by logarithmic transforms of odd order at one fiber. The resulting orbifolds V^1_q then admit anorexic sequences as before, as do the 4-manifolds M^1_q obtained by replacing their singular points by 2-spheres of self-intersection -2. These manifolds are homeomorphic to $K3$ surfaces, but as smooth manifolds they are distinct, not only from $K3$, but also from each other. Proposition 10.3 thus tells us that none of them admits optimal metrics, even though they are homeomorphic to $K3$, which *does* admit optimal metrics. We have thus proved Theorem B:

Theorem 10.12. *The existence or nonexistence of optimal metrics depends on the choice of smooth structure. In particular, the topological 4-manifolds $K3$ and $\mathbb{CP}_2\#k\overline{\mathbb{CP}}_2$, $k \geqslant 14$, admit infinitely many exotic smooth structures for which no optimal metric exists, even though each also admits a "standard" smooth structure for which optimal metrics do exist.*

A similar construction [**46**] yields anorexic sequences on many exotic manifolds homeomorphic to $j\mathbb{CP}_2\#k\overline{\mathbb{CP}}_2$. Consider the complex orbifold $[\Sigma \times E]/\mathbb{Z}_2$, where Σ is a hyperelliptic Riemann surface of genus $\mathbf{g}$:

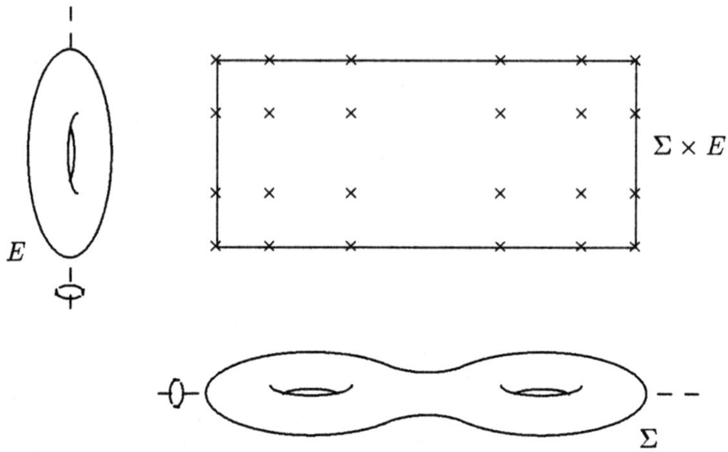

Giving E its compatible flat metric of various areas, products with a fixed metric on Σ give us anorexic sequences on the orbifold $V^{\mathbf{g}} = [\Sigma \times E]/\mathbb{Z}_2$, and we also get anorexic sequences on the orbifold $V^{\mathbf{g}}_q$ obtained by performing a logarithmic transform to one fiber by the previous gluing argument. If $M^{\mathbf{g}}_q$ denotes the complex surface obtained by replacing the singularities of $V^{\mathbf{g}}_q$ by (-2)-curves, then Lemma 10.4 guarantees that $M^{\mathbf{g}}_q$ also admits anorexic sequences, and Lemma 10.10 then tells us that $M^{\mathbf{g}}_q \# \ell \overline{\mathbb{CP}}_2$ admits anorexic sequences, too. Now $M^{\mathbf{g}}_q$ is a simply connected complex surface with $p_g = \mathbf{g}$ and $c_1^2 = 0$; and it is non-spin if either $\mathbf{g}$ or q is even. For any $\ell > 0$, Theorem 2.3 tells us that $M^{\mathbf{g}}_q \# \ell \overline{\mathbb{CP}}_2$ is homeomorphic to $(2\mathbf{g}+1)\mathbb{CP}_2 \# (10\mathbf{g} + 9 + \ell)\overline{\mathbb{CP}}_2$; and we also get the analogous statement for $\ell = 0$ if q is even. However, by varying q, gauge theory can be used to show **[23, 27]** that we obtain infinitely many distinct smooth structures in this way for any fixed $\mathbf{g}$ and ℓ. By Proposition 10.3, we therefore have the following result:

Theorem 10.13. *For any odd $j \geqslant 1$ and any $k \geqslant 5j+4$, the topological manifold $j\mathbb{CP}_2 \# k\overline{\mathbb{CP}}_2$ admits infinitely distinct smooth structures for which no optimal metric exists.*

Lemma 10.9 now allows us to produce more examples of smooth 4-manifolds with anorexic sequences by taking connected sums of our previous examples. Determining whether we obtain distinct differentiable structures in this way is a cutting-edge problem in gauge-theory, however, and it is only by applying the sophisticated new machinery of Bauer and Furuta [**9, 8**] that a result of this type can be obtained. Specifically, if $\mathbf{g}$ is *odd*, then the bandwidth argument of [**35**] shows that the $M^{\mathbf{g}}\#M^1_q\#\ell\overline{\mathbb{CP}}_2$ run through infinitely many differentiable structures on $(2\mathbf{g}+4)\mathbb{CP}_2\#(10\mathbf{g}+28+\ell)\overline{\mathbb{CP}}_2$ as we vary the even integer $q \geqslant 2$, and that $2M^{\mathbf{g}}\#2M^1_q\#\ell\overline{\mathbb{CP}}_2$ runs through infinitely many differentiable structures on $(4\mathbf{g}+8)\mathbb{CP}_2\#(20\mathbf{g}+56+\ell)\overline{\mathbb{CP}}_2$. Since Proposition 10.3 tells us that none of these spaces can admit optimal metrics, we thus obtain the second part of Theorem C:

Theorem 10.14. *If j and k are integers, with $j \geqslant 5$, $j \not\equiv 0 \bmod 8$, and $k \geqslant 9j$, then the topological manifold $j\mathbb{CP}_2\#k\overline{\mathbb{CP}}_2$ admits infinitely distinct smooth structures for which no optimal metric exists.*

11. Concluding Remarks

While we have seen that many topological 4-manifolds fail to admit optimal metrics for many choices of smooth structure, the techniques developed here do not by any means allow us to determine whether or not an optimal metric exists for an *arbitrary* smooth structure. The reason is that the arguments deployed in §10 are heavily dependent on the existence of anorexic sequences of metrics, whereas such sequence simply do not exist for many smooth structures. For example, if X is a minimal complex surface of general type, and if $M = X\#\ell\overline{\mathbb{CP}}_2$ is the complex surface obtained from it by blowing up ℓ points, then Seiberg–Witten theory can be used to show [**47**] that any metric on M satisfies

$$\int_M \left(\frac{s^2}{24}+2|W_+|^2\right)d\mu \geqslant \frac{8\pi^2}{3}c_1^2(X), \tag{13}$$

so there are certainly no anorexic metrics on such an M. If X is simply connected and ℓ is sufficiently large, however, this 4-manifold is homeomorphic to one of the manifolds treated by Theorem C, and so represents an exotic smooth structure on this topological 4-manifold which is simply not amenable to treatment with the current technology.

The main obstacle to progress on this front is that the estimate (13) does not appear to be sharp; if we compare it with known minimizing sequences for $\int s^2 d\mu$ on a complex surface M of general type with minimal model X, we only obtain upper and lower bounds

$$\frac{16\pi^2}{3}c_1^2(X) - 8\pi^2(\chi + 3\tau)(M) \leqslant \mathcal{I}_{\mathcal{R}}(M) \leqslant 8\pi^2 c_1^2(X) - 8\pi^2(\chi + 3\tau)(M)$$

for $\mathcal{I}_{\mathcal{R}}(M)$. These bounds certainly do not allow one to compute $\mathcal{I}_{\mathcal{R}}$, but they do allow us to estimate with sufficient accuracy to be able to know that at least several different values of $\mathcal{I}_{\mathcal{R}}$ must occur for many fixed homeotypes. An exact computation of $\mathcal{I}_{\mathcal{R}}$ for such examples would have many interesting ramifications, and should be considered as an outstanding open problem in the subject.

While current technology does not suffice to compute $\mathcal{I}_{\mathcal{R}}$ for many of the most interesting 4-manifolds, the analogous invariants

$$\begin{aligned} \mathcal{I}_s(M) &= \inf_g \int_M |s_g|^{n/2} d\mu_g \\ \mathcal{I}_r(M) &= \inf_g \int_M |r|_g^{n/2} d\mu_g \end{aligned}$$

arising from the scalar and Ricci curvatures *do* turn out to be exactly computable for complex surfaces of general type and many of their connect sums [**47, 36**]. It may therefore come as a surprise that one key trick used in computations of $\mathcal{I}_r$ is closely related to the techniques developed here. Indeed, the Gauss–Bonnet and signature formulæ tell one that

$$\int_M |r|^2 d\mu = -8\pi^2(2\chi + 3\tau)(M) + 8\int_M \left(\frac{s^2}{24} + \frac{1}{2}|W_+|^2\right) d\mu$$

so that the existence of an anorexic sequence is certainly quite sufficient to allow one to calculate $\mathcal{I}_r$. However, the curious difference is that the available Seiberg–Witten estimate analogous to (13) for this particular combination of s and W_+ turns out to typically be sharp. For example, if $M = X\#\ell\overline{\mathbb{CP}}_2$ is a complex surface of general type with minimal model X, one obtains the estimate

$$\int_M \left(\frac{s^2}{24} + \frac{1}{2}|W_+|^2\right) d\mu \geqslant 2\pi^2 c_1^2(X),$$

and one can actually find sequences of metrics for which the left-hand side approaches the expression on the right; thus, an exact formula

$$\mathcal{I}_r(M) = 8\pi^2[c_1^2(X) + \ell]$$

emerges from the discussion. Moreover, related arguments show that this infimum is unattained whenever $\ell > 0$. For details and further applications, see [**47, 36**].

While Theorem A may shed a fair amount of light on the existence of SFASD metrics on compact 4-manifolds, it by no means closes the book on the subject. For example, we still do not know whether such metrics exist on $5\overline{\mathbb{CP}}_2$. The existence of SFASD metrics on $\mathbb{CP}_2\#k\overline{\mathbb{CP}}_2$ for $10 \leqslant k \leqslant 13$ is also not covered by Theorem A, although Yann Rollin and Michael Singer seem to have recently made considerable progress concerning these manifolds. The global structure of the moduli space of SFASD metrics still remains a mystery. And it would obviously be of the greatest interest to sharpen Proposition 3.5 so as to say something about diffeotype when $b_+ = 0$, or to definitively handle the non-simply connected case.

We now have many examples of optimal metrics on compact 4-manifolds, essentially falling into two classes: the Einstein metrics, and the scalar-flat anti-self-dual metrics. Of course, we can cheaply obtain further examples by reversing the orientation of SFASD manifolds to make them self-dual instead of anti-self-dual. But, such trickery aside, there do not really seem to be any other known examples of optimal metrics on compact 4-manifolds. In particular, it seems that all known examples of optimal metrics are critical points of $\int s^2 d\mu$, and so, by optimality, also of $\int |W_+|^2 d\mu$. Now a metric on a 4-manifold is a critical point of $\int |W_+|^2 d\mu$ iff it has vanishing Bach tensor [**11**]. Are there any optimal metrics on compact 4-manifolds that are not Bach-flat? Are there scalar-flat optimal metrics which are Bach-flat, but neither self-dual nor anti-self-dual? Both of these questions should illustrate the degree to which we still remain fundamentally ignorant as to the true nature of general optimal metrics, even in dimension four.

Acknowledgments. I would like to express my profound gratitude to Dominic Joyce, who originally convinced me, by an entirely different line of reasoning, that Theorems 6.3 and 6.4 ought to be true. I would also like to warmly thank Mike Eastwood and Michael Singer for granting me permission to present some of their unpublished results in §8. Finally, I would like to thank Chris Bishop, Mark de Cataldo, Sorin Popescu, and Michael Taylor for helpful suggestions concerning a number of important technical details.

References

1. C. Allendoerfer and A. Weil, *The Gauss-Bonnet theorem for Riemannian polyhedra*, Trans. Am. Math. Soc. **53** (1943), 101–129.
2. A. Altman and S. Kleiman, *Introduction to Grothendieck Duality Theory*, Lect. Notes Math. **146**, 1970.
3. M. F. Atiyah, *Green's functions for self-dual four-manifolds*, in: Mathematical Analysis and Applications, Part A, vol. 7 of Adv. in Math. Suppl. Stud., Academic Press, New York, 1981, pp. 129–158.
4. M. F. Atiyah, N. J. Hitchin, and I. M. Singer, *Self-duality in four-dimensional Riemannian geometry*, Proc. Roy. Soc. London Ser. A **362** (1978), 425–461.
5. T. N. Bailey and M. A. Singer, *Twistors, massless fields and the Penrose transform*, in: Twistors in Mathematics and Physics, London Math. Soc. Lect. Notes **156**, Cambridge Univ. Press, Cambridge, 1990, pp. 299–338.
6. C. Bănică and O. Stănăşilă, *Algebraic Methods in the Global Theory of Complex Spaces*, Editura Academiei, Bucharest, 1976.
7. W. Barth, C. Peters, and A. V. de Ven, *Compact Complex Surfaces*, Springer Verlag, 1984.
8. S. Bauer, *A stable cohomotopy refinement of Seiberg-Witten invariants. II*, Invent. Math. **155** (2004), 21–40.
9. S. Bauer and M. Furuta, *A stable cohomotopy refinement of Seiberg-Witten invariants. I*, Invent. Math. **155** (2004), 1–19.
10. M. Berger, *Riemannian Geometry During the Second Half of the Twentieth Century*, Am. Math. Soc., Providence, RI, 2000. Reprint of the 1998 original.
11. A. Besse, *Einstein Manifolds*, Springer Verlag, 1987.
12. C. J. Bishop and P. W. Jones, *Hausdorff dimension and Kleinian groups*, Acta Math. **179** (1997), 1–39.
13. J.-P. Bourguignon, *Les variétés de dimension 4 à signature non nulle dont la courbure est harmonique sont d'Einstein*, Invent. Math. **63** (1981), 263–286.
14. J. Cheeger and M. Gromov, *Collapsing Riemannian manifolds while keeping their curvature bounded. I*, J. Differ. Geom. **23** (1986), 309–346.
15. S. Donaldson and R. Friedman, *Connected sums of self-dual manifolds and deformations of singular spaces*, Nonlinearity **2** (1989), 197–239.
16. S. K. Donaldson, *An application of gauge theory to four-dimensional topology*, J. Differ. Geom. **18** (1983), 279–315.
17. M. G. Eastwood and J. W. Rice, *Conformally invariant differential operators on Minkowski space and their curved analogues*, Comm. Math. Phys. **109** (1987), 207–228.
18. M. G. Eastwood and M. A. Singer, *On the geometry of twistor spaces*, unpublished manuscript, 1991.
19. T. Eguchi and A. J. Hanson, *Self-dual solutions to Euclidean gravity*, Ann. Phys. **120** (1979), 82–106.

20. M. Freedman, *On the topology of 4-manifolds*, J. Differ. Geom. **17** (1982), 357–454.
21. R. Friedman, *Global smoothings of varieties with normal crossings*, Ann. Math. (2) **118** (1983), 75–114.
22. R. Friedman, *Vector bundles and* SO(3)*-invariants for elliptic surfaces*, J. Am. Math. Soc. **8** (1995), 29–139.
23. R. Friedman and J. W. Morgan, *On the diffeomorphism types of certain algebraic surfaces. I*, J. Differ. Geom. **27** (1988), 297–369.
24. J. Gasqui and H. Goldschmidt, *Théorèmes de dualité en géométrie conforme*, C. R. Acad. Sci. Paris Sér. I Math. **294** (1982), 99–102.
25. J. Gasqui and H. Goldschmidt, *Déformations Infinitésimales des Structures Conformes Plates*, Progress in Mathematics **52**, Birkhäuser Boston Inc., Boston, MA, 1984.
26. D. Gilbarg and N. S. Trudinger, *Elliptic Partial Differential Equations of Second Order*, Springer Verlag, Berlin, second ed., 1983.
27. R. E. Gompf and A. I. Stipsicz, *4-Manifolds and Kirby Calculus*, Am. Math. Soc., Providence, RI, 1999.
28. P. Griffiths and J. Harris, *Principles of Algebraic Geometry*, Wiley-Interscience [John Wiley & Sons], New York, 1978.
29. M. Gromov, *Structures métriques pour les variétés riemanniennes*, CEDIC, Paris, 1981. Edited by J. Lafontaine and P. Pansu.
30. R. C. Gunning and H. Rossi, *Analytic Functions of Several Complex Variables*, Prentice-Hall Inc., Englewood Cliffs, N.J., 1965.
31. F. Hirzebruch, *Topological Methods in Algebraic Geometry*, Springer Verlag, 1966.
32. N. J. Hitchin, *On compact four-dimensional Einstein manifolds*, J. Differ. Geom. **9** (1974), 435–442.
33. N. J. Hitchin, *Kählerian twistor spaces*, Proc. London Math. Soc. (3), **43** (1981), 133–150.
34. G. Horrocks, *A construction for locally free sheaves*, Topology **7** (1968), 117–120.
35. M. Ishida and C. LeBrun, *Spin manifolds, Einstein metrics, and differential topology*, Math. Res. Lett. **9** (2002), 229–240.
36. M. Ishida and C. LeBrun, *Curvature, connected sums, and Seiberg-Witten theory*, Comm. Anal. Geom. **11** (2003), 809–836.
37. D. Joyce, *Constant scalar curvature metrics on connected sums*, Int. J. Math. Math. Sci. (2003), 405–450.
38. T. Kato, *Perturbation Theory for Linear Operators*, Springer Verlag, New York, 1966.
39. J. Kim, *On the scalar curvature of self-dual manifolds*, Math. Ann. **297** (1993), 235–251.
40. J. Kim, C. LeBrun, and M. Pontecorvo, *Scalar-flat Kähler surfaces of all genera*, J. Reine Angew. Math. **486** (1997), 69–95.
41. N. H. Kuiper, *On conformally-flat spaces in the large*, Ann. Math . (2) **50** (1949), 916–924.
42. J. Lafontaine, *Remarques sur les variétés conformément plates*, Math. Ann. **259** (1982), 313–319.

43. C. LeBrun, *On the topology of self-dual 4-manifolds*, Proc. Am. Math. Soc. **98** (1986), 637–640.
44. C. LeBrun, *Counter-examples to the generalized positive action conjecture*, Comm. Math. Phys. **118** (1988), 591–596.
45. C. LeBrun, *Anti-self-dual metrics and Kähler geometry*, in: Proc. Intern. Congress Mathematicians, Vol. 1, 2 (Zürich, 1994), Basel, 1995, Birkhäuser, pp. 498–507.
46. C. LeBrun, *Kodaira dimension and the Yamabe problem*, Comm. Anal. Geom. **7** (1999), 133–156.
47. C. LeBrun, *Ricci curvature, minimal volumes, and Seiberg-Witten theory*, Inv. Math. **145** (2001), 279–316.
48. C. LeBrun and M. Singer, *A Kummer-type construction of self-dual 4-manifolds*, Math. Ann. **300** (1994), 165–180.
49. J. Lee and T. Parker, *The Yamabe problem*, Bull. Am. Math. Soc. **17** (1987), 37–91.
50. B. Maskit, *Kleinian Groups*, Springer Verlag, Berlin, 1988.
51. S. Nayatani, *Patterson-Sullivan measure and conformally flat metrics*, Math. Z. **225** (1997), 115–131.
52. A. Newlander and L. Nirenberg, *Complex analytic coordinates in almost complex manifolds*, Ann. Math. (2) **65** (1957), 391–404.
53. R. Penrose, *Nonlinear gravitons and curved twistor theory*, General Relativity and Gravitation **7** (1976), 31–52.
54. M. H. Protter and H. F. Weinberger, *Maximum Principles in Differential Equations*, Prentice-Hall Inc., Englewood Cliffs, N.J., 1967.
55. R. Schoen and S.-T. Yau, *Conformally flat manifolds, Kleinian groups and scalar curvature*, Invent. Math. **92** (1988), 47–71.
56. J.-P. Serre, *Modules projectifs et espaces fibrés à fibre vectorielle*, in: Séminaire P. Dubreil, M.-L. Dubreil-Jacotin et C. Pisot, 1957/58, Fasc. 2, Exposé 23, Secrétariat mathématique, Paris, 1958, p. 18.
57. J. A. Thorpe, *Some remarks on the Gauss-Bonnet formula*, J. Math. Mech. **18** (1969), 779–786.
58. R. O. Wells, Jr., *Differential Analysis on Complex Manifolds*, Springer Verlag, New York, second ed., 1980.
59. S. T. Yau, *On the curvature of compact Hermitian manifolds*, Invent. Math. **25** (1974), 213–239.
60. S. T. Yau, *On the Ricci curvature of a compact Kähler manifold and the complex Monge-Ampère equation. I*, Comm. Pure Appl. Math. **31** (1978), 339–411.
61. S. T. Yau, *Open problems in geometry*, in: Differential Geometry: Partial Differential Equations on Manifolds (Los Angeles, CA, 1990), Am. Math. Soc., Providence, RI, 1993, pp. 1–28.

Amoebas of Algebraic Varieties and Tropical Geometry

Grigory Mikhalkin[†]

University of Utah
Utah, USA

University of Toronto
Toronto, Canada

POMI RAS
St. Petersburg, Russia

This survey consists of two parts. Part 1 is devoted to amoebas. These are images of algebraic subvarieties in $\mathbb{C}^n \supset (\mathbb{C}^*)^n$ under the logarithmic moment map. The amoebas have essentially piecewise-linear shape if viewed *at large*. Furthermore, they degenerate to certain piecewise-linear objects called *tropical varieties* whose behavior is governed by algebraic geometry over the so-called tropical semifield. Geometric aspects of tropical algebraic geometry is the content of Part 2. We pay special attention to tropical curves. Both parts also include relevant applications of the theories. Part 1 of this survey is a revised and updated version of the report [**28**].

[†]Supported in part by the NSF.

***Different Faces of Geometry*, Edited by Donaldson *et al.*,**
Kluwer Academic/Plenum Publishers, New York, 2004

Part 1. AMOEBAS

1. Definition and Basic Properties of Amoebas

1.1. Definitions. Let $V \subset (\mathbb{C}^*)^n$ be an algebraic variety. Recall that $\mathbb{C}^* = \mathbb{C} \setminus 0$ is the group of complex numbers under multiplication. Let $\mathrm{Log} : (\mathbb{C}^*)^n \to \mathbb{R}^n$ be defined by $\mathrm{Log}(z_1, \dots, z_n) \to (\log|z_1|, \dots, \log|z_n|)$.

Definition 1.1 (Gelfand–Kapranov–Zelevinski [**11**]). The *amoeba* of V is $\mathcal{A} = \mathrm{Log}(V) \subset \mathbb{R}^n$.

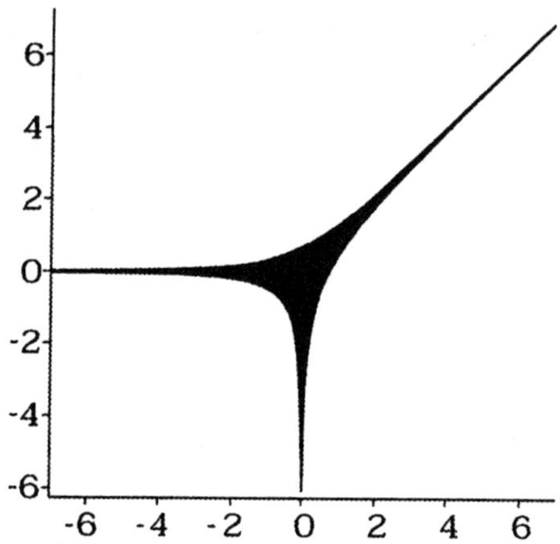

FIGURE 1. The amoeba of the line $\{x + y + 1 = 0\} \subset (\mathbb{C}^*)^2$.

Proposition 1.2 ([**11**]). *The amoeba* $\mathcal{A} \subset \mathbb{R}^n$ *is a closed set with a nonempty complement.*

If $\mathbb{C}T \supset (\mathbb{C}^*)^n$ is a closed n-dimensional toric variety and $\overline{V} \subset \mathbb{C}T$ is a compactification of V then we say that $\mathcal{A}$ is the amoeba of $\overline{V}$ (recall that $\mathcal{A}$ is also the amoeba of $V = \overline{V} \cap (\mathbb{C}^*)^n$). Thus we can speak about amoebas of projective varieties once the coordinates in $\mathbb{CP}^n$, or at least an action of $(\mathbb{C}^*)^n$, is chosen.

If $\mathbb{C}T$ is equipped with a $(\mathbb{C}^*)^n$-invariant symplectic form then we can also consider the corresponding moment map $\overline{\mu} : \mathbb{C}T \to \Delta$ (see

[**4**],[**11**]), where Δ is the convex polyhedron associated to the toric variety $\mathbb{C}T$ with the given symplectic form. The polyhedron Δ is a subset of $\mathbb{R}^n$ but it is well defined only up to a translation. In this case we can also define the *compactified amoeba* of $\overline{V}$.

Definition 1.3 ([**11**]). The *compactified amoeba* of V is

$$\overline{\mathcal{A}} = \overline{\mu}(V) \subset \Delta.$$

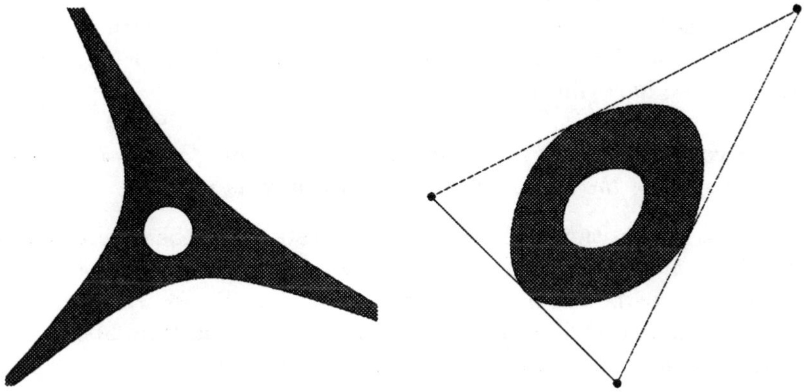

FIGURE 2. [**26**] An amoeba $\mathcal{A}$ and a compactified amoeba $\overline{\mathcal{A}}$.

Remark 1.4. Maps $\overline{\mu}|_{(\mathbb{C}^*)^n}$ and Log are submersions and have the same real n-tori as fibers. Thus $\mathcal{A}$ is mapped diffeomorphically onto $\overline{\mathcal{A}} \cap \operatorname{Int}\Delta$ under a reparameterization of $\mathbb{R}^n$ onto $\operatorname{Int}\Delta$.

Using the compactified amoeba we can describe the behavior of $\mathcal{A}$ near infinity. Note that each face Δ' of Δ determines a toric variety $\mathbb{C}T' \subset \mathbb{C}T$. Consider $\overline{V}' = \overline{V} \cap \mathbb{C}T'$. Let $\overline{\mathcal{A}}'$ be the compactified amoeba of $\overline{V}'$.

Proposition 1.5 ([**11**]). *We have* $\overline{\mathcal{A}}' = \overline{\mathcal{A}} \cap \Delta'$.

This proposition can be used to describe the behavior of $\mathcal{A} \subset \mathbb{R}^n$ near infinity.

1.2. Amoebas at infinity. Consider a linear subspace $L \subset \mathbb{R}^n$ parallel to Δ' and with $\dim L = \dim \Delta'$. Let $H \subset \mathbb{R}^n$ be a supporting hyperplane for the convex polyhedron Δ at the face Δ', i.e., a hyperplane such that $\Delta \cap H = \Delta'$. Let $\vec{v}$ be an outwards normal vector to H. Let $\mathcal{A}_t^{\Delta'}$, $t > 0$, be the intersection of L with the result of translation of $\mathcal{A}$ by $-t\,\vec{v}$.

Recall that the *Hausdorff distance* between two closed sets $A, B \subset \mathbb{R}^n$ is

$$d_{\mathrm{Haus}}(A, B) = \max\{\sup_{a \in A} d(a, B), \sup_{b \in B} d(b, A)\},$$

where $d(a, B)$ is the Euclidean distance between a point a and a set B. We say that a sequence $A_t \subset \mathbb{R}^n$ converges to a set A' when $t \to \infty$ *with respect to the Hausdorff metric on compacts in* $\mathbb{R}^n$ if for any compact $K \subset \mathbb{R}^n$ we have $\lim_{t \to \infty} d_{\mathrm{Haus}}(A_t \cap K, A' \cap K) = 0$.

Proposition 1.6. *The subsets $\mathcal{A}_t^{\Delta'}$ converge to $\mathcal{A}'$ when $t \to \infty$ with respect to the Hausdorff metric on compacts in $\mathbb{R}^n$.*

This proposition can be informally restated in the case $n = 2$ and $\dim V = 1$. In this case Δ is a polygon and the amoeba $\mathcal{A}$ develops "tentacles" perpendicular to the sides of Δ (see Figure 3). The number of tentacles perpendicular to a side of Δ is bounded from above by the integer length of this side, i.e., one plus the number of the lattice points in the interior of the side.

Corollary 1.7. *For a generic choice of the slope of a line ℓ in $\mathbb{R}^n$ the intersection $\mathcal{A} \cap \ell$ is compact.*

1.3. Amoebas of hypersurfaces: concavity and topology of the complement. Forsberg, Passare, and Tsikh treated amoebas of hypersurfaces in [**10**]. In this case V is a zero set of a single polynomial $f(z) = \sum_j a_j z^j, a_j \in \mathbb{C}$. Here we use the multiindex notations $z = (z_1, \dots, z_n)$, $j = (j_1, \dots, j_n) \in \mathbb{Z}^n$ and $z^j = z_1^{j_1} \dots z_n^{j_n}$. Let

$$\Delta = \text{Convex hull}\{j \mid a_j \neq 0\} \subset \mathbb{R}^n \tag{1}$$

be the Newton polyhedron of f.

Theorem 1.8 (Forsberg–Passare–Tsikh [**10**]). *Each component of $\mathbb{R}^n \setminus \mathcal{A}$ is a convex domain in $\mathbb{R}^n$. There exists a locally constant function*

$$\mathrm{ind} : \mathbb{R}^n \setminus \mathcal{A} \to \Delta \cap \mathbb{Z}^n$$

which maps different components of the complement of $\mathcal{A}$ to different lattice points of Δ.

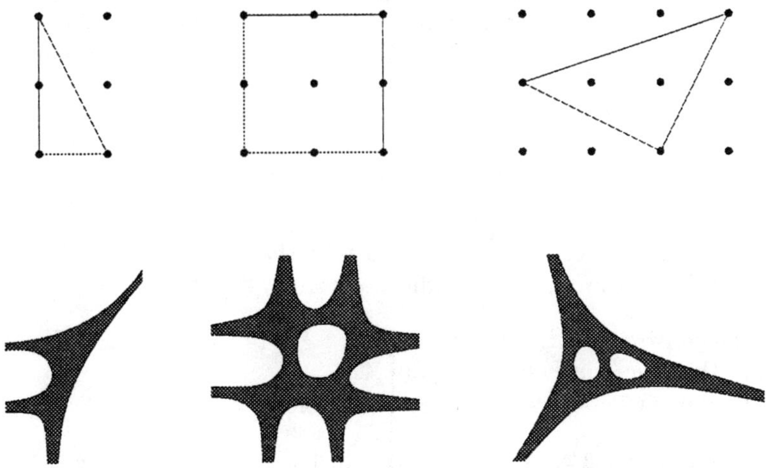

FIGURE 3. Amoebas together with their Newton polyhedra.

Corollary 1.9 ([**10**]). *The number of components of $\mathbb{R}^n \setminus \mathcal{A}$ is never greater then the number of lattice points of Δ.*

Theorem 1.8 and Proposition 1.6 indicate the dependence of the amoeba on the Newton polyhedron.

The inequality of Corollary 1.9 is sharp. This sharpness is a special case of Theorem 2.8. Also examples of amoebas with the maximal number of the components of the complement are supplied by Theorem 4.6.

The concavity of $\mathcal{A}$ is equivalent to concavity of its boundary. The boundary $\partial\mathcal{A}$ is contained in the critical value locus of $\mathrm{Log}|_V$. The following proposition also takes care of some interior branches of this locus.

Proposition 1.10 ([**26**]). *Let $D \subset \mathbb{R}^n$ be an open convex domain and V' be a connected component of $\mathrm{Log}^{-1}(D) \cap V$. Then $D \setminus \mathrm{Log}(V')$ is convex.*

1.4. Amoebas in higher codimension: concavity. The amoeba of a hypersurface is of full dimension in $\mathbb{R}^n$, $n > 1$, unless its Newton polyhedron Δ is contained in a line. The boundary $\partial\mathcal{A}$ at its generic point is a smooth $(n-1)$-dimensional submanifold. Its normal curvature form has no negative squares with respect to the outwards normal (because of convexity of components of $\mathbb{R}^n \setminus \mathcal{A}$). This property can be generalized to the nonsmooth points in the following way.

Definition 1.11. An open interval $D^1 \subset L$, where L is a straight line in $\mathbb{R}^n$, is called a *supporting* 1-*cap* for $\mathcal{A}$ if

- $D^1 \cap \mathcal{A}$ is nonempty and compact;
- there exists a vector $\vec{v} \in \mathbb{R}^n$ such that the translation of D^1 by $\varepsilon\vec{v}$ is disjoint from $\mathcal{A}$ for all sufficiently small $\varepsilon > 0$.

The convexity of the components of $\mathbb{R}^n \setminus \mathcal{A}$ can be reformulated as stating that *there are no 1-caps for* $\mathcal{A}$.

Similarly we may define higher-dimensional caps.

Definition 1.12. An open round disk $D^k \subset L$ of radius $\delta > 0$ in a k-plane $L \subset \mathbb{R}^n$ is called a *supporting k-cap* for $\mathcal{A}$ if

- $D^k \cap \mathcal{A}$ is non -empty and compact;
- there exists a vector $\vec{v} \in \mathbb{R}^n$ such that the translation of D^k by $\varepsilon\vec{v}$ is disjoint from $\mathcal{A}$ for all sufficiently small $\varepsilon > 0$.

Consider now the general case, where $V \subset (\mathbb{C}^*)^n$ is l-dimensional. Let $k = n-l$ be the codimension of V. The amoeba $\mathcal{A}$ is of full dimension in $\mathbb{R}^n$ if $2l \geqslant n$. The boundary $\partial\mathcal{A}$ at its generic point is a smooth $(n-1)$-dimensional submanifold. Its normal curvature form may not have more than $k-1$ negative squares with respect to the outwards normal. To see that note that a composition of $\mathrm{Log}|_V : V \to \mathbb{R}^n$ and any linear projection $\mathbb{R}^n \to \mathbb{R}$ is a pluriharmonic function.

Note that this implies that there are no k-caps for $\mathcal{A}$ at its smooth points. It turns out that there are no k-caps for $\mathcal{A}$ at the nonsmooth points as well and also in the case of $2l < n$ when $\mathcal{A}$ is $2l$-dimensional.

Proposition 1.13 (Local higher-dimensional concavity of $\mathcal{A}$). *If $V \subset (\mathbb{C}^*)^n$ is of codimension k then $\mathcal{A}$ does not have supporting k-caps.*

A global formulation of convexity was treated by André Henriques **[13]**.

Definition 1.14 (Henriques [**13**]). A subset $\mathcal{A} \subset \mathbb{R}^n$ is called *k-convex* if for any k-plane $L \subset \mathbb{R}^n$ the induced homomorphism $H_{k-1}(L \setminus \mathcal{A}) \to H_{k-1}(\mathbb{R}^n \setminus \mathcal{A})$ is injective.

Conjecturally the amoeba of a codimension k variety in $(\mathbb{C}^*)^n$ is k-convex. A proof of a somewhat weaker version of this statement is contained in [**13**].

1.5. Amoebas in higher codimension: topology of the complement. Recall that in the hypersurface case each component of $\mathbb{R}^n \setminus \mathcal{A}$ is connected and that there are not more than $\#(\Delta \cap \mathbb{Z}^n)$ such components. The correspondence between the components of the complement and the lattice points of Δ can be viewed as a cohomology class $\alpha \in H^0(\mathbb{R}^n \setminus \mathcal{A}; \mathbb{Z}^n)$ whose evaluation on a point in each component of $\mathbb{R}^n \setminus \mathcal{A}$ is the corresponding lattice point.

Similarly, when V is of codimension k there exists a natural class (cf. [**40**])

$$\alpha \in H^{k-1}(\mathbb{R}^n \setminus \mathcal{A}; H^k(T^n)),$$

where T^n is the real n-torus, the fiber of Log, $H^k(T^n) = H^k((\mathbb{C}^*)^n)$. The value of α on each $(k-1)$-cycle C in $\mathbb{R}^n \setminus \mathcal{A}$ and k-cycle C' in T^n is the linking number in $\mathbb{C}^n \supset (\mathbb{C}^*)^n$ of $C \times C'$ and the closure of V.

The cohomology class α corresponds to the linking with the fundamental class of V. Consider now the linking with smaller-dimensional homology of V.

Note that for an l-dimensional variety $V \subset (\mathbb{C}^*)^n$ we have $H_j(V) = 0$, $j > l$. Similarly, $H^c_j(V) = 0$, $j < l$, where H^c stands for homology with closed support. The linking number in $\mathbb{R}^n$ composed with Log : $(\mathbb{C}^*)^n \to \mathbb{R}^n$ defines the following pairing

$$H^c_l(V) \times H_{k-1}(\mathbb{R}^n \setminus \mathcal{A}) \to \mathbb{Z}.$$

Together with the Poincaré duality between $H^c_l(V)$ and $H_l(V)$ this pairing defines the homomorphism

$$\iota : H_{k-1}(\mathbb{R}^n \setminus \mathcal{A}) \to H_l(V).$$

Question 1.15. Is ι injective?

Recall that a subspace $L \subset H_l(V)$ is called *isotropic* if the restriction of the intersection form to L is trivial.

Proposition 1.16. *The image $\iota(H_{k-1}(\mathbb{R}^n \setminus \mathcal{A}))$ is isotropic in $H_l(V)$.*

Remark 1.17. A positive answer to Question 1.15 together with Proposition 1.16 would produce an upper bound for the dimension of $H_{k-1}(\mathbb{R}^n \setminus \mathcal{A})$.

One may also define similar linking forms for $H_j(\mathbb{R}^n \setminus \mathcal{A})$, $j \neq k-1$ (if $j > k-1$ then we can use ordinary homology $H_{n-j-1}(V)$ instead of homology with closed support).

The answer to Question 1.15 is currently unknown even in the case when $V \subset (\mathbb{C}^*)^2$ is a curve. In this case V is a Riemann surface and it is defined by a single polynomial. Let Δ be the Newton polygon of V. The genus of V is equal to the number of lattice points strictly inside Δ (see **[22]**) while the number of punctures is equal to the number of lattice points on the boundary of Δ). Thus the dimension of a maximal isotropic subspace of $H_1(V)$ is equal to $\#(\Delta \cap \mathbb{Z}^2)$ and Question 1.15 agrees with Corollary 1.9 for this case.

2. Analytic Treatment of Amoebas

This section outlines the results obtained by Passare and Rullgård in **[33]**, **[39]** and **[40]**.

We assume that $V \subset (\mathbb{C}^*)^n$ is a hypersurface in this section. Thus $V = \{f = 0\}$ for a polynomial $f : (\mathbb{C}^*)^n \to \mathbb{C}$ and we can consider $\Delta \subset \mathbb{R}^n$, the Newton polyhedron of V (see 1.3).

2.1. The Ronkin function N_f. Since f is a holomorphic function, $\log|f| : (\mathbb{C}^*)^n \setminus V \to \mathbb{R}$ is a pluriharmonic function. Furthermore, if we set $\log(0) = -\infty$ then we have a plurisubharmonic function

$$\log|f| : (\mathbb{C}^*)^n \to \mathbb{R} \cup \{-\infty\},$$

which is, obviously, strictly plurisubharmonic over V. Recall that a function F in a domain $\Omega \subset \mathbb{C}^n$ is called plurisubharmonic if its restriction to any complex line L is subharmonic, i.e., the value of F at each point $z \in L$ is smaller or equal than the average of the value of F along a small circle in L around z.

Let $N_f : \mathbb{R}^n \to \mathbb{R}$ be the push-forward of $\log|f|$ under the map $\mathrm{Log} : (\mathbb{C}^*)^n \to \mathbb{R}^n$, i.e.,

$$N_f(x_1, \ldots, x_n) = \frac{1}{(2\pi i)^n} \int\limits_{\mathrm{Log}^{-1}(x_1,\ldots,x_n)} \log|f(z_1, \ldots, z_n)| \frac{dz_1}{z_1} \wedge \cdots \wedge \frac{dz_n}{z_n}$$

(cf. [**38**]). This function was called the *Ronkin function* in [**33**]. It is easy to see that it takes real (finite) values even over $\mathcal{A} = \mathrm{Log}(V)$ where the integral is singular.

Proposition 2.1 (Ronkin–Passare–Rullgård [**33**], [**38**]). *The function $N_f : \mathbb{R}^n \to \mathbb{R}$ is convex. It is strictly convex over $\mathcal{A}$ and linear over each component of $\mathbb{R}^n \setminus \mathcal{A}$.*

This follows from plurisubharmonicity of $\log|f| : (\mathbb{C}^*)^n \to \mathbb{R}$, its strict plurisubharmonicity over V and its pluriharmonicity in $(\mathbb{C}^*)^n \setminus V$. Indeed the convexity of a function in a connected real domain is just a real counterpart of plurisubharmonicity. A harmonic function of one real variable has to be linear and thus a function of several real variables is real-plurisubharmonic if and only if it is convex. Over each connected component of $\mathbb{R}^n \setminus \mathcal{A}$ the function is linear as the push-forward of a pluriharmonic function.

Remark 2.2. Note that just the existence of a convex function N_f, which is strictly convex over $\mathcal{A}$ and linear over components of $\mathbb{R}^n \setminus \mathcal{A}$, implies that each component of $\mathbb{R}^n \setminus \mathcal{A}$ is convex.

Thus the gradient $\nabla N_f : \mathbb{R}^n \to \mathbb{R}^n$ is constant over each component E of $\mathbb{R}^n \setminus \mathcal{A}$. Recall the classical Jensen's formula in complex analysis

$$\frac{1}{2\pi i} \int\limits_{|z|=e^x} \log|f(z)| \frac{dz}{z} = Nx + \log|f(0)| - \sum_{k=1}^{N} \log|a_k|,$$

where $a_1, \ldots, a_N$ are the zeroes of f in $|z| < e^x$, if $f(0) \neq 0$ and $f(z) \neq 0$ if $|z| = e^x$. This formula implies that $\nabla N_f(E) \in \mathbb{Z}^n \cap \Delta$.

Proposition 2.3 (Passare–Rullgård [**33**]). *We have*

$$\mathrm{Int}\,\Delta \subset \nabla N_f(\mathbb{R}^n) \subset \Delta,$$

where $\mathrm{Int}\,\Delta$ *is the interior of the Newton polyhedron.*

Recall that Theorem 1.8 associates a lattice point to each component of $\mathbb{R}^n \setminus \mathcal{A}$.

Proposition 2.4 ([**33**]). *We have*

$$\nabla N_f(E) = \mathrm{ind}(E)$$

for each component E of $\mathbb{R}^n \setminus \mathcal{A}$.

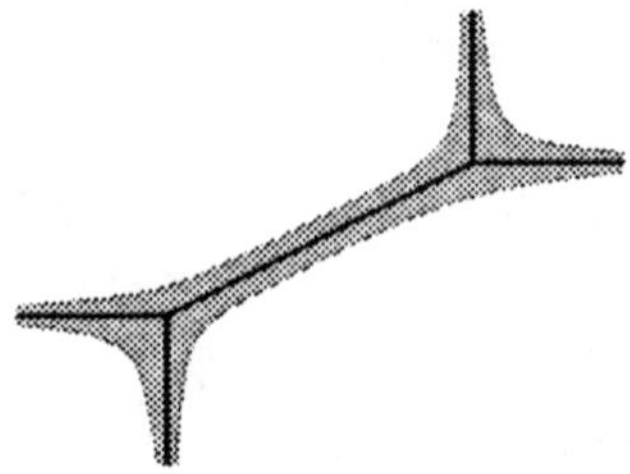

FIGURE 4. An amoeba and its spine.

2.2. The spine of amoeba. Passare and Rullgård [**33**] used N_f to define the *spine* of amoeba. Recall that N_f is piecewise-linear on $\mathbb{R}^n \smallsetminus \mathcal{A}$ and convex in $\mathbb{R}^n$. Thus we may define a superscribed convex linear function $N_f^{\inf ty}$ by letting

$$N_f^\infty = \max_E N_E,$$

where E runs over all components of $\mathbb{R}^n \smallsetminus E$ and $N_E : \mathbb{R}^n \to \mathbb{R}$ is the linear function obtained by extending $N_f|_E$ to $\mathbb{R}^n$ by linearity.

Definition 2.5 ([**33**]). The spine $\mathcal{S}$ of amoeba is the corner locus of N_f^∞, i.e., the set of points in $\mathbb{R}^n$ where N_f^∞ is not locally linear.

Note that $\mathcal{S} \subset \mathcal{A}$ and that s is a piecewise-linear polyhedral complex. The following theorem shows that $\mathcal{S}$ is indeed a spine of $\mathcal{A}$ in the topological sense.

Theorem 2.6 ([**33**], [**40**]). *The spine $\mathcal{S}$ is a strong deformational retract of the amoeba $\mathcal{A}$.*

Thus each component of $\mathbb{R}^n \smallsetminus \mathcal{S}$ (i.e., each maximal open domain where N_f^∞ is linear) contains a unique component of $\mathbb{R}^n \smallsetminus \mathcal{A}$.

2.3. Spine of amoebas and some functions on the space of complex polynomials. Now we return to the study of the spine $\mathcal{S} \subset \mathcal{A}$ of a complex amoeba. The spine $\mathcal{S}$ itself a certain amoeba over a non-Archimedean field K. It does not matter what is the field K as long as the corresponding hypersurface over K has the coefficients $a_j \in K$ with the correct valuations. We can find these valuations from N_f^∞ by taking its Legendre transform. Since N_f^∞ is obtained as a maximum of a finite number of linear function with integer slopes its Legendre transform has a support on a convex lattice polyhedron $\Delta \subset \mathbb{R}^n$. Let $c_\alpha \in \mathbb{R}$, $\alpha \in \Delta \cap \mathbb{Z}^n$ be the value of the Legendre transform of N_f^∞ at α. To present $\mathcal{S}$ as a non-Archimedean amoeba we choose $a_j \in K$ such that $v(a_j) = c_\alpha$.

For each $\alpha \in \Delta \cap \mathbb{Z}^n$ let U_α be the space of all polynomials whose Newton polyhedron is contained in Δ and whose amoeba contains a component of the complement of index α. The space of all polynomials whose Newton polyhedron is contained in Δ is isomorphic to $\mathbb{C}^N$, where $N = \#(\Delta \cap \mathbb{Z}^n)$. The subset $U_\alpha \subset \mathbb{C}^N$ is an open domain. Note that c_α defines a real-valued function on U_α. This function was used by Rullgård [**39**], [**40**] for the study of geometry of U_α.

2.4. Geometry of U_α. Fix $\alpha \in \Delta \cap \mathbb{Z}^n$. Consider the following function in the space $\mathbb{C}^N$ of all polynomials f whose Newton polyhedron is contained in Δ

$$u_\alpha(f) = \inf_{x \in \mathbb{R}^n} \frac{1}{(2\pi i)^n} \int\limits_{\mathrm{Log}^{-1}(x)} \log\left|\frac{f(z)}{z^\alpha}\right| \frac{dz_1}{z_1} \wedge \cdots \wedge \frac{dz_n}{z_n}, z \in (\mathbb{C}^*)^n.$$

Rullgård [**39**] observed that this function is plurisubharmonic in $\mathbb{C}^N$ while pluriharmonic over U_α. Indeed, over U_α there is a component $E_\alpha \subset \mathbb{R}^n \setminus \mathcal{A}$ corresponding to α and $u_\alpha = \operatorname{Re} \Phi_\alpha$, where

$$\Phi_\alpha = \frac{1}{(2\pi i)^n} \int\limits_{\mathrm{Log}^{-1}(x)} \log\left(\frac{f(z)}{z^\alpha}\right) \frac{dz_1}{z_1} \wedge \cdots \wedge \frac{dz_n}{z_n}, x \in E_\alpha$$

is a $(\mathbb{C}/2\pi i\mathbb{Z})$-valued holomorphic function. Note that over $\mathrm{Log}^{-1}(E_\alpha)$ we can choose a holomorphic branch of $\log(\frac{f(z)}{z^\alpha})$ and that Φ_α does not depend on the choice of $x \in E_\alpha$. Therefore, U_α is pseudo-convex.

Note that U_α is invariant under the natural $\mathbb{C}^*$-action in $\mathbb{C}^N$. Let $\mathcal{C} \subset \mathbb{CP}^{N-1}$ be the complement of the image of U_α under the projection $\mathbb{C}^N \to \mathbb{CP}^{N-1}$.

Theorem 2.7 (Rullgård [**39**]). *For any line $L \subset \mathbb{CP}^{n-1}$ the set $L \cap \mathcal{C}$ is nonempty and connected.*

The next theorem describes how the sets U_α with different $\alpha \in \Delta \cap \mathbb{Z}^n$ intersect. It turns out that for any choice of subdivision $\Delta \cap \mathbb{Z}^n = A \cup B$ with $A \cap B = \varnothing$ the sets $\bigcup\limits_{\alpha \in A} U_\alpha$ and $\mathbb{C}^N \setminus \bigcup\limits_{\beta \in B} U_\beta$ intersect. A stronger statement was found by Rullgård. Let $A, B \subset \Delta \cap \mathbb{Z}^n$ be disjoint sets. The set $A \cup B \subset \Delta \cap \mathbb{Z}^n$ defines a subspace $\mathbb{C}^{\#(A\cup B)} \subset \mathbb{C}^N$.

Theorem 2.8 ([39]). *For any $\#(A \cup B)$-dimensional space L parallel to $\mathbb{C}^{\#(A\cup B)}$ the intersection $L \cap \bigcup_{\alpha\in A} U_\alpha \cap \mathbb{C}^N \setminus \bigcup_{\beta\in B} U_\beta$ is nonempty.*

2.5. The Monge–Ampère measure and the symplectic volume.

Definition 2.9 (Passare–Rullgård [33]). The Monge–Ampère measure on $\mathcal{A}$ is the pull-back of the Lebesgue measure on $\Delta \subset \mathbb{R}^n$ under ∇N_f.

Indeed by Proposition 2.1 the Monge–Ampère measure is well-defined. Furthermore, we have the following proposition.

Proposition 2.10 ([33]). *The Monge–Ampère measure has its support on $\mathcal{A}$. The total Monge–Ampère measure of $\mathcal{A}$ is* $\operatorname{Vol}\Delta$.

By Definition 2.9 the Monge–Ampère measure is given by the determinant of the Hessian of N_f. By convexity of N_f its Hessian $\operatorname{Hess} N_f$ is a nonnegatively defined matrix-valued function. The trace of $\operatorname{Hess} N_f$ is the Laplacian of N_f, it gives another natural measure supported on $\mathcal{A}$. Note that $\omega = \sum_{k=1}^{n} \frac{dz_k}{z_k} \wedge \frac{d\bar{z}_k}{\bar{z}_k}$ is a symplectic form on $(\mathbb{C}^*)^n$ invariant with respect to the group structure. The restriction $\omega|_V$ is a symplectic form on V. Its $(n-1)$-th power divided by $(n-1)!$ is a volume form called the *symplectic volume* on the $(n-1)$-manifold V.

Theorem 2.11 ([33]). *The measure on $\mathcal{A}$ defined by the Laplacian of N_f coincides with the push-forward of the symplectic volume on V, i.e., for any Borel set A*

$$\int_A \Delta N_f = \int_{\operatorname{Log}^{-1}(A)\cap V} \omega^{n-1}.$$

This theorem appears in [33] as a particular case of a computation for the *mixed Monge–Ampère operator*, the symmetric multilinear operator associating a measure to n functions $f_1, \ldots, f_n$ (recall that by our convention n is the number of variables) and such that its value on $f, \ldots, f$ is the Monge–Ampère measure from Definition 2.9. The total mixed Monge–Ampère measure for $f_1, \ldots, f_n$ is equal to the mixed volume of the Newton polyhedra of $f_1, \ldots, f_n$ divided by $n!$.

Recall that this mixed volume divided by $n!$ appears in the Bernstein formula [6] which counts the number of common solutions of the system of equations $f_k = 0$ (assuming that the corresponding hypersurfaces intersect transversely). Passare and Rullgård found the following local analogue of the Bernstein formula which also serves as a geometric interpretation of the mixed Monge–Ampère measure. Note that the complex torus $(\mathbb{C}^*)^n$ acts on polynomials of n variables. The value of $t \in (\mathbb{C}^*)^n$ on $f : (\mathbb{C}^*)^n \to \mathbb{C}$ is the composition $f \circ t$ of the multiplication by t followed by application of f. In particular, the real torus $T^n = \mathrm{Log}^{-1}(0) \subset (\mathbb{C}^*)^n$ acts on polynomials of n variables.

Theorem 2.12 ([**33**]). *The mixed Monge–Ampère measure for $f_1, \ldots, f_n$ of a Borel set $A \subset \mathbb{R}^n$ is equal to the average number of solutions of the system of equations $f_k \circ t_k = 0$ in $\mathrm{Log}^{-1}(E) \subset (\mathbb{C}^*)^n$, $t_k \in T^n$, $k = 1, \ldots, n$.*

The number of solution of this system of equations does not depend on t_k as long as the choice of t_k is generic. Thus Theorem 2.12 produces the Bernstein formula when $E = \mathbb{R}^n$.

2.6. The area of a planar amoeba. The computations of the previous subsection can be used to obtain an upper bound on amoeba's area in the case when $V \subset (\mathbb{C}^*)^2$ is a curve. With the help of Theorem 2.12 Passare and Rullgård [**33**] showed that in this case the Lebesgue measure on $\mathcal{A}$ is not greater than π^2 times the Monge–Ampère measure. In particular, we have the following theorem.

Theorem 2.13 ([**33**]). *If $V \subset (\mathbb{C}^*)^2$ is an algebraic curve then*

$$\mathrm{Area}\,\mathcal{A} \leqslant \pi^2\,\mathrm{Area}\,\Delta.$$

This theorem is specific for the case $\mathcal{A} \subset \mathbb{R}^2$. Nondegenerate higher-dimensional amoebas of hypersurfaces have infinite volume. This follows from Proposition 1.6 since the area of the cross-section at infinity must be separated from zero.

3. Some Applications of Amoebas

3.1. The first part of Hilbert's 16th problem. Most applications considered here are in the framework of Hilbert's 16th problem. Consider the classical setup of its first part ([**14**]). Let $\mathbb{R}\overline{V} \subset \mathbb{RP}^2$ be a smooth algebraic curve of degree d. *What are the possible topological types of pairs* $(\mathbb{RP}^2, \mathbb{R}\overline{V})$ *for a given* d*?*

Since $\mathbb{R}\overline{V}$ is smooth it is homeomorphic to a disjoint union of circles. All of these circles must be contractible in $\mathbb{R}P^2$ (such circles are called the *ovals*) if d is even. If d is odd then exactly one of these circles is noncontractible. Therefore, the topological type of $(\mathbb{RP}^2, \mathbb{R}\overline{V})$ (also called the *topological arrangement* of $\mathbb{R}\overline{V}$ in $\mathbb{RP}^2$) is determined by the number of components of $\mathbb{R}\overline{V}$ together with the information on the mutual position of the ovals.

The possible number of components of $\mathbb{R}\overline{V}$ was determined by Harnack [**12**]. He proved that it cannot be greater than $\frac{(d-1)(d-2)}{2} + 1$. Furthermore he proved that for any number

$$l \leqslant \frac{(d-1)(d-2)}{2} + 1$$

there exists a curve of degree d with exactly l components as long as $l > 0$ in the case of odd d (recall that for odd d we always have to have a noncontractible component).

Note that each oval separates $\mathbb{RP}^2$ into its *interior*, which is homeomorphic to a disk, and its *exterior*, which is homeomorphic to a Möbius band. If the interiors of the ovals intersect then the ovals are called *nested.* Otherwise the ovals are called *disjoint.* Hilbert's problem started from a question whether a curve of degree 6 which has 11 ovals (the maximal number according to Harnack) can have all of the ovals disjoint. This question was answered negatively by Petrovsky [**34**] who showed that at least two ovals of a sextic must be nested if the total number of ovals is 11.

In general the number of topological arrangements of curves of degree d grows exponentially with d. Even for small d the number of the possible types is enormous. Many powerful theorems restricting possible topological arrangements were found for over 100 years of history of this problem, see, in particular, [**34**], [**3**], [**37**], [**44**]. A powerful *patchworking* construction technique [**42**] counters these theorems. The complete classifications is currently known for $d \leqslant 7$ (see [**42**]).

The most restricted turn out to be curves with the maximal numbers of components, i.e., with $l = \frac{(d-1)(d-2)}{2} + 1$. Such curves were called *M-curves* by Petrovsky. However, even for M-curves, the number of topological arrangements grows exponentially with d.

The situation becomes different if we consider $\mathbb{RP}^2$ as a toric surface, i.e., as a compactification of $(\mathbb{R}^*)^2$. Recall that $\mathbb{RP}^2 \setminus (\mathbb{R}^*)^2$ consists of three lines l_0, l_1 and l_2 which can be viewed as coordinate axes for homogeneous coordinates in $\mathbb{RP}^2$. Thus we have three affine charts for $\mathbb{RP}^2$. The intersection of all three charts is $(\mathbb{R}^*)^2 \subset \mathbb{RP}^2$. We denote $\mathbb{R}V = \mathbb{R}\overline{V} \cap (\mathbb{R}^*)^2$. The complexification $V \subset (\mathbb{C}^*)^2$ is the complex hypersurface defined by the same equation as $\mathbb{R}V$. Thus we are in position to apply the content of the previous sections of the paper to the amoeba of V.

In [**26**] it was shown (with the help of amoebas) that for each d the topological type of the pair $(\mathbb{RP}^2, \mathbb{R}\overline{V})$ is unique as long as the curve $\mathbb{R}\overline{V}$ is maximal in each of the three affine charts of $\mathbb{RP}^2$. Furthermore, the diffeomorphism type of the triad $(\mathbb{RP}^2; \mathbb{R}\overline{V}, l_0 \cup l_1 \cup l_2)$ is unique. In subsection 3.5 we formulate this maximality condition and sketch the proof of uniqueness. A similar statement holds for curves in other toric surfaces. The Newton polygon Δ plays then the rôle of the degree d.

3.2. Relation to amoebas: the real part $\mathbb{R}V$ as a subset of the critical locus of $\mathrm{Log}\,|_V$ and the logarithmic Gauss map. Suppose that the hypersurface $V \subset (\mathbb{C}^*)^n$ is defined over real numbers (i.e., by a polynomial with real coefficients). Denote its real part via $\mathbb{R}V = V \cap (\mathbb{R}^*)^n$. We also assume that V is nonsingular. Let $F \subset V$ be the critical locus of the map $\mathrm{Log}\,|_V : V \to \mathbb{R}^n$. It turns out that the real part $\mathbb{R}V$ is always contained in F.

Proposition 3.1 (Mikhalkin [**26**]). *$\mathbb{R}V \subset F$.*

This proposition indicates that the amoeba must carry some information about $\mathbb{R}V$.

The proof of this proposition makes use of the *logarithmic Gauss map.*

Note that since $(\mathbb{C}^*)^n$ is a Lie group there is a canonical trivialization of its tangent bundle. If $z \in (\mathbb{C}^*)^n$ then the multiplication by z^{-1} induces an isomorphism $T_z(\mathbb{C}^*)^n \approx T_1(\mathbb{C}^*)^n$ of the tangent bundles at z and $1 = (1, \ldots, 1) \in (\mathbb{C}^*)^n$.

Definition 3.2 (Kapranov [**17**]). The *logarithmic Gauss map* is a map

$$\gamma : V \to \mathbb{CP}^{n-1}.$$

It sends each point $z \in V$ to the image of the hyperplane $T_z V \subset T_z(\mathbb{C}^*)^n$ under the canonical isomorphism $T_z(\mathbb{C}^*)^n \approx T_1(\mathbb{C}^*)^n = \mathbb{C}^n$.

The map γ is a composition of a branch of a holomorphic logarithm $(\mathbb{C}^*)^n \to \mathbb{C}^n$ defined locally up to translation by $2\pi i$ with the usual Gauss map of the image of V. We may define γ explicitly in terms of the defining polynomial f for V by logarithmic differentiation formula. If $z = (z_1, \dots, z_n) \in V$ then

$$\gamma(z) = [\langle \nabla f, z\rangle] = \left[\frac{\partial f}{\partial z_1} z_1 : \dots : \frac{\partial f}{\partial z_n} z_n\right] \in \mathbb{CP}^{n-1}.$$

Lemma 3.3 ([**26**]). $F = \gamma^{-1}(\mathbb{RP}^{n-1})$

To justify this lemma we recall that $\mathrm{Log} : (\mathbb{C}^*)^n \to \mathbb{R}^n$ is a smooth fibration and V is nonsingular. Thus $z \in V$ is critical for $\mathrm{Log}|_V$ if and only if the tangent vector space to V and the tangent vector space to the fiber torus $\gamma^{-1}(\gamma(z))$ intersect along an $(n-1)$-dimensional subspace. Such points are mapped to real points of $\mathbb{CP}^{n-1}$ by γ.

Note that this lemma implies Proposition 3.1. If V is defined over $\mathbb{R}$ then γ is equivariant with respect to the complex conjugation and maps $\mathbb{R}V$ to $\mathbb{RP}^{n-1}$

3.3. Compactification: a toric variety associated to a hypersurface in $(\mathbb{C}^*)^n$. A hypersurface $V \subset (\mathbb{C}^*)^n$ is defined by a polynomial $f : \mathbb{C}^n \to C$. If the coefficients of f are real then we define the real part of V by $\mathbb{R}V = V \cap (\mathbb{R}^*)^n$. Recall that the Newton polyhedron $\Delta \subset \mathbb{R}^n$ of V is an integer convex polyhedron obtained as the convex hull of the indices of monomials participating in f, see (1) in subsection 1.3.

Let $\mathbb{C}\,T_\Delta \supset (\mathbb{C}^*)^n$ be the toric variety corresponding to Δ (see, for example, [**11**]), and let $\mathbb{R}T_\Delta \supset (\mathbb{R}^*)^n$ be its real part. We define $\overline{V} \subset \mathbb{C}\,T_\Delta$ as the closure of V in $\mathbb{C}\,T_\Delta$ and we denote via $\mathbb{R}\overline{V}$ its real part.

Note that $\overline{V}$ may be singular even if V is not. Nevertheless $\mathbb{C}\,T_\Delta$ is, in some sense, the best toric compactification of $(\mathbb{C}^*)^n$ for V. Namely, $\overline{V}$ does not pass via the points of $\mathbb{C}\,T_\Delta$ corresponding to the vertices of Δ and therefore it does not have singularities there. Furthermore, $\mathbb{C}\,T_\Delta$ is minimal among such toric varieties, since $\overline{V}$ intersect any line in $\mathbb{C}\,T_\Delta$ corresponding to an edge of Δ.

Thus we may naturally compactify the pair $((\mathbb{C}^*)^n, V)$ to the pair $(\mathbb{C}\,T_\Delta, \overline{V})$. In such a setup the polyhedron Δ plays the rôle of the degree in $\mathbb{C}\,T_\Delta$. Indeed, two integer polyhedra Δ define the same toric variety $\mathbb{C}\,T_\Delta$ if their corresponding faces are parallel. But the choice of Δ also fixes the homology class of $\overline{V}$ in $H_{2n-2}(\mathbb{C}\,T_\Delta)$.

The simplest example is the projective space $\mathbb{CP}^n$. The corresponding Δ is, up to translation and the action of $SL_n(\mathbb{Z})$ the simplex defined by equations $z_j > 0$, $z_1 + \cdots + z_n < d$. Thus in this case Δ is parameterized by a single natural number d which is the degree of $\overline{V} \subset \mathbb{CP}^n$.

3.4. Maximality condition for $\mathbb{R}V$. The inequality $l \leqslant \frac{(d-1)(d-2)}{2}$ discovered by Harnack for the number l of components of a curve $\mathbb{R}\overline{V}$ is a part of a more general *Harnack–Smith inequality.* Let X be a topological space and let Y be the fixed point set of a of a continuous involution on X. Denote by $b_*(X;\mathbb{Z}_2) = \dim H_*(X;\mathbb{Z}_2)$ the total $\mathbb{Z}_2$-Betti number of X.

Theorem 3.4 (P. A. Smith (see, e.g., the appendix in [**44**])).

$$b_*(Y;\mathbb{Z}_2) \leqslant b_*(X;\mathbb{Z}_2).$$

Corollary 3.5. $b_*(\mathbb{R}\overline{V};\mathbb{Z}_2) \leqslant b_*(\overline{V};\mathbb{Z}_2)$, $b_*(\mathbb{R}V;\mathbb{Z}_2) \leqslant b_*(V;\mathbb{Z}_2)$.

Note that Theorem 3.4 can also be applied to pairs which consist of a real variety and real subvariety and other similar objects.

Definition 3.6 (Rokhlin [**37**]). A variety $\mathbb{R}\overline{V}$ is called an *M-variety* if

$$b_*(\mathbb{R}\overline{V};\mathbb{Z}_2) = b_*(\overline{V};\mathbb{Z}_2).$$

For example, if $\overline{V} \subset \mathbb{CP}^2$ is a smooth curve of degree d then $\overline{V}$ is a Riemann surface of genus $g = \frac{(d-1)(d-2)}{2}$. Thus $b_*(\overline{V};\mathbb{Z}_2) = 2 + 2g$. On the other hand, $b_*(\mathbb{R}\overline{V};\mathbb{Z}_2) = 2l$, where l is the number of (circle) components of $\mathbb{R}\overline{V}$.

Let $\mathbb{R}V \subset (\mathbb{R}^*)^n$ be an algebraic hypersurface, Δ be its Newton polyhedron, $\mathbb{R}T_\Delta$ be the toric variety corresponding to Δ and $\mathbb{R}\overline{V} \subset \mathbb{R}T_\Delta$ the closure of $\mathbb{R}V$ in $\mathbb{R}T_\Delta$. We denote with $V \subset (\mathbb{C}^*)^n$ and $\overline{V} \subset \mathbb{C}\,T_\Delta$ the complexifications of these objects. Recall (see, for example, [**11**]) that each (closed) k-dimensional face Δ' of Δ corresponds to a closed k-dimensional toric variety $\mathbb{R}T_{\Delta'} \subset \mathbb{R}T_\Delta$ (and, similarly, $\mathbb{C}\,T_{\Delta'} \subset \mathbb{C}\,T_\Delta$). The intersection $V_{\Delta'} = \overline{V} \cap \mathbb{C}\,T_{\Delta'}$ is itself a hypersurface in the k-dimensional toric variety $\mathbb{C}\,T_{\Delta'}$ with the Newton polyhedron Δ'. Its real part is $\mathbb{R}V_{\Delta'} = V_{\Delta'} \cap \mathbb{R}\overline{V}$.

Denote with $\operatorname{St}\Delta' \subset \partial\Delta$ the union of all the closed faces of Δ containing Δ'. Denote $V_{\operatorname{St}\Delta'} = \bigcup\limits_{\Delta'' \subset \operatorname{St}\Delta'} V_{\Delta'}$ and $\mathbb{R}V_{\operatorname{St}\Delta'} = V_{\operatorname{St}\Delta'} \cap \mathbb{R}T_\Delta$.

Definition 3.7. A hypersurface $\mathbb{R}\overline{V} \subset \mathbb{C}\,T_\Delta$ is called *torically maximal* if the following conditions hold

- $\mathbb{R}\overline{V}$ is an M-variety, i.e., $b_*(\mathbb{R}\overline{V};\mathbb{Z}_2) = b_*(\overline{V};\mathbb{Z}_2)$;
- the hypersurface $\overline{V} \cap \mathbb{C}\,T_{\Delta'} \subset \mathbb{C}\,T_\Delta$ is torically maximal for each face $\Delta' \subset \Delta$ (inductively we assume that this notion is already defined in smaller dimensions);
- for each face $\Delta' \subset \Delta$ we have $b_*(\mathbb{R}V \cup \mathbb{R}V_{\mathrm{St}\,\Delta'}, \mathbb{R}V_{\mathrm{St}\,\Delta'};\mathbb{Z}_2) = b_*(V \cup V_{\mathrm{St}\,\Delta'}, V_{\mathrm{St}\,\Delta'};\mathbb{Z}_2)$.

Consider a linear function $h : \mathbb{R}^n \to \mathbb{R}$. A facet $\Delta' \subset \Delta$ is called *negative* with respect to h if the image of its outward normal vector under h is negative. We define $\mathbb{C}\,T^- = \bigcup_{\text{negative } \Delta'} \mathbb{C}\,T_{\Delta'}$. In these formula we take the union over all the closed facets Δ' negative with respect to h. Let $V^- = \overline{V} \cap \mathbb{C}\,T^-$ and $\mathbb{R}V^- = V^- \cap \mathbb{R}\overline{V}$.

We call a linear function $h : \mathbb{R}^n \to \mathbb{R}$ generic if its kernel does not contain vectors orthogonal to facets of Δ.

Proposition 3.8. *If a hypersurface $\mathbb{R}\overline{V} \subset \mathbb{R}T_\Delta$ is torically maximal then for any generic linear function h we have*

$$b_*(\mathbb{R}V \cup \mathbb{R}V^-, \mathbb{R}V^-;\mathbb{Z}_2) = b_*(V \cup V^-, V^-;\mathbb{Z}_2).$$

3.5. Curves in the plane.

3.5.1. *Curves in $\mathbb{R}\mathbb{P}^2$ and their bases.* Note that if $\mathbb{R}V \subset (\mathbb{R}^*)^2$ is a torically maximal curve then the number of components of $\mathbb{R}\overline{V}$ coincides with the genus of $\mathbb{C}\overline{V}$. In other words (cf. 3.1) $\mathbb{R}\overline{V}$ is an M-curve.

We start by reformulating the maximality condition of Definition 3.7 for the case of curves in the projective plane. Let $\mathbb{R}C \subset \mathbb{R}\mathbb{P}^2$ be a nonsingular curve of degree d.

FIGURE 5. Possible bases for a real quartic curve.

Definition 3.9 (Brusotti [**8**]). Let α be an arc (i.e., an embedded closed interval) in $\mathbb{R}C$. The arc α is called a *base* (or a *base of rank 1* (see [**8**])) if there exists a line $L \subset \mathbb{R}\mathbb{P}^2$ such that the intersection $L \cap \alpha$ consists of d distinct points.

Note if three lines L_1, L_2, L_3 in $\mathbb{RP}^2$ are generic, i.e., they do not pass through the same point, then $=\mathbb{RP}^2 \smallsetminus (L_1 \cup L_2 \cup L_3) = (\mathbb{R}^*)^2$. We call such $(\mathbb{R}^*)^2$ a *toric chart* of $\mathbb{RP}^2$. Thus $\mathbb{R}V = \mathbb{R}C \smallsetminus (L_1 \cup L_2 \cup L_3)$ is a curve in $(\mathbb{R}^*)^2$. If $\mathbb{R}C$ does not pass via $L_j \cap L_k$ then the Newton polygon of $\mathbb{R}V$ (for any choice of coordinates (x, y) in $(\mathbb{R}^*)^2$ extendable to affine coordinates in $\mathbb{R}^2 = \mathbb{RP}^2 \smallsetminus L_j$ for some j) is the triangle $\Delta_d = \{x \geqslant 0\} \cap \{y \geqslant 0\} \cap \{x + y \leqslant d\}$.

Proposition 3.10 (Mikhalkin [**32**]). *The curve $\mathbb{R}C \subset \mathbb{RP}^2$ is maximal in some toric chart of $\mathbb{RP}^2$ if and only if $\mathbb{R}C$ is an M-curve with three disjoint bases.*

Many M-curves with one or two disjoint bases are known (see, for example, [**8**]).

However there is (topologically) only one known example of curve with three disjoint bases, namely the first M-curve constructed by Harnack [**12**]. Theorem 3.12 asserts that this example is the only possible.

Definition 3.11 (simple Harnack curve in $\mathbb{RP}^2$, see [**12**] and [**27**]). A nonsingular curve $\mathbb{R}C \subset \mathbb{RP}^2$ of degree d is called a (smooth) simple Harnack curve if it is an M-curve and

- all ovals of $\mathbb{R}C$ are disjoint (i.e., have disjoint interiors, see 3.1) if $d = 2k - 1$ is odd;
- one oval of $\mathbb{R}C$ contains $\frac{(k-1)(k-2)}{2}$ ovals in its interior while all other ovals are disjoint if $d = 2k$ is even.

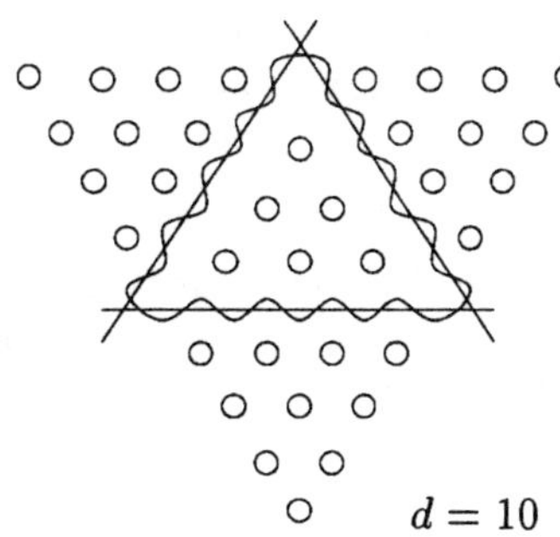

FIGURE 6. [**26**] A simple Harnack curve.

Theorem 3.12 ([**26**]). *Any smooth M-curve $\mathbb{R}C \subset \mathbb{R}\mathbb{P}^2$ with at least three base is a simple Harnack curve.*

There are several topological arrangements of M-curves with fewer than 3 bases for each d (in fact, their number grows exponentially with d). There is a unique (Harnack) topological arrangement of an M-curve with 3 bases by Theorem 3.12. In the same time 3 is the highest number of bases an M-curve of sufficiently high degree can have as the next theorem shows.

Theorem 3.13 ([**26**]). *No M-curve in $\mathbb{R}\mathbb{P}^2$ can have more than 3 bases if $d \geqslant 3$.*

3.5.2. *Curves in real toric surfaces.* Theorem 3.12 has a generalization applicable to other toric surfaces. Let $\mathbb{R}V \subset (\mathbb{R}^*)^2$ be a curve with the Newton polygon Δ. The sides of Δ correspond to lines $L_1, \ldots, L_n$ in $\mathbb{R}T_\Delta$. We have $\mathbb{R}V = \mathbb{R}\overline{V} \setminus (L_1 \cup \cdots \cup L_n$.

Theorem 3.14 ([**26**]). *The topological arrangement of a torically maximal curve is unique for each Δ. More precisely, the topological type of the triad $(\mathbb{R}T_\Delta; \mathbb{R}\overline{V}, L_1 \cup \cdots \cup L_n)$ and, in particular, the topological type of the pair $((\mathbb{R}^*)^2, \mathbb{R}V)$ depends only on Δ as long as $\mathbb{R}V$ is a torically maximal curve.*

A torically maximal curve $\mathbb{R}\overline{V}$ is a counterpart of a simple Harnack curve for $\mathbb{R}T_\Delta$. All of its components except for one are ovals with disjoint interiors. The remaining component is not homologous to zero unless Δ is even (i.e., obtained from another lattice polygon by a homotetehy with coefficient 2). If Δ is even the remaining component is also an oval whose interior contains $g(V)$ ovals of $\mathbb{R}V$. Recall that, by Khovanskii's formula [**22**], $g(V)$ coincides with the number of lattice points in the interior of Δ.

Theorem 3.15 (Harnack, Itenberg–Viro [**12**], [**16**]). *For any Δ there exists a curve $\mathbb{R}V \subset (\mathbb{R}^*)^2$ which is torically maximal and has Δ as its Newton polygon.*

As in Definition 3.11 we call such curves *simple Harnack curves* (see [**27**]).

3.5.3. *Geometric properties of algebraic curves in $(\mathbb{R}^*)^2$.* It turns out that the simple Harnack curves have peculiar geometric properties, but they are better seen after a logarithmic reparameterization $\operatorname{Log}|_{(\mathbb{R}^*)^2}$:

$(\mathbb{R}^*)^2 \to \mathbb{R}^2$. A point of $\mathbb{R}V$ is called a logarithmic inflection point if it corresponds to an inflection point of $\mathrm{Log}(\mathbb{R}V) Subset \mathbb{R}^2$ under Log.

Theorem 3.16 ([26]). *The following conditions are equivalent.*

- $\mathbb{R}V \subset (\mathbb{R}^*)^2$ *is a simple Harnack curve.*
- $\mathbb{R}V \subset (\mathbb{R}^*)^2$ *has no real logarithmic inflection points.*

Remark 3.17. Recall that by Proposition 3.1 $\mathrm{Log}(\mathbb{R}V)$ is contained in the critical value locus of $\mathrm{Log}|_V$. The map $\mathrm{Log}|_V : V \to \mathbb{R}^2$ is a surface-to-surface map in our case and its most generic singularities are folds. By Proposition 1.10 the folds are convex. Thus a logarithmic inflection point of $\mathbb{R}V$ must correspond to a higher singularity of $\mathrm{Log}|_V$.

FIGURE 7. A junction point.

In [26] it was stated that there are two types of stable (surviving small deformations of $\mathbb{R}V$) logarithmic inflection points of $\mathbb{R}V$. Here we'd like to correct this statement. Only one of these two types is genuinely stable. The first type (see Figure 7), called *junction*, corresponds to an intersection of $\mathbb{R}V$ with a branch of imaginary folding curve. A junction logarithmic inflection point can be found at the curve $y = (x-1)^2 + 1$. Note that the image of the imaginary folding curve under the complex conjugation is also a folding curve. Thus over its image we have a double fold.

The second type, called *pinching*, corresponds to intersection of $\mathbb{R}V$ with a circle $E \subset V$ that gets contracted by Log. The circle E intersect $\mathbb{R}V$ at two points. These points belong to different quadrants of $(\mathbb{R}^*)^2$, but have the same absolute values of their coordinates. Both of these points are logarithmic inflection points. The pinching is not stable even in the class of real deformations. A small perturbation breaks it to two junctions with a corner of two branches of the amoeba as in Figure 8.

Proposition 3.18. *The logarithmic image* $\mathrm{Log}(\mathbb{R}V)$ *is trivial in the closed support homology group* $H_1^c(\mathbb{R}^2)$.

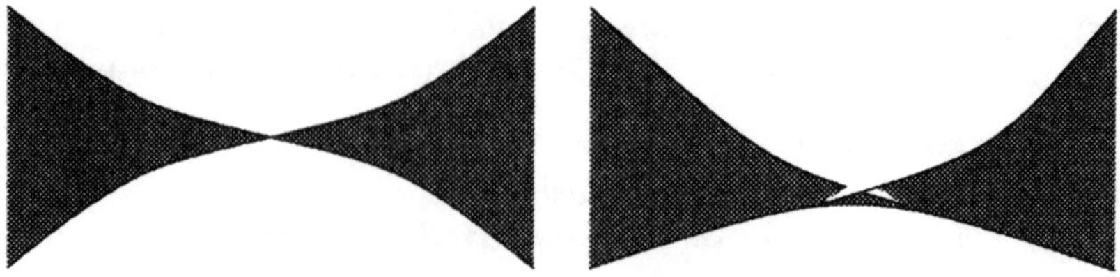

FIGURE 8. Deformation of a pinching point into two junction points.

Thus the curve $\mathrm{Log}(\mathbb{R}V)$ spans a surface in $(\mathbb{R}^*)^2$. Theorem 2.13 has the following corollary.

Corollary 3.19. *The area of any region spanned by branches of* $\mathrm{Log}(\mathbb{R}V)$ *is smaller than* $\mathrm{Area}\,\Delta$.

The situation is especially simple for the logarithmic image of a simple Harnack curve.

Proposition 3.20 ([26]). *If* $\mathbb{R}V$ *is a simple Harnack curve then* $\mathrm{Log}\,|_{\mathbb{R}V}$ *is an embedding and* $\mathrm{Log}\,\mathbb{R}V = \partial\mathcal{A}$.

Thus in this case $\mathcal{A}$ coincides with the region spanned by the whole curve $\mathrm{Log}(\mathbb{R}V)$. Furthermore, in [**27**] it was shown that simple Harnack curves maximize the area of this region.

Theorem 3.21 (Mikhalkin–Rullgård, [**27**]). *If* $\mathbb{R}V$ *is a simple Harnack curve then* $\mathrm{Area}\,\mathcal{A} = \mathrm{Area}\,\Delta$.

In the opposite direction we have the following theorem. We say that a curve $V \subset (\mathbb{C}^*)^2$ is real up to translation if there exists $a \in (\mathbb{C}^*)^2$ such that aV is defined by a polynomial with real coefficients. We denote the corresponding real part with $\mathbb{R}V$. (Note that in general this real part might depend on the choice of translation.)

Theorem 3.22 ([**27**]). *If* $\mathrm{Area}\,\mathcal{A} = \mathrm{Area}\,\Delta > 0$ *and* V *is nonsingular and transverse to the lines (coordinate axes) in* $\mathbb{C}T_\Delta$ *corresponding to the sides of* Δ *then* V *is real up to translation in a unique way and* $\mathbb{R}V$ *is a simple Harnack curve.*

Furthermore, in [**27**] it was shown that the only singularities that V can have in the case $\mathrm{Area}\,\mathcal{A} = \mathrm{Area}\,\Delta > 0$ are ordinary real isolated double points.

3.6. A higher-dimensional case.

3.6.1. *Surfaces in* $(\mathbb{R}^*)^3$. Let $\mathbb{R}V \subset (\mathbb{R}^*)^3$ be an algebraic surface with the Newton polyhedron $\Delta \subset \mathbb{R}^3$. Let $\mathbb{R}\overline{V} \subset \mathbb{R}T_\Delta$ be its compactification.

Recall (see Definition 3.7) that if $\mathbb{R}V$ is a torically maximal surface then $b_*(\mathbb{R}\overline{V};\mathbb{Z}_2) = b_*(\overline{V};\mathbb{Z}_2)$, i.e., $\mathbb{R}\overline{V}$ is an M-surface.

Theorem 3.23 ([32]). *Given a Newton polyhedron Δ the topological type of a torically maximal surface $\mathbb{R}\overline{V} \subset \mathbb{R}T_\Delta$ is unique.*

To describe the topological type of $\mathbb{R}\overline{V}$ it is useful to compute the total Betti number $b_*(\overline{V};\mathbb{Z}_2)$ in terms of Δ. Note that by the Lefschetz hyperplane theorem $b_*(\overline{V};\mathbb{Z}_2) = \chi(\overline{V})$.

We denote by Area $\partial\Delta$ the total area of the faces of Δ. Each of these faces sits in a plane $P \subset \mathbb{R}^3$. The intersection $P \cap \mathbb{Z}^3$ determines the area form on P. This area form is translation invariant and such that the area of the smallest lattice parallelogram is 1.

Similarly we denote by Length $\mathrm{Sk}^1\,\Delta$ the total length of all the edges of Δ. Again, each edge sits in a line $L \subset \mathbb{R}^3$. The intersection $L \cap \mathbb{Z}^3$ determines the length on L by setting the length of the smallest lattice interval 1.

Proposition 3.24. $b_*(V;\mathbb{Z}_2) = 6\,\mathrm{Vol}\,\Delta - 2\,\mathrm{Area}\,\partial\Delta + \mathrm{Length}\,\mathrm{Sk}^1\,\Delta.$

This proposition follows from Khovanskii's formula [**22**].

Theorem 3.25 ([32]). *A torically maximal surface $\mathbb{R}\overline{V}$ consists of $p_g + 1$ components, where p_g is the number of points in the interior of Δ. There are p_g components homeomorphic to 2-spheres and contained in $(\mathbb{R}^*)^3$. These spheres bound disjoint spheres in $(\mathbb{R}^*)^3$. The remaining component is homeomorphic to*

- *a sphere with $b_*(V;\mathbb{Z}_2) - 2p_g(V) - 2$ Möbius bands in the case when Δ is odd (i.e., cannot be presented as $2\Delta'$ for some lattice polyhedron Δ');*
- *a sphere with $\frac{1}{2}b_*(V;\mathbb{Z}_2) - p_g(V) - 1$ handles in the case Δ is even.*

Remark 3.26. Not for every Newton polyhedron Δ a torically maximal surface $\mathbb{R}V \subset (\mathbb{R}^*)^3$ exists. The following example is due to B. Bertrand. Let $\Delta \subset \mathbb{R}^3$ be the convex hull of $(1,0,0)$, $(0,1,0)$, $(1,1,0)$ and $(0,0,2k+1)$. If $k > 0$ then there is no M-surface $\mathbb{R}\overline{V}$ with the Newton polyhedron Δ. In particular, there is no torically maximal surface $\mathbb{R}V$ for Δ.

Example 3.27. There are 3 different topological types of smooth M-quartics in $\mathbb{RP}^3$ (see [**21**]). They realize all topological possibilities for maximal real structures on abstract K3-surfaces. Namely, such real surface may be homeomorphic to

- the disjoint union of 9 spheres and a surface of genus 2;
- the disjoint union of 5 spheres and a surface of genus 6;
- the disjoint union of a sphere and a surface of genus 10.

Theorem 3.25 asserts that only the last type can be a torically maximal quartic in $\mathbb{RP}^3$. More generally, only the last type can be a torically maximal surface is a toric 3-fold $\mathbb{R}T_\Delta$.

3.6.2. *Geometric properties of maximal algebraic surfaces in* $(\mathbb{R}^*)^3$. Recall the classical geometric terminology. Let $S \subset \mathbb{R}^3$ be a smooth surface. We call a point $x \in S$ *elliptic*, *hyperbolic* or *parabolic* if the Gauss curvature of S at x is positive, negative or zero.

Remark 3.28. Of course we do not actually need to use the Riemannian metric on S do define these points. Here is an equivalent definition without referring to the curvature. Locally near x we can present S as the graph of a function $\mathbb{R}^2 \to \mathbb{R}$. If the Hessian form of this function at x is degenerate then we call x parabolic. If not, the intersection of S with the tangent plane at x is a real curve with an ordinary double point in x. If this point is isolated we call x elliptic. If it is an intersection of two real branches of the curve we call it hyperbolic.

We say that a point $x \in \mathbb{R}V \subset (\mathbb{R}^*)^3$ is *logarithmically* elliptic, hyperbolic or parabolic if it maps to such point under $\mathrm{Log}|_{(\mathbb{R}^*)^3} : (\mathbb{R}^*)^3 \to \mathbb{R}^3$.

Generically for a smooth surface in $\mathbb{R}^3$ the parabolic locus, i.e., the set of parabolic points, is a 1-dimensional curve. So is the logarithmic parabolic locus for a surface in $(\mathbb{R}^*)^3$. In a contrast to this we have the following theorem for torically maximal surfaces. Note that torically maximal surfaces form an open subset in the space of all surfaces with a given Newton polyhedron.

Theorem 3.29 ([**32**]). *The logarithmic parabolic locus of a torically maximal surface consists of a finite number of points.*

Note that such a zero-dimensional locus cannot separate the surface $\mathbb{R}V$. Thus each component of $\mathbb{R}V$ is either logarithmically elliptic (all its points except finitely many are logarithmically elliptic) or logarithmically hyperbolic (all its points except finitely many are logarithmically hyperbolic).

Corollary 3.30 ([**32**]). *Every compact component of* $\mathbb{R}V$ *is diffeomorphic to a sphere.*

This corollary is a part of Theorem 3.25.

Remark 3.31 (logarithmic monkey saddles of $\mathbb{R}V$)**.** The Hessian at the isolated parabolic points $\mathrm{Log}(\mathbb{R}V)$ vanishes. Generic parabolic points sitting on hyperbolic components of $\mathrm{Log}(\mathbb{R}V)$ look like so-called monkey saddles (given in some local coordinates (x, y, z) by $z = x(y^2 - x^2)$).

Logarithmic monkey saddles do not appear on generic *smooth* surfaces in $(\mathbb{R}^*)^3$. But they do appear on generic *real algebraic* surfaces in $(\mathbb{R}^*)^3$. In particular, they appear on every torically maximal surface of sufficiently high degree.

The counterpart on the elliptic components of $\mathrm{Log}(\mathbb{R}V)$, the *imaginary monkey saddles*, are locally given by $z = x(y^2 + x^2)$.

3.6.3. *General case.* Let $\mathbb{R}V \subset (\mathbb{R}^*)^n$ be a hypersurface. Theorems 3.14 and 3.23 have a weaker version that holds for an arbitrary n.

Theorem 3.32 ([**32**]). *If* $\mathbb{R}V$ *is torically maximal then every compact component of* $\mathbb{R}V$ *is a sphere. All these* $(n-1)$*-spheres bound disjoint* n*-balls in* $(\mathbb{R}^*)^n$.

The following theorem is a counterpart of Theorem 3.29 and a weaker version of Theorem 3.16.

Theorem 3.33 ([**32**]). *The parabolic locus of* $\mathrm{Log}(\mathbb{R}V) \subset \mathbb{R}^n$ *is of codimension 2 if* $\mathbb{R}V$ *is torically maximal.*

Existence of torically maximal hypersurfaces for a given polyhedron Δ seems to be a challenging question if $n > 2$.

3.7. Amoebas and dimers. Amoebas and, in particular, the amoebas of simple Harnack curves have appeared in a recent work of Kenyon, Okounkov and Sheffield on dimers (see [**20**] and [**19**]). In particular, Figure 1 of [**20**] sketches a probabilistic approximation of the amoeba of a line in the plane.

One starts from the negative octant

$$O = \{(x, y, z) \in \mathbb{R}^3 \mid x < 0, y < 0, z < 0\}.$$

Its projection onto $\mathbb{R}^2$ along the vector $(1,1,1)$ defines a fan with 3 corners (see Figure 9). For each $(x_0, y_0, z_0) \in \mathbb{R}^3$ let

$$Q_{(x_0,y_0,z_0)} = \{(x,y,z) \in \mathbb{R}^3 \mid x_0 - 1 < x \leqslant x_0, y_0 - 1 < y \leqslant y_0, z_0 - 1 < z \leqslant z_0\}$$

be the unit cube with the "outer" vertex (x_0, y_0, z_0). Let us fix a large natural number N and remove N such unit cubes from O according to the following procedure.

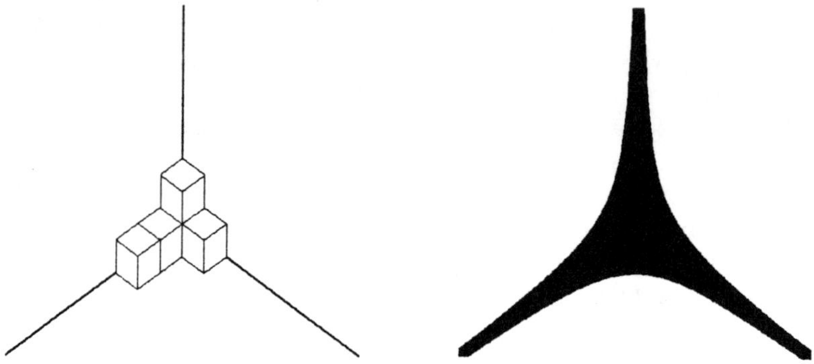

FIGURE 9. The fan with the dimer tiling and the statistical accumulation R.

At the first step we remove $Q_{(0,0,0)}$. The region $O \setminus Q_{(0,0,0)}$ has three outer vertices, namely $(-1,0,0)$, $(0,-1,0)$ and $(0,0,-1)$. At the second step we remove a unit cube whose outer vertex is one of these three and proceed inductively. For each N we have a finite number of possible resulting regions O'.

The projection of such region defines a tiling by diamond-shaped figures (dimers) as in Figure 9. Clearly there is no more than $3N$ dimers in the tiling. Each dimer in $\mathbb{R}^2$ is assigned a weight in a double-periodic fashion with some integer period vectors. The probability of a tiling is determined by these weights.

It is shown in **[20]** that after some rescaling the union of the dimer tiles converges to some limiting region $R \subset \mathbb{R}^2$ that depends only on the choice of the (periodic) choice of weights of the dimers when $N \to \infty$.

Furthermore, according to **[20]** there exists a simple Harnack curve V with the amoeba $\mathcal{A} \subset \mathbb{R}^2$ such that

$$R = T(\mathcal{A})$$

for the linear transformation $T = \begin{pmatrix} \frac{\sqrt{3}}{2} & -\frac{\sqrt{3}}{2} \\ \frac{1}{2} & \frac{1}{2} \end{pmatrix}$ in $\mathbb{R}^2$. The curve V is a line (as in Figure 9) if all the dimer weights are the same. For other periodic weight choices any simple Harnack curve can appear.

Using such dimer interpretation Kenyon and Okounkov [**19**] have constructed an explicit parameterization for the set of all simple Harnack curves of the same degree. It is shown in [**19**] that this set is contractible.

Part 2. TROPICAL GEOMETRY

4. Tropical Degeneration and the Limits of Amoebas

4.1. Tropical algebra.

Definition 4.1. The *tropical semifield* $\mathbb{R}_{\text{trop}}$ is the set of real numbers $\mathbb{R}$ equipped with the following two operation called *tropical addition* and *tropical multiplication*. We use quotation marks to distinguish tropical arithmetical operations from the standard ones. For $x, y \in \mathbb{R}_{\text{trop}}$ we set "$x + y$" $= \max\{x, y\}$ and "xy" $= x + y$.

This definition appeared in Computer Science. The term "tropical" was given in honor of Imre Simon who resides in São Paolo, Brazil (see [**35**]). Strictly speaking, the tropical addition in Computer Science is usually taken to be the minimum (instead of the maximum), but, clearly, the minimum generates an isomorphic semifield.

The semifield $\mathbb{R}_{\text{trop}}$ lacks the subtraction. However it is not needed to define polynomials. Indeed the tropical polynomial is defined as

$$\text{“}\sum_j a_j x^j\text{”} = \max_j \langle j, x\rangle + a_j$$

for any finite collections of coefficients $a_j \in \mathbb{R}_{\text{trop}}$ parameterized by indices $j = (j_1, \dots, j_n) \in \mathbb{Z}^n$. Here $x = (x_1, \dots, x_n) \in \mathbb{R}^n$, $x^j = x_1^{j_1} \dots x_n^{j_n}$ and $\langle j, x\rangle = j_1 x_1 + \dots + j_n x_n$.

Thus the tropical polynomials are piecewise-linear functions. They are simply the Legendre transforms of the function $j \mapsto -a_j$ (this function is defined only on finitely many points, but its Legendre transform is defined everywhere on $\mathbb{R}^n$).

It turns out that these polynomials are responsible for some piecewise-linear geometry in $\mathbb{R}^n$ that is similar in many ways to the classical algebraic geometry defined by the polynomials with complex coefficients. Furthermore, this tropical geometry can be obtained as the result of a certain degeneration of the (conventional) complex geometry in the torus $(\mathbb{C}^*)^n$.

4.2. Patchworking as tropical degeneration. In 1979 Viro discovered a *patchworking* technique for construction of real algebraic hypersurfaces (see [**42**]). Fix a convex lattice polyhedron $\Delta \in \mathbb{R}^n$. Choose a function $v : \Delta \cap \mathbb{Z}^n \to \mathbb{R}$. The graph of v is a discrete set of points in $\mathbb{R}^n \times \mathbb{R}$. The overgraph is a family of parallel rays. Thus the convex hull of the overgraph is a semi-infinite polyhedron $\tilde{\Delta}$. The facets of $\tilde{\Delta}$ which project isomorphically to $\mathbb{R}^n$ define a subdivision of Δ into smaller convex lattice polyhedra Δ_k.

Let $F(z) = \sum\limits_{j \in \Delta} a_j z^j$ be a generic polynomial in the class of polynomial whose Newton polyhedron is Δ. The *truncation* of F to Δ_k is $F_{\Delta_k} = \sum\limits_{j \in \Delta_k} a_j z^j$. The *patchworking polynomial* f is defined by formula

$$f_t^v(z) = \sum_j a_j t^{v(j)} z^j, \tag{2}$$

$z \in \mathbb{R}^n$, $t > 1$ and $j \in \mathbb{Z}^n$.

Consider the hypersurfaces V_{Δ_k} and V_t in $(\mathbb{C}^*)^n$ defined by F_{Δ_k} and f_t^v. If F has real coefficients then we denote $\mathbb{R}V_{\Delta_k} = V_{\Delta_k} \cap (\mathbb{R}^*)^n$ and $\mathbb{R}V_t = V_t \cap (\mathbb{R}^*)^n$. Viro's patchworking theorem [**42**] asserts that for large values of t the hypersurface $\mathbb{R}V_t$ can be obtained from $\mathbb{R}V_{\Delta_k}$ by a certain patchworking procedure. The same holds for amoebas of the hypersurfaces V_t and $\mathbb{R}V_{\Delta_k}$. In fact patchworking of real hypersurfaces can be interpreted as the real version of patchworking of amoebas (see Appendix in [**26**]). It was noted by Viro in [**43**] that patchworking is related to so-called *Maslov's dequantization* of positive real numbers.

Recall that a *quantization* of a semiring R is a family of semirings R_h, $h \geqslant 0$ such that $R_0 = R$ and $R_t \approx R_s$ as long as $s, t > 0$, but R_0 is not isomorphic to R_t. The semiring R_h with $h > 0$ is called a *quantized* version of R_0.

Maslov (see [**25**]) observed that the "classical" semiring $\mathbb{R}_+$ of real positive number is a quantized version of some other ring in this sense Let R_h be the set of positive numbers with the usual multiplication and with the addition operation $z \oplus_h w = (z^{\frac{1}{h}} + w^{\frac{1}{h}})^h$ for $h > 0$ and $z \oplus_h w = \max\{z, w\}$ for $h = 0$. Note that

$$\lim_{h \to 0} (z^{\frac{1}{h}} + w^{\frac{1}{h}})^h = \max\{z, w\}$$

and thus this is a continuous family of arithmetic operations.

The semiring R_1 coincides with the standard semiring $\mathbb{R}_+$. The isomorphism between $\mathbb{R}_+$ and R_h with $h > 0$ is given by $z \mapsto z^h$. On the other hand the semiring R_0 is not isomorphic to $\mathbb{R}_+$ since it is idempotent, indeed $z + z = \max\{z, z\} = z$.

Alternatively we may define the dequantization deformation with the help of the logarithm. The logarithm $\log_t$, $t > 1$, induces a semiring structure on $\mathbb{R}$ from $\mathbb{R}_+$,

$$x \oplus_t y = \log_t(t^x + t^y), \;\; x \otimes_t y = x + y, \;\; x, y \in \mathbb{R}.$$

Similarly we have $x \oplus_\infty y = \max\{x, y\}$. Let $R_t^{\log}$ be the resulting semiring.

Proposition 4.2. *The map* $\log : R_h \to R_t^{\log}$, *where* $t = e^{\frac{1}{h}}$, *is an isomorphism.*

The patchworking polynomial (2) can be viewed as a deformation of the polynomial f_1^v. We define a similar deformation with the help of Maslov's dequantization. Instead of deforming the coefficients we keep the coefficients but deform the arithmetic operations.

Choose any coefficients α_j, $j \in \Delta$. Let $\varphi_t : (R_t^{\log})^n \to R_t^{\log}$, $t \geqslant e$, be a polynomial whose coefficients are α, i.e.,

$$\varphi_t(x) = \bigoplus_t (\alpha_j + jx), \;\; x \in \mathbb{R}^n.$$

Let $\mathrm{Log}_t : (\mathbb{C}^*)^n \to \mathbb{R}^n$ be defined by $(x_1, \ldots, x_n) = (\log|z_1|, \ldots, \log|z_n|)$.

Proposition 4.3 (Maslov [**25**],Viro [**43**]). *The function* $f_t = (\log_t)^{-1} \circ \varphi_t \circ \mathrm{Log}_t : (\mathbb{R}_+)^n \to \mathbb{R}_+$ *is a polynomial with respect to the standard arithmetic operations in* $\mathbb{R}_+$, *namely we have*

$$f_t(z) = \sum_j t^{\alpha_j} z^j.$$

This is a speci al case of the patchworking polynomial (2). The coefficients α_j define the function $v : \Delta \cap \mathbb{Z}^n \to \mathbb{R}$.

4.3. Limit set of amoebas. Let $V_t \subset (\mathbb{C}^*)^n$ be the zero set of f_t and let $\mathcal{A}_t = \operatorname{Log}_t(V_t) \subset \mathbb{R}^n$. Note that $\mathcal{A}_t$ is the amoeba of V_t scaled $\log t$ times. Note also that the family $f_t = \sum_j t^{\alpha_j} z^j$ can be considered as a single polynomial whose coefficients are powers of t. Such coefficients are a very simple instance of the so-called *Puiseux series.*

The field K of the real-power Puiseux series is obtained from the field of the Laurent series in t by taking the algebraic closure first and then taking the metric completion with respect to the ultranorm

$$||\sum a_j t^j|| = \min\{j \in \mathbb{R} \mid a_j \neq 0\}.$$

The logarithm $\operatorname{val} : K^* \to \mathbb{R}$ of this norm is an example of the so-called non-Archimedean valuation as $\operatorname{val}(a+b) \leqslant \max\{\operatorname{val}(a) + \operatorname{val}(b)\}$ and $\operatorname{val}(ab) = \operatorname{val}(a) + \operatorname{val}(b)$ for any $a, b \in K^* = K \setminus \{0\}$.

Definition 4.4 (Kapranov [**18**]). Let $V_K \subset (K^*)^n$ be an algebraic variety. Its *(non-Archimedean) amoeba* is

$$\mathcal{A}_K = \operatorname{Val}(V_k) \subset \mathbb{R}^n,$$

where $\operatorname{Val}(z_1, \dots, z_n) = (\operatorname{val}(z_1), \dots, \operatorname{val}(z_n))$.

We have a uniform convergence of the addition operation in $R_t^{\log}$ to the addition operation in $R_\infty^{\log}$. As it was observed by Viro it follows from the following inequality

$$\max\{x, y\} \leqslant x \oplus_t y = \log_t(t^x + t^y) \leqslant \max\{x, y\} + \log_t 2.$$

More generally, we have the following lemma.

Lemma 4.5.

$$\max_{j \in \Delta}(\alpha_j + jx) \leqslant \varphi_t(x) \leqslant \max_{j \in \Delta}(\alpha_j + jx) + \log N,$$

where N is the number of lattice points in Δ.

Recall that the *Hausdorff metric* is defined on closed subsets $A, B \subset \mathbb{R}^n$ by

$$d_{\text{Hausdorff}}(A, B) = \max\{\sup_{a \in A} d(a, B), \sup_{b \in B} d(A, b)\},$$

where d is the Euclidean distance in $\mathbb{R}^n$. The following theorem is a corollary of Lemma 4.5.

Theorem 4.6 (Mikhalkin [**29**], Rullgård [**40**]). *The subsets $\mathcal{A}_t \subset \mathbb{R}^n$ tend in the Hausdorff metric to $\mathcal{A}_K$ when $t \to 0$.*

Recall that in our setup $t > 0$. Alternatively we may replace t with $\frac{1}{t}$ to get a limit with $t \to +\infty$.

4.4. Tropical varieties and non-Archimedean amoebas. We start by defining tropical hypersurfaces. The semiring $\mathbb{R}_{\text{trop}}$ lacks (additive) zero so the tropical hypersurfaces are defined as singular loci and not as zero loci. Let $F : \mathbb{R}^n \to \mathbb{R}$ be a tropical polynomial. It is a continuous convex piecewise-linear function. Unless F is linear it is not everywhere smooth.

Definition 4.7. The *tropical variety* $V_F \subset \mathbb{R}^n$ *of* F is the set of all points in $\mathbb{R}^n$ where F is not smooth.

Equivalently we may define V_F as the set of points where more than one monomial of $F(x) =$ "$\sum a_j x^j$" reaches the maximum.

Let us go back to the non-Archimedean field K of Puiseux series. Let

$$f(z) = \sum_j \alpha_j z^j,$$

$\alpha_j \in K$, $j \in \mathbb{Z}^n$, $z \in K^n$, be a polynomial that defines a hypersurface $V_K \subset (K^*)^n$ and let $\mathcal{A}_K \subset \mathbb{R}^n$ be the corresponding non-Archimedean amoeba. We form a tropical polynomial

$$F(x) = \sum_j \text{val}(\alpha_j) x^j,$$

$x \in \mathbb{R}^n$.

Kapranov's description **[18]** of the non-Archimedean amoebas can be restated in the following way.

Theorem 4.8 ([**18**]). *The amoeba $\mathcal{A}_K$ coincides with the tropical hypersurface V_F.*

Definition of tropical varieties in higher codimension in $\mathbb{R}^n$ gets somewhat tricky as intersections of tropical hypersurfaces are not always tropical. As is was suggested in **[36]** non-Archimedean amoebas provide a byway for such definition as tropical varieties can be simply defined as non-Archimedean amoebas for algebraic varieties in $(K^*)^n$.

In the next section we concentrate on the study of tropical curves. References to some higher-dimensional tropical varieties treatments include **[29]** for the case of hypersurfaces and **[41]** for the case of the Grassmanian varieties.

5. Calculus of Tropical Curves in $\mathbb{R}^n$

5.1. Definitions. Let $\overline{\Gamma}$ be a finite graph whose edges are weighted by natural numbers. Let $\mathcal{V}_1$ be the set of 1-valent vertices of Γ. We set

$$\Gamma = \overline{\Gamma} \smallsetminus \mathcal{V}_1.$$

Definition 5.1 (Mikhalkin [**31**]). A proper map $h : \Gamma \to \mathbb{R}^n$ is called a *parameterized tropical curve* if it satisfies to the following two conditions.

- For every edge $E \subset \Gamma$ the restriction $h|_E$ is an embedding. The image $h(E)$ is contained in a line $l \subset \mathbb{R}^n$ such that the slope of l is rational.
- For every vertex $V \in \Gamma$ we have the following property. Let $E_1, \ldots, E_m \subset \Gamma$ be the edges adjacent to V, let $w_1, \ldots, w_m \in \mathbb{N}$ be their weights and let $v_1, \ldots, v_m \in \mathbb{Z}^n$ be the primitive integer vectors from V in the direction of the edges. We have

$$\sum_{j=1}^{m} w_j v_j = 0. \tag{3}$$

Two parameterized tropical curves $h : \Gamma \to \mathbb{R}^n$ and $h' : \Gamma' \to \mathbb{R}^n$ are called *equivalent* if there exists a homeomorphism $\Phi : \Gamma \to \Gamma'$ which respects the weights of the edges and such that $h = h' \circ \Phi$. We do not distinguish equivalent parameterized tropical curves.

The image

$$C = h(\Gamma) \subset \mathbb{R}^n$$

is called the (unparameterized) tropical curve. It is a weighted piecewise-linear graph in $\mathbb{R}^n$. Note that the same curve $C \subset \mathbb{R}^2$ may admit nonequivalent parameterizations. The curve C is called *irreducible* if Γ is connected for any parameterization. Otherwise the curve is called reducible.

Remark 5.2. In dimension 2 the notion of tropical curve coincides with the notion of (p, q)-webs introduced by Aharony, Hanany and Kol in [**2**] (see also [**1**]).

It is convenient to prescribe a multiplicity to a 3-valent vertex $A \in \Gamma$ of the tropical curve $h : \Gamma \to \mathbb{R}^n$ as in [**31**]. As in Definition 5.1 let w_1, w_2, w_3 be their weights of the edges of $h(\Gamma)$ adjacent to A and let v_1, v_2, v_3 be the primitive integer vectors in the direction of the edges.

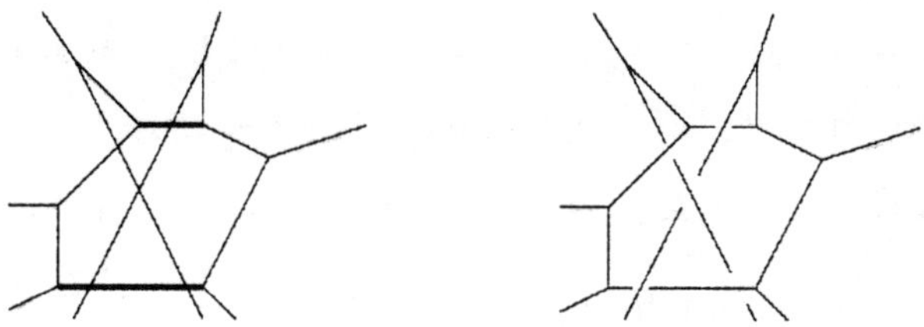

FIGURE 10. A tropical curve in $\mathbb{R}^2$ and its possible lift to $\mathbb{R}^3$. The edges of weight 2 are bold (at the left picture). Note that lifts of such edges can have weight 1.

Definition 5.3. The *multiplicity* of a 3-valent vertex A in $h(\Gamma)$ is $w_1 w_2 |v_1 \times v_2|$. Here $|v_1 \times v_2|$ is the "length of the vector product of v_1 and v_2" in $\mathbb{R}^n$ being interpreted as the area of the parallelogram spanned by v_1 and v_2. Note that

$$w_1 w_2 |v_1 \times v_2| = w_2 w_3 |v_2 \times v_3| = w_3 w_1 |v_3 \times v_1|$$

since $v_1 w_1 + v_2 w_2 + v_3 w_3 = 0$ by Definition 5.1.

If the multiplicity of a vertex is greater than 1 then it is possible to deform it with an appearance of a new cycle as in Figure 11.

FIGURE 11. Deformation of a multiple 3-valent vertex.

5.2. Degree, genus and the tropical Riemann–Roch formula. Heuristically, the degree of a tropical curve $C \subset \mathbb{R}^n$ is the set of its asymptotic directions. For each end of a tropical curve $C = h(\Gamma)$ we fix a primitive integer vector parallel to this ray in the outward direction and multiply it by the weight of the corresponding (half-infinite) edge. Doing this for every end of C we get a collection $\mathcal{C}$ of integer vectors in $\mathbb{Z}^n$.

Let us add all vectors in $\mathcal{C}$ that are positive multiples of each other. The result is a set $\mathcal{T} = \{\tau_1, \ldots, \tau_q\} \subset \mathbb{Z}^n$ of nonzero integer vectors such that $\sum_{j=1}^{q} \tau_j = 0$. Note that in this set we do not have positive multiples of each other, i.e. if $\tau_j = m\tau_k$ for $m \in \mathbb{N}$ then $\tau_j = \tau_k$.

Definition 5.4 (**[31]**). The set $\mathcal{T}$ is called the *degree of the tropical curve* $C \subset \mathbb{R}^n$. The *genus* of a parameterized tropical curve $h : \Gamma \to \mathbb{R}^n$ is $\dim(H_1(\Gamma)) + 1 - \dim(H_0(\Gamma))$ so that if Γ is connected then it coincides with the number of cycles $\dim(H_1(\Gamma))$ in Γ. The genus of a tropical curve $C \subset \mathbb{R}^n$ is the minimal genus among all the parameterization $C = h(\Gamma)$.

There is an important class of tropical curves that behaves especially nice with respect to a genus-preserving deformation.

Definition 5.5 (**[31]**). A parameterized tropical curve $h : \Gamma \to \mathbb{R}^n$ is called *simple* if

- Γ is 3-valent,
- h is an immersion,
- if $a, b \in \Gamma$ are such that $h(a) = h(b)$ then neither a nor b can be a vertex of Γ.

In this case the image $h(\Gamma)$ is called a *simple tropical curve.*

Simple curves locally deform in a linear space.

Theorem 5.6 (Tropical Riemann–Roch, **[31]**). *Let $h : \Gamma \to \mathbb{R}^n$ be a simple tropical curve, where Γ is a graph with x ends. Non-equivalent tropical curves of the same genus and with the same number of ends close to h locally form a k-dimensional real vector space, where*

$$k \geqslant x + (n-3)(1-g).$$

If the curve is nonsimple then its space of deformation is locally piecewise-linear.

5.3. Enumerative tropical geometry in $\mathbb{R}^2$. It was suggested by Kontsevich (see **[24]** and **[28]**) that tropical curves can be used for enumeration of holomorphic curves. This suggestion was realized in **[31]** for the case of toric surfaces. Let us review the problem of curve counting for the complex torus $(\mathbb{C}^*)^2$.

Any algebraic curve $V \subset (\mathbb{C}^*)^2$ is defined by a polynomial

$$f(z,w) = \sum_{j,k} a_{jk} z^j w^k.$$

Recall that from the topological viewpoint the degree of a variety is its homology class in the ambient variety. Here we have a difficulty caused by noncompactness of $(\mathbb{C}^*)^2$.

Help is provided by the *Newton polygon*

$$\Delta(f) = \text{ConvexHull}\{(j,k) \mid a_{jk} \neq 0\}$$

of f. The polygon $\Delta = \Delta(f)$ can be interpreted as the (*toric*) *degree* of V. Indeed being a compact lattice polygon Δ defines a compact toric surface $\mathbb{C}T_\Delta \supset (\mathbb{C}^*)^2$, for example, by taking the closure of the image under the *Veronese embedding* $(\mathbb{C}^*)^2 \to \mathbb{CP}^{\#(\Delta\cap\mathbb{Z}^2)}$ (see, for example, [**11**]). The closure of V in $\mathbb{C}T_\Delta$ defines a homology class induced from the hyperplane section by the Veronese embedding.

Note that the definition of the toric degree agrees with its tropical counterpart in Definition 5.4. Indeed, for each side Δ' of Δ we can take the primitive integer normal vector in the outward direction and multiply it by the lattice length $\#(\Delta' \cap \mathbb{Z}^2) - 1$ of the side. The result is a tropical degree set $\mathcal{T}(\Delta)$. Accordingly we define

$$x = \#(\partial\Delta \cap \mathbb{Z}^2)$$

which is the number of ends of a general curve of degree Δ in $(\mathbb{C}^*)^2$.

An irreducible curve V has *geometric genus* which is the genus of its normalization $\widetilde{V} \to V$. In the case when V is not necessarily irreducible it is convenient to define the genus as the sum of the genera of all irreducible components minus the number of such components plus one.

Let us fix the genus (i.e. a number $g \in \mathbb{Z}$) and the toric degree (i.e., a polygon $\Delta \subset \mathbb{R}^2$). Let

$$\mathcal{P} = \{p_1, \dots, p_{x+g-1}\} \subset (\mathbb{C}^*)^2$$

be an configuration of $x+g-1$ general points in $(\mathbb{C}^*)^2$. We set $N(g,\Delta)$ to be equal to be the number of curves in $(\mathbb{C}^*)^2$ of genus g and degree Δ passing through $\mathcal{P}$. Similarly we set $N^{\text{irr}}(g,\Delta)$ to be the number of irreducible curves among them.

These numbers are close relatives of the Gromov–Witten invariants of $\mathbb{C}T_\Delta$ (see [**23**] for the definition). In the case when $\mathbb{C}T_\Delta$ is smooth Fano they coincide with the corresponding Gromov–Witten invariants. The numbers $N(g,\Delta)$ and $N^{\text{irr}}(g,\Delta)$ have tropical counterparts.

For a fixed genus g and a toric degree Δ we fix a configuration

$$\mathcal{R} = \{r_1, \dots, r_{x+g-1}\} \subset \mathbb{R}^2$$

of $x+g-1$ general points in the tropical plane $\mathbb{R}^2$ (for a rigorous definition of tropical general position see [**31**]). We have a finite number of tropical curves of genus g and degree $\mathcal{T}(\Delta)$ passing through $\mathcal{R}$ (see [**31**]). Generically all such curves are simple (see Definition 5.5. However unlike the situation in $(\mathbb{C}^*)^2$ the number of such curves is different for different configurations of $x+g-1$ general point.

Definition 5.7 ([**31**]). The multiplicity mult(h) of a simple tropical curve $h:\Gamma\to\mathbb{R}^2$ of degree Δ and genus g passing via $\mathcal{R}$ equals to the product of the multiplicities of the (3-valent) vertices of Γ. (see Definition 5.3).

Theorem 5.8 ([**31**]). *The number of irreducible tropical curves of genus g and degree Δ passing via $\mathcal{R}$ and counted with multiplicity from Definition* 5.7 *equals to $N^{\mathrm{irr}}(g,\Delta)$.*

The number of all tropical curves of genus g and degree Δ passing via $\mathcal{R}$ and counted with multiplicity from Definition 5.7 equals to $N(g,\Delta)$.

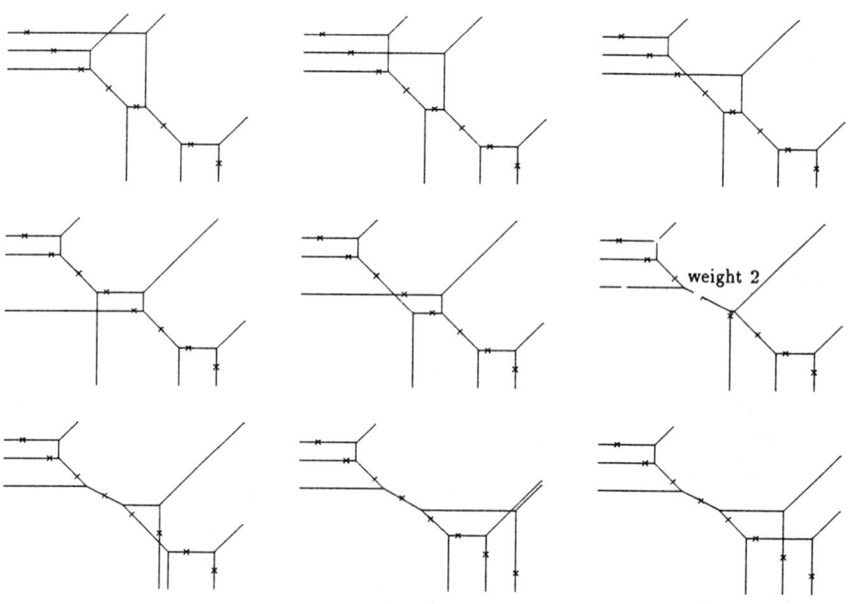

FIGURE 12. Tropical projective rational cubics via 8 points.

Example 5.9. Figure 12 shows a (generic) configuration of 8 points $\mathcal{R} \subset \mathbb{R}^2$ and all curves of genus 0 and of projective degree 3 passing through $\mathcal{R}$. Out of these nine curves eight have multiplicity 1 and one (with a weight 2 edge) has multiplicity 4. All the curves are irreducible. Thus $N^{\text{irr}}(g, \Delta) = N(g, \Delta) = 12$.

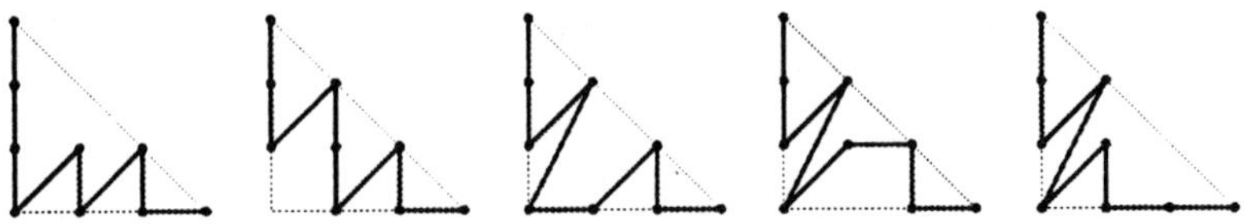

FIGURE 13. The lattice paths describing the tropical curves from Figure 12 and the path multiplicities.

Theorem 5.8 thus reduces the problem of finding $N^{\text{irr}}(g, \Delta)$ and $N(g, \Delta)$ to the corresponding tropical problems. Furthermore, it allows us to use any general configuration $\mathcal{R}$ in the tropical plane $\mathbb{R}^2$ (as it implies that the answer is independent of $\mathcal{R}$). We can take the configuration $\mathcal{R}$ on the same affine (not tropical) line $L \subset \mathbb{R}^2$ and still insure tropical general position as long as the slope of L is irrational. It was shown in **[31]** that such curves are encoded by lattice paths of length $x + g - 1$ connecting a pair of vertices in Δ.

Namely, the slope of L determines a linear function $\lambda : \mathbb{R}^2 \to \mathbb{R}$ such that $\lambda|_{\Delta \cap \mathbb{Z}^2}$ is injective and thus a linear order on the lattice points of Δ. There is a combinatorial rule (see **[30]** or **[31]**) that associates a nonnegative integer multiplicity to every λ-increasing lattice path of length $x + g - 1$, i.e., to every order-increasing sequence of lattice points of Δ that contains $x + g$ points. This multiplicity is only nonzero if the first and the last points of the sequence are the points where $\lambda|_\Delta$ reaches its minimum and maximum.

Example 5.10. The tropical curves from Figure 12 are described by the lattice paths from Figure 13 shown together with their multiplicities. Here the first path describes the first 3 tropical curve from Figure 12, the second — the next two paths, the third — the next curve (which itself corresponds to 4 distinct holomorphic curves), the fourth — the next curve and the fifth — the last two tropical curves from Figure 12. These paths are λ-increasing for $\lambda(x, y) = y - (1 + \varepsilon x)$, where $\varepsilon > 0$ is very small.

5.4. Enumerative tropical geometry in $\mathbb{R}^3$ (and higher dimension). The results of previous subsections can be established with the help of the following restatement of Theorem 4.6 in the case of $\mathbb{R}^2$.

Lemma 5.11. *If $C = h(\Gamma) \subset \mathbb{R}^2$ is a tropical curve then there exists a family $V_t \subset (\mathbb{C}^*)^2$ of holomorphic curves for $t > 0$ such that $\mathrm{Log}_t(V_t) = C$. Here the degree of C coincides with the degree of V_t.*

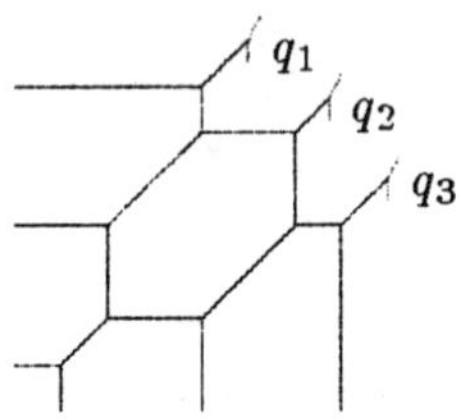

FIGURE 14. A planar part of a superabundant spatial cubic.

The situation is more complicated if $n > 2$ as such statement is no longer true for *all* tropical curves in $\mathbb{R}^n$.

Example 5.12. Consider the graph $C' \subset \mathbb{R}^2 \subset \mathbb{R}^3$ depicted on Figure 14. This set can be obtained by removing three rays from a planar projective cubic curve. Let $q_1, q_2, q_3 \in \mathbb{R}^2$ be the end points of these rays. Consider the curve

$$C = C' \cup \bigcup_{j=1}^{n} (\{(q_j, t) \mid t \leqslant 0\} \cup \{(q_j + t, t) \mid t \geqslant 0\}).$$

It is easy to check that $C \subset \mathbb{R}^3$ is a (spatial) projective curve of degree 3 and genus 1. Suppose that q_1, q_2, q_3 are not tropically collinear, i.e., are not lying on the same tropical line in $\mathbb{R}^2$ (for example, we may choose q_1, q_2, q_3 to be in tropically general position). Then C cannot be obtained as the limit of $\mathrm{Log}_t(V_t)$ for cubic curves $V_t \subset \mathbb{CP}^3$ (since $\mathrm{Log}_t(V_t)$ is not everywhere defined $\mathrm{Log}_t(V_t)$ stands for $\mathrm{Log}_t(V_t \cap (\mathbb{C}^*)^3)$.

Indeed, any cubic curve $V_t \subset \mathbb{CP}^3$ of genus 1 is planar, i.e., is contained in a plane $H_t \subset \mathbb{CP}^3$. It is easy to see (after passing to a subsequence, see **[29]** and **[31]**) that there has to exist a limiting set H for $\mathrm{Log}_t(H_t)$. Furthermore, H is a tropical hypersurface in $\mathbb{R}^3$ whose Newton polyhedron is contained in the polyhedron of a hyperplane. Since $C \subset H$ we can deduce that H has to be a hyperplane. But then the intersection of H with $\mathbb{R}^2$ is (up to a translation in $\mathbb{R}^2$) a union of the negative quadrant $\{(x,y) \mid x \leqslant 0, y \leqslant 0\}$ and the ray $\{(t,t) \mid t \geqslant 0$. The points p_j have to sit on the boundary of the quadrant (which is impossible unless they are tropically collinear).

Note that the tropical Riemann–Roch formula (Theorem 5.6) is a strict inequality for the curve C. In accordance with the classical terminology such curves are called *superabundant*. Conversely, a tropical curve is called *regular* if the Riemann–Roch formula turns into equality. It is easy to see that all rational curves are regular and that the superabundancy of C is caused by the cycle contained in an affine plane in $\mathbb{R}^3$. Conjecturally all regular curves are limits of the corresponding complex amoebas. Hopefully the technique developed in the Symplectic Field Theory, see **[9]** and **[7]** can help to verify this conjecture.

Let us formulate a tropical enumerative tropical problem in $\mathbb{R}^n$. We fix the genus g and the degree $\mathcal{T} = \{\tau_1, \dots, \tau_q\} \subset \mathbb{Z}^n$. In addition we fix a configuration $\mathcal{R}$ which consists of some points and some higher dimensional tropical varieties in $\mathbb{R}^n$ in general position. Let k be the sum of the codimensions of all varieties in $\mathcal{R}$. For each τ_j let $x_j \in \mathbb{N}$ be the maximal integer that divides it. Let $x = \sum_{j=1}^{g} x_j$.

If $k = x + (1-g)(n-3)$ then the expected number of tropical curves of genus g and degree $\mathcal{T}$ passing through $\mathcal{R}$ is finite. However there may exist positive-dimensional families of superabundant curves of genus g and degree $\mathcal{T}$ through $\mathcal{R}$.

One way to avoid this (higher-dimensional) difficulty is to restrict ourselves to the genus zero case. In this case one can assign multiplicities to tropical rational curves passing through $\mathcal{R}$ so that the total number of tropical curves counted with these multiplicities agrees with the number of curves in the corresponding complex enumerative problem (details are subject to a future paper).

5.5. Complex and real tropical curves. Tropical curve $C \subset \mathbb{R}^n$ can be presented as images $C = \mathrm{Log}(B)$ for certain objects $B \subset (\mathbb{C}^*)^n$ called *complex tropical curves.* (Recall that $\mathrm{Log} : (\mathbb{C}^*)^n \to \mathbb{R}^n$ is the coordinatewise logarithm of the absolute value.) Let $H_t : (\mathbb{C}^*)^n \to (\mathbb{C}^*)^n$ be the self-diffeomorphism defined by

$$H_t(z_1, \dots, z_n) = (|z_1|^{\log(t)-1} z^1, \dots, |z_n|^{\log(t)-1} z_n).$$

We have $\mathrm{Log}_t(z) = \mathrm{Log}(H_t(z))$.

Definition 5.13. The set $B \subset (\mathbb{C}^*)^n$ is called a complex tropical curve if it satisfies to the following condition.

- For every $x \in \mathbb{R}^n$ there exist a neighborhood $U \ni x$ and a family $V_t \subset (\mathbb{C}^*)^n$, $t > 1$ of holomorphic curves such that
$$B \cap \mathrm{Log}^{-1}(U) = \lim_{t \to +\infty} (H_t^{-1}(V_t) \cap U),$$
where the limit is taken with respect to the Hausdorff metric.
- For every open set $U \subset \mathbb{R}^n$ for every component B' of $B \cap \mathrm{Log}^{-1}(U)$ there exists a tropical curve $C' \subset \mathbb{R}^n$ such that projection $\mathrm{Log}(B') = C' \cap U$.

It is easy to see that for every open edge of $E \subset C$ the inverse image $\mathrm{Log}^{-1}(E) \cap B$ is a disjoint union of holomorphic cylinders. We can prescribe the weights to this cylinder so that the sum is equal to the weight of E. (In fact the second condition in Definition 5.13 is needed only to insure that the cylinder weights in different neighborhoods are consistent.)

Complex tropical curves can be viewed as curves "holomorphic" with respect to a (maximally) degenerate complex structure in $(\mathbb{C}^*)^n$. Consider a family of almost complex structures J_t induced from the standard structure on $(\mathbb{C}^*)^n$ by the self-diffeomorphism H_t, $t > 1$. For every finite t it is an integrable complex structure (isomorphic to the standard one by H_t). The curves $H^1(V_t)$ are J_t-holomorphic as long as V_t is holomorphic (with respect to the standard, i.e., J_e-holomorphic structure). The limiting J_∞-structure is no longer complex or almost complex, but it is convenient to view B as a "J_∞-holomorphic curve."

If $C = \mathrm{Log}(B)$ admits a parameterization by a 3-valent graph Γ then one can equip the edges of Γ with some extra data called the *phases* that determine B. Let E be a phase of weight w and parallel to a primitive integer vector $v \in \mathbb{Z}^n$. The vector v determines an equivalence relation $\sim_v$ in the torus T^n. We have $a \sim_v b$ for $a, b \in T^n$ if $a - b$ is proportional to v. Clearly, $T^n / \sim_v$ is an $(n-1)$-dimensional torus. The *phase* of E is a multiset $\Phi = \{\varphi_1, \dots, \varphi_w\}$, $\varphi_j \in T^n / \sim_v$ (recall that w is the weight of E). Alternatively, φ_j may be viewed as a geodesic circle in T^n. We orient this geodesic by choosing v going away from A along E. A phase determines a collection of holomorphic cylinders in $\mathrm{Log}^{-1}(E) \subset (\mathbb{C}^*)^n$. If some of φ_j coincide then some of these cylinders have multiple weight.

Let A be a 3-valent vertex of Γ and E, E', E'' are the three adjacent edges. to A with phases Φ, Φ', Φ''. The phases are called *compatible* at

A if the geodesics of $\Phi \cup \Phi' \cup \Phi''$ can be divided into subcollections Ψ such that for every $\Psi = \{\psi_1, \ldots, \psi_k\}$ there exists a subtorus $T^2 \subset T^n$ containing all geodesics ψ_j and these (oriented) geodesics bound a region of zero area in this T^2.

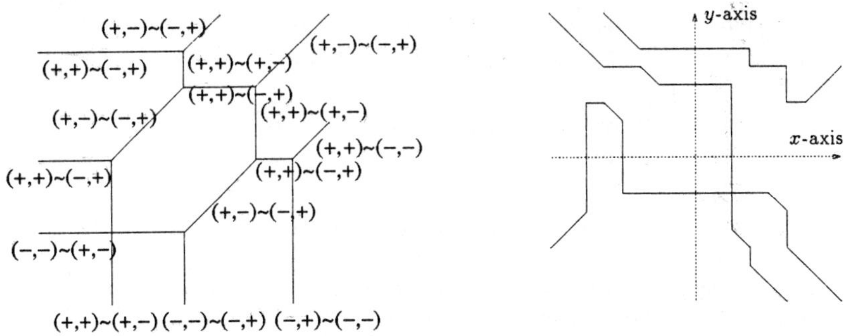

FIGURE 15. A real tropical projective cubic curve.

Definition 5.14. A *simple complex tropical curve* is a simple tropical curve $h : \Gamma \to \mathbb{R}^n$ (see Definition 5.5) whose edges are equipped with admissible phases such that for every edge $E \subset \Gamma$ the phase $\Phi = \{\varphi_1, \ldots, \varphi_w\}$ consists of the same geodesic $\varphi_1 = \cdots = \varphi_w$.

Note that a simple complex tropical curve defines a complex tropical curve $B \subset (\mathbb{C}^*)^n$ of the same genus as $h : \Gamma \to \mathbb{R}^n$. If the phase of a bounded edge of Γ consists of distinct geodesics then the genus of B is strictly greater than that of C.

In a similar way one can define *real tropical curves* by requiring all curves V_t in Definition 5.13 to be real. Our next purpose is to define simple real tropical curves. Let $h : \Gamma \to \mathbb{R}^n$ be a simple tropical curve. Consider an edge $E \subset \Gamma$ of weight w parallel to a primitive vector $v \in \mathbb{Z}^n$. The scalar multiple wv defines an equivalence relation $\sim_{wv}$ in $\mathbb{Z}_2^n$. We have $a \sim_{wv} b$ if $a - b \in \mathbb{Z}_2^n$ is a multiple of $wv \mod 2$. The equivalence is trivial if w is even. Otherwise $\mathbb{Z}_2^n / \sim_{wv} \approx \mathbb{Z}_2^{n-1}$. The *sign* of E is an element $\mathbb{Z}_2^n / \sim_{wv}$. The choice of signs has to be compatible at the vertices of Γ. Let A be a vertex of Γ and E_1, E_2, E_3 be the adjacent edges of weight w_1, w_2, w_3 parallel to the pri mitive vectors $v_j \in \mathbb{Z}^n$. Let σ_j be the sign of E_j. We say that the sign choice is *compatible* at A if every element in the equivalence class σ_j, $j = 1, 2, 3$, is contained in another equivalence class σ_k, $k = 1, 2, 3$, $k \neq j$.

Definition 5.15. A *simple real tropical curve* is a tropical curve $h : \Gamma \to \mathbb{R}^n$ whose edges are equipped with signs compatible at every vertex of Γ.

If all edges of Γ have weight 1 then this definition agrees with *combinatorial patchworking* (see [**16**]). Simple real tropical curves can be used in real enumerative problems (see [**31**] and [**15**] for details in the case of $\mathbb{R}^2$).

Figure 15 sketches a tropical curve equipped with admissible signs and the corresponding real tropical curve.

References

1. O. Aharony and A. Hanany, *Branes, superpotentials and superconformal fixed points*, http://arxiv.org hep-th/9704170.
2. O. Aharony, A. Hanany, and B. Kol, *Webs of (p,q) 5-branes, five dimensional field theories and grid diagrams*, http://arxiv.org hep-th/9710116.
3. V. I. Arnold, *The situation of ovals of real plane algebraic curves, the involutions of four-dimensional smooth manifolds, and the arithmetic of integral quadratic forms* [in Russian], Funk. Anal. Pril. **5** (1971) no. 3, 1-9.
4. M. F. Atiyah, *Angular momentum, convex polyhedra and algebraic geometry*, Proc. Edinburgh Math. Soc. **26** (1983) 121-133.
5. G. M. Bergman, *The logarithmic limit set of an algebraic variety*, Trans. Am. Math. Soc. **157** (1971), 459-469.
6. D. Bernstein, *The number of roots of a system of equations*, Funct. Anal. Appl. **9** (1975), 183-185.
7. F. Bourgeois, *A Morse–Bott Approach to Contact Homology*, Dissertation, Stanford University, 2002.
8. L. Brusotti, *Curve generatrici e curve aggregate nella costruzione di curve piane d'ordine assegnato dotate del massimo numero di circuiti*, Rend. Circ. Mat. Palermo **42** (1917), 138-144.
9. Y. Eliashberg, A. Givental, and H. Hofer, *Introduction to symplectic field theory*, GAFA 2000, Special Volume, Part II, 560–673.
10. M. Forsberg, M. Passare, and A. Tsikh, *Laurent determinants and arangements of hyperplane amoebas*, Adv. Math. **151** (2000), 45–70.
11. I. M. Gelfand, M. M. Kapranov, A. V. Zelevinsky, *Discriminants, resultants, and multidimensional determinants*. Mathematics: Theory & Applications. Birkhäuser Boston, Inc., Boston, MA, 1994.
12. A. Harnack, *Über Vieltheiligkeit der ebenen algebraischen Curven*, Math. Ann. **10** (1876), 189-199.
13. A. Henriques, *An Analogue of Convexity for Complements of Amoebas of Varieties of Higher Codimensions*, Preprint, Berkeley, May 2001.
14. D. Hilbert, *Mathematische Probleme*, Arch. Math. Phys. (3) **1** (1901), 213-237.
15. I. Itenberg, V. Kharlamov, E. Shustin, *Welschinger invariant and enumeration of real plane rational curves*, http://arxiv.org/abs/math.AG/0303378.

16. I. Itenberg and O. Viro, *Patchworking real algebraic curves disproves the Ragsdale conjecture*, Math. Intel. **18** (1996), 19–28.
17. M. M. Kapranov, *A characterization of A-discriminantal hypersurfaces in terms of the logarithmic Gauss map*, Math. Ann. **290** (1991), 277-285
18. M. M. Kapranov, *Amoebas over Non-Archimedian Fields*, Preprint, 2000.
19. R. Kenyon and A. Okounkov, *Planar dimers and Harnack curves*, http://arxiv.org math.AG/03 11062.
20. R. Kenyon, A. Okounkov, and S. Sheffield, *Dimers and amoebae*, http://arxiv.org math-ph/0311005.
21. V. M. Kharlamov, *On the classification of nonsingular surfaces of degree 4 in* $\mathbb{RP}^3$ *with respect to rigid isotopies* [in Russian], Funkt. Anal. Pril. **18** (1984), no. 1, 49-56.
22. A. G. Khovanskii, *Newton polyhedra and toric varieties,* [in Russian], Funkt. Anal. Pril. **11** (1977), no. 4, 56 - 64.
23. M. Kontsevich and Yu. Manin, *Gromov–Witten classes, quantum cohomology and enumerative geometry*, Comm. Math. Phys. **164** (1994), 525-562.
24. M. Kontsevich and Y. Soibelman, *Homological mirror symmetry and torus fibrations,* Symplectic geometry and mirror symmetry (Seoul, 2000), 203–263, World Sci. Publishing, River Edge, NJ, 2001.
25. V. P. Maslov, *New superposition principle for optimization problems*, in Seminaire sur les Equations auc Dérivées Partielles 1985/6, Centre Mathématiques de l'École Polytechnique, Palaiseau, 1986, exposé 24.
26. G. Mikhalkin, *Real algebraic curves, moment map and amoebas*, Ann. Math. **151** (2000), 309 - 326.
27. G. Mikhalkin and H. Rullgård, *Amoebas of maximal area*, Intern. Math. Res. Notices **9** (2001), 441-451.
28. G. Mikhalkin, *Amoebas of algebraic varieties,* a report for the Real Algebraic and Analytic Geometry congress, June 2001, Rennes, France, http://arxiv.org math.AG/0108225.
29. G. Mikhalkin, *Decomposition into Pairs-of-Pants for Complex Algebraic Hypersurfaces*, http://arxiv.org math.GT/0205011 Preprint 2002 [To appear in Topology.]
30. G. Mikhalkin, *Counting curves via lattice paths in polygons*, C. R. Math. Acad. Sci. Paris **336** (2003), no. 8, 629-634.
31. G. Mikhalkin, *Enumerative tropical geometry in* $\mathbb{R}^2$, http://arxiv.org math.AG/0312530.
32. G. Mikhalkin, *Maximal real algebraic hypersurfaces* (in preparation)
33. M. Passare and H. Rullgård, *Amoebas, Monge–Ampère Measures, and Triangulations of the Newton Polytope.* Preprint, Stockholm University, 2000.
34. I. G. Petrovsky, *On the topology of real plane algebraic curves*, Ann. Math. **39** (1938), 187–209.
35. J.-E. Pin, *Tropical semirings,* Idempotency (Bristol, 1994), 50–69, Publ. Newton Inst., 11, Cambridge Univ. Press, Cambridge, 1998.
36. J. Richter-Gebert, B. Sturmfels, Th. Theobald, *First steps in tropical geometry,* http://arxiv.org math.AG/0306366.

37. V. A. Rohlin, *Congruences modulo* 16 *in Hilbert's sixteenth problem*, [in Russian], Funkt. Anal. Pril. **6** (1972), no. 4, 58-64.
38. L. Ronkin, *On zeroes of almost periodic functions generated by holomorphic functions in a multicircular domain* [To appear in "Complex Analysis in Modern Mathematics," Fazis, Moscow, 2000, 243-256].
39. H. Rullgård, *Stratification des espaces de polynômes de Laurent et la structure de leurs amibes*, C. R. Acad. Sci. Paris, Série I, **331** (2000), 355-358.
40. H. Rullgård, *Polynomial Amoebas and Convexity*, Preprint, Stockholm Univerity, 2001.
41. D. Speyer, B. Sturmfels, *The tropical Grassmanian*, http://arxiv.org math.AG/0304218.
42. O. Ya. Viro, *Real plane algebraic curves: constructions with controlled topology*, Leningrad Math. J. **1** (1990), no. 5, 1059–1134.
43. O. Ya. Viro, *Dequantization of Real Algebraic Geometry on a Logarithmic Paper*, Proceedings of the European Congress of Mathematicians (2000).
44. G. Wilson, *Hilbert's sixteenth problem*, Topology **17** (1978), 53-73.

Heegaard Diagrams and Holomorphic Disks

Peter Ozsváth†
Columbia University, New York, USA
Institute for Advanced Study, Princeton, USA

Zoltán Szabó††
Princeton University, New Jersey, USA

In this paper, we describe some invariants for low-dimensional manifolds that are constructed using suitable counts of pseudo-holomorphic disks.

1. Introduction

Gromov's theory of pseudo-holomorphic disks [**40**] has wide-reaching consequences in symplectic geometry and low-dimensional topology. Our aim here is to describe certain invariants for low-dimensional manifolds built on this theory.

The invariants we describe here associate a graded Abelian group to each closed, oriented three-manifold Y, the Heegaard Floer homology of Y. These invariants also have a four-dimensional counterpart, which

†Peter Ozsváth was partially supported by NSF grant numbers DMS-0234311, DMS-0111298, and FRG-0244663.

†† Zoltán Szabó was partially supported by NSF grant numbers DMS-0107792 and FRG-0244663, and a Packard Fellowship.

associates to each smooth cobordism between two such three-manifolds a map between the corresponding Floer homology groups. In another direction, there is a variant which gives rise to an invariant of knots in Y.

1.1. Some background on Floer homology. To place Heegaard Floer homology into a wider context, we begin with Casson's invariant. Starting with a Heegaard decomposition of an integer homology three-sphere Y, Casson constructs a numerical invariant which roughly speaking gives an obstruction to disjoining the $SU(2)$ character varieties of the two handlebodies inside the character variety for the Heegaard surface Σ (cf. [**1**] and [**95**]).

During the time when Casson introduced his invariants to three-dimensional topology, smooth four-dimensional topology was being revolutionized by the work of Donaldson [**16**], who showed that the moduli spaces of solutions to certain nonlinear, elliptic PDEs – gauge theory equations which were first written down by physicists – revealed a great deal about the underlying smooth four-manifold topology. Indeed, he constructed certain diffeomorphism invariants, called Donaldson polynomials, defined by counting (in a suitable sense) solutions to these PDEs, the anti-self-dual Yang-Mills equations [**13**], [**14**], [**10**], [**33**], [**52**].

It was proved by Taubes in [**100**] that Casson's invariant admits a gauge-theoretic interpretation. This interpretation was carried further by Floer [**28**], who constructed a homology theory whose Euler characteristic is Casson's invariant. The construction of this instanton Floer homology proceeds by defining a chain complex whose generators are equivalence classes of flat $SU(2)$ connections over Y (or, more precisely, a suitably perturbed notion of flat connections, as required for transversality), and whose differentials count solutions to the anti-self-dual Yang-Mills equations. In fact, Floer's instanton homology quickly became a central tool in the calculation of Donaldson's invariants, see for example [**68**], [**26**], [**12**]. More specifically, under suitable conditions, the Donaldson invariant of a four-manifold X separated along a three-manifold Y could be viewed as a pairing, in the Floer homology of Y, of relative Donaldson invariants coming from the two sides.

Floer's construction seemed closely related to an earlier construction Floer gave in the context of Hamiltonian dynamics, known as "Lagrangian Floer homology" [**27**]. That theory – which is also very closely related to Gromov's invariants for symplectic manifolds (cf. [**40**]) — associates to a symplectic manifold V, equipped with a pair of Lagrangian submanifolds L_0 and L_1 (that are in generic position, and satisfy certain topological restrictions), a homology theory whose Euler characteristic is the algebraic intersection number of L_0 and L_1, but which gives a refined symplectic obstruction to disjoining the Lagrangians through exact Hamiltonian isotopies. More specifically, the generators for this chain complex are intersection points for L_0 and L_1, and its differentials count holomorphic Whitney disks which interpolate between these intersection points, see also [**34**].

The close parallel between Floer's two constructions, which take us back to Casson's original picture, were further explored by Atiyah [**4**]. Atiyah conjectured that Floer's instanton theory coincides with a suitable version of Floer's Lagrangian theory, where one considers the $SU(2)$ character variety of Σ as the ambient symplectic manifold, equipped with the Lagrangian submanifolds which are the character varieties of the two handlebodies. The Atiyah-Floer conjecture remains open to this day. For related results, see [**17**], [**94**], [**106**].

In 1994, there was another drastic turn of events in gauge theory and its interaction with smooth four-manifold topology, namely, the introduction of a new set of equations coming from physics, the Seiberg–Witten monopole equations [**107**]. These are a novel system of nonlinear, elliptic, first-order equations which one can associate to a smooth four-manifold equipped with a Riemannian metric. Just as the Yang-Mills equations lead to Donaldson polynomials, the Seiberg–Witten equations lead to another smooth four-manifold invariant, the Seiberg–Witten invariant (cf. [**107**], [**67**], [**11**], [**53**], [**101**]). These two theories seem very closely related. In fact, Witten conjectured a precise relationship between the two four-manifold invariants, see [**107**], [**24**], see also [**51**]. Moreover, many of the formal aspects of Donaldson theory have their analogues in Seiberg–Witten theory. In particular, it was natural to expect a similar relationship between their three-dimensional counterparts [**65**], [**69**], [**64**].

But now, a question arises in studying the three-dimensional theory: what is the geometric picture playing the role of character varieties in this new context? In attempting to formulate an answer to this question, we came upon a construction which has, as its starting point a Heegaard diagram $(\Sigma, \boldsymbol{\alpha}, \boldsymbol{\beta})$ for a three-manifold Y [**85**]. That is, Σ is an oriented surface of genus g, and $\boldsymbol{\alpha} = \{\alpha_1, ..., \alpha_g\}$ and $\boldsymbol{\beta} = \{\beta_1, ..., \beta_g\}$ are a pair of g-tuples of embedded, homologically linearly independent, mutually disjoint, closed curves. Thus, the $\boldsymbol{\alpha}$ and $\boldsymbol{\beta}$ specify a pair of handlebodies

U_α and U_β which bound Σ, so that $Y \cong U_\alpha \cup_\Sigma U_\beta$. Note that any oriented, closed three-manifold can be described by a Heegaard diagram. We associate to Σ its g-fold symmetric product $\mathrm{Sym}^g(\Sigma)$, the space of unordered g-tuples of points in Σ. This space is equipped with a pair of g-dimensional tori

$$\mathbb{T}_\alpha = \alpha_1 \times ... \times \alpha_g \quad \text{and} \quad \mathbb{T}_\beta = \beta_1 \times ... \times \beta_g.$$

The most naive numerical invariant in this context – the oriented intersection number of $\mathbb{T}_\alpha$ and $\mathbb{T}_\beta$ – depends only on $H_1(Y;\mathbb{Z})$. However, by using the holomorphic disk techniques of Lagrangian Floer homology, we obtain a nontrivial invariant for three-manifolds, $\widehat{HF}(Y)$, whose Euler characteristic is this intersection number. In fact, there are some additional elaborations of this construction which give other variants of Heegaard Floer homology (denoted HF^-, HF^∞, and HF^+, discussed below).

This geometric construction gives rise to invariants whose definition is quite different in flavor than its gauge-theoretic predecessors. And yet, it is natural to conjecture that certain variants give the same information as Seiberg–Witten theory. This conjecture, in turn, can be viewed as an analogue of the Atiyah-Floer conjecture in the Seiberg–Witten context. With this said, it is also fruitful to study Heegaard Floer homology and its structure independently from its gauge-theoretical origins.

1.2. Structure of this paper. Our aim in this article is to give a leisurely introduction to Heegaard Floer homology. We begin by recalling some of the details of the construction in Section 2. In Section 3, we describe some of the properties. Broader summaries can be found in some of our other papers (cf. [**84**], [**82**]). In Section 4 we describe in further detail the relationship between Heegaard Floer homology and knots (cf. [**76**], [**78**] and also the work of Rasmussen [**90**], [**88**]). We conclude in Section 5 with some problems and questions raised by these investigations.

We have not attempted to give a full account of the state of Heegaard Floer homology. In particular, we have said very little about the four-manifold invariants. We do not discuss here the Dehn surgery characterization of the unknot which follows from properties of Floer homology (see Corollary 1.3 of [**77**], and also [**57**] for the original proof using Seiberg-Witten monopole Floer homology; compare also [**39**], [**8**], [**36**]). Another topic to which we have paid only fleeting attention is the close relationship between Heegaard Floer homology and contact geometry [**80**]. As a result, we do not have the opportunity to describe the recent results of Lisca and Stipsicz in contact geometry which result from this interplay, see for example [**60**].

1.3. Further remarks. The conjectured relationship between Heegaard Floer homology and Seiberg–Witten theory can be put on a more precise footing with the help of some more recent developments in gauge theory. For example, Kronheimer and Mrowka [**55**] have given a complete construction of a Seiberg–Witten–Floer package, which associates to each closed, oriented three-manifold a triple of Floer homology groups $\check{\mathrm{HM}}(Y)$, $\overline{\mathrm{HM}}(Y)$, and $\widehat{\mathrm{HM}}(Y)$ which are functorial under cobordisms between three-manifolds. They conjecture that the three functors in this sequence are isomorphic to (suitable completions of) $HF^-(Y)$, $HF^\infty(Y)$ and $HF^+(Y)$ respectively. A different approach is taken in papers by Manolescu and Kronheimer (cf. [**63**], [**50**]).

It should also be pointed out that a different approach to understanding gauge theory from a geometrical point of view has been adopted by Taubes [**102**], building on his fundamental earlier work relating the Seiberg–Witten and Gromov invariants of symplectic four-manifolds [**103**], [**104**], [**101**].

2. The Construction

We recall the construction of Heegaard Floer homology. In Subsection 2.1, we explain the construction for rational homology three-spheres (i.e., those three-manifolds whose first Betti number vanishes). In Subsection 2.2, we give an example which illustrates some of the subtleties involved in the Floer complex. In Subsection 2.4, we outline how the construction can be generalized for arbitrary closed, oriented three-manifolds. In Subsection 2.5, we sketch the construction of the maps induced by cobordisms, and in Subsection 2.6 we give preliminaries on the construction of the invariants for knots in S^3. The material in Sections 2.1-2.4 is derived from [**85**]. The material from Subsection 2.5 is an account of the material starting in Section 8 of [**85**] and continued in [**83**]. The material from Subsection 2.6 can be found in [**76**], see also [**88**].

2.1. Heegaard Floer homology for rational homology three-spheres. A *genus g handlebody U* is the three-manifold-with-boundary obtained by attaching g one-handles to a zero-handle. More informally, a genus g handlebody is homeomorphic to a regular neighborhood of a bouquet of g circles in $\mathbb{R}^3$. The boundary of U is a two-manifold with genus g. If Y is any oriented three-manifold, for some g we can write Y as a union of two genus g handlebodies U_0 and U_1, glued together along their boundary. A natural way of thinking about Heegaard decompositions is to consider self-indexing Morse functions

$$f\colon Y \longrightarrow [0,3]$$

with one index 0 critical point, one index three critical point, and g index one (hence also index two) critical points. The space U_0, then, is the preimage of the interval $[0,3/2]$ (with boundary the preimage of $3/2$); while U_1 is the preimage of $[3/2,3]$. We orient Σ as the boundary of U_0.

Heegaard decompositions give rise to a combinatorial description of three-manifolds. Specifically, let Σ be a closed, oriented surface of genus g. A *set of attaching circles* for Σ is a g-tuple of homologically linearly independent, pairwise disjoint, embedded curves $\boldsymbol{\gamma} = \{\gamma_1, ..., \gamma_g\}$. A *Heegaard diagram* is a triple consisting of $(\Sigma, \boldsymbol{\alpha}, \boldsymbol{\beta})$, where $\boldsymbol{\alpha}$ and $\boldsymbol{\beta}$ are both sets of attaching circles for Σ. From the Morse-theoretic point of view, the points in $\boldsymbol{\alpha}$ can be thought of as the points on Σ which flow out of the index one critical points under the gradient flow equations for f (taken with respect to a suitable metric on Y), and the points in $\boldsymbol{\beta}$ are points in Σ which flow into the index two critical points.

In the opposite direction, a set of attaching circles for Σ specifies a handlebody which bounds Σ, and hence a Heegaard diagram specifies an oriented three-manifold Y. It is a classical theorem of Singer [**99**] that every closed, oriented three-manifold Y admits a Heegaard diagram, and if two Heegaard diagrams describe the same three-manifold, then they can be connected by a sequence of moves of the following type:

- *isotopies:* replace α_i by a curve α_i' which is isotopic through isotopies which are disjoint from the other α_j' $(j \neq i)$; or, the same moves amongst the $\boldsymbol{\beta}$
- *handleslides:* replace α_i by α_i', which is a curve with the property that $\alpha_i \cup \alpha_i' \cup \alpha_j$ bound a pair of pants which is disjoint from the remaining α_k $(k \neq i,j)$; or, the same moves amongst the $\boldsymbol{\beta}$
- *stabilizations/destabilizations:* A stabilization replaces Σ by its connected sum with a genus one surface $\Sigma' = \Sigma \# E$, and replaces $\{\alpha_1, ..., \alpha_g\}$ and $\{\beta_1, ..., \beta_g\}$ by $\{\alpha_1, \ldots, \alpha_{g+1}\}$ and $\{\beta_1, \ldots, \beta_{g+1}\}$

respectively, where α_{g+1} and β_{g+1} are a pair of curves supported in E, meeting transversally in a single point.

Our goal is to associate a group to each Heegaard diagram, which is unchanged by the above three operations, and hence an invariant of the underlying three-manifold.

To this end, we will use a variant of Floer homology in the g-fold symmetric product of a genus g Heegaard surface Σ, relative to the pair of totally real subspaces

$$\mathbb{T}_\alpha = \alpha_1 \times \ldots \times \alpha_g \quad \text{and} \quad \mathbb{T}_\beta = \beta_1 \times \ldots \times \beta_g.$$

That is to say, we define a chain complex generated by intersection points between $\mathbb{T}_\alpha$ and $\mathbb{T}_\beta$, and whose boundary maps count pseudo-holomorphic disks in $\mathrm{Sym}^g(\Sigma)$. Again, it is useful to bear in mind the Morse-theoretic interpretation: an intersection point $\mathbf{x}$ between $\mathbb{T}_\alpha$ and $\mathbb{T}_\beta$ can be viewed as a g-tuple of gradient flow-lines which connect all the index two and index one critical points. We denote the corresponding one-chain in Y by $\gamma_{\mathbf{x}}$ and call it a *simultaneous trajectory.*

Before studying the space of pseudo-holomorphic Whitney disks, we turn our attention to the algebraic topology of Whitney disks. Specifically, we consider the unit disk $\mathbb{D}$ in $\mathbb{C}$, and let $e_1 \subset \partial\mathbb{D}$ denote the arc where $\mathrm{Re}(z) \geqslant 0$, and $e_2 \subset \partial\mathbb{D}$ denote the arc where $\mathrm{Re}(z) \leqslant 0$. Let $\pi_2(\mathbf{x}, \mathbf{y})$ denote the set of homotopy classes of Whitney disks, i.e., maps

$$\left\{ u\colon \mathbb{D} \longrightarrow \mathrm{Sym}^g(\Sigma) \middle| \begin{array}{l} u(-i) = \mathbf{x}, u(i) = \mathbf{y} \\ u(e_1) \subset \mathbb{T}_\alpha, u(e_2) \subset \mathbb{T}_\beta \end{array} \right\}.$$

Fix intersection points $\mathbf{x}, \mathbf{y} \in \mathbb{T}_\alpha \cap \mathbb{T}_\beta$. There is an obvious obstruction to the existence of a Whitney disk which lives in $H_1(Y;\mathbb{Z})$. It is obtained as follows. Given $\mathbf{x}$ and $\mathbf{y}$, consider the corresponding simultaneous trajectories $\gamma_{\mathbf{x}}$ and $\gamma_{\mathbf{y}}$. The difference $\gamma_{\mathbf{x}} - \gamma_{\mathbf{y}}$ gives a closed loop in Y, whose homology class is trivial if there exists a Whitney disk connecting $\mathbf{x}$ and $\mathbf{y}$. (Indeed, the condition is sufficient to guarantee existence of such a Whitney disk when $g > 1$.)

What this shows is that the space of intersection points between $\mathbb{T}_\alpha$ and $\mathbb{T}_\beta$ naturally fall into equivalence classes labeled by elements in an affine space over $H_1(Y;\mathbb{Z})$.

There is another very familiar such affine space over $H_1(Y;\mathbb{Z})$: the space of Spin^c structures over Y. Following Turaev **[105]**, one can think of this space concretely as the space of equivalence classes of vector fields. Specifically, we say that two vector fields v and v' over Y are *homologous* if they agree outside a Euclidean ball in Y. The space of homology classes of vector fields (which in turn are identified with the more standard definitions of Spin^c structures, see [**105**], [**38**], [**56**]) is also an affine space over $H_1(Y;\mathbb{Z})$.

In order to link these two concepts, we fix a basepoint $z \in \Sigma - \alpha_1 - \ldots - \alpha_g - \beta_1 - \ldots - \beta_g$. Thinking of the Heegaard decomposition as a Morse function as described earlier, the base point z describes a flow from the index zero to the index three critical point (for generic metric on Y). Now, each tuple $\mathbf{x} \in \mathbb{T}_\alpha \cap \mathbb{T}_\beta$ specifies a g-tuple of connecting flows between the index one and index two critical points. Modifying the gradient vector field $\vec{\nabla} f$ in a tubular neighborhood of these $g+1$ flow-lines so that it does not vanish there, we obtain a nowhere vanishing vector field over Y, whose homology class gives us a Spin^c structure, depending on $\mathbf{x}$ and z. This gives rise to an assignment

$$\mathfrak{s}_z : \mathbb{T}_\alpha \cap \mathbb{T}_\beta \longrightarrow \mathrm{Spin}^c(Y).$$

It follows from the previous discussion that $\mathbf{x}$ and $\mathbf{y}$ have $\mathfrak{s}_z(\mathbf{x}) = \mathfrak{s}_z(\mathbf{y})$ if and only if (when $g > 1$) $\mathbf{x}$ and $\mathbf{y}$ can be connected by a Whitney disk.

Having answered the existence problem for Whitney disks, we turn to questions of its uniqueness (up to homotopy). For this, we turn once again to our fixed base-point $z \in \Sigma - \alpha_1 - \ldots - \alpha_g - \beta_1 - \ldots - \beta_g$. Given a Whitney disk u connecting $\mathbf{x}$ with $\mathbf{y}$, we can consider the intersection number between u and the submanifold $\{z\} \times \mathrm{Sym}^{g-1}(\Sigma) \subset \mathrm{Sym}^g(\Sigma)$. This descends to homotopy classes to give a map

$$n_z : \pi_2(\mathbf{x}, \mathbf{y}) \longrightarrow \mathbb{Z}$$

which is additive in the sense that $n_z(\varphi * \psi) = n_z(\varphi) + n_z(\psi)$, where here $*$ denotes the natural juxtaposition operation. The multiplicity n_z can be modified by splicing in a copy of the two-sphere which generates $\pi_2(\mathrm{Sym}^g(\Sigma)) \cong \mathbb{Z}$ (when $g > 2$). Moreover, when Y is a rational homology sphere and $g > 2$, $n_z(\varphi)$ uniquely determines the homotopy class of φ – more generally, we have an identification $\pi_2(\mathbf{x}, \mathbf{y}) \cong \mathbb{Z} \oplus H^1(Y;\mathbb{Z})$. For the present discussion, though, we focus attention to the rational homology sphere case.

The most naive application of Floer's theory would then give a $\mathbb{Z}/2\mathbb{Z}$-graded theory. However, the calculation of $\pi_2(\mathbf{x}, \mathbf{y})$ suggests that if we count each intersection point infinitely many times, we obtain a relatively $\mathbb{Z}$-graded theory. Specifically, for a fixed Spin^c structure $\mathfrak{s}$ over Y, let the set $\mathfrak{S} \subset \mathbb{T}_\alpha \cap \mathbb{T}_\beta$ consist of intersection points which induce $\mathfrak{s}$, with respect to the fixed base-point z. Now we can consider the Abelian group $CF^\infty(Y, \mathfrak{s})$ freely generated by the set of pairs $[\mathbf{x}, i] \in \mathfrak{S} \times \mathbb{Z}$. We can give this space a natural relative $\mathbb{Z}$-grading, by

$$\operatorname{gr}([\mathbf{x}, i], [\mathbf{y}, j]) = \mu(\varphi) - 2(i - j + n_z(\varphi)), \tag{1}$$

where here φ is any homotopy class of Whitney disks which connects $\mathbf{x}$ and $\mathbf{y}$, and $\mu(\varphi)$ denotes the Maslov index of φ, that is, the expected dimension of the moduli space of pseudo-holomorphic representatives of φ, see also [**91**]. If S denotes the generator of $\pi_2(\operatorname{Sym}^g(\Sigma)) \cong \mathbb{Z}$ (when $g > 2$), then $\mu(\varphi * S) = \mu(\varphi) + 2$ while $n_z(\varphi * S) = n_z(\varphi) + 1$, and hence gr is independent of the choice of φ. (It is easy to see that gr extends also to the cases where $g \leqslant 2$.)

Our aim will be to count pseudo-holomorphic disks. For this to make sense, we need to have a sufficiently generic situation, so that the moduli spaces are cut out transversally and, in particular,

$$\dim \mathcal{M}(\varphi) = \mu(\varphi). \tag{2}$$

To achieve this, we need to introduce a suitable perturbation of the notion of pseudo-holomorphic disks, see for instance Section 3 of [**85**], see also [**31**], [**34**]. Specifically, for such a perturbation, we can arrange Equation (2) to hold for all homotopy classes φ with $\mu(\varphi) \leqslant 2$.

Indeed, since there is a one-parameter family of holomorphic automorphisms of the disks which preserve $\pm i$ and the boundary arcs e_1 and e_2, the moduli space $\mathcal{M}(\varphi)$ admits a free action by $\mathbb{R}$, provided that φ is nontrivial. In particular, if φ has $\mu(\varphi) = 1$, then $\mathcal{M}(\varphi)/\mathbb{R}$ is a zero-dimensional manifold.

We then define a boundary map on $CF^\infty(Y, \mathfrak{s})$ by the formula

$$\partial^\infty[\mathbf{x}, i] = \sum_{\mathbf{y}} \sum_{\{\varphi \in \pi_2(\mathbf{x}, \mathbf{y}) | \mu(\varphi) = 1\}} \# \left(\frac{\mathcal{M}(\varphi)}{\mathbb{R}} \right) [\mathbf{y}, i - n_z(\varphi)]. \tag{3}$$

Here, $\# \left(\frac{\mathcal{M}(\varphi)}{\mathbb{R}} \right)$ can be thought of as either a count modulo 2 of the number of points in the moduli space, in the case where we consider Floer homology with coefficients in $\mathbb{Z}/2\mathbb{Z}$ or, in a more general case, to be an appropriately signed count of the number of points in the moduli space. For a discussion on signs, see [**85**], see also [**30**], [**34**].

By analyzing Gromov limits of pseudo-holomorphic disks, one can prove that $(\partial^\infty)^2 = 0$, i.e., that $CF^\infty(Y, \mathfrak{s})$ is a chain complex.

It is easy to see that if a given homotopy class φ contains a holomorphic representative, then the intersection number $n_z(\varphi)$ is nonnegative. This observation ensures that the subset $CF^-(Y,\mathfrak{s}) \subset CF^\infty(Y,\mathfrak{s})$ generated by $[\mathbf{x},i]$ with $i<0$ is a subcomplex. It is also interesting to consider the quotient complex $CF^+(Y,\mathfrak{s})$ (which we can think of as generated by pairs $[\mathbf{x},i]$ with $i \geqslant 0$). Note that all three complexes can be thought of as $\mathbb{Z}[U]$-modules, where

$$U \cdot [\mathbf{x},i] = [\mathbf{x},i-1];$$

i.e., multiplication by U lowers grading by two. Similarly, we can define a complex $\widehat{CF}(Y,\mathfrak{s})$ which is generated by the kernel of the U-action on $CF^+(Y,\mathfrak{s})$. We can think of $\widehat{CF}(Y)$ directly as generated by intersection points between $\mathbb{T}_\alpha$ and $\mathbb{T}_\beta$, endowed with the differential

$$\widehat{\partial}\mathbf{x} = \sum_{\{\varphi\in\pi_2(\mathbf{x},\mathbf{y})\,\big|\,\mu(\varphi)=1,\, n_z(\varphi)=0\}} \#\left(\frac{\mathcal{M}(\varphi)}{\mathbb{R}}\right)\cdot \mathbf{y}. \tag{4}$$

We now define Floer homology theories $HF^-(Y,\mathfrak{s})$, $HF^\infty(Y,\mathfrak{s})$, $HF^+(Y,\mathfrak{s})$, $\widehat{HF}(Y,\mathfrak{s})$, which are the homologies of the chain complexes $CF^-(Y,\mathfrak{s})$, $CF^\infty(Y,\mathfrak{s})$, $CF^+(Y,\mathfrak{s})$, and $\widehat{CF}(Y,\mathfrak{s})$ respectively. Note that all of these groups are $\mathbb{Z}[U]$ modules (where the U action is trivial on $\widehat{HF}(Y,\mathfrak{s})$).

The main result of [**85**], then, is the following topological invariance of these theories:

Theorem 2.1. *The relatively $\mathbb{Z}$-graded theories $HF^-(Y,\mathfrak{s})$, $HF^\infty(Y,\mathfrak{s})$, $HF^+(Y,\mathfrak{s})$, and $\widehat{HF}(Y,\mathfrak{s})$ are topological invariants of the underlying three-manifold Y and its* Spin^c *structure $\mathfrak{s}$.*

The content of the above result is that the invariants are independent of the various choices going into the definition of the homology theories. It can be broken up into parts, where one shows that the homology groups are identified as the Heegaard diagram undergoes the following changes:

1. the complex structure over Σ is varied,
2. the attaching circles are moved by isotopies (in the complement of z),
3. the attaching circles are moved by handle-slides (in the complement of z),
4. the Heegaard diagram is stabilized.

The first step is a direct adaptation of the corresponding fact from Lagrangian Floer theory (independence of the particular compatible almost-complex structure). To see the second step, we observe that any isotopy of the $\boldsymbol{\alpha}$ and $\boldsymbol{\beta}$ can be realized as a sequence of exact Hamiltonian isotopies and metric changes over Σ. The third step follows from naturality properties of the Floer homology theories (using a holomorphic triangle construction which we return to in Subsection 2.5), and a direct calculation in a special case (where handle-slides are made over a g-fold connected sum of $S^1 \times S^2$). The final step can be seen as an invariance of the theory under a natural degeneration of the $(g+1)$-fold symmetric product of the connected sum of Σ with E, as the connected sum neck is stretched, compare also [**59**], [**42**].

Although the study of holomorphic disks in general is a daunting task, holomorphic disks in symmetric products in a Riemann surface admit a particularly nice interpretation in terms of the underlying Riemann surface: indeed, holomorphic disks in the g-fold symmetric product correspond to g-fold branched coverings of the disk by a Riemann surface-with-boundary, together with a holomorphic map from the Riemann-surface-with-boundary into Σ. This gives a geometric grasp of the objects under study, and hence, in many special cases, a way to calculate the boundary maps. In particular, it makes the model calculations used in the proof of handle-slide invariance mentioned above possible. We use it also freely in the following example.

2.2. An example. We give a concrete example to illustrate some of the familiar subtleties arising in the Floer complex. For simplicity, we stick to the case of $\widehat{HF}(S^3)$. Moreover, since we have made no attempt to explain sign conventions, we consider the group with coefficients in $\mathbb{Z}/2\mathbb{Z}$

Of course, S^3 can be given a genus one Heegaard diagram, with two attaching circles α_1 and β_1, which meet in a unique transverse intersection point. Correspondingly, the complex $\widehat{CF}(S^3)$ in this case has a single generator, and there are no differentials. Hence, $\widehat{HF}(S^3) \cong \mathbb{Z}/2\mathbb{Z}$. The diagram from Figure 1 is isotopic to this diagram, but now there are three intersections between α_1 and α_2, x_1, x_2, and x_3. By the Riemann mapping theorem, it is easy to see that $\widehat{\partial} x_1 = x_2 = \widehat{\partial} x_3$. Thus, $x_1 + x_3$ generates $\widehat{HF}(S^3)$. Clearly, the chain complex changed under the isotopy since the combinatorics of the new Heegaard diagram is different (but, of course, its homology stayed the same).

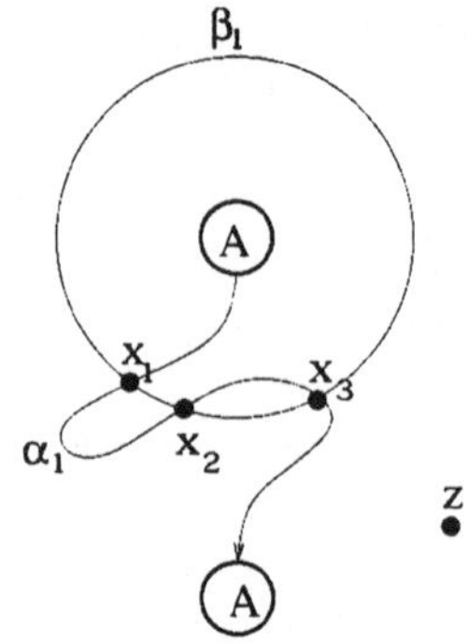

FIGURE 1. *A genus one Heegaard diagram for S^3.* In this diagram, the two circles labeled A are to be identified, to obtain a torus.

But the chain complex can change for reasons more subtle than combinatorics. Consider the Heegaard diagram for S^3 illustrated in Figure 2.

For this diagram, there are two different chain complexes, depending on the choice of complex structure over Σ (and the geometry of the attaching circles). We sketch the argument below.

First, it is easy to see that there are nine generators, corresponding to the points $x_i \times y_j \in \mathrm{Sym}^2(\Sigma)$ for $i, j = 1, ..., 3$. Again, by the Riemann mapping theorem applied to the region Γ, there are holomorphic disks connecting $x_i \times y_3$ to $x_i \times y_2$ for for all $i = 1, ..., 3$. In a similar way, an inspection of Figure 2 reveals disks connecting $x_1 \times y_j$ to $x_2 \times y_j$ and $x_i \times y_1$ to $x_i \times y_2$. However, the question of whether or not there is a holomorphic disk in $\mathrm{Sym}^2(\Sigma)$ (with $n_z(\varphi) = 0$) connecting $x_3 \times y_i$ to $x_2 \times y_j$ is dictated by the conformal structures of the annuli in the diagram.

More precisely, consider the annular region Δ illustrated in Figure 2. Δ has a uniformization as a standard annulus with four points marked on its boundary, corresponding to the points x_1, x_2, y_2, and y_3. Let a denote the angle of the arc in the boundary connecting x_1 and x_2 which is the image of the corresponding segment in α_1 under this uniformization; let b denote the angle of the arc in the boundary connecting y_2 and

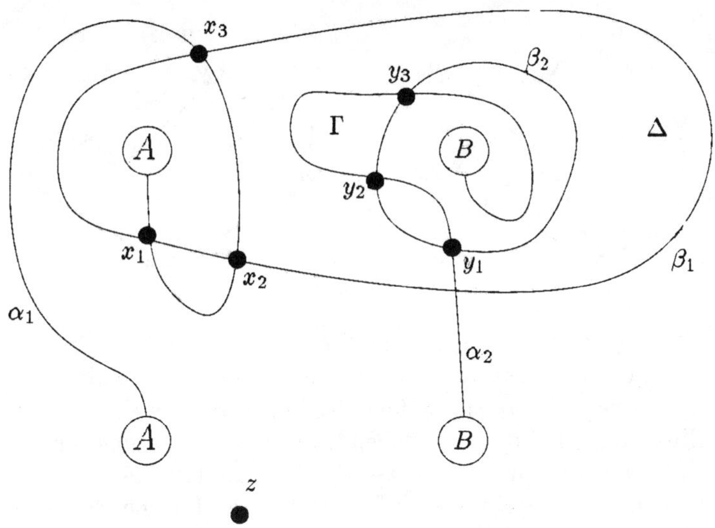

FIGURE 2. *A genus two Heegaard diagram for S^3.* In this diagram, the two circles labeled A are to be identified, as are the circles labeled by B. The resulting surface Σ of genus two is divided into connected components by the union $\alpha_1 \cup \alpha_2 \cup \beta_1 \cup \beta_2$. Let Δ be the component annular region indicated by taking the closure of the component indicated.

y_3 which is the image of the corresponding segment in α_2 under the uniformization. Now, the question of whether there is a holomorphic disk in $\mathrm{Sym}^2(\Sigma)$ connecting $x_3 \times y_3$ to $x_2 \times y_3$ admits the following conformal reformulation. Consider the one-parameter family of conformal annuli with four marked boundary points obtained from $\Delta \cup \Gamma$ by cutting a slit along α_2 starting at y_3. The four boundary points are the images of x_3, x_2, and y_3 (counted twice) under a uniformization map. A four-times marked annulus which admits an involution (interchanging the two α-arcs on the boundary) gives rise to a holomorphic disk connecting $x_3 \times y_3$ to $x_2 \times y_3$. By analyzing the conformal angles of the α arcs in this one-parameter family, one can prove that the mod 2 count of the holomorphic is 1 iff $a < b$.

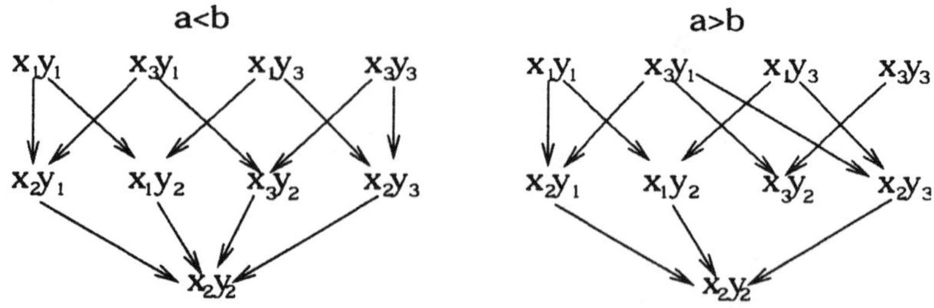

FIGURE 3. *Complexes for $\widehat{CF}(S^3)$ coming from the Heegaard diagram in Figure 2.* The above two complexes can be realized as $\widehat{CF}$ for a Heegaard diagram for S^3 illustrated in Figure 2, depending on the relations for the conformal parameter described in the text. Arrows here indicate nontrivial differentials; e.g. for the complex on the left, we have that $\widehat{\partial} x_1 \times y_1 = x_2 \times y_1 + x_1 \times y_2$.

Proceeding in the like manner for the other homotopy classes, we see that in the regime where $a < b$ resp. $a > b$, the chain complex $\widehat{CF}$ has differentials listed on the left resp. right in Figure 3. These two complexes are different, but of course, they are chain homotopic.

2.3. Algebra. The reason for this zoo of groups HF^-, HF^∞, HF^+, $\widehat{HF}$ can be traced to a simple algebraic reason: $CF^-(Y,\mathfrak{s})$ (whose chain homotopy type is an invariant of Y) is a finitely-generated chain complex of free $\mathbb{Z}[U]$-modules. All of the other groups are obtained from this from canonical algebraic operations. $CF^\infty(Y,\mathfrak{s})$ is the "localization" $CF^-(Y,\mathfrak{s}) \otimes_{\mathbb{Z}[U]} \mathbb{Z}[U,U^{-1}]$, $CF^+(Y,\mathfrak{s})$ is the cokernel of the localization map, and $\widehat{CF}(Y,\mathfrak{s})$ is the quotient $CF^-(Y,\mathfrak{s})/U \cdot CF^-(Y,\mathfrak{s})$.

Correspondingly, the various Floer homology groups are related by natural long exact sequences

$$\begin{array}{c} \cdots \longrightarrow HF^-(Y,\mathfrak{s}) \xrightarrow{i} HF^\infty(Y,\mathfrak{s}) \xrightarrow{\pi} HF^+(Y,\mathfrak{s}) \xrightarrow{\delta} \cdots \\ \cdots \longrightarrow \widehat{HF}(Y,\mathfrak{s}) \xrightarrow{j} HF^+(Y,\mathfrak{s}) \xrightarrow{U} HF^+(Y,\mathfrak{s}) \longrightarrow \cdots, \end{array} \quad (5)$$

and a more precise version of Theorem 2.1 states that both of the above diagrams are topological invariants of Y. The interrelationships between these groups is essential in the study of four-manifold invariants, as we shall see.

We can form another topological invariant, $HF^+_{\mathrm{red}}(Y, \mathfrak{s})$, which is the cokernel of π appearing in Diagram (5).

2.4. Manifolds with $b_1(Y) > 0$. When $b_1(Y) > 0$, there are a number of additional technical issues which arise in the definition of Heegaard Floer homology. The crux of the matter is that there are homotopically nontrivial cylinders connecting $\mathbb{T}_\alpha$ and $\mathbb{T}_\beta$. Specifically, given any point $\mathbf{x}$, we have a subgroup of $\pi_2(\mathbf{x}, \mathbf{x})$ consisting of classes of Whitney disks φ with $n_z(\varphi) = 0$. This group, the group of "periodic classes," is naturally identified with the cohomology group $H^1(Y;\mathbb{Z})$, and hence (provided $g > 2$) $\pi_2(\mathbf{x}, \mathbf{y}) \cong \mathbb{Z} \oplus H^2(Y;\mathbb{Z})$. In particular, there are infinitely many homotopy classes of Whitney disks with a fixed multiplicity at a given point z; thus, the coefficients appearing in Equation (3) might *a priori* be infinite. One way to remedy this situation is to work with special Heegaard diagrams for Y. For example, in defining $\widehat{HF}$ and HF^+, one can use Heegaard diagrams with the property that each nontrivial periodic class has a negative multiplicity at some $z' \in \Sigma - \alpha_1 - \ldots - \alpha_g - \beta_1 - \ldots - \beta_g$. Such diagrams are called *weakly admissible*, (cf. Section 4.2.2 of [**85**] for a detailed account, and also for a discussion of the stronger hypotheses needed for the construction of HF^- and HF^∞).

Another related issue is that now it is no longer true that the dimension of the space of holomorphic disks connecting $\mathbf{x}, \mathbf{y}$ depends only on the multiplicity at z. Specifically, given a one-dimensional cohomology class $\gamma \in H^1(Y;\mathbb{Z})$, if $\varphi \in \pi_2(\mathbf{x}, \mathbf{y})$ and $\gamma * \varphi$ denotes the new element of $\pi_2(\mathbf{x}, \mathbf{y})$ obtained by letting the periodic class associated to γ act on φ, the Maslov classes of φ and $\gamma * \varphi$ are related by the formula:

$$\mu(\gamma * \varphi) - \mu(\varphi) = \langle c_1(\mathfrak{s}_z(\mathbf{x})) \cup \gamma, [Y] \rangle,$$

where the right-hand-side is, of course, calculated over the three-manifold Y. Letting $\delta(\mathfrak{s})$ be the greatest common divisor of the integers of the form $c_1(\mathfrak{s}) \cup H^1(Y;\mathbb{Z})$, the above discussion shows that the grading defined in Equation (1) gives rise to a relatively $\mathbb{Z}/\delta(\mathfrak{s})\mathbb{Z}$-graded theory.

With this said, there is an analogue of Theorem 2.1: when $b_1(Y) > 0$, the homology theories $HF^+(Y, \mathfrak{s})$, $HF^-(Y;\mathfrak{s})$, and $HF^+_{\mathrm{red}}(Y;\mathfrak{s})$ (as calculated for special Heegaard diagrams) are relatively $\mathbb{Z}/\delta(\mathfrak{s})$-graded topological invariants. Note that, there are only finitely many Spin^c structures $\mathfrak{s}$ over Y for which $HF^+(Y,\mathfrak{s})$ is nonzero.

2.5. Maps induced by cobordisms. Cobordisms between three-manifolds give rise to maps between their Floer homology groups. The construction of these maps relies on the holomorphic triangle construction from symplectic geometry (cf. [**7**], [**34**]).

A bridge between the symplectic geometry construction and the four-manifold picture can be given as follows.

A *Heegaard triple-diagram of genus* g is an oriented two-manifold and three g-tuples $\boldsymbol{\alpha}$, $\boldsymbol{\beta}$, and $\boldsymbol{\gamma}$ which are sets of attaching circles for handlebodies U_α, U_β, and U_γ respectively. Let $Y_{\alpha,\beta}=U_\alpha\cup U_\beta$, $Y_{\beta,\gamma}=U_\beta\cup U_\gamma$, and $Y_{\alpha,\gamma}=U_\alpha\cup U_\gamma$ denote the three induced three-manifolds. A Heegaard triple-diagram naturally specifies a cobordism $X_{\alpha,\beta,\gamma}$ between these three-manifolds. The cobordism is constructed as follows.

Let Δ denote the two-simplex, with vertices $v_\alpha, v_\beta, v_\gamma$ labeled clockwise, and let e_i denote the edge from v_j to v_k, where $\{i,j,k\}=\{\alpha,\beta,\gamma\}$. Then, we form the identification space

$$X_{\alpha,\beta,\gamma}=\frac{(\Delta\times\Sigma)\coprod(e_\alpha\times U_\alpha)\coprod(e_\beta\times U_\beta)\coprod(e_\gamma\times U_\gamma)}{(e_\alpha\times\Sigma)\sim(e_\alpha\times\partial U_\alpha),(e_\beta\times\Sigma)\sim(e_\beta\times\partial U_\beta),(e_\gamma\times\Sigma)\sim(e_\gamma\times\partial U_\gamma)}.$$

Over the vertices of Δ, this space has corners, which can be naturally smoothed out to obtain a smooth, oriented, four-dimensional cobordism between the three-manifolds $Y_{\alpha,\beta}$, $Y_{\beta,\gamma}$, and $Y_{\alpha,\gamma}$ as claimed.

We will call the cobordism $X_{\alpha,\beta,\gamma}$ described above a *pair of pants connecting* $Y_{\alpha,\beta}$, $Y_{\beta,\gamma}$, *and* $Y_{\alpha,\gamma}$. Note that if we give $X_{\alpha,\beta,\gamma}$ its natural orientation, then $\partial X_{\alpha,\beta,\gamma}=-Y_{\alpha,\beta}-Y_{\beta,\gamma}+Y_{\alpha,\gamma}$.

Fix $\mathbf{x}\in\mathbb{T}_\alpha\cap\mathbb{T}_\beta$, $\mathbf{y}\in\mathbb{T}_\beta\cap\mathbb{T}_\gamma$, $\mathbf{w}\in\mathbb{T}_\alpha\cap\mathbb{T}_\gamma$. A map

$$u\colon\Delta\longrightarrow\text{Sym}^g(\Sigma)$$

with the boundary conditions that $u(v_\gamma)=\mathbf{x}$, $u(v_\alpha)=\mathbf{y}$, and $u(v_\beta)=\mathbf{w}$, and $u(e_\alpha)\subset\mathbb{T}_\alpha$, $u(e_\beta)\subset\mathbb{T}_\beta$, $u(e_\gamma)\subset\mathbb{T}_\gamma$ is called a *Whitney triangle connecting* $\mathbf{x}$, $\mathbf{y}$, *and* $\mathbf{w}$. Two Whitney triangles are homotopic if the maps are homotopic through maps which are all Whitney triangles. We let $\pi_2(\mathbf{x},\mathbf{y},\mathbf{w})$ denote the space of homotopy classes of Whitney triangles connecting $\mathbf{x}$, $\mathbf{y}$, and $\mathbf{w}$.

Using a base-point $z\in\Sigma-\alpha_1-\ldots-\alpha_g-\beta_1-\ldots-\beta_g-\gamma_1-\ldots-\gamma_g$, we obtain an intersection number

$$n_z\colon\pi_2(\mathbf{x},\mathbf{y},\mathbf{w})\longrightarrow\mathbb{Z}$$

between the Whitney triangle and the subspace $\{z\} \times \mathrm{Sym}^{g-1}(\Sigma)$. If the space of homotopy classes of Whitney triangles $\pi_2(\mathbf{x}, \mathbf{y}, \mathbf{w})$ is nonempty, then it can be identified with $\mathbb{Z} \oplus H_2(X_{\alpha,\beta,\gamma}; \mathbb{Z})$, in the case where $g > 2$.

As explained in Section 8 of [**85**], the choice of base-point $z \in \Sigma - \alpha_1 - \ldots - \alpha_g - \beta_1 - \ldots - \beta_g - \gamma_1 - \ldots - \gamma_g$ gives rise to a map

$$\mathfrak{s}_z : \pi_2(\mathbf{x}, \mathbf{y}, \mathbf{w}) \longrightarrow \mathrm{Spin}^c(X_{\alpha,\beta,\gamma}).$$

A Spin^c structure over X gives rise to a map

$$f^\infty(\,\cdot\,;\mathfrak{s})\colon CF^\infty(Y_{\alpha,\beta}, \mathfrak{s}_{\alpha,\beta}) \otimes CF^\infty(Y_{\beta,\gamma}, \mathfrak{s}_{\beta,\gamma}) \longrightarrow CF^\infty(Y_{\alpha,\gamma}, \mathfrak{s}_{\alpha,\gamma})$$

by the formula:

$$f^\infty_{\alpha,\beta,\gamma}([\mathbf{x}, i] \otimes [\mathbf{y}, j]; \mathfrak{s}) = \sum_{\mathbf{w} \in \mathbb{T}_\alpha \cap \mathbb{T}_\gamma} \sum_{\{\psi \in \pi_2(\mathbf{x},\mathbf{y},\mathbf{w}) | \mathfrak{s}_z(\psi) = \mathfrak{s}, \mu(\psi) = 0\}} (\#\mathcal{M}(\psi)) \cdot [\mathbf{w}, i+j-n_z(\psi)]. \quad (6)$$

Under suitable admissibility hypotheses on the Heegaard diagrams, these sums are finite (cf. Section 8 of [**85**]). Indeed, there are induced maps on some of the other variants of Floer homology, and again, we refer the interested reader to that discussion for a more detailed account.

Let X be a smooth, connected, oriented four-manifold with boundary given by $\partial X = -Y_0 \cup Y_1$ where Y_0 and Y_1 are connected, oriented three-manifolds. We call such a four-manifold a cobordism from Y_0 to Y_1. If X is a cobordism from Y_0 to Y_1, and $\mathfrak{s} \in \mathrm{Spin}^c(X)$ is a Spin^c structure, then there is a naturally induced map

$$F^\infty_{X,\mathfrak{s}}\colon HF^\infty(Y_0, \mathfrak{s}_i) \longrightarrow HF^\infty(Y_1, \mathfrak{s}_i)$$

where here $\mathfrak{s}_i$ denotes the restriction of $\mathfrak{s}$ to Y_i. This map is constructed as follows. First assume that X is given as a collection of two-handles. Then we claim that in the complement of the regular neighborhood of a one-complex, X can be realized as a pair-of-pants cobordism, one of whose boundary components is $-Y_0$, the other which is Y_1, and the third of which is a connected sum of copies of $S^2 \times S^1$. Next, pairing Floer homology classes coming from Y_0 with a certain canonically associated Floer homology class on the connected sum of $S^2 \times S^1$, we obtain a map using the holomorphic triangle construction as defined in Equation (6) to obtain a the desired map to $HF^\infty(Y_1)$. For the cases of one- and three-handles, the associated maps are defined in a more formal manner. The fact that these maps are independent (modulo an overall multiplication by ± 1) of the many choices which go into their construction is established in [**83**].

Indeed, variants of this construction can be extended to the following situation (again, see [**83**]): if X is a smooth, oriented cobordism from Y_0 to Y_1, then there are induced maps (of $\mathbb{Z}[U]$ modules) between the corresponding Heegaard Floer homology groups, which make the squares in the following diagrams commutate:

$$\begin{array}{ccccccccc}
\cdots & \longrightarrow & HF^-(Y_0,\mathfrak{s}_0) & \longrightarrow & HF^\infty(Y_0,\mathfrak{s}_0) & \xrightarrow{\pi} & HF^+(Y_0,\mathfrak{s}_0) & \xrightarrow{\delta} & \cdots \\
 & & \downarrow F^-_{X,\mathfrak{s}} & & \downarrow F^\infty_{X,\mathfrak{s}} & & \downarrow F^+_{X,\mathfrak{s}} & & \\
\cdots & \longrightarrow & HF^-(Y_1,\mathfrak{s}_1) & \longrightarrow & HF^\infty(Y_1,\mathfrak{s}_1) & \xrightarrow{\pi} & HF^+(Y_1,\mathfrak{s}_1) & \xrightarrow{\delta} & \cdots
\end{array}$$

$$\begin{array}{ccccccccc}
\cdots & \longrightarrow & \widehat{HF}(Y_0,\mathfrak{s}_0) & \longrightarrow & HF^+(Y_0,\mathfrak{s}_0) & \xrightarrow{U} & HF^+(Y_0,\mathfrak{s}_0) & \longrightarrow & \cdots, \\
 & & \downarrow \widehat{F}_{X,\mathfrak{s}} & & \downarrow F^+_{X,\mathfrak{s}} & & \downarrow F^+_{X,\mathfrak{s}} & & \\
\cdots & \longrightarrow & \widehat{HF}(Y_1,\mathfrak{s}_1) & \longrightarrow & HF^+(Y_1,\mathfrak{s}_1) & \xrightarrow{U} & HF^+(Y_1,\mathfrak{s}_1) & \longrightarrow & \cdots,
\end{array} \tag{7}$$

Naturality of the maps induced by cobordisms can be phrased as follows. Suppose that W_0 is a smooth cobordism from Y_0 to Y_1 and W_1 is a cobordism from Y_1 to Y_2, then for fixed Spin^c structures $\mathfrak{s}_i$ over W_i which agree over Y_1, we have that

$$\sum_{\{\mathfrak{s}\in \mathrm{Spin}^c(W_0\cup_{Y_1} W_1) \,|\, \mathfrak{s}|_{W_i}=\mathfrak{s}_i\}} F^\circ_{W_0\cup_{Y_1}W_1,\mathfrak{s}} = F^\circ_{W_1,\mathfrak{s}_1}\circ F^\circ_{W_0,\mathfrak{s}_0},$$

where here $F^\circ = F^-$, F^∞, F^+, $\widehat{F}$ (cf. Theorem 3.4 of [**83**]).

Sometimes, it is convenient to obtain topological invariants by summing over all Spin^c structures. To this end, we write, for example,

$$HF^+(Y) \cong \bigoplus_{\mathfrak{s}\in \mathrm{Spin}^c(Y)} HF^+(Y,\mathfrak{s}).$$

It is convenient to have a corresponding notion for cobordisms, only in that case a little more care must be taken. For fixed X and $\xi \in HF^+(Y_0,\mathfrak{s}_0)$, we have that $F^+_{X,\mathfrak{s}}(\xi) = 0$ for all but finitely many $\mathfrak{s} \in \mathrm{Spin}^c(X)$ (cf. Theorem 3.3 of [**83**]), and hence there is a well-defined map

$$F^+_X \colon HF^+(Y_0) \longrightarrow HF^+(Y_1),$$

defined by

$$F_X^+ = \sum_{\mathfrak{s}\in \mathrm{Spin}^c(X)} F_{X,\mathfrak{s}}^+.$$

(Note that the same construction works for $\widehat{HF}$, but it does not work for HF^-, HF^∞: for a given $\xi \in HF^\infty(Y_0)$, there might be infinitely many different $\mathfrak{s} \in \mathrm{Spin}^c(X)$ for which $F_{X,\mathfrak{s}}^\infty(\xi)$ is nonzero.)

2.6. Doubly-pointed Heegaard diagrams and knot invariants. Additional basepoints give rise to additional filtrations on Floer homology. These additional filtrations can be given topological interpretations. We consider the case of two basepoints.

Specifically, a Heegaard diagram $(\Sigma, \boldsymbol{\alpha}, \boldsymbol{\beta})$ for Y equipped with two basepoints w and z in $\Sigma - \alpha_1 - ... - \alpha_g - \beta_1 - ... - \beta_g$ gives rise to a knot in Y as follows. We connect w and z by a curve a in $\Sigma - \alpha_1 - ... - \alpha_g$ and also by another curve b in $\Sigma - \beta_1 - ... - \beta_g$. By pushing a and b into U_α and U_β respectively, we obtain a knot $K \subset Y$. We call the data $(\Sigma, \boldsymbol{\alpha}, \boldsymbol{\beta}, w, z)$ a *doubly-pointed Heegaard diagram compatible with the knot* $K \subset Y$. Given a knot K in Y, one can always find such a Heegaard diagram.

This can be thought of from the following Morse-theoretic point of view. Let Y be an oriented three-manifold, equipped with a Riemannian metric and a self-indexing Morse function

$$f\colon Y \longrightarrow [0,3]$$

with one index 0 critical point, one index three critical point, and g index one (hence also g index two) critical points. The knot K now is obtained from the union of the two flows connecting the index 0 to the index 3 critical points which pass through w and z. We call a Morse function as in the above construction one which is compatible with K. Note also that an ordering of w and z is equivalent to an orientation on K. However, the invariants we construct can be shown to be independent of the orientation of K, see **[76]**, **[88]**.

The simplest construction now is to consider a differential on $\mathbb{T}_\alpha \cap \mathbb{T}_\beta$ defined analogously to Equation (4), only now we count holomorphic disks for which $n_z(\varphi) = n_w(\varphi) = 0$. More generally, we use the reference point w to construct the Heegaard Floer complex for Y, and then use the additional basepoint z to induce a filtration on this complex. We describe this in detail for the case of knots in S^3, and using the chain complex $\widehat{CF}(S^3)$.

There is a unique function $\mathcal{F}\colon \mathbb{T}_\alpha \cap \mathbb{T}_\beta \longrightarrow \mathbb{Z}$ satisfying the relation

$$\mathcal{F}(\mathbf{x}) - \mathcal{F}(\mathbf{y}) = n_z(\varphi) - n_w(\varphi), \tag{8}$$

for any $\varphi \in \pi_2(\mathbf{x}, \mathbf{y})$, and the additional symmetry

$$\#\{\mathbf{x} \in \mathbb{T}_\alpha \cap \mathbb{T}_\beta \big| \mathcal{F}(\mathbf{x}) = i\} \equiv \#\{\mathbf{x} \in \mathbb{T}_\alpha \cap \mathbb{T}_\beta \big| \mathcal{F}(\mathbf{x}) = -i\} \pmod 2$$

for all i (compare, more generally, Equation (12)). (Alternatively, a more intrinsic characterization can be given in terms of relative Spin^c structures on the knot complement.) Clearly, if $\mathbf{y}$ appears in $\widehat{\partial}(\mathbf{x})$ with nonzero multiplicity, then the homotopy class $\varphi \in \pi_2(\mathbf{x}, \mathbf{y})$ with $n_w(\varphi) = 0$ admits a holomorphic representative, and hence $\mathcal{F}(\mathbf{x}) - \mathcal{F}(\mathbf{y}) \geqslant 0$. Thus, any filtration satisfying Equation (8) induces a filtration on the complex $\widehat{CF}(S^3)$, by the rule that $\mathcal{F}(K, i) \subset \widehat{CF}(S^3)$ is the subcomplex generated by $\mathbf{x} \in \mathbb{T}_\alpha \cap \mathbb{T}_\beta$ with $\mathcal{F}(\mathbf{x}) \leqslant i$.

It is shown in [**76**] and [**88**] that the chain homotopy type of this filtration is a knot invariant. More precisely, recall that a *filtered chain complex* is a chain complex C, together with a sequence of subcomplexes $X(C, i)$ indexed by $i \in \mathbb{Z}$, where $X(C, i) \subseteq X(C, i+1) \subset C$. Our filtered complexes are always bounded, meaning that for all sufficiently large i, $X(C, -i) = 0$ and $X(C, i) = C$. A *filtered map* between chain complexes $\Phi\colon C \longrightarrow C'$ is one whose restriction to $X(C, i) \subset C$ is contained in $X(C', i)$. Two filtered chain complexes are said to be have the same *filtered chain type* if there are maps $f\colon C \longrightarrow C'$ and $f'\colon C' \longrightarrow C$ and $H\colon C \longrightarrow C'$ and $H'\colon C' \longrightarrow C'$, all four of which are filtered maps, f and f' are chain maps, and also we have that

$$f \circ f' - \mathrm{Id} = \partial' \circ H' + H' \circ \partial' \quad \text{and} \quad f' \circ f - \mathrm{Id} = \partial \circ H + H \circ \partial.$$

The construction we mentioned earlier – counting holomorphic disks with $n_w(\varphi) = n_z(\varphi) = 0$ can be thought of as the chain complex of the associated graded object

$$\bigoplus_i \mathcal{F}(K, i)/\mathcal{F}(K, i-1).$$

The homology of this is also a knot invariant. We return to properties of this invariant in Section 4.

3. Basic Properties

We outline here some of the basic properties of Heegaard Floer homology, to give a flavor for its structure. We have not attempted to summarize all of its properties; for additional properties, see [**84**], [**82**], [**83**]. We focus on material which is useful for calculations: an exact sequence and rational gradings. We then turn briefly to properties of the maps induced on HF^∞, which have some important consequences explained later, but they also shed light on the special role played by $b_2^+(X)$ in Heegaard Floer homology. In Section 3.4 we give a few sample calculations. In Section 3.5, we describe one of the first applications of the rational gradings: a constraint on the intersection forms of four-manifolds which bound a given three-manifold, compare the gauge-theoretic analogue of Frøyshov [**32**]. Finally, in Subsection 3.6, we sketch how the maps induced by cobordisms give rise to an interesting invariant of closed, smooth four-manifolds X with $b_2^+(X) > 1$, which are conjectured to agree with the Seiberg–Witten invariants (cf. [**107**]).

3.1. Long exact sequences. An important calculational device is provided by the surgery long exact sequence. Long exact sequences of this type were first explored by Floer in the context of instanton Floer homology [**29**], [**7**], see also [**98**], [**57**].

Heegaard Floer homology satisfies a surgery long exact sequence, which we state presently. Suppose that M is a three-manifold with torus boundary, and fix three simple, closed curves γ_0, γ_1, and γ_2 in ∂M with

$$\#(\gamma_0 \cap \gamma_1) = \#(\gamma_1 \cap \gamma_2) = \#(\gamma_2 \cap \gamma_0) = -1 \tag{9}$$

(where here the algebraic intersection number is calculated in ∂M, oriented as the boundary of M), so that Y_0 resp. Y_1 resp. Y_2 are obtained from M by attaching a solid torus along the boundary with meridian γ_0 resp. γ_1 resp. γ_2.

Theorem 3.1. *Let Y_0, Y_1, and Y_2 be related as above. Then, there is a long exact sequence relating the Heegaard Floer homology groups:*

$$\ldots \longrightarrow HF^+(Y_0) \longrightarrow HF^+(Y_1) \longrightarrow HF^+(Y_2) \longrightarrow \ldots$$

The above theorem is proved in Theorem 9.12 of [**84**]. A variant for Seiberg–Witten monopole Floer homology, with coefficients in $\mathbb{Z}/2\mathbb{Z}$ is proved in [**57**].

The maps in the long exact sequence have a four-dimensional interpretation. To this end, note that there are two-handle cobordisms W_i connecting Y_i to Y_{i+1} (where we view $i \in \mathbb{Z}/3\mathbb{Z}$). When we work with Heegaard Floer homology over the field $\mathbb{Z}/2\mathbb{Z}$, the map from $HF^+(Y_i)$ to $HF^+(Y_{i+1})$ in the above exact sequence is the map induced by the corresponding cobordism W_i, $F^+_{W_i}$ (i.e., obtained by summing the maps induced by all Spin^c structures over W_i). When working over $\mathbb{Z}$, though, one must make additional choices of signs to ensure that exactness holds.

3.2. Gradings. It is proved in Section 10 of [**84**] that if Y is a rational homology three-sphere and $\mathfrak{s}$ is any Spin^c structure over it, then $HF^\infty(Y,\mathfrak{s}) \cong \mathbb{Z}[U,U^{-1}]$, thus, this invariant is not a very subtle invariant of three-manifolds. However, extra information can still be gleaned from the interplay between HF^∞ and HF^+, with the help of some additional structure on Floer homology.

It is shown in [**83**] that when Y is an oriented rational homology three-sphere and $\mathfrak{s}$ is a Spin^c structure over Y, the relative $\mathbb{Z}$ grading on the Heegaard Floer homology described earlier can be lifted to an absolute $\mathbb{Q}$-grading. This gives $HF^\circ(Y,\mathfrak{s})$ is a $\mathbb{Q}$-graded module over the polynomial algebra $\mathbb{Z}[U]$ (where here $HF^\circ(Y,\mathfrak{s})$ is any of $HF^-(Y,\mathfrak{s})$, $HF^\infty(Y,\mathfrak{s})$, $HF^+(Y,\mathfrak{s})$, or $\widehat{HF}(Y,\mathfrak{s})$),

$$HF^\circ(Y,\mathfrak{s}) = \bigoplus_{d\in\mathbb{Q}} HF^\circ_d(Y,\mathfrak{s}),$$

where multiplication by U lowers degree by two. In each grading, $i \in \mathbb{Q}$, $HF^\circ_i(Y,\mathfrak{s})$ is a finitely generated Abelian group.

The maps i, π, and j in Diagram (5) preserve this $\mathbb{Q}$-grading, and moreover, maps induced by cobordisms $F^\circ_{X,\mathfrak{s}}$ (again, $F^\circ_{X,\mathfrak{s}}$ denotes any of $F^-_{X,\mathfrak{s}}$, $F^\infty_{X,\mathfrak{s}}$, $F^+_{X,\mathfrak{s}}$ or $\widehat{F}_{X,\mathfrak{s}}$) respect the $\mathbb{Q}$-grading in the following sense. If Y_0 and Y_1 are rational homology three-spheres, and X is a cobordism from Y_0 to Y_1, with Spin^c structure $\mathfrak{s}$, the map induced by the cobordism maps

$$F^\circ_{X,\mathfrak{s}} \colon HF^\circ_d(Y_0,\mathfrak{s}_0) \longrightarrow HF^\circ_{d+\Delta}(Y_1,\mathfrak{s}_1),$$

for

$$\Delta = \frac{c_1(\mathfrak{s})^2 - 2\chi(X) - 3\sigma(X)}{4}, \tag{10}$$

where here $\chi(X)$ denotes the Euler characteristic of X, and $\sigma(X)$ denotes its signature. In fact (cf. Theorem 7.1 of [**83**]) the $\mathbb{Q}$ grading is uniquely characterized by the above property, together with the convention that $\widehat{HF}(S^3)$ is supported in degree zero.

The image of π determines a function

$$d\colon \mathrm{Spin}^c(Y) \longrightarrow \mathbb{Q}$$

(the "correction terms" of [**82**]) which associates to each Spin^c structure the minimal $\mathbb{Q}$-grading of any (nonzero) homogeneous element in $HF^+(Y,\mathfrak{s}) \otimes_{\mathbb{Z}} \mathbb{Q}$ in the image of π.

Certain properties of the correction terms can be neatly summarized, with the help of the following definitions.

The three-dimensional Spin^c homology bordism group θ^c is the set of equivalence classes of pairs $(Y,\mathfrak{t})$ where Y is a rational homology three-sphere, and $\mathfrak{t}$ is a Spin^c structure over Y, and the equivalence relation identifies $(Y_1,\mathfrak{t}_1) \sim (Y_2,\mathfrak{t}_2)$ if there is a (connected, oriented, smooth) cobordism W from Y_1 to Y_2 with $H_i(W,\mathbb{Q}) = 0$ for $i = 1$ and 2, which can be endowed with a Spin^c structure $\mathfrak{s}$ whose restrictions to Y_1 and Y_2 are $\mathfrak{t}_1$ and $\mathfrak{t}_2$ respectively. The connected sum operation endows this set with the structure of an Abelian group (whose unit is S^3 endowed with its unique Spin^c structure).

There is a classical homomorphism

$$\rho\colon \theta^c \longrightarrow \mathbb{Q}/2\mathbb{Z}$$

(see for instance [**3**]), defined as follows. Consider a rational homology three-sphere $(Y,\mathfrak{t})$, and let X be any four-manifold equipped with a Spin^c structure $\mathfrak{s}$ with $\partial X \cong Y$ and $\mathfrak{s}|\partial X \cong \mathfrak{t}$. Then

$$\rho(Y,\mathfrak{t}) \equiv \frac{c_1(\mathfrak{s})^2 - \sigma(X)}{4} \pmod{2\mathbb{Z}}$$

where $\sigma(X)$ denotes the signature of the intersection form of X.

It is shown in [**82**] that the numerical invariant $d(Y,\mathfrak{t})$ descends to give a group homomorphism

$$d\colon \theta^c \longrightarrow \mathbb{Q}$$

which is a lift of ρ. Moreover, d is invariant under conjugation; i.e., $d(Y,\mathfrak{t}) = d(Y,\bar{\mathfrak{t}})$.

The rational gradings can be introduced for three-manifolds with $b_1(Y) > 0$, as well, only there one must restrict to Spin^c structures whose first Chern class is trivial. In this case, the gradings are fixed so that Equation (10) still holds. With these conventions, for example, for a three-manifold with $H_1(Y;\mathbb{Z}) \cong \mathbb{Z}$, the Heegaard Floer homologies of

$HF^{\circ}(Y, \mathfrak{s})$ for the Spin^c structure with $c_1(\mathfrak{s}) = 0$ have a grading which takes its values in $\frac{1}{2} + \mathbb{Z}$.

3.3. Maps on HF^{∞}. As we have seen, for rational homology three-spheres, the structure of HF^{∞} is rather simple. There are corresponding statements for the maps on HF^{∞} induced by cobordisms.

Indeed, if W is a cobordism from Y_1 to Y_2 with $b_2^+(W) > 0$, the induced map $F^{\infty}_{W,\mathfrak{s}} = 0$ for any $\mathfrak{s} \in \mathrm{Spin}^c(W)$ (cf. Lemma 8.2 of [**83**]). Moreover, if W is a cobordism from Y_1 to Y_2 (both of which are rational homology three-spheres), and W satisfies $b_2^+(W) = b_1(W) = 0$, then $F^{\infty}_{W,\mathfrak{s}}$ is an isomorphism, as proved in Propositions 9.3 and 9.4 of [**82**].

3.4. Examples. We begin with some algebraic notions for describing Heegaard Floer homology groups. Let $\mathcal{T}_{(d)}$ denote the graded $\mathbb{Z}[U]$-module $\mathbb{Z}[U, U^{-1}]/U \cdot \mathbb{Z}[U]$, graded so that the element 1 has grading d.

A rational homology three-sphere Y is called an *L-space* if $HF^+(Y)$ has no torsion and the map from $HF^{\infty}(Y)$ to $HF^+(Y)$ is surjective. The Floer homology of an L-space can be uniquely specified by its correction terms. That is, if Y is an L-space, then

$$\widehat{HF}(Y, \mathfrak{s}) \cong \mathbb{Z}_{(d(Y,\mathfrak{s}))},$$

where here (and indeed throughout this subsection) the subscript denotes absolute grading, and

$$HF^+(Y, \mathfrak{s}) \cong \mathcal{T}_{(d)}.$$

By a direct inspection of the corresponding genus one Heegaard diagrams, one can see that S^3 is an L-space. Indeed, by a similar picture, all lens spaces are L-spaces.

The absolute $\mathbb{Q}$ grading can also be calculated for lens spaces [**82**]. For example, for $L(2,1) \cong \mathbb{RP}^3$, there are two Spin^c structures $\mathfrak{s}$ and $\mathfrak{s}'$ with correction terms $1/4$ and $-1/4$ respectively.

The Brieskorn homology sphere $\Sigma(2,3,5)$ is also an L-space, and it has $d(\Sigma(2,3,5)) = 2$.

However, $\Sigma(2,3,7)$ is not an L-space. Its Heegaard Floer homology is determined by

$$HF^+(\Sigma(2,3,7)) \cong \mathcal{T}_{(0)} \oplus \mathbb{Z}_{(-1)}.$$

A combinatorial description of the Heegaard Floer homology of Brieskorn spheres and some other plumbings can be found in [**73**]; see also [**82**], [**70**], [**93**].

3.5. Intersection form bounds. The correction terms of a rational homology three-sphere Y constrain the intersection forms of smooth four-manifolds which bound Y, according to the following result, which is analogous to a gauge-theoretic result of Frøyshov [**32**]:

Theorem 3.2. *Let Y be a rational homology and W be a smooth four-manifold which bounds Y with negative-definite intersection form. Then, for each* Spin^c *structure $\mathfrak{s}$ over W, we have that*

$$c_1(\mathfrak{s})^2 + b_2(W) \leqslant 4d(Y, \mathfrak{s}|_Y). \tag{11}$$

The above theorem gives strong restrictions on the intersection forms of four-manifolds which bound a given three-manifold Y. In particular, if Y is an integral homology three-sphere, following a standard argument from Seiberg–Witten theory, compare [**32**], one can combine the above theorem with a number-theoretic result of Elkies [**22**] to show that if Y can be realized as the boundary of a smooth, negative-definite four-manifold, then $d(Y) \geqslant 0$; moreover if $d(Y) = 0$, then if X has negative-definite intersection form, then it must be diagonalizable.

3.6. Four-manifold invariants. The invariants associated to cobordisms can be used to construct an invariant for smooth, closed four-manifolds which is very similar in spirit to the Seiberg–Witten invariant for four-manifolds. Indeed, all known calculations support the conjecture that the two smooth four-manifold invariants agree.

Suppose that X is a four-manifold with $b_2^+(X) > 1$. We delete four-ball neighborhoods of two points in X, and view the result as a cobordism from S^3 to S^3, which we can further subdivide along a separating hypersurface N into a union $W_1 \cup_N W_2$, with the following properties:

- W_1 is a cobordism from S^3 to N with $b_2^+(W_1) > 0$,
- W_2 is a cobordism from N to S^3 with $b_2^+(W_2) > 0$,
- restriction map $H^2(W_1 \cup_N W_2) \longrightarrow H^2(W_1) \oplus H^2(W_2)$ is injective.

Such a separating hypersurface is called an *admissible cut* for X.

Let $HF^-_{\mathrm{red}}(Y)$ denote the kernel of the map $HF^-(Y) \longrightarrow HF^\infty(Y)$. Of course, this is isomorphic to the group $HF^+_{\mathrm{red}}(Y)$, via an identification coming from the homomorphism δ from Equation (7). Since $b_2^+(W_i) > 0$, the maps on HF^∞ induced by cobordisms are trivial (see Lemma 8.2 of [**83**]), and in particular the image of the map

$$F^-_{W_1,\mathfrak{s}|W_1} : HF^-(S^3) \longrightarrow HF^-(N, \mathfrak{s}|N)$$

lies in the kernel $HF^-_{\mathrm{red}}(N,\mathfrak{s}|N)$ of the map i (cf. Diagram (5). Moreover, the map

$$F^+_{W_2,\mathfrak{s}|W_2}\colon HF^+(N,\mathfrak{s}|N)\longrightarrow HF^+(S^3).$$

factors through the projection of $HF^+(N,\mathfrak{s}|N)$ to $HF^+_{\mathrm{red}}(N,\mathfrak{s}|N)$ (the cokernel of the map π from Diagram (5)). Thus, we can define

$$\Phi_{X,\mathfrak{s}}\colon HF^-(S^3)\longrightarrow HF^+(S^3)$$

to be the composite:

$$F^+_{W_2,\mathfrak{s}|W_2}\circ(\delta')^{-1}\circ F^-_{W_1,\mathfrak{s}|W_1},$$

where

$$\delta'\colon HF^+_{\mathrm{red}}(N,\mathfrak{s}|N)\longrightarrow HF^-_{\mathrm{red}}(N,\mathfrak{s}|N)$$

is the natural isomorphism induced from δ.

The definition of $\Phi_{X,\mathfrak{s}}$ depends on a choice of admissible cut for X, but it is not difficult to verify [**83**] that $\Phi_{X,\mathfrak{s}}$ is independent of this choice, giving a well-defined four-manifold invariant.

The element $\Phi_{X,\mathfrak{s}}$ is nontrivial for only finitely many Spin^c structures over X. It vanishes for connected sums of four-manifolds with $b_2^+(X)>0$; cf. Theorem 1.3 of [**83**] (compare [**14**] and [**107**] for corresponding results for Donaldson polynomials and Seiberg–Witten invariants respectively). In fact, according to [**81**], if (X,ω) is a symplectic four-manifold $\Phi_{X,k}\neq 0$ for the canonical Spin^c structure k associated to the symplectic structure. This can be seen as an analogue of a theorem of Taubes [**103**] in the Seiberg–Witten context. Whereas Taubes' theorem is proved by perturbing the Seiberg–Witten equations using a symplectic two-form, the nonvanishing theorem of Φ is proved by first associating to (X,ω) a compatible Lefschetz pencil, which can be done according to a theorem of Donaldson (cf. [**15**]) blowing up to obtain a Lefschetz fibration, and then analyzing maps between Floer homology induced by two-handles coming from the singular fibers in the Lefschetz fibration, with the help of Theorem 3.1.

4. Knots in S^3

We describe here constructions of Heegaard Floer homology applicable to knots. For simplicity, we restrict attention to knots in S^3. This knot invariant was introduced in [**76**] and also independently by Rasmussen in [**90**], [**88**]. The calculations in Subsection 4.2 are based on the results of **[78]**, **[74]**, **[75]**. In Subsection 4.3, we discuss the fact that knot Floer homology detects the Seifert genus of a knot. This result is proved in **[77]**. The relationship with the four-ball genus is discussed in 4.4, where we discuss the concordance invariant of **[86]** and **[88]**, and also the method of Owens and Strle **[71]**. Finally, in Subsection 4.5, we discuss an application to the problem of knots with unknotting number one from **[72]**. This application uses the Heegaard Floer homology of the branched double-cover associated to a knot.

4.1. Knot Floer homology. In Subsection 2.6, we described a construction which associates to an oriented knot in a three-manifold Y a filtration of the chain complex $\widehat{CF}(Y)$. Our aim here is to describe properties of this invariant when the ambient three-manifold is S^3 (although we will be forced to generalize to the case of knots in a connected sum of copies of $S^2 \times S^1$, as we shall see later). In this case, a knot $K \subset S^3$ induces a filtration $\mathcal{F}(K,i)$ indexed by $i \in \mathbb{Z}$ of the chain complex $\widehat{CF}(S^3)$, whose homology is a single $\mathbb{Z}$. With some loss of information, we can take the homology of the associated graded object, to obtain the "knot Floer homology"

$$\widehat{HFK}_*(K,i) = H_*(\mathcal{F}(K,i)/\mathcal{F}(K,i-1)).$$

Note that this can be viewed as one bigraded Abelian group

$$\widehat{HFK}(K) = \bigoplus_{d,i\in\mathbb{Z}} \widehat{HFK}_d(K,i).$$

We call here i the filtration level and d (the grading induced from the Heegaard Floer complex $\widehat{CF}(S^3)$) the Maslov grading.

These homology groups satisfy a number of basic properties, which we outline presently. Sometimes, it is simplest to state these properties for $\widehat{HFK}(K,i,\mathbb{Q})$, the homology with rational coefficients:

$$\widehat{HFK}(K,i,\mathbb{Q}) \cong \widehat{HFK}(K,i) \otimes_{\mathbb{Z}} \mathbb{Q}.$$

The Euler characteristic is related to the Alexander polynomial of K, $\Delta_K(T)$ by the following formula:

$$\sum \chi(\widehat{HFK}(K,i,\mathbb{Q})) \cdot T^i = \Delta_K(T) \tag{12}$$

(it is interesting to compare this with [**1**], [**65**], and [**25**]). The sign conventions on the Euler characteristic here are given by

$$\chi(\widehat{HFK}(K,i,\mathbb{Q})) = \sum_{d=-\infty}^{+\infty} (-1)^d \cdot \mathrm{rk}\left(\widehat{HFK}_d(K,i,\mathbb{Q})\right).$$

Unlike the Alexander polynomial, the knot Floer homology is sensitive to the chirality of the knot. Specifically, if $\overline{K}$ denotes the mirror of K (i.e., switch over- and under-crossings in a projection for K), then

$$\widehat{HFK}_d(K,i,\mathbb{Q}) \cong \widehat{HFK}_{-d}(\overline{K},-i,\mathbb{Q}). \tag{13}$$

Another symmetry these invariants enjoy is the following conjugation symmetry:

$$\widehat{HFK}_d(K,i,\mathbb{Q}) \cong \widehat{HFK}_{d-2i}(K,-i,\mathbb{Q}), \tag{14}$$

refining the symmetry of the Alexander polynomial.

These groups also satisfy a Künneth principle for connected sums. Specifically, let K_1 and K_2 be a pair of knots, and let $K_1 \# K_2$ denote their connected sum. Then,

$$\widehat{HFK}(K_1\#K_2,i,\mathbb{Q}) \cong \bigoplus_{i_1+i_2=i} \widehat{HFK}(K_1,i_1,\mathbb{Q}) \otimes_{\mathbb{Q}} \widehat{HFK}(K_2,i_2,\mathbb{Q}) \tag{15}$$

(see Corollary 7.2 of [**76**], and [**88**]). Of course, this can be seen as a refinement of the fact that the Alexander polynomial is multiplicative under connected sums of knots. These invariants also satisfy a "skein exact sequence" (compare [**29**], [**7**], [**82**], [**48**]). To state it, we must generalize to the case of oriented links in S^3. This can be done in the following manner: an n-component oriented link in S^3 gives rise, in a natural way, to an n-component oriented knot in $\#^{n-1}(S^2 \times S^1)$. Specifically, we attach $n-1$ one-handles to S^3, so that the two feet of each one-handle lie on different components of the link, and so that each link component meets at least one foot. Next, we form the connected sum of the various components of the link via standard strips which pass through the one-handles. In this way, we view the link invariant for an n-component link $L \subset S^3$ as a knot invariant for the associated knot in $\#^{n-1}(S^2 \times S^1)$.

In this manner, the homology of the associated graded object – the link Floer homology – is a sequence of graded Abelian groups $\widehat{HFK}_*(L,i)$, where here $i \in \mathbb{Z}$. If L has an odd number of components, the Maslov grading is a $\mathbb{Z}$-grading, while if it has an even number of components,

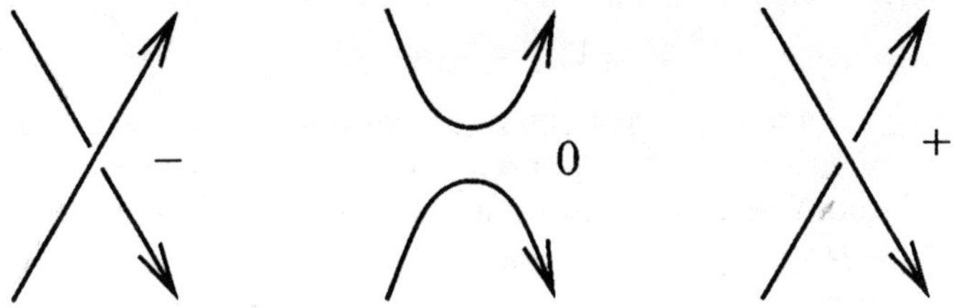

FIGURE 4. *Skein moves at a double-point.*

the Maslov grading takes values in $\frac{1}{2}+\mathbb{Z}$. As a justification for this convention, observe that the reflection formula, Equation (13), remains true in the context of links.

Suppose that L is a link, and suppose that p is a positive crossing of some projection of L. Following the usual conventions from skein theory, there are two other associated links, L_0 and L_-, where here L_- agrees with L_+, except that the crossing at p is changed, while L_0 agrees with L_+, except that here the crossing p is resolved in a manner consistent with orientations, as illustrated in Figure 4. There are two cases of the skein exact sequence, according to whether or not the two strands of L_+ which project to p belong to the same component of L_+.

Suppose first that the two strands which project to p belong to the same component of L_+. In this case, the skein exact sequence reads:

$$\ldots \longrightarrow \widehat{HFK}(L_-) \longrightarrow \widehat{HFK}(L_0) \longrightarrow \widehat{HFK}(L_+) \longrightarrow \ldots, \tag{16}$$

where all the maps above respect the splitting of $\widehat{HFK}(L)$ into summands (e.g. $\widehat{HFK}(L_-,i)$ is mapped to $\widehat{HFK}(L_0,i)$). Furthermore, the maps to and from $\widehat{HFK}(L_0)$ drop degree by $\frac{1}{2}$. The remaining map from $\widehat{HFK}(L_+)$ to $\widehat{HFK}(L_-)$ does not necessarily respect the absolute grading; however, it can be expressed as a sum of homogeneous maps, none of which increases absolute grading. When the two strands belong to different components, we obtain the following:

$$\ldots \longrightarrow \widehat{HFK}(L_-) \longrightarrow \widehat{HFK}(L_0)\otimes V \longrightarrow \widehat{HFK}(L_+) \longrightarrow \ldots, \tag{17}$$

where V denotes the four-dimensional vector space

$$V = V_{-1} \oplus V_0 \oplus V_1,$$

where here $V_{\pm 1}$ are one-dimensional pieces supported in degree ± 1, while V_0 is a two-dimensional piece supported in degree 0. Moreover, the maps respect the decomposition into summands, where the ith summand of the middle piece $\widehat{HFK}(L_0) \otimes V$ is given by

$$(\widehat{HFK}(L_0, i-1) \otimes V_1) \oplus (\widehat{HFK}(L_0, i) \otimes V_0) \oplus (\widehat{HFK}(L_0, i+1) \otimes V_{-1}).$$

The shifts in the absolute gradings work just as they did in the previous case.

The skein exact sequence is, of course, very closely related to Theorem 3.1. Indeed, its proof proceeds by using the surgery long exact sequence associated to an unknot which links the crossing one is considering, and analyzing the behaviour of the induced maps (cf. Section 8 of [**76**]).

4.2. Calculations of knot Floer homology. It is useful to have a concrete description of the generators of the knot Floer complex in terms of the combinatorics of a knot projection. In fact, the data we fix at first is an oriented knot projection (with at most double-point singularities), equipped with a choice of distinguished edge e which appears in the closure of the unbounded region A in the planar projection. We call this data a *decorated projection* for K. We denote the planar graph of the projection by G.

We can construct a doubly-pointed Heegaard diagram compatible with K from a decorated projection of K, as follows.

Let B denote the other region which contains the edge e in its closure, and let Σ be the boundary of a regular neighborhood of G, thought of as a one-complex in S^3 (i.e., if our projection has n double-points, then Σ has genus $n+1$); we orient Σ as $\partial(S^3 - \mathrm{nd}(G))$. We associate to each region $r \in R(G) - A$, an attaching circle α_r (which follows along the boundary of r). To each crossing v in G we associate an attaching circle β_v as indicated in Figure 5. In addition, we let μ denote the meridian of the knot, chosen to be supported in a neighborhood of the distinguished edge e.

Each vertex v is contained in four (not necessarily distinct) regions. Indeed, it is clear from Figure 5, that in a neighborhood of each vertex v, there are at most four intersection points of β_v with circles corresponding to the four regions which contain v. (There are fewer than four

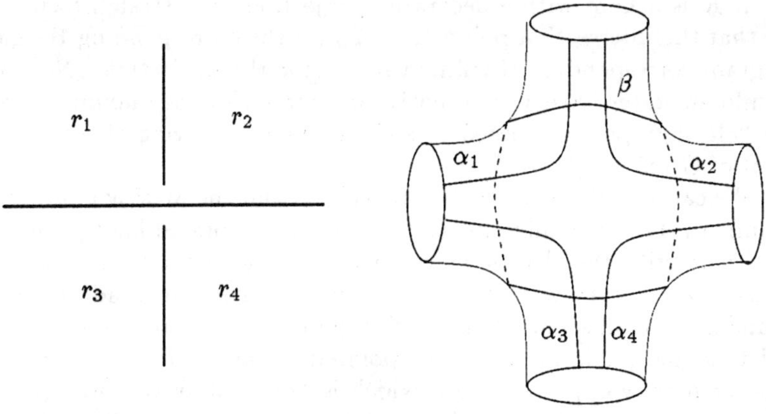

FIGURE 5. *Special Heegaard diagram for knot crossings.* At each crossing as pictured on the left, we construct a piece of the Heegaard surface on the right (which is topologically a four-punctured sphere). The curve β is the one corresponding to the crossing on the left; the four arcs $\alpha_1, \ldots, \alpha_4$ will close up. (Note that if one of the four regions $r_1, \ldots, r_4$ contains the distinguished edge e, its corresponding α-curve should *not* be included.) Note that the Heegaard surface is oriented from the outside.

intersection points with β_v if v is a corner for the unbounded region A.) Moreover, the circle corresponding to μ meets the circle α_B in a single point (and is disjoint from the other circles). Placing one reference point w and z on each side of μ, we obtain a doubly-pointed Heegaard diagram for S^3 compatible with K.

We can now describe the generators $\mathbb{T}_\alpha \cap \mathbb{T}_\beta$ for the knot Floer homology in terms of the planar graph G of the projection.

Definition 4.1. A *Kauffman state* (cf. [**45**]) for a decorated knot projection of K is a map which associates to each vertex of G one of the four in-coming quadrants, so that:

- the quadrants associated to distinct vertices are subsets of distinct regions in $S^2 - G$
- none of the quadrants is a corner of the distinguished regions A or B (whose closure contains the edge e).

If K is a knot with a decorated projection, it is straightforward to see that the intersection points $\mathbb{T}_\alpha \cap \mathbb{T}_\beta$ for the corresponding Heegaard diagram correspond to Kauffman states for the projection. Note that Kauffman states have an alternative interpretation, as maximal trees in the "black graph" associated to a checkerboard coloring of the complement of G (cf. [**45**]).

We can also describe the filtration level and the Maslov grading of a Kauffman state in combinatorial terms of the decorated knot projection.

To describe the filtration level, note that the orientation on the knot K associates to each vertex $v \in G$ a distinguished quadrant whose boundary contains both edges which point towards the vertex v. We call this the quadrant which is "pointed towards" at v. There is also a diagonally opposite region which is "pointed away from" (i.e., its boundary contains the two edges pointing away from v). We define the local filtration contribution of x at v, denoted $s(x,v)$, by the following rule (illustrated in Figure 7), where $\epsilon(v)$ denotes the sign of the crossing (which we recall in Figure 6):

$$2\epsilon(v)s(x,v) = \begin{cases} 1 & x(v) \text{ is the quadrant pointed towards at } v \\ -1 & x(v) \text{ is the quadrant away from at } v \\ 0 & \text{otherwise.} \end{cases}$$

The filtration level associated to a Kauffman state, then, is given by the sum

$$\mathcal{F}(x) = \sum_{v \in \mathrm{Vert}(G)} s(x,v).$$

Note that the function $\mathcal{F}(x)$ is the T-power appearing for the contribution of x to the symmetrized Alexander polynomial, see [**2**], [**44**].

The Maslov grading $\mathrm{gr}(x)$ is defined analogously. First, at each vertex v, we define the local grading contribution $m(x,v)$. This local contributions is nonzero on only one of the four quadrants – the one which is pointed away from at v. At this quadrant, the grading contribution is minus the sign $\epsilon(v)$ of the crossing, as illustrated in Figure 8. Now, the grading $\mathrm{gr}(x)$ of a Kauffman state x is defined by the formula

$$\mathrm{gr}(x) = \sum_{v \in \mathrm{Vert}(G)} m(x,v).$$

A verification of these formulas can be found in Theorem 1.2 of [**78**].

It is clear from the above formulas that if K has an alternating proj ection, then $\mathcal{F}(\mathbf{x}) - \mathrm{gr}(\mathbf{x})$ is independent of the choice of state $\mathbf{x}$. It

FIGURE 6. *Crossing conventions.* Crossings of the first kind are assigned +1, and those of the second kind are assigned −1.

FIGURE 7. *Local filtration level contributions* $s(x, v)$. We have illustrated the local contributions of $s(x, v)$ for both kinds of crossings. (In both pictures, "upwards" region is the one which the two edges point towards.).

follows that if we use the chain complex associated to this Heegaard diagram, then there are no differentials in the knot Floer homology, and indeed, its rank is determined by its Euler characteristic. Indeed, by calculating the constant, we get the following result, proved in Theorem 1.3 of [**78**]:

FIGURE 8. *Local grading contributions $m(x, v)$.* We have illustrated the local contribution of $m(x, v)$.

Theorem 4.2. *Let $K \subset S^3$ be an alternating knot in the three-sphere, and write its symmetrized Alexander polynomial as*

$$\Delta_K(T) = a_0 + \sum_{s>0} a_s(T^s + T^{-s}),$$

and let $\sigma(K)$ denote its signature. Then, $\widehat{HFK}(S^3, K, s)$ is supported entirely in dimension $s + \frac{\sigma(K)}{2}$, and indeed

$$\widehat{HFK}(S^3, K, s) \cong \mathbb{Z}^{|a_s|}.$$

Thus, for alternating knots, this choice of Heegaard diagram is remarkably successful. However, in general, there are differentials one must grapple with, and these admit, at present, no combinatorial description in terms of Kauffman states. However, they do respect certain additional filtrations which can be described in terms of states, and this property, together with some additional tricks, can be used to give calculations of knot Floer homology groups in certain cases (cf. [**75**], [**18**]). As a particular example, these filtrations are used in [**75**] to show that knot Floer homology of the eleven-crossing Kinoshita–Terasaka knot (a knot whose Alexander polynomial is trivial) differs from that of its Conway mutant.

In a different direction, some knots admit Heegaard diagrams on a genus one surface. For these knots, calculation of the differentials becomes a purely combinatorial matter (cf. Section 6 of [**76**] and also [**90**], [**88**], [**37**]).

Sometimes, it is more convenient to use more abstract methods to calculate knot Floer homology. In particular, there is a relationship between knot Floer homology and the Heegaard Floer homology three-manifolds obtained by surgery along K (cf. [**76**], [**88**]). With the help of this relationship, we obtain the following structure for the knot Floer homology of a knot for which some positive surgery is an L-space (proved in Theorem 1.2 of [**74**]):

Theorem 4.3. *Suppose that $K \subset S^3$ is a knot for which there is a positive integer p for which $S^3_p(K)$ is an L-space. Then, there is an increasing sequence of nonnegative integers*

$$n_{-m} < \dots < n_m$$

with the property that $n_i = -n_{-i}$, with the following significance. If we let

$$\delta_i = \begin{cases} 0 & \text{if } i = m \\ \delta_{i+1} - 2(n_{i+1} - n_i) + 1 & \text{if } m - i \text{ is odd} \\ \delta_{i+1} - 1 & \text{if } m - i > 0 \text{ is even,} \end{cases}$$

then $\widehat{HFK}(K, j) = 0$ unless $j = n_i$ for some i, in which case $\widehat{HFK}(K, j) \cong \mathbb{Z}$ and it is supported entirely in dimension δ_i.

For example, all (right-handed) torus knots satisfy the hypothesis of this theorem. (Recal that if $T_{p,q}$ denotes the right-handed (p, q) torus knot, then $S^3_{pq\pm1}(T_{p,q})$ is a lens space.) The knot Floer homology of the $(3, 4)$ torus knot is illustrated in Figure 9. The above theorem can be fruitfully thought of from three perspectives: as a source of examples of knot Floer homology calculations (for example, a calculation of the knot Floer homology of torus knots), as a restriction on knots which admit L-space surgeries (for example, it shows that if $K \subset S^3$ admits a lens space surgery, then all the coefficients of its Alexander polynomial satisfy $|a_i| \leqslant 1$), or as a restriction on L-spaces which can arise as surgeries on knots in S^3 (cf. [**74**]).

4.3. Knot Floer homology and the Seifert genus. A knot $K \subset S^3$ can be realized as the boundary of an embedded, orientable surface in S^3. Such a surface is called a Seifert surface for K, and the minimal genus of any Seifert surface for K is called its *Seifert genus*, denoted $g(K)$. Of course, a knot has $g(K) = 0$ if and only if it is the unknot.

The knot Floer homology of K detects the Seifert genus, and in particular it distinguishes the unknot, according to the following result

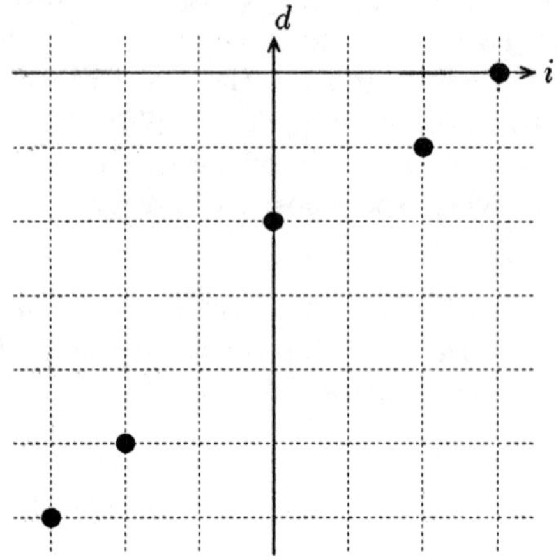

FIGURE 9. *Knot Floer homology for the* (3,4) *torus knot.* The dots represent $\mathbb{Z}$ summands, and the bigrading is specified by the d and i coordinates.

proved in [**77**]. To state it, we first define the degree of the knot Floer homology to be the integer

$$\deg \widehat{HFK}(K) = \max\{i \in \mathbb{Z} \big| \widehat{HFK}(K,i) \neq 0\}.$$

Theorem 4.4. *For any knot $K \subset S^3$, $g(K) = \deg \widehat{HFK}(K)$.*

Given a Seifert surface of genus g for K, one can construct a Heegaard diagram for which all the points in $\mathbb{T}_\alpha \cap \mathbb{T}_\beta$ have filtration level $\leqslant g$. This gives at once the bound

$$\deg \widehat{HFK}(K) \leqslant g(K)$$

(this result is analogous to a classical bound on the genus of a knot in terms of the degree of its Alexander polynomial).

The inequality in the other direction is much more subtle, involving much of the theory described so far. First, one relates the degree of the knot Floer homology by a similar quantity defined using the Floer homology of the zero-surgery $S^3_0(K)$. Next, one appeals to a theorem of Gabai [**36**], according to which if K is a knot with Seifert genus $g > 0$, then $S^3_0(K)$ admits a taut foliation $\mathcal{F}$ whose first Chern class is $g-1$ times a generator for $H^2(S^3_0(K);\mathbb{Z})$. The taut foliation naturally induces a symplectic structure on $[-1,1]\times S^3_0(K)$, according to a result of Eliashberg and Thurston [**21**], which, according to a recent result of Eliashberg [**20**], [**23**] can be embedded in a closed symplectic four-manifold X (indeed, one can arrange for $S^3_0(K)$ to divide the four-manifold X into two pieces with $b_2^+(X_i)>0$). The nonvanishing of the four-manifold invariant $\Phi_{X,k}$ for a symplectic four-manifold can then be used to prove that the Heegaard Floer homology of $S^3_0(K)$ is nontrivial in the Spin^c structure gotten by restricting the canonical Spin^c structure k of the ambient symplectic four-manifold – i.e., this is the Spin^c structure belonging to the foliation $\mathcal{F}$. The details of this argument are given in [**77**].

Since the generators of knot Floer homology can be thought of from a Morse-theoretic point of view as simultaneous trajectories of gradient flow-lines, Theorem 4.4 immediately gives a curious Morse-theoretic characterization of the Seifert genus of K, as the minimum over all Morse functions compatible with K of the maximal filtration level of any simultaneous trajectory.

4.4. The four-ball genus. A knot $K\subset S^3$ can be viewed as a knot in the boundary of the four-ball, and as such, it can be realized as the boundary of a smoothly embedded oriented surface in the four-ball. The minimal genus of any such surface is called the *four-ball genus* of the knot, and it is denoted $g^*(K)$. Obviously, $g^*(K)\leqslant g(K)$. In general, $g^*(K)$ is quite difficult to calculate.

Lower bounds on $g^*(K)$ can be obtained from Heegaard Floer homology. The construction involves going deeper into the knot filtration. Specifically, as explained explained in Subsection 2.6, the filtered chain homotopy type of the sequence of inclusions

$$.....\subseteq \mathcal{F}(K,i)\subseteq \mathcal{F}(K,i+1)\subseteq ...\subseteq \widehat{CF}(S^3)$$

is a knot invariant; passing to the homology of the associated graded object constitutes some loss of information. There is a quantity associated to the filtered complex which goes beyond knot Floer homology, and that is the integer $\tau(K)$ which is defined by

$$\tau(K)=\min\{i\in\mathbb{Z}\,|\,H_*(\mathcal{F}(K,i))\longrightarrow \widehat{HF}(S^3)\text{ is nontrivial}\}.$$

It is proved in [**86**], [**88**] that

$$|\tau(K)|\leqslant g^*(K). \tag{18}$$

The above inequality can be used to prove a property of τ which underscores its analogy with the correction terms $d(Y, \mathfrak{s})$ described earlier. To put this result into context, we give a definition. Two knots K_1 and K_2 are said to be *concordant* if there is a smoothly embedded cylinder C in $[1,2] \times S^3$ with $C \cap \{i\} \times S^3 = K_i$ for $i = 1,2$. The set of concordance classes of knots can be made into an Abelian group, under the connected sum operation. It follows from Inequality (18), together with the additivity of τ under connected sums, that τ gives a homomorphism from the concordance group of knots to $\mathbb{Z}$.

Another such homomorphism is provided by $\sigma(K)/2$. In fact, according to Theorem 4.2,

$$2\tau(K) = -\sigma(K) \tag{19}$$

when K is alternating. By contrast, Theorem 4.3 gives many examples where Equation (19) fails. Indeed, combining Theorem 4.3 with Theorem 4.4 and Equation (12) and (18), we see that if K is a knot which admits a positive surgery which is an L-space surgery, then

$$\tau(K) = g(K) = g^*(K) = \deg \Delta_K;$$

and in particular, if $K = T_{p,q}$, then

$$\tau(K) = \frac{(p-1)(q-1)}{2}. \tag{20}$$

Note that the fact that $g^*(T_{p,q})$ is given by the above formula was conjectured by Milnor and first proved by Kronheimer and Mrowka using gauge theory [**54**].

In [**61**], see also [**62**] Livingston shows that properties of the concordance invariant $\tau(K)$ (specifically, the fact that it is a homomorphism whose absolute value bounds the four-ball genus of K, and satisfies Equation (20)) leads to the result that if K is the closure of a positive on k strands with n crossings, then

$$\tau(K) = \frac{n-k+1}{2} = g^*(K).$$

The second of these equations was proved first by Rudolph [**92**] using the local Thom conjecture proved by Kronheimer and Mrowka [**54**]. Further links between the Thurston-Bennequin invariant and τ are explored by Plamenevskaya, see [**87**].

A different method for bounding $g^*(K)$ is given by Owens and Strle in [**71**], where they describe a method using the correction terms for the

FIGURE 10. *A knot with* $u(K) = 2$.

branched double-cover of S^3 along K, $\Sigma(K)$. Under favorable circumstances, their method gives an obstruction for Murasugi's bound

$$|\sigma(K)| \leqslant 2g^*(K). \tag{21}$$

to being sharp. Specifically, taking the branched double-cover $\Sigma(F)$ of a surface F in B^4 which bounds K, one obtains a four-manifold which bounds $\Sigma(K)$. When F is a surface with $2g(F) = \sigma(K)$, the branched double $\Sigma(F)$ is a four-manifold with definite intersection form, whose second Betti number is $2g(F)$. Comparing this construction with the inequality from Theorem 3.2, one obtains restrictions on the correction terms of $\Sigma(K)$, which sometimes can rule out the existence of such surfaces F.

4.5. Unknotting numbers. Recall that the *unknotting number* of a knot K, denoted $u(K)$, is the the minimal number of crossing-changes required to unknot K. An unknotting of K can be realized as an immersed disk in B^4 which bounds S^3. Resolving the self-intersections, one gets the inequality $g^*(K) \leqslant u(K)$. However, there are circumstances where one needs better bounds (most strikingly, for any nontrivial slice knot). In [**72**], we describe an obstruction to knots K having $u(K) = 1$, in terms of the correction terms of the branched double-cover of K.

This construction works best in the case where K is alternating. In this case, the branched double-cover $\Sigma(K)$ is an L-space (cf. [**79**]). A classical construction of Montesinos [**66**] shows that if $u(K) = 1$, then $\Sigma(K)$ can be obtained as $\pm D/2$ surgery on another knot C in S^3, where here D denotes the determinant of the knot K

(i.e., $D = |\Delta_K(-1)|$). On the one hand, correction terms for an L-space which is realized as $n/2$ surgery (for some integer n) on a knot in S^3 satisfy certain symmetries (cf. Theorem 4.1 of [**72**]); on the other hand, the correction terms of the branched double-cover of an alternating knot can be calculated explicitly by classical data associated to an alternating projection of K (cf. [**79**]). Rather than recalling the result here, we content ourselves with illustrating an alternating knot K (listed as 8_{10} in the Alexander-Briggs notation) whose unknotting number was previously unknown, but which can now be shown to have $u(K) = 2$ using these techniques.

5. Problems and Questions

The investigation of Heegaard Floer homology naturally leads us to the following problems and questions.

Perhaps the most important problem in this circle of ideas is the following:

Problem 1. *Give a purely combinatorial calculation of Heegaard Floer homology or, more generally, the Heegaard Floer functor.*

In certain special cases, combinatorial calculations can be given, for example [**73**], [**78**]. This problem would be very interesting to solve even for certain restricted classes of three-manifolds, for example for those which fiber over the circle, compare. [**41**], [**97**], [**96**] [**9**].

In a related direction, it is simpler to consider the case of knots in S^3. Recall that in Section 4 we showed that the generators of the knot Floer complex can be thought of as Kauffman states.

Question 2. *Is there a combinatorial description of the differential on Kauffman states whose homology gives the knot Floer homology of K.*

It is intriguing to compare this with Khovanov's new invariants for links, see [**48**]. These invariants have a very similar structure to the knot Floer homology considered here, except that their Euler characteristic gives the Jones polynomial, see also [**43**], [**47**], [**46**], [**49**] [**5**]. Indeed, the similarities are further underscored by the work of E-S. Lee [**58**], who describes a spectral sequence which converges to a vector space of dimension 2^n, where n is the number of components of L, see also [**89**].

In a similar vein, it would be interesting to give combinatorial calculations of the numerical invariants arising from Heegaard Floer homology, specifically, the correction terms $d(Y, \mathfrak{s})$ or the concordance invariant $\tau(K)$. An intriguing conjecture of Rasmussen [**89**] relates τ with a numerical invariant coming from Khovanov homology.

Problem 3. *Establish the conjectured relationship between Heegaard Floer homology and Seiberg–Witten theory.*

There are two approaches one might take to this problem. One direct, analytical approach would be to analyze moduli spaces of solutions to the Seiberg–Witten equations over a three-manifold equipped with a Heegaard decomposition. Another approach would involve an affirmative answer to the following question:

Question 4. *Is there an axiomatic characterization of Heegaard Floer homology?*

A *Floer functor* is a map which associates to any closed, oriented three-manifold Y a $\mathbb{Z}/2\mathbb{Z}$-graded Abelian group $\mathcal{H}(Y)$ and to any cobordism W from Y_1 to Y_2 a homorphism $\mathcal{D}_W : \mathcal{H}(Y_1) \longrightarrow \mathcal{H}(Y_2)$, which is natural under composition of cobordisms, and which induce exact sequences for triples of three-manifolds (Y_0, Y_1, Y_2) which are related as in the hypotheses of Theorem 3.1. It is interesting to observe that if $\mathcal{T}$ is a natural transformation between Floer functors $\mathcal{H}$ and $\mathcal{D}$ to $\mathcal{H}'$ and $\mathcal{D}'$, then if $\mathcal{T}$ induces an isomorphism $\mathcal{T}(S^3)\colon \mathcal{H}(S^3) \longrightarrow \mathcal{H}'(S^3)$, then $\mathcal{T}$ induces isomorphisms for all three-manifolds $\mathcal{T}(Y)\colon \mathcal{H}(Y) \longrightarrow \mathcal{H}'(Y)$. This can be proved from Kirby calculus, following the outline laid out in [**7**]. We know from [**57**] that monopole Floer homology (taken with $\mathbb{Z}/2\mathbb{Z}$ coefficients) is a Floer functor in this sense, and also (Theorem 3.1) that Heegaard Floer homology is a Floer functor. Unfortunately, this still falls short of giving an axiomatic characterization: one needs axioms which are sufficient to assemble a natural transformation $\mathcal{T}$.

Problem 5. *Develop cut-and-paste techniques for calculating the Heegaard Floer homology of Y in terms of data associated to its pieces.*

As a special case, one can ask how the knot Floer homology of a satellite knot can be calculated from data associated to the companion and the pattern. Of course, the Künneth principle for connected sums can be viewed as an example of this. Another example, of Whitehead doubling, has been studied by Eftekhary [**19**].

We have seen that the set of Spin^c structures for which the Heegaard Floer homology of Y is nontrivial determines the Thurston norm of Y. It is natural to ask what additional topological information is contained in the groups themselves. It is possible that these groups contain further information about foliations over Y.

We specialize to the case of knot Floer homology. If $K \subset S^3$ is a fibered knot of genus g, then it is shown in [**80**] that $\widehat{HFK}(K,g) \cong \mathbb{Z}$.

Question 6. *If $K \subset S^3$ is a knot with genus g and $\widehat{HFK}(K,g) \cong \mathbb{Z}$, does it follow that K is fibered?*

Calculations give some evidence that the answer to the above question is positive.

Question 7. *If $K \subset S^3$ is a knot, is there an explicit relationship between the fundamental group of $S^3 - K$ and the knot Floer homology $\widehat{HFK}(K)$?*

The above question is, of course, very closely related to the following:

Question 8. *Is there an explicit relationship between the Heegaard Floer homology and the fundamental group of Y?*

For example, one could try to relate the Heegaard Floer homology with the instanton Floer homology of Y. (Note, though, that presently instanton Floer homology is defined only for a restricted class of three-manifolds, cf. [**7**].) A link between Seiberg–Witten theory and instanton Floer homology is given by Pidstrygach and Tyurin's $PU(2)$ monopole equations [**24**]. This connection has been successfully exploited in Kronheimer and Mrowka's recent proof that all knots in S^3 have Property P [**51**].

The conjectured relationship with Seiberg–Witten invariants raises further questions. Specifically, Bauer and Furuta [**35**], [**6**] have constructed refinements of the Seiberg–Witten invariant which use properties of the Seiberg–Witten equations beyond merely their solution counts. Correspondingly, these invariants carry topological information about four-manifolds beyond their usual Seiberg–Witten invariants. A three-dimensional analogue is studied in work of Manolescu and Kronheimer, see [**63**], [**50**]

Question 9. *Is there a refinement of the four-manifold invariant Φ defined using Heegaard Floer homology which captures the information in the Bauer-Furuta construction?*

In the opposite direction, it is natural to study the following:

Problem 10. *Construct a gauge-theoretic analogue of knot Floer homology.*

Recall from Section 3 that there is a class of three-manifolds whose Heegaard Floer homology is as simple as possible, the so-called L-spaces. This class of three-manifolds includes all lens spaces, and more generally branched double-covers of alternating knots. The set of L-spaces is closed under connected sums. According to [**77**], L-spaces admit no (coorientable) taut foliations. A striking theorem of Némethi [**70**] characterizes those L-spaces which are boundaries of negative-definite plumbings of spheres: they are the links of rational surface singularities. Note also that there is an analogous class of three-manifolds in the context of Seiberg–Witten monopole Floer homology (cf. [**57**]).

Question 11. *Is there a topological characterization of L-spaces (i.e., which makes no reference to Floer homology)?*

References

1. S. Akbulut and J. D. McCarthy, *Casson's invariant for oriented homology 3-spheres*, Math. Notes **36**, Princeton University Press, 1990,
2. J. W. Alexander, *Topological invariants of knots and links*, Trans. Am. Math. Soc. **30** (1928), no. 2, 275–306.
3. M. F. Atiyah, V. K. Patodi, and I. M. Singer, *Spectral asymmetry and Riemannian geometry. II*, Math. Proc. Cambridge Philos. Soc. **78** (1975), no. 3, 405–432.
4. M. F. Atiyah, *Floer homology*, in: The Floer Memorial Volume, Birkhäuser, 1995, pp. 105–108.
5. D. Bar-Natan, *On Khovanov's categorification of the Jones polynomial*, Algebr. Geom. Topol. **2** (2002), 337-370.
6. S. Bauer and M. Furuta, *A Stable Cohomotopy Refinement of Seiberg–Witten Invariants: I*, Preprint, math.DG/0204340.
7. P. Braam and S. K. Donaldson, *Floer's work on instanton homology, knots, and surgery*, in: The Floer Memorial Volume, Birkhäuser, 1995, pp. 1995–256.
8. M. Culler, C. McA. Gordon, J. Luecke, and P. B. Shalen. Dehn surgery on knots. *Ann. of Math*, 125(2):237–300, 1987.
9. S. Donaldson and I. Smith, *Lefschetz pencils and the canonical class for symplectic four-manifolds*, Topology **42** (2003), no. 4, 743–785.
10. S. K. Donaldson and P. B. Kronheimer, *The Geometry of Four-Manifolds*, Oxford University Press, 1990.
11. S. K. Donaldson, *The Seiberg-Witten equations and 4-manifold topology*, Bull. Am. Math. Soc. (N.S.) **33** (1966), no. 1, 45–70.
12. S. K. Donaldson, *Floer Homology Groups in Yang-Mills Theory*, Cambridge University Press, Cambridge, 2002.
13. S. K. Donaldson, *Irrationality and the h-cobordism conjecture*, J. Differ. Geom. **26** (1987), no. 1, 141–168.

14. S. K. Donaldson, *Polynomial invariants for smooth four-manifolds*, Topology **29** (1990), no. 3, 257–315.
15. S. K. Donaldson, *Lefschetz pencils on symplectic manifolds*, J. Differ. Geom. **53** (1999), no. 2, 205–236.
16. S. K. Donaldson, *An application of gauge theory to four-dimensional topology*, J. Differ. Geom. **18** (1983), no. 2, 279–315.
17. S. Dostoglou and D. A. Salamon, *Self-dual instantons and holomorphic curves*, Ann. of Math. (2) **139** (1994), no. 3, 581–640.
18. E. Eftekhary, *Knot Floer homologies for pretzel knots*, math.GT/0311419.
19. E. Eftekhary, in preparation, 2004.
20. Ya. M. Eliashberg, *Few Remarks about Symplectic Filling*, Preprint, math.SG/0311459.
21. Ya. M. Eliashberg and W. P. Thurston, *Confoliations*, Am. Math. Soc., 1998.
22. N. D. Elkies, *A characterization of the* Z^n *lattice*, Math. Res. Lett. **2** (1995), no. 3, 321–326.
23. J. B. Etnyre, *On Symplectic Fillings*, Preprint, math.SG/0312091, 2003.
24. P. M. N. Feehan and T. G. Leness, SO(3) *monopoles, level-one Seiberg-Witten moduli spaces, and Witten's conjecture in low degrees*, in: Proc. the 1999 Georgia Topology Conference (Athens, GA), Topology Appl., **124** (2002), no. 2, 221–326.
25. R. Fintushel and R. J. Stern, *Knots, links, and 4-manifolds*, Invent. Math. **134** (1998), no. 2, 363–400.
26. R. Fintushel and R. J. Stern, *Seifert fibered homology three-spheres*, Proc. London Math. Soc. **61** (1990), 109–137.
27. A. Floer, *Morse theory for Lagrangian intersections*, J. Differ. Geometry, **28**, (1988) 513–547.
28. A. Floer, *An instanton-invariant for 3-manifolds*, Comm. Math. Phys. **119** (1988), 215-240.
29. A. Floer, *Instanton homology and Dehn surgery*, in: The Floer Memorial Volume, Birkhäuser, 1995, pp. 77–97.
30. A. Floer and H. Hofer, *Coherent orientations for periodic orbit problems in symplectic geometry*, Math. Z. **212** (1993), no. 1, 13–38.
31. A. Floer, H. Hofer, and D. Salamon, *Transversality in elliptic Morse theory for the symplectic action*, Duke Math. J. **80** (1995), no. 1, 251–290.
32. K. A. Frøyshov, *The Seiberg-Witten equations and four-manifolds with boundary*, Math. Res. Lett. **3** (1996), 373–390.
33. R. Friedman and J. W. Morgan, *Smooth Four-Manifolds and Complex Surfaces*, Springer-Verlag, 1994.
34. K. Fukaya, Y-G. Oh, K. Ono, and H. Ohta, *Lagrangian Intersection Floer Theory—Anomaly and Obstruction*, Kyoto University, 2000.
35. M. Furuta, *Finite dimensional approximations in geometry*, in: Proc. Intern. Congress Math. Vol. II (Beijing, 2002), Higher Ed. Press, Beijing, 2002, pp. 395–403.
36. D. Gabai, *Foliations and the topology of* 3*-manifolds III*, J. Differ. Geom. **26** (1987), no. 3, 479–536.
37. H. Goda, H. Matsuda, and T. Morifuji, *Knot Floer Homology of* (1,1)*-Knots*, Preprint, math.GT/0311084.

38. R. E. Gompf and A. I. Stipsicz, *4-manifolds and Kirby calculus*, Am. Math. Soc., 1999.
39. C. McA. Gordon and J. Luecke. Knots are determined by their complements. *J. Amer. Math. Soc.*, 2(2):371–415, 1989.
40. M. Gromov, *Pseudo holomorphic curves in symplectic manifolds*, Invent. Math. **82** (1985), 307–347.
41. M. Hutchings and M. Sullivan, *The periodic Floer homology of a Dehn twist*, http://math.berkeley.edu/ hutching/pub/index.htm, 2002.
42. E. Ionel and T. H. Parker, *Relative Gromov-Witten invariants*, Ann. Math. (2), **157**, (2003), no. 1, 45–96.
43. M. Jacobsson, *An Invariant of Link Cobordisms from Khovanov's Homology Theory*, Preprint, math.GT/0206303, 2002.
44. L. H. Kauffman, *On Knots*, Princeton University Press, 1987.
45. L. H. Kauffman, *Formal Knot Theory*, Princeton University Press, 1983.
46. M. Khovanov, *sl*(3) *Link Homology I*, Preprint, math.QA/0304375.
47. M. Khovanov, *An invariant of Tangle Cobordisms*, Preprint, math.QA/0207264, 2002.
48. M. Khovanov, *A categorification of the Jones polynomial*, Duke Math. J. **101** (2000), no. 3, 359–426.
49. M. Khovanov and L. Rozansky, *Matrix Factorizations and Link Homology*, Preprint, math.QA/0401268.
50. P. B. Kronheimer and C. Manolescu, *Periodic Floer Pro-spectra from the Seiberg-Witten Equations*, Preprint, math.GT/0203243.
51. P. B. Kronheimer and T. S. Mrowka, *Witten's Conjecture and Property P*, Preprint, math.GT/0311489, 2003.
52. P. B. Kronheimer and T. S. Mrowka, *Embedded surfaces and the structure of Donaldson's polynomial invariants*, J. Differ. Geom. **3** (1995), 573–734.
53. P. B. Kronheimer and T. S. Mrowka, *The genus of embedded surfaces in the projective plane*, Math. Res. Lett. **1** (1994), 797–808.
54. P. B. Kronheimer and T. S. Mrowka, *Gauge theory for embedded surfaces. I*, Topology **32** (1993), no. 4, 773–826.
55. P. B. Kronheimer and T. S. Mrowka, *Floer homology for Seiberg-Witten Monopoles* [In preparation.]
56. P. B. Kronheimer and T. S. Mrowka, *Monopoles and contact structures*, Invent. Math. **130** (1997) no. 2, 209–255.
57. P. B. Kronheimer, T. S. Mrowka, P. S. Ozsváth, and Z. Szabó, *Monopoles and Lens Space Surgeries*, Preprint, math.GT/0310164.
58. E. S. Lee, *The Support of the Khovanov's Invariants for Alternating Knots*, Preprint, math.GT/0201105, 2002.
59. A-M. Li and Y. Ruan, *Symplectic surgery and Gromov-Witten invariants of Calabi-Yau 3-folds*, Invent. Math. **145** =200. no. 1, 151–218.
60. P. Lisca and A. I. Stipsicz, *Heegaard Floer Invariants and Tight Contact Three-Manifolds*, Preprint, math.SG/0303280, 2003.
61. C. Livingston, *Computations of the Ozsváth-Szabó Concordance Invariant*, Preprint, math.GT/0311036.

62. C. Livingston, *Splitting the concordance group of Algebraically slice knots*, Geom. Topol. **7** (2003), 641-463.
63. C. Manolescu, *Seiberg-Witten-Floer stable homotopy type of three-manifolds with $b_1 = 0$*, Geom. Topol. **7** (2003), 889–932.
64. M. Marcolli and B-L. Wang, *Equivariant Seiberg-Witten Floer homology*, Comm. Anal. Geom. **9** (2001), no. 3, 451–639.
65. G. Meng and C. H. Taubes, *SW=Milnor torsion*, Math. Res. Lett. **3** (1996), 661–674.
66. J. M. Montesinos, *Surgery on links and double branched covers of S^3*, in: Knots, Groups, and 3-Manifolds (Papers dedicated to the memory of R. H. Fox), Princeton Univ. Press, 1975, pp. 227–259.
67. J. W. Morgan, *The Seiberg-Witten Equations and Applications to the Topology of Smooth Four-Manifold*, Princeton University Press, 1996.
68. J. W. Morgan, T. S. Mrowka, and D. Ruberman, *The L^2-Moduli Space and a Vanishing Theorem for Donaldson Polynomial Invariants*, International Press, 1994.
69. T. S. Mrowka, P. S. Ozsváth, and B. Yu, *Seiberg-Witten Monopoles on Seifert Fibered Spaces*, Comm. Analysis and Geometry, **5** (1997), no. 4, 685-793.
70. A. Némethi, *On the Ozsváth-Szabó Invariant of Negative Definite Plumbed 3-Manifolds*, Preprint, math.GT/0310083.
71. B. Owens and S. Strle, *Rational Homology Spheres and Four-Ball Genus*, Preprint, math.GT/0308073, 2003.
72. P. S. Ozsváth and Z. Szabó, *Knots with Unknotting Number One and Heegaard Floer Homology*, Preprint, math.GT/0401426.
73. P. S. Ozsváth and Z. Szabó, *On the Floer homology of plumbed three-manifolds*, Geom. Topol. **7** (2003), 185–224.
74. P. S. Ozsváth and Z. Szabó, *On knot Floer Homology and Lens Space Surgeries*, Preprint, math.GT/0303017.
75. P. S. Ozsváth and Z. Szabó" *Knot Floer Homology, Genus Bounds, and Mutation*, Preprint, math.GT/0303225, 2003.
76. P. S. Ozsváth and Z. Szabó, *Holomorphic Disks and Knot Invariants*, Preprint, math.GT/0209056, 2002.
77. P. S. Ozsváth and Z. Szabó, *Holomorphic Disks and Genus Bounds*, Preprint, math.GT/0311496, 2003.
78. P. S. Ozsváth and Z. Szabó, *Heegaard Floer homology and alternating knots*, Geom. Topol. **7** (2003), 225–254 (electronic).
79. P. S. Ozsváth and Z. Szabó, *On the Heegaard Floer Homology of Branched Double-Covers*, Preprint, math.GT/0309170, 2003.
80. P. S. Ozsváth and Z. Szabó, *Heegaard Floer Homologies and Contact Structures*, Preprint, math.SG/0210127, 2002.
81. P. S. Ozsváth and Z. Szabó, *Holomorphic Disk Invariants for Symplectic Four-Manifolds*, Preprint, math.SG/0201059, 2002.
82. P. S. Ozsváth and Z. Szabó, *Absolutely graded Floer homologies and intersection forms for four-manifolds with boundary*, Adv. Math. **173** (2003), no. 2, 179–261.
83. P. S. Ozsváth and Z. Szabó, *Holomorphic Triangles and Invariants for Smooth Four-Manifolds*, Preprint, math.SG/0110169.

84. P. S. Ozsváth and Z. Szabó, *Holomorphic disks and three-manifold invariants: properties and applications*, math.SG/0105202. [To appear in *Ann. Math.*]
85. P. S. Ozsváth and Z. Szabó, *Holomorphic disks and topological invariants for closed three-manifolds*, math.SG/0101206. [To appear in *Ann. Math.*]
86. P. S. Ozsváth and Z. Szabó, *Knot Floer homology and the four-ball genus*, Geom. Topol. **7** (2003), 615-639.
87. O. Plamenevskaya, *Bounds for Thurston–Bennequin Number from Floer Homology*, Preprint, math.SG/0311090, 2003.
88. J. A. Rasmussen, *Floer Homology and Knot Complements*, Preprint, math.GT/0306378, 2003.
89. J. A. Rasmussen, *Khovanov Homology and the Slice Genus*, Preprint, 2003.
90. J. A. Rasmussen, *Floer homology of surgeries on two-bridge knots*, Algebr. Geom. Topol. **2** (2002), 757–789 (electronic).
91. J. Robbin and D. Salamon, *The Maslov index for paths*, Topology **32** (1993), no. 4, 827–844.
92. L. Rudolph, *Quasipositivity as an obstruction to sliceness*, Bull. Am. Math. Soc. (N.S.), **29** (1993), no. 1, 51–59.
93. R. Rustamov, *Calculation of Heegaard Floer Homology for a Class of Brieskorn Spheres*, Preprint, math.SG/0312071, 2003.
94. D. Salamon, *Lagrangian intersections, 3-manifold with boundary, and the Atiyah-Floer conjecture*, in Proc. Intern. Congress Math., Birkhäuser, 1994, pp. 526–536.
95. N. Saveliev, *Lectures on the Topology of 3-Manifolds*, Walter de Gruyter & Co., Berlin, 1999.
96. P. Seidel, *Vanishing cycles and mutation*, in: European Congress Math. Vol. II (Barcelona, 2000), Birkhäuser, Basel, 2001, pp. 65–85,
97. P. Seidel, *The symplectic Floer homology of a Dehn twist*, Math. Res. Lett. **3**, (1996), no. 6, 829–834.
98. P. Seidel, *A long exact sequence for symplectic Floer cohomology*, Topology **42** (2003), no. 5, 1003–1063.
99. J. Singer, *Three-dimensional manifolds and their Heegaard diagrams*, Trans. Am. Math. Soc., **35** (1933), no. 1, 88–111.
100. C. H. Taubes, *Casson's invariant and gauge theory*, J. Differ. Geom., **31** (1990), no. 2, 547–599.
101. C. H. Taubes, *Seiberg Witten and Gromov invariants for symplectic 4-manifolds*, International Press, 2000.
102. C. H. Taubes, *Seiberg-Witten invariants, self-dual harmonic 2-forms and the Hofer-Wysocki-Zehnder formalism*, in: Surveys in Differential Geometry, Int. Press, Somerville, MA, 2000, pp. 625–672.
103. C. H. Taubes, *Seiberg-Witten invariants and symplectic forms*, Math. Res. Lett. **1** (1994), no. 6, 809–822.
104. C. H. Taubes, *More constraints on symplectic forms from Seiberg-Witten invariants*, Math. Res. Lett. **2** (1995), no. 1, 9–13.
105. V. Turaev, *Torsion invariants of Spin^c-structures on 3-manifolds*, Math. Res. Lett. **4** (1997), 679–695.

106. K. Wehrheim, *Anti-self-dual instantonts with Lagrangian boundary conditions*, Swiss Federal Institute of Technology, Zürich, 2002.
107. E. Witten, *Monopoles and four-manifolds*, Math. Res. Lett. **1** (1994), 769–796.

Exact Lagrangian Submanifolds in T^*S^n and the Graded Kronecker Quiver

Paul Seidel
S. University Ave
Chicago, USA

The topology of Lagrangian submanifolds gives rise to some of the most basic, and also most difficult, questions in symplectic geometry. We have many tools that can be brought to bear on these questions, but each one is effective only in a special class of situations, and their interrelation is by no means clear. The present paper is a piece of shamelessly biased propaganda for a relatively obscure approach, using Fukaya categories. We will test-ride this machinery in a particularly simple case, where the computations are very explicit, and where the outcome can be nicely compared to known results. For clarity, the level of technical sophistication has been damped down a little, and therefore the resulting theorem is not quite the best one can get. Also, in view of the expository nature of the paper, we do not take the most direct path to the conclusion, but instead choose a more scenic route bringing the reader past some classical questions from linear algebra.

***Different Faces of Geometry*, Edited by Donaldson *et al.*,**
Kluwer Academic/Plenum Publishers, New York, 2004

1. Introduction

Take $M = T^*S^n$, $n \geqslant 2$, with its canonical symplectic structure. We will consider compact connected Lagrangian submanifolds $L \subset M$ which satisfy $H^1(L) = 0$ and whose second Stiefel-Whitney class $w_2(L)$ vanishes. Note that such an L is automatically orientable (because $w_1(L)$ is the reduction of the integer-valued Maslov class, which must vanish).

Theorem 1. *For such L,*

1. $[L] = \pm[S^n] \in H_n(M)$;
2. $H^*(L;\mathbb{C}) \cong H^*(S^n;\mathbb{C})$;
3. $\pi_1(L)$ *has no nontrivial finite-dimensional complex representations;*
4. *if $L' \subset M$ is another Lagrangian submanifold satisfying the same conditions, then $L \cap L' \neq \varnothing$.*

As mentioned above, the problem of exact Lagrangian submanifolds in M and more generally in cotangent bundles has been addressed previously by several authors, and their results overlap substantially with Theorem 1. Part (4), with somewhat weakened assumptions, was proved by Lalonde-Sikorav [**12**] in one of the first papers on the subject, which is still well worth reading today. For odd n, Buhovsky [**3**] proves a statement essentially corresponding to (2) (he does not assume $w_2(L) = 0$, and works with cohomology with $\mathbb{Z}/2$ coefficients; in fact, his result can be used to remove that assumption from our theorem for odd n). Viterbo's work [**19, 18**] has some relevant implications, for instance he proved that there is no exact Lagrangian embedding of a $K(\pi,1)$-manifold in T^*S^n. Finally, in the lowest nontrivial dimension $n = 2$ much sharper results are known. Viterbo's theorem implies that any exact $L \subset T^*S^2$ must be a two-sphere; Eliashberg-Polterovich [**4**] showed that such a sphere is necessarily differentiably isotopic to the zero-section, and this was improved to Lagrangian isotopy by Hind [**9**]. Therefore, the new parts of Theorem 1 would seem to be (1), (2) for even $n \geqslant 4$, and (3) in so far as it goes beyond Viterbo's results. These overlaps are quite intriguing, as the arguments used in proving them are widely different (even though all of the higher-dimensional methods involve Floer homology in some way).

We close with some forward-looking observations. Both Buhovsky's and Viterbo's approaches should still admit some technical refinements, as does ours, and one can eventually hope to prove that any exact

$L \subset T^*S^n$ is homeomorphic to S^n. In contrast, the question of whether L needs to be diffeomorphic to the standard sphere is hard to attack, and issues about its Lagrangian isotopy class seem to be entirely beyond current capabilities (for $n > 2$). In the course of proving Theorem 1, we will show that any submanifold L satisfying its assumptions is "Floer-theoretically equivalent" (in rigorous language, isomorphic in the Donaldson-Fukaya category) to the zero-section. This is a much weaker equivalence relation than Lagrangian isotopy, but still sufficiently strong to imply properties (1) and (4) above.

2. The Donaldson–Fukaya Category

In the first place, Fukaya categories provide a convenient way of packaging the information obtained from Lagrangian Floer cohomology groups. We will find it convenient to include noncompact Lagrangian submanifolds with a fixed behaviour at infinity, so we choose a point on S^n and consider the corresponding cotangent fibre $F_0 \subset M$. The objects of the Donaldson-Fukaya category $H^0\mathfrak{F}(M)$ are constructed from connected Lagrangian submanifolds $L \subset M$ subject to the following restrictions:

- L is either compact, or else agrees with F_0 outside a compact subset.
- For any smooth disc with boundary on L, $u : (D, \partial D) \to (M, L)$, the symplectic area $\int u^*\omega$ vanishes;
- Similarly, the Maslov number of any disc u, which is a relative Chern class $\int u^*(2c_1(M, L)) \in \mathbb{Z}$, must vanish. As remarked above, this implies orientability of L;
- $w_2(L) = 0$.

To be more precise, the objects are Lagrangian branes $L^\flat$, which means Lagrangian submanifolds L as described above coming with certain choices of additional data:

- A grading $\tilde{\alpha}_L$, which is a lift of the Lagrangian phase function $\alpha_L : L \to S^1$ to a real-valued function. This exists because of the zero Maslov condition, but clearly there are infinitely many different choices, differing by integer constants. We write $L^\flat[k]$ for the brane obtained by shifting the grading down by k.
- A spin structure on L, and a flat complex vector bundle ξ_L. These are actually coupled, in the sense that changing the spin structure by an element of $H^1(L; \mathbb{Z}/2)$, and simultaneously tensoring ξ_L with

the corresponding complex line bundle (with monodromy ± 1), does not change our Lagrangian brane.

This list is no doubt baffling to the non-specialist reader, but its purpose is just to make the Lagrangian Floer cohomology groups well-defined and as nicely behaved as possibly. We need these to provide the rest of the category structure, namely, the morphisms between two objects are the Floer cohomology groups in degree zero, with twisted coefficients in our flat vector bundles:

$$\mathrm{Hom}_{H^0\mathcal{F}(M)}(L_0^\flat, L_1^\flat) = HF^0(L_0^\flat, L_1^\flat),$$

and composition of morphisms is provided by the pair-of-pants (or more appropriately pseudo-holomorphic triangle) product

$$HF^0(L_1^\flat, L_2^\flat) \otimes HF^0(L_0^\flat, L_1^\flat) \longrightarrow HF^0(L_0^\flat, L_2^\flat). \tag{1}$$

Some basic reminders about Floer cohomology are in order. First of all, one can recover the full Floer groups by considering morphisms into shifted objects, $\mathrm{Hom}\,(L_0^\flat, L_1^\flat[k]) = HF^k(L_0^\flat, L_1^\flat)$. The endomorphism group of any brane equipped with the trivial line bundle is its usual cohomology

$$HF^*(L^\flat, L^\flat) \cong H^*(L;\mathbb{C}) \quad \text{for} \quad \xi_L \cong \mathbb{C} \times L. \tag{2}$$

More generally, if we take L with a fixed grading and spin structure, and equip it with two different flat vector bundles ξ_L, ξ_L', the resulting two branes satisfy

$$HF^*(L^\flat, L^{\flat\prime}) \cong H^*(L; \xi_L^* \otimes \xi_L'). \tag{3}$$

In our arguments, three simple Lagrangian submanifolds in M will be prominent. One is the zero-section Z, which we make into a brane by choosing some grading and equipping it with the trivial line bundle. The second is the fibre F_0, treated in the same way, and the third is the image $F_1 = \tau_Z(F_0)$ under the Dehn twist τ_Z, which inherits a brane structure from F_0.

Lemma 2. 1. *The groups $HF^*(Z^\flat, F_k^\flat)$, $HF^*(F_k^\flat, Z^\flat)$ for $k = 0, 1$ are all one-dimensional;*
2. $HF^*(F_0^\flat, F_1^\flat) \cong H^*(S^{n-1};\mathbb{C})$;
3. $HF^*(F_1^\flat, F_0^\flat) = 0$.

Part (1) is obvious, because the Lagrangian submanifolds intersect in a single point. For the rest, one needs to remember that Floer cohomology of a pair of noncompact Lagrangian submanifolds is defined by moving the first one slightly by the normalized geodesic flow φ_t, which is the Hamiltonian flow of the function H with $H(\xi) = |\xi|$ outside a compact subset. As a consequence, the intersections

$$\varphi_t(L_0) \cap L_1, \quad \text{for } t > 0 \text{ small}$$

are relevant for computing $HF^*(L_0^\flat, L_1^\flat)$. In the case of (F_0, F_1), the effect of this perturbation is make the submanifolds intersect cleanly along an S^{n-1}, and standard Bott-Morse methods yield (2). For (F_1, F_0), in contrast, the perturbation will make them disjoint, so (3) follows.

References. All that we have discussed belongs to the basics of Floer homology theory. Gradings and twisted coefficients were both introduced in [**10**]; for the use of spin structures see [**6**]; for the product (1) see [**5**] (these are by no means the only possible references).

3. Triangulated Categories

A triangulated category lurks wherever long exact sequences of (any kind of) cohomology groups appear. The axioms of exact triangles formalize some nontrivial properties of such sequences, thus allowing one to manipulate them more efficiently. We will give an informal description of the axioms, which is neither exhaustive nor totally rigorous, but which suffices for our purpose. Let $\mathcal{C}$ be a category in which the morphism spaces $\mathrm{Hom}\,(X_0, X_1)$ between any two objects are complex vector spaces. Having a triangulated structure on $\mathcal{C}$ gives one certain ways of constructing new objects out of old ones. First of all, we assume that $\mathcal{C}$ is additive, which means that one can form the direct sum $X_0 \oplus X_1$ of two objects, with the expected properties. Secondly, one can shift or translate an object by an integer amount, $X \mapsto X[k]$, which allows one to define higher degree morphism spaces $\mathrm{Hom}^k(X_0, X_1) = \mathrm{Hom}\,(X_0, X_1[k])$. We will denote the direct sum of all these spaces by $\mathrm{Hom}^*(X_0, X_1)$. As a side-remark which will be useful later, note that by using direct sums and shifts, one can define the tensor product $V \otimes X$ of any object X with a finite-dimensional graded complex vector space V: choose a homogeneous basis v_j for V, and set

$$V \otimes X = \bigoplus_j V[-\deg(v_j)], \tag{4}$$

which is easily seen to be independent of the choice of basis up to isomorphism. The final and most important construction procedure for objects is the following: to any morphism $a : X_0 \to X_1$ one can associate a new object, the mapping cone $\mathrm{Cone}\,(a)$, which is unique up to isomorphism. This can be interpreted as measuring the failure of a to be an isomorphism: at one extreme, if $a = 0$ then $\mathrm{Cone}\,(a) = X_0[1] \oplus X_1$, and on the other hand, a is an isomorphism iff $\mathrm{Cone}\,(a)$ is the zero object.

Mapping cones come with canonical maps $i : X_1 \to \mathrm{Cone}\,(a)$, $\pi : \mathrm{Cone}\,(a) \to X_0[1]$, such that the composition of any two arrows in the diagram

$$X_0 \xrightarrow{a} X_1 \xrightarrow{i} \mathrm{Cone}\,(a) \xrightarrow{\pi} X_0[1] \xrightarrow{a[1]} X_1[1] \tag{5}$$

is zero. By applying $\mathrm{Hom}\,(Y,-)$ or $\mathrm{Hom}\,(-,Y)$ for some object Y one gets long exact sequences of vector spaces,

$$\begin{aligned} &\cdots \mathrm{Hom}^k(Y, X_0) \to \mathrm{Hom}^k(Y, X_1) \to \mathrm{Hom}^k(Y, \mathrm{Cone}\,(a)) \cdots \\ &\cdots \mathrm{Hom}^k(X_0, Y) \leftarrow \mathrm{Hom}^k(X_1, Y) \leftarrow \mathrm{Hom}^k(\mathrm{Cone}\,(a, Y)) \cdots \end{aligned} \tag{6}$$

which is what we were talking about at the beginning of the section. Diagrams of the form (5) (and isomorphic ones) are called exact triangles, and drawn rolled up like this:

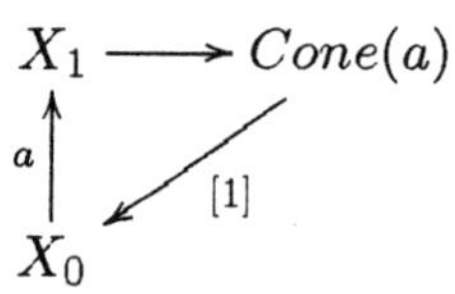

where the [1] reminds us that this arrow is really a morphism to $X_0[1]$. One remarkable thing is that triangles can be rotated:

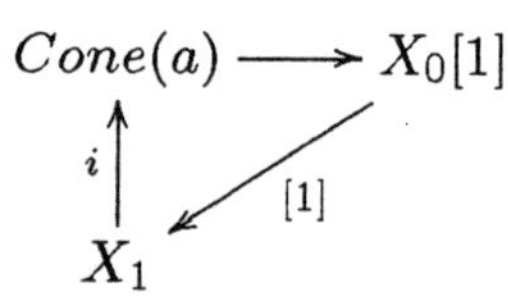

is again an exact triangle, which means that $X_0[1] \cong \mathrm{Cone}\,(i)$, and similarly $X_1[1] \cong \mathrm{Cone}\,(\pi)$. We know that the cone of the zero morphism is a direct sum, and therefore in (5)

$$\begin{aligned} a = 0 &\Longrightarrow \mathrm{Cone}\,(a) \cong X_0[1] \oplus X_1, \\ i = 0 &\Longrightarrow X_0 \cong X_1 \oplus \mathrm{Cone}\,(a)[-1], \\ \pi = 0 &\Longrightarrow X_1 \cong \mathrm{Cone}\,(a) \oplus X_0. \end{aligned} \tag{7}$$

Loosely speaking, the formalism of triangulated categories puts the boundary operator in long exact sequences on the same footing as the other two maps. To round off our picture, we look at the situation where one has two morphisms

$$X_0 \xrightarrow{a} X_1 \xrightarrow{b} X_2.$$

Since the mapping cone measures the failure of a map to be an isomorphism, it seems intuitive that these defects should somehow add up under composition, and indeed there is an exact triangle

$$\begin{array}{ccc} Cone(a) & \longrightarrow & Cone(ba) \\ \uparrow & \swarrow_{[1]} & \\ Cone(b)[-1]. & & \end{array} \tag{8}$$

This is part of the "octahedral axiom" of triangulated structures (the name comes from a more complicated diagram, which describes various compatibility conditions between the maps in (8) and those in the triangles defining Cone (a), Cone (b)).

The following simple construction arose first in algebraic geometry as part of the theory of mutations. Suppose that $\mathcal{C}$ is a triangulated category in which the graded vector spaces $V = \mathrm{Hom}^*(X,Y)$ are finite-dimensional. One can define the tensor products $V \otimes X$ and $V^\vee \otimes Y$ as in (4), and these come with canonical evaluation morphisms $ev : \mathrm{Hom}^*(X,Y) \otimes X \to Y$, $ev' : X \to \mathrm{Hom}^*(X,Y)^\vee \otimes Y$. We define $T_X(Y) = \mathrm{Cone}\,(ev)$, $T'_Y(X) = \mathrm{Cone}\,(ev')[-1]$. By definition, this means that $T_X(Y)$ sits in an exact triangle

$$\begin{array}{ccc} Y & \longrightarrow & T_X(Y) \\ {\scriptstyle ev}\uparrow & \swarrow_{[1]} & \\ Hom^*(X,Y) \otimes X & & \end{array}$$

and similarly for T'_X. Unfortunately, the axioms of triangulated categories are not quite strong enough to make T_X, T'_X into actual functors. Still, these operations can be shown to have nice behaviour, such as taking direct sums to direct sums, and cones to cones (up to isomorphism).

We will now come to the geometric interpretation of T, T'. Contrary to what the reader may have hoped, the Donaldson-Fukaya category $H^0\mathcal{F}(M)$ is not triangulated. It does have a shift operation with the correct properties, namely the change of grading for a Lagrangian brane, but neither direct sums not cones exist in general. It is an open question whether one can define a triangulated version of this category by geometric means, such as including Lagrangian submanifolds with self-intersections. Meanwhile, there is a purely algebraic construction in terms of twisted complexes, which yields a triangulated category $D^b\mathcal{F}(M)$ containing $H^0\mathcal{F}(M)$ as a full subcategory. The gist of this is simply to add on new objects in a formal way, so that the requirements of a triangulated category are satisfied. The details are slightly less straightforward than this description may suggest, and involve the use of Fukaya's higher order (A_∞) product structures on Floer cohomology, but for the purposes of this paper, all we need is the knowledge that $D^b\mathcal{F}(M)$ exists, and the following fact:

Theorem 3. *For any object $L^\flat$ of $H^0\mathcal{F}(M)$, the "algebraic twist" $T_{Z^\flat}(L^\flat)$ and the "geometric (Dehn) twist" $\tau_Z(L^\flat)$ are isomorphic objects of $D^b\mathcal{F}(M)$. In the same vein, one has $T'_{Z^\flat}(L^\flat) \cong \tau_Z^{-1}(L^\flat)$.* □

References. [**7**] is an accessible presentation of the abstract theory of triangulated categories. For mutations see the papers in [**14**]. The construction of $D^b\mathcal{F}(M)$ is outlined in [**10**]. The long exact sequence in Floer cohomology which is a consequence of Theorem 3 was introduced in [**17**]. The argument given there can easily be adapted to prove the result as stated, see [**16**] for an explanation.

4. The Graded Kronecker Quiver

We will now switch gears slightly. The representation theory of quivers is a subject with deceptively humble appearance. Superficially, it is no more than a convenient way of reformulating certain questions in linear algebra, but in many cases these reveal themselves to be equivalent to much less elementary problems in other areas, such as algebraic geometry. We will only need a very simple instance of the theory, namely the following graded Kronecker quiver, with nonzero $d \in \mathbb{Z}$:

$$\bullet \underset{d}{\overset{0}{\rightrightarrows}} \bullet \tag{9}$$

By definition, a representation of (9) consists of two finite-dimensional graded $\mathbb{C}$-vector spaces V, W, together with linear maps $\alpha : V \rightarrow W$ of degree zero and $\beta : V \rightarrow W[d]$ of degree d, respectively. Two representations (V, W, α, β) and $(V', W', \alpha', \beta')$ are isomorphic if there are graded linear isomorphisms $\varphi : V \rightarrow V'$, $\psi : W \rightarrow W'$ such that $\alpha'\varphi = \psi\alpha$, $\beta'\varphi = \psi\beta$. Of course, any representation splits into a direct sum of indecomposable ones. The Kronecker quiver is nice in that the latter can be classified explicitly:

Proposition 4. *Any indecomposable representation is isomorphic, up to a common shift in the gradings of both vector spaces, to one of the following:*

- $(\mathcal{O}_X(k),\ k < 0)$*:* $\dim(V) = -k$, $\dim(W) = -k - 1$,

$$V = \mathbb{C} \oplus \mathbb{C}[-d] \oplus \cdots \oplus \mathbb{C}[(k-1)d],$$

$$\alpha = \begin{pmatrix} 0 & 1 & & \\ & & 1 & \\ & & & \ddots & \\ & & & & 1 \end{pmatrix} \Bigg\downarrow\Bigg\downarrow \beta = \begin{pmatrix} 1 & & & \\ & 1 & & \\ & & \ddots & \\ & & & 1 & 0 \end{pmatrix}$$

$$W = \mathbb{C}[-d] \oplus \mathbb{C}[-2d] \oplus \cdots \oplus \mathbb{C}[(k-1)d]$$

- $(\mathcal{O}_X(k),\ k \geqslant 0)$*:* $\dim(V) = k$, $\dim(W) = k + 1$,

$$V = \mathbb{C}[d] \oplus \mathbb{C}[2d] \oplus \cdots \oplus \mathbb{C}[kd]$$

$$\alpha = \begin{pmatrix} 0 & & & \\ 1 & & & \\ & 1 & & \\ & & \ddots & \\ & & & 1 \end{pmatrix} \Bigg\downarrow\Bigg\downarrow \beta = \begin{pmatrix} 1 & & & \\ & 1 & & \\ & & \ddots & \\ & & & 1 \\ & & & 0 \end{pmatrix}$$

$$W = \mathbb{C} \oplus \mathbb{C}[d] \oplus \cdots \oplus \mathbb{C}[kd]$$

- $(\mathcal{O}_X/\mathcal{J}^k_{X,0},\ k \geqslant 1)$: $\dim(V) = \dim(W) = k$,

$$V = \mathbb{C}[d] \oplus \mathbb{C}[2d] \oplus \cdots \oplus \mathbb{C}[kd]$$

$$\alpha=\begin{pmatrix}0&&&\\1&&&\\&1&&\\&&\ddots&\\&&1&0\end{pmatrix}\ \Big\downarrow\Big\downarrow\ \beta=\begin{pmatrix}1&&&\\&1&&\\&&\ddots&\\&&&1\end{pmatrix}$$

$$W = \mathbb{C} \oplus \mathbb{C}[d] \oplus \cdots \oplus \mathbb{C}[(k-1)d]$$

- $(\mathcal{O}_X/\mathcal{J}^k_{X,\infty},\ k \geqslant 1)$: $\dim(V) = \dim(W) = k$,

$$V = \mathbb{C} \oplus \mathbb{C}[d] \oplus \cdots \oplus \mathbb{C}[(k-1)d],$$

$$\alpha=\begin{pmatrix}1&&&\\&1&&\\&&\ddots&\\&&&1\end{pmatrix}\ \Big\downarrow\Big\downarrow\ \beta=\begin{pmatrix}0&1&&\\&&1&\\&&&\ddots\\&&&1\\&&&0\end{pmatrix}$$

$$W = \mathbb{C} \oplus \mathbb{C}[d] \oplus \cdots \oplus \mathbb{C}[(k-1)d]$$

This is actually simple enough to be tackled using only linear algebra, but the meaning of the classification becomes much clearer in algebro-geometric terms. To any representation (V, W, α, β) of (9) one can associate a map of algebraic vector bundles on $X = \mathbb{CP}^1$,

$$\mathcal{O}_X(-1) \otimes V \xrightarrow{\alpha x + \beta y} \mathcal{O}_X \otimes W. \tag{10}$$

Take the $\mathbb{C}^*$-action on $\mathbb{C}^2$ given by $\zeta \cdot (x, y) = (x, \zeta^{-d} y)$, and the induced action on X. Then the sheaves $\mathcal{O}_X, \mathcal{O}_X(-1)$ are naturally equivariant, and if one equips V, W with the $\mathbb{C}^*$-actions whose weights are given by the grading, $\alpha x + \beta y$ becomes an equivariant map. At this point, the usual procedure would be to look at the kernel and cokernel of (10), and to use Grothendieck's splitting theorem for holomorphic vector bundles on $\mathbb{CP}^1$ to derive Proposition 4, but we prefer a sleeker and more high-tech approach using triangulated categories. The category $\mathrm{Coh}_{\mathbb{C}^*}(X)$ of equivariant coherent sheaves admits a full embedding into a triangulated category, its derived category $D^b\,\mathrm{Coh}_{\mathbb{C}^*}(X)$. One can look at the cone of (10) as an object $\mathcal{E}$ in that category. A computation using the long exact sequences (6) shows that the morphisms between two such cones are precisely the homomorphisms of quiver representations, which means that the cone construction embeds the category of representations of our quiver as a full subcategory into $D^b \operatorname{Coh}_{\mathbb{C}^*}(X)$. In particular, indecomposable representations must give rise to indecomposable objects $\mathcal{E}$.

Grothendieck's theorem extends easily to the derived category and to the equivariant case: each indecomposable object of $D^b \operatorname{Coh}_{\mathbb{C}^*}(X)$ is, up to a shift, either a line bundle $\mathcal{O}_X(k)$, or else a torsion sheaf of the form $\mathcal{O}_X/\mathcal{J}^k_{X,p}$, where $p \in \{0, \infty\}$ is one of the two fixed points $0 = [0:1]$ or $\infty = [1:0]$ of the $\mathbb{C}^*$-action. Going back to the objects $\mathcal{E}$ constructed above, one finds that only two essentially different possibilities can occur. One is that $\mathcal{E}[-1] \cong \mathcal{O}_X(k)$ with $k < 0$, in which case the map (10) is surjective, with kernel $\mathcal{E}[-1]$. The usual long exact sequence in sheaf cohomology shows that one can recover the representation from $\mathcal{E}$ as follows:

$$V \cong H^1(\mathcal{E}[-1] \otimes \mathcal{O}_X(-1)), \quad W \cong H^1(\mathcal{E}[-1]),$$

and the maps α, β are the (Yoneda) products with the standard generators of $H^0(\mathcal{O}_X(1))$. The other case is where $\mathcal{E}$ is either $\mathcal{O}_X(k)$ with $k \geqslant 0$, or a torsion sheaf; then (10) is surjective, with cokernel $\mathcal{E}$, and this time one finds that

$$V \cong H^0(\mathcal{E} \otimes \mathcal{O}_X(-1)), \quad W \cong H^0(\mathcal{E}).$$

A straightforward computation of cohomology groups identifies the various $\mathcal{E}$ with the corresponding cases in Proposition 4, thereby concluding our proof of that result.

References. The analogue of Proposition 4 for the ungraded quiver is due to Kronecker, and is explained in textbooks on the representation theory of finite-dimensional algebras [**1, 2**]. The connection with coherent sheaves is well-known. The fact that an indecomposable object of the derived category is actually a single sheaf is a general property of abelian categories of homological dimension one, and is used extensively in papers about mirror symmetry on elliptic curves [**13, 11**]. Grothendieck's paper is [**8**].

5. Proof of Theorem 1

Consider the three basic Lagrangian branes $Z^\flat$, $F_0^\flat$ and $F_1^\flat$ as objects in $D^b\mathcal{F}(M)$. Because F_1 is the image of F_0 under τ_Z, Theorem 3 can be applied, and we use this to prove:

Lemma 5. *$T_{F_0^\flat} T_{F_1^\flat}(Z^\flat)$ is the zero object.*

PROOF. From Lemma 2(1) we know that $HF^*(F_1^\flat, Z^\flat)$ is one-dimensional. For simplicity, assume that the gradings have been chosen in such a way that the nontrivial Floer group lies in degree zero, and denote a nonzero element of it by c. Using the definitions of T, T', and Theorem 3, one gets

$$T_{F_1^\flat}(Z^\flat) = \text{Cone}\,\left(F_1^\flat \xrightarrow{c} Z^\flat\right) = T'_{Z^\flat}(F_1^\flat)[1] \cong \tau_Z^{-1}(F_1^\flat)[1] \cong F_0^\flat[1].$$

By (2) $HF^*(F_0^\flat, F_0^\flat) \cong H^*(F_0; \mathbb{C}) \cong \mathbb{C}$, and therefore

$$T_{F_0^\flat}(F_0^\flat) = \text{Cone}\,\left(F_0^\flat \xrightarrow{id} F_0^\flat\right) = 0. \quad \square$$

Take a Lagrangian submanifold $L \subset M$ satisfying the assumptions of Theorem 1. Because $H^1(L; \mathbb{C}) = 0$, L is automatically exact and has zero Maslov class, so we can make it into an object $L^\flat$ of $H^0\mathcal{F}(M)$ by choosing a grading and spin structure, as well as the trivial line bundle. Near Z, τ_Z is equal to the antipodal involution, and so τ_Z^2 is equal to the identity. By an isotopy along the Liouville (compressing) flow, one can move L arbitrarily close to Z, so $\tau_Z^2(L)$ is Lagrangian isotopic to L. When one considers the gradings on both submanifolds, however, there is a difference:

$$\tau_Z^2(L^\flat) \sim L^\flat[2-2n]. \tag{11}$$

From Theorem 3 and the definition of T as a cone, we get exact triangles

$$\begin{array}{ccc} L^\flat & \longrightarrow & \tau_Z(L^\flat) \\ \uparrow & \overset{[1]}{\swarrow} & \\ HF^*(Z^\flat, L^\flat) \otimes Z^\flat & & \end{array}$$

and

$$\begin{array}{ccc} \tau_Z(L^\flat) & \longrightarrow & \tau_Z^2(L^\flat) \cong L^\flat[2-2n] \\ \uparrow & \overset{[1]}{\swarrow} & \\ HF^*(Z^\flat, \tau_Z(L^\flat)) \otimes Z^\flat & & \end{array}$$

The bottom term can be simplified by noticing that $HF^*(Z^\flat, \tau_Z(L^\flat)) \cong HF^*(\tau_Z^{-1}(Z^\flat), L^\flat) \cong HF^*(Z^\flat[n-1], L^\flat) \cong HF^{*+1-n}(Z^\flat, L^\flat)$. Using the octahedral axiom (8), the two exact triangles can be spliced together to a single one,

$$\begin{array}{ccc} L^\flat & \longrightarrow & L^\flat[2-2n] \\ \uparrow & \overset{[1]}{\swarrow} & \\ Cone\big(HF^*(Z^\flat, L^\flat)[-n] \otimes Z^\flat \to HF^*(Z^\flat, L^\flat) \otimes Z^\flat\big) & & \end{array}$$

The $\rightarrow$ must be given by an element of $HF^{2-2n}(L^\flat, L^\flat)$, but because of (2) the Floer cohomology groups vanish in negative degrees. Hence the morphism is necessarily zero, and as explained in (7),

$$L^\flat \oplus L^\flat[1-2n] \cong \text{Cone}\ \big(HF^*(Z^\flat, L^\flat)[-n] \otimes Z^\flat \rightarrow HF^*(Z^\flat, L^\flat) \otimes Z^\flat\big). \tag{12}$$

Lemma 6. *$T_{F_0^\flat} T_{F_1^\flat}(L^\flat)$ is the zero object.*

PROOF. From Lemma 5 we know that $T_{F_0^\flat} T_{F_1^\flat}$ annihilates $Z^\flat$. Since it carries direct sums to direct sums and cones to cones, this operation also annihilates the right hand side of (12), hence $L^\flat$. □

Lemma 6 gives us a powerful hold on the a priori unknown object $L^\flat$. Namely, by putting together the two exact triangles coming from the definition of T (we omit the details, since they are parallel to the computation carried out above) one finds that

$$L^\flat \cong \text{Cone}\ \big(\text{Hom}^*(F_0^\flat, T_{F_1^\flat}(L^\flat))[-1] \otimes F_0^\flat \rightarrow \text{Hom}^*(F_1^\flat, L^\flat) \otimes F_1^\flat\big). \tag{13}$$

Hence, the isomorphism class of $L^\flat$ as an object of the Fukaya category is determined by two finite-dimensional graded vector spaces

$$V = \text{Hom}^*(F_0^\flat, T_{F_1^\flat}(L^\flat))[-1], \quad W = \text{Hom}^*(F_0^\flat, L^\flat)$$

and the arrow in (13), which is an element $x \in \text{Hom}^*_{\mathbb{C}}(V, W) \otimes HF^*(F_0^\flat, F_1^\flat)$ of degree zero. We computed the relevant Floer cohomology group in Lemma 2(2). After choosing generators a of degree zero and b of degree $n-1$, one writes $x = \alpha \otimes a + \beta \otimes b$ where $\alpha : V \rightarrow W$ is a linear map of degree zero, and $\beta : V \rightarrow W$ one of degree $1-n$. We see that (V, W, α, β) is a representation of the graded Kronecker quiver with

$$d = 1 - n < 0. \tag{14}$$

If this representation was decomposable, it would give rise to a corresponding decomposition of the object $L^\flat$ into direct summands in $D^b\mathcal{F}(M)$, but that is impossible since $\mathrm{Hom}(L^\flat, L^\flat) = HF^0(L^\flat, L^\flat) = H^0(L^\flat; \mathbb{C}) = \mathbb{C}$. More generally, the long exact sequences (6) applied to the cone (13) show that $H^*(L;\mathbb{C}) \cong HF^*(L^\flat, L^\flat)$ is the cohomology of the two-step complex

$$C = \Big\{0 \to \mathrm{Hom}_{\mathbb{C}}(V,V) \oplus \mathrm{Hom}_{\mathbb{C}}(W,W) \longrightarrow \xrightarrow{\begin{pmatrix} -\alpha\circ\cdot & \cdot\circ\alpha \\ -\beta\circ\cdot & \cdot\circ\beta \end{pmatrix}} \mathrm{Hom}_{\mathbb{C}}(V,W) \oplus \mathrm{Hom}_{\mathbb{C}}(V,W)[d] \to 0\Big\} \quad (15)$$

This is actually a complex of graded vector spaces, so its cohomology $H^*(C)$ is bigraded, and one obtains $H^*(L;\mathbb{C})$ by summing up the two gradings. One can compute $H^*(C)$ directly for each of the cases in Proposition 4, or alternatively, one can go back to our algebro-geometric proof of that result and note that $H^*(C)$ is equal to $Ext^*_X(\mathcal{E},\mathcal{E})$ for the associated sheaf $\mathcal{E}$ (which is bigraded by the cohomological degree and $\mathbb{C}^*$-action). In either way, one sees that

- in the case where the representation is of type $\mathcal{O}_X(k)$, $H^*(C)$ is one-dimensional and concentrated in degree zero;
- for $\mathcal{O}_X/\mathcal{J}^k_{X,0}$, $H^*(C)$ is $2k$-dimensional, with generators in degrees $0, -d, \dots, (1-k)d$ and $d+1, 2d+1, \dots, kd+1$.
- for $\mathcal{O}_X/\mathcal{J}^k_{X,\infty}$, $H^*(C)$ is $2k$-dimensional, with generators in degrees $0, d, \dots, (k-1)d$ and $1-d, 1-2d, \dots, 1-kd$.

Bearing in mind (14), one sees that $H^*(C)$ cannot be the cohomology of an n-dimensional oriented manifold except in one case, $\mathcal{O}_X/\mathcal{J}_{X,\infty}$. We have proved:

Lemma 7. *Up to a shift, $L^\flat$ is isomorphic to* $\mathrm{Cone}(a : F_0^\flat \to F_1^\flat)$ *in the derived Fukaya category.* □

Part (2) of Theorem 1 follows immediately, since in the case of $\mathcal{O}_X/\mathcal{J}_{X,\infty}$, $H^*(C) \cong H^*(S^n;\mathbb{C})$. Lemma 7 also implies that there is a long exact sequence

$$\cdots HF^*(F_1^\flat, F_1^\flat) \to HF^*(F_1^\flat, L^\flat) \to HF^{*+1}(F_1^\flat, F_0^\flat) \cdots$$

which in view of Lemma 2 shows that $HF^*(F_1^\flat, L^\flat)$ is one-dimensional. Because the Euler characteristic of Floer cohomology is the ordinary intersection number, it follows that $F_1 \cdot L = \pm 1$, which implies part (1) of Theorem 1. Next, if we had two Lagrangian submanifolds satisfying the conditions of that theorem, they would give rise to isomorphic objects in the Fukaya category by Lemma 7, hence their Floer cohomology would be equal to the ordinary cohomology of each. Since that is nonzero, the two Lagrangian submanifolds necessarily intersect, which is part (4).

The remaining statement (3) about the fundamental group is slightly more complicated. Take an indecomposable flat complex vector bundle ξ_L on L, and use that to define a brane $L^{\flat,'}$. From (3) we see that $HF^0(L^{\flat,'}, L^{\flat,'}) = H^0(\xi_L^* \otimes \xi_L)$ cannot contain any nontrivial idempotents; therefore the representation of (9) associated to $L^{\flat,'}$ must still be indecomposable. Again, one finds that $\mathcal{O}_X/\mathcal{I}_{X,\infty}$ is the only possibility, so $L^{\flat,'}$ is isomorphic to the previously considered brane $L^\flat$. Now applying (3) to this pair of branes, one finds that ξ_L must be the trivial line bundle.

References. Equation (11) is taken from [**15**].

Acknowledgments. I would like to thank Paul Biran for fruitful conversations, and Dan Quillen whose lectures introduced me to the Kronecker quiver.

References

1. M. Auslander, I. Reiten, and S. Smalo, *Representation theory of Artin algebras*, Cambridge Stud. Advanc. Math. **36**, Cambridge University Press, Cambridge, 1995.
2. D. Benson. *Representations and cohomology. I*, Cambridge Stud. Advanc. Math. **30**, Cambridge University Press, 1998.
3. L. Buhovsky, *Homology of Lagrangian Submanifolds of Cotangent Bundles*, Preprint, math.SG/0312265.
4. Ya. Eliashberg and L. Polterovich, *Unknottedkness of Lagrangian surfaces in symplectic 4-manifolds*, *Int. Math. Res. Notices* **11** (1993), 295–301.
5. K. Fukaya and Y.-G. Oh, *Zero-loop open strings in the cotangent bundle and Morse homotopy*, Asian J. Math. **1** (1998), 96–180.
6. K. Fukaya, Y.-G. Oh, H. Ohta, and K. Ono, *Lagrangian Intersection Floer Theory – Anomaly and Obstruction*, Preprint, 2000.
7. S. Gelfand and Yu. Manin, *Methods of Homological Algebra*, Springer, 1996.
8. A. Grothendieck, *Sur la classification des fibrés holomorphes sur la sphère de Riemann*, Amer. J. Math. **79** (1957), 121–138.
9. R. Hind, *Lagrangian Isotopies in Stein Manifolds*, Preprint, math.SG/0311093.
10. M. Kontsevich, *Homological algebra of mirror symmetry*, in: Proceedings of the International Congress of Mathematicians (Zürich, 1994), Birkhäuser, 1995, pp. 120–139.

11. B. Kreussler, *Homological Mirror Symmetry in Dimension One*, Preprint, math.AG/0012018.
12. F. Lalonde and J.-C. Sikorav, *Sous-variétés lagrangiennes et lagrangiennes exactes des fibrés cotangents*, Commun. Math. Helv. **66** (1991), 18–33.
13. A. Polishchuk and E. Zaslow, *Categorical mirror symmetry: the elliptic curve*, Adv. Theor. Math. Phys. **2** (1998), 443–470.
14. A. Rudakov et al. *Helices and vector bundles: Seminaire Rudakov*, LMS Lecture Note Series **148**, Cambridge University Press, 1990.
15. P. Seidel, *Graded Lagrangian submanifolds*, Bull. Soc. Math. France **128** (2000), 103–146.
16. P. Seidel, *Homological Mirror Symmetry for the Quartic Surface*, Preprint, math.SG/0310414, 2003.
17. P. Seidel, *A long exact sequence for symplectic Floer cohomology*, Topology **42** (2003), 1003–1063.
18. C. Viterbo, *Functors and Computations in Floer Homology with Applications, Part II,* Preprint, 1996.
19. C. Viterbo, *Exact Lagrangian submanifolds, periodic orbits and the cohomology of free loop space*, J. Differ. Geom. **47** (1997), 420–468.

The Construction Problem in Kähler Geometry

Carlos Simpson
Université de Nice-Sophia Antipolis
Nice, France

Dedicated to Alexander Reznikov

One of the most surprising things in algebraic geometry is the fact that algebraic varieties over the complex numbers benefit from a collection of metric properties which strongly influence their topological and geometric shapes. The existence of a Kähler metric leads to all sorts of Hodge theoretical restrictions on the homotopy types of algebraic varieties. On the other hand, a sparse collection of examples shows that the remaining liberty is nontrivially large. Paradoxically, with all of this information, the research field remains as wide open as it was many decades ago, because the gap between the known restrictions, and the known examples of what can occur, only seems to grow wider and wider the more closely we look at it.

In spite of the differential-geometric nature of the questions and methods, the origins of the situation are very algebraic. We look at subvarieties of projective space over the complex numbers. The main

***Different Faces of Geometry*, Edited by Donaldson *et al.*,**
Kluwer Academic/Plenum Publishers, New York, 2004

over-arching problem in algebraic geometry is to understand the classification of algebro-geometric objects. The topology of the usual complex-valued points of a variety plays an important role, because the topological type is a locally constant function on any classifying space. Thus the partitioning of the classification problem according to topological type provides a coarse, and more calculable, alternative to the partition by connected components. Furthermore, the topology of a variety strongly influences its geometric properties. A sociological observation is that the quest for understanding the topology of algebraic varieties has led to a rich set of techniques which found applications elsewhere, even in physics.

Perhaps a word about the choice of ground field is appropriate. One could also look at the shapes of the real points of real algebraic varieties. However, by Weierstrass approximation, pretty much anything can arise if you let the degree get big enough. Thus the classical question in this case is "which shapes can occur for a given degree?" This is so much more difficult that it constitutes a separate subject.[1] Taking the algebraically closed field of complexes means that we do not unwittingly leave any points out of the picture, so it seems reasonable to regard this as one possible canonical choice. Historically at least, and probably also for some philosophical reasons, investigations into the topology of complex varieties have in turn been mirrored in arithmetic algebraic geometry over other fields. A lot of what we say for complex Hodge theory could also apply to the p-adic versions.

It is a standard norm for mathematical papers, to discuss positive contributions to the collective knowledge. I would like to take advantage of the present opportunity, to focus on a zone where we have little or no knowledge, and where I have nothing new to report, in any case. I apologize therefore for the very vague nature of most of the discussion. I hope that the references[2] will give readers a place to look for whatever details are available. With our purpose stated this way, we should normally try to cover most of modern Kähler geometry. However, this would be unreasonable, because for many parts of the subject I would

---

[1] We can hope, however, that the intuition developed by workers in that field (do a recursive search on the references in [**28**]) could be of help in the complex case we are discussing here.

[2] In compiling the references, the online version of AMS Math Reviews was a great help in remembering, finding, organizing and writing them up (often by cut-and-paste). Readers having access to this service can greatly expand the amount

not have anything further to say than just repeating what would be in the references. So, toning down our ambitions a little bit, we get to the description of what I really would like to talk about, which is to isolate a certain family of questions which seem important, and about which very little seems to be known: how to construct algebraic (or Kähler) varieties with interesting topological behavior.

To get going, recall that projective space is endowed with the Fubini–Study metric, a hermitian metric which has the Kähler property which we now explain in two ways. The first way is to note that if you write the matrix for the hermitian metric in terms of a local system of holomorphic coordinates, then transform the tensor product between dz_i and $d\overline{z}_j$ to a wedge product, you get an alternating form ω. It is well-defined independently of the choice of coordinates. The Kähler condition is that the two-form ω is closed, i.e. it gives a symplectic structure. The second definition is to say that a metric is Kähler if at each point there exists a holomorphic system of coordinates in which the matrix for the metric is equal to the identity, up to an error term in $|z|^2$ (rather than just $|z|$ as would be the case for an arbitrary choice of coordinates). The good system is called an *osculating system of coordinates.* This second point of view leads directly to the famous *Kähler identities*

$$(D')^* = \sqrt{-1}[\Lambda, D''], \quad (D'')^* = -\sqrt{-1}[\Lambda, D']$$

where $D' = \partial$ and $D'' = \overline{\partial}$ are the $(1,0)$ and $(0,1)$-components of the de Rham differential, and Λ is the pointwise adjoint of the operation $\wedge\omega$.

The correspondence between hermitian metrics and $(1,1)$-forms is compatible with restriction to subvarieties, as is the Kähler condition $d\omega = 0$, so any smooth subvariety $X \subset \mathbf{P}^n$ inherits a Kähler metric too. This extrinsic description can be modified so as to seem more intrinsic, by saying that the Kähler metric is obtained as the Chern $(1,1)$-form of a positively curved metric on a line bundle L over X: in the projective case L is the pullback of the tautological bundle $\mathcal{O}_{\mathbf{P}^n}(1)$ on projective space. In a philosophical sense, a Kähler metric should

---

of referential information available, by searching starting with the authors cited in our bibliography. A crucial recent improvement has been the inclusion of forward searching to get future papers referring to a given paper. This results in such a vast amount of information that even our rather long reference list can only be considered as a sampling.

be thought of as corresponding to a curvature of a connection on a line bundle; this will only be precise if the cohomology class $[\omega]$ is integral. On the other hand, there may be algebraic varieties, or other more exotic things such as algebraic spaces, which do not have projective embeddings and so are not Kähler. Nonetheless, the main intention of the theory is to investigate the properties of projective varieties. There are some significant differences between the Kähler and the projective situations **[194]**, but in spite of those we will often mix things up, so that questions posed for one class of varieties might well be reinterpreted for any other related class.

When a variety X has a Kähler metric ω, the Kähler identities imply that the Laplacians for d and D'' are proportional. This leads[3] to the Hodge decomposition for cohomology (certainly when X is compact, and some results are possible in the noncompact case too **[63]** **[156]** **[183]** **[201]**). It gives us a first Hodge-theoretic result about the homotopy of X: if i is odd then $b_i(X) := \dim H^i(X, \mathbf{C})$ is even.[4] Johnson and Rees point out that this result carries over easily to the case of cohomology with coefficients in a local system with finite monodromy group **[104]**, and they use it to show that certain free products cannot occur. This was extended by Arapura **[11]** to cover certain amalgamated products, in a more elementary version of Gromov' general theorem that $\pi_1(X)$ can never be a nontrivial amalgamated product **[88]**.

Treating the noncompact case by compactifying, resolving singularities,[5] and then using the logarithmic de Rham complex, leads to mixed Hodge structures on the cohomology in this case **[63]** **[183]** **[201]** **[173]** **[172]** **[149]**. Combining with simplicial techniques gives mixed Hodge structures on singular varieties too.

A classical technique for analyzing the topology of algebraic varieties is *Lefschetz devissage* **[63]**. One can arrange (by a birational transformation) to have a fibration $X \to \mathbf{P}^1$ whose fibers are $n-1$-dimensional varieties, a finite number of them singular (that is to say more singular than the others). The topological properties of X are determined by the

---

[3] Some good references for the theory starting from the beginning are **[87]**, **[194]**, **[195]**, **[27]**.

[4] This apparently characterizes exactly those algebraic surfaces which are projective, as was stated in Simanca's Math Review MR 1723831 of **[10]** but I did not find the actual reference for it.

[5] Hironaka's result **[98]** has recently been revisited by several people **[1]** **[24]** **[36]**.

topological properties of the general fiber, the way its topology is transformed under the monodromy representation of $\pi_1(\mathbf{P}^1 - \{s_1, \ldots, s_k\})$, and the way the singular fibers fill things in. This led to things such as Zariski's calculations of fundamental groups, as well as proofs of certain parts of the Lefschetz theorems (fuller and easier proofs being obtained using the Hodge decomposition for cohomology). Griffiths proposed to throw Hodge theory into this situation, namely to look at how the Hodge decomposition of the cohomology of the fibers varies. He discovered the fundamental *Griffiths transversality relation* governing the variance of the Hodge filtration, and developed much of the basic theory of *variations of Hodge structure* which shows how this family of Hodge structures integrates to give the Hodge structure on the total space. This left open the question of what happens near the singularities, the answers furnished by [**174**] [**51**]. Similarly, what happens when the total space, and hence the fibers, are noncompact and/or singular—in this case we obtain *variations of mixed Hodge structure*, and the combined situation where this degenerates leads to a mountain of linear algebra [**183**] [**201**] which very few can grasp (anyone out there?).

To wrap up this historical perspective a little faster, basically the remaining theoretical directions have been to look at Hodge theory for the full homotopy type of X. This can mean looking at the homotopy groups, which obtain mixed Hodge structures[6] [**141**], idem for the pronilpotent completion of the fundamental group [**141**] [**93**]; looking at representations of $\pi_1(X)$ in nonabelian groups such as $GL(n)$, which have harmonic representatives [**53**]; and combining various of these. Much of what we shall say below issues from these basic themes.

In a family of varieties, all of the Hodge-theoretic invariants give rise to *variations of mixed Hodge structure* over the base. This imposes restrictions on the monodromy. We have things like the invariant cycle theorem stating that a global section which is of type (p, q) at one point is also of type (p, q) at other points.

---

[6] One notable exception to the principle that homotopic invariants which are finite dimensional complex vector spaces, have mixed Hodge structures, is the case of the group cohomology of π_1. This was a question that was raised by many workers such as Toledo, Carlson, Hain, long ago. I do not recall whether I have heard of any progress (or counterexample) in the meantime, and Google did not help either.

A crucial property is the *formality theorem* of **[65]**, saying that the cohomology ring of a simply connected smooth projective variety determines its rational homotopy type. This has been extended to the pronilpotent completion of $\pi_1(X)$ **[141]** **[93]**, to the spaces of representations **[144]** **[81]**, and to relative Malcev completions at other representations in **[93]**. In the case of open varieties, the quadratic formality condition is replaced by a weaker but still restrictive condition of generation in low degrees **[141]**.

Any representation $\pi_1(X) \to GL(n, \mathbf{C})$ with reductive Zariski closure, admits a structure of *harmonic bundle* **[53]**. These behave formally much like variations of Hodge structure except that they lack a notion of "Hodge type." The space of representations of $\pi_1(X)$ has a *hyperkähler structure*,**[99]** **[78]** where the other complex structures come from the interpretation as a moduli space of *Higgs bundles* **[99]** **[151]** **[177]** **[78]**. The cohomologies of X with coefficients in the different representations give a nicely varying family of complexes indexed by the representation space. The support loci are hyperkähler subvarieties, and even have certain rationality properties. An important distinction occurs between representations which are rigid (i.e. isolated points of the space of representations) and those which are not. Rigid representations of the fundamental group are variations of Hodge structure, and one could conjecture that they must be *motivic* i.e. that they should come from families of varieties (this is a special case of a slightly more general conjecture **[177]** (i)). On the other hand, nonrigid representations fit into a bigger *factorization theory* **[178]** **[47]** **[202]** **[111]** **[105]** **[75]** **[117]** **[147]** **[59]** which also encompasses results such as the fact (due to L^2 cohomology on infinite covers of X) that if π_1 decomposes into a nontrivial amalgamated sum then there must be a map from X to a Riemann surface of genus $\geqslant 2$ **[88]**. We now know the *Shafarevich holomorphic convexity conjecture* relative to reductive representations in a linear group **[111]** **[75]** **[124]**. There are a number of directions of investigation running which look at cohomological invariants associated to representations, specially work of Reznikov **[170]** which has been applied in **[117]**. We are starting to get a good picture of the theory of representations of the fundamental group for quasiprojective open varieties **[53]** **[105]** **[138]**.

The above results lead to important classes of restrictions of the form, certain homotopy types or homotopical invariants, cannot occur for (projective) algebraic and/or Kähler varieties. In fact, it becomes possible to write down explicit large families of topological types which cannot occur, see **[47]** **[178]** **[45]** **[110]** **[177]** .... Reznikov has proven some of the deepest versions of these restrictions, including a theorem covering many cases of a conjecture due to him, and independantly Goldman and Donaldson, saying that irreducible 3-manifold groups are not Kähler groups **[170]**.

Intrinsic global methods appeared in the work of Calabi and Yau giving certain types of special Kähler metrics. This hooked up with a sector of the subject of interest to the physicists generating the subject of "Mirror symmetry" which can no longer be considered as just a subfield of Hodge theory or Kähler geometry. These important subjects are underrepresented in present article.

Investigations of the Hodge-theoretic properties of algebraic cycles have led very far in the direction of K-theory. In particular, this leads to the introduction of new invariants such as regulators **[23]** **[167]** **[170]**, behaving in a way which is related to Hodge theory via the Beilinson conjectures and Deligne cohomology.

Recent work of Voisin has underlined the difference between the Kähler and the algebraic categories. For example, the $\mathbf{Q}$-polarization of the Hodge structure of an algebraic variety is a distinguishing characteristic, the existence of which can be ruled out for certain Kähler homotopy types; this leads to Kähler homotopy types whose cohomology ring cannot be the cohomology ring of an algebraic variety (**[194]** (v)). Philosophically going in the same direction is her result that the Hodge conjecture for representing cycles by Chern classes of bundles, cannot be true for arbitrary Kähler manifolds. Of course we do not know whether it is true for algebraic varieties, but Voisin's proof **[194]** (iv) clearly distinguishes these cases since it makes use of a Kähler variety (a complex torus) which has no rational Hodge classes in degrees 2 or 4 (in particular no hyperplane class). In view of these results, the questions of realization of homotopy types in the Kähler and projective categories are clearly distinct and can be treated as separate problems. The reader is invited to double-up our questions in this way.

It seems safe to say (if nothing else, in view of the length of the reference list which is only partial) that a relatively complete picture of the Hodge-theoretic properties of algebraic varieties, is in the making. It would be good to understand how Lefschetz devissage works in the context of Hodge structures on homotopy, or nonabelian cohomology, see Navarro Aznar **[148]** for example. The theory should be extended to apply to algebraic stacks and other exotic objects. One detail for the curious, is the variance of the formality equivalence on the homotopy type given by [**65**]. And more globally, it seems unclear whether the full complicated version of the theory as sketched above has really been explicitly applied in many examples. Aside from these brief remarks, it is best to consult the references for a good picture of where the current development of the theory stands, and in which directions it will (or should) be developed in the near future.

Now we come to the main point of the present exposition, which is to try to isolate some of the important questions in the subject. It should be stressed that these are not at all new, but rather come from the folklore explicit or implicit in most of the work that has been done. Still, it seems like a good idea to look at them from an analytical viewpoint. A historical overview brings out an inescapable fact: we know a lot more about what (projective) algebraic, or Kähler, varieties *do not* look like, than about what they *do*. This is because the most subtle tool we have found yet is Hodge theory, and this is essentially geared toward additional properties, or equivalently, restrictions on the topology and/or geometry of Kähler manifolds. This type of application determines the style of numerous theorems in the subject. In some hypothetical alternative world, this might have been enough to gain a complete understanding of the subject, if we could prove so many properties that there were very few possibilities left and we would get a complete classification of what could happen. Unfortunately, or perhaps fortunately for the interest of the subject at least, this is not the case. In spite of the large and varied collection of properties given by Hodge theory, it seems nowhere close to giving any kind of classification of the topological behavior which could arise.

On the other side of the story, this situation leaves wide open the whole problem of how to construct varieties exhibiting topological behavior that is allowed by Hodge theory. Theoretical machinery is not in the fore here, although it is present insofar as it facilitates the *calculation* of the topological behavior of a variety which for other reasons we have at hand. Unfortunately (in all senses) on this side of things, we do not have very many deep techniques of construction.[7] I will try to discuss below the possible techniques of construction we know about. For the purposes of the present paragraph, let's just look at what might be seen

[7] The only full exception is the Lefschetz (1, 1)-theorem.

as a good candidate for a "deep technique of construction:" the notion of quotient of a symmetric domain by a group action. The associated variety comes out of a transcendental construction, so it is at least very indirect as a construction. However, the geographical place occupied by the varieties obtained in this way is quite infinitesimal, so in spite of a number of successful applications of this construction, we are mostly left wondering how come there is not a way of twiddling it to cover more nearby cases.[8] To give a precise version of these statements, look at the case of surfaces: those which are uniformized by the ball have $c_1^2 = 3c_2$, whereas there exist other surfaces uniformized by a product of discs, with $c_1^2 = 2c_2$. Siu mentioned as a folkloric problem, the question of whether or not every variety with c_1^2 lying between $2c_2$ and $3c_2$, would have to have infinite fundamental group. I do not know whether it is reasonable to expect such a thing or not, but it does seem quite clear that it is silly to have such a wide swath where we think non-simply connected varieties should be lying, but to have a nice general construction only on the boundary of the region. This example is pretty emblematic of the whole problem of constructing interesting varieties: we have the impression that there is a huge mass of stuff out there, waiting to be constructed or seen, but we have no idea how to get there.

The construction question can take many forms, depending on what type of thing you want to construct. Among the infinite variations on this question, we can try to isolate some exemplary ones. Start with the simply connected case. In all of the questions, we will for simplicity ask for smooth projective varieties but the same question could be asked for proper smooth algebraic spaces, compact Kähler manifolds, etc.; and we stick to the compact smooth case again for simplicity, but more subtle questions involving mixed Hodge structures are available for the open and/or singular case.

The easiest version of the question would be the following: given n and a vector $(b_0, \ldots, b_{2n})$ satisfying the known constraints such as $b_k = b_{2n-k}$ and that b_{2k+1} is even, plus the constraint given by the Lefschetz decomposition, does there exist a smooth projective variety X with $\dim H^i(X, \mathbf{C}) = b_i$? A recent investigation of the very first case $b_i = 0$ is [**200**]. Take note of the obvious (and in this case pretty useless) observation that there exists an algorithm which lists all of the vectors which occur: just run through all possible varieties defined over finite

[8] Actually, this has been done in some cases such as Toledo's examples [**191**], see also [**3**]

extensions of $\mathbf{Q}$ and for each one, calculate the Betti numbers. This does not imply that there is a decision algorithm for a given vector. To get even an impractical decision algorithm we would have to get a bound on the degree and height of varieties with a given set of Betti numbers. Of course the algorithmic question does not really represent what we would like, which is a clean statement saying, a vector of Betti numbers occurs if and only if it satisfies a given simple constraint.

We can refine the above question by asking for the full collection of Hodge numbers $h^{p,q}$. See [**74**] and [**64**] for example. In this case, it is interesting even to look at only one cohomology group: given a collection $(h^{k,0},\ldots,h^{0,k})$ does there exist a variety X such that the Hodge structure on $H^k(X,\mathbf{C})$ has these Hodge numbers? In a sort of vague philosophical sense, Mirror symmetry says that there may be some relationship between this Hodge number question and the Betti number question in the previous paragraph.

At the very least, there is a question which intuitively suggests itself in both cases. This was the subject of a conversation with J. Kollar a long time ago. Is it possible to have big numbers on the ends of the vector, and small numbers in the middle? Or, on the other hand, is there some type of bound that says the middle Hodge (or Betti) numbers have to be bigger than the outer ones? Our intuition says that the middle numbers probably should be bigger than the outer ones, but there does not seem to be any result in this direction. This can be made into a very small and specific question, for example do there exist surfaces with Hodge numbers $h^{2,0} = a$, $h^{1,1} = b$, with $a \gg b$? Or say with a arbitrarily large and $b = 1$ or $b = 2$? The same concrete question could be posed for threefolds, or more generally for the Hodge structures on H^2 or H^3. One can weaken the requirement by looking at motives. In this case, we just mean the image of an algebraically defined correspondence. Can the image Hodge structure have Hodge numbers with $a \gg b$?

In formulating the same question for Betti numbers, care must be taken with the Lefschetz decomposition, which already provides the type of bound we are looking for in a certain sense. One could ask, do there exist varieties with large primitive cohomology P^i and small P^j for $i < j \leqslant (\dim X)$. Again here it might be convenient to distinguish odd and even cohomology.

When we see that we do not even know what to say about these reasonably simple questions, it is apparent that more complicated things such as the possible structure of the cohomology ring or just of certain fixed cup product operations, are completely open problems too. Without trying to list these questions exhaustively, we can formulate an example. If X is a 3-fold with a nontrivial vector space $V := H^2(X, \mathbf{Q})$ then the cup product gives a cubic form $\eta \in (V^*)^{\otimes 3}$. What are the possibilities for the equivalence class of (V, η)? Even in the simplest case $\dim V = 2$, the space of forms is 8-dimensional, and it is acted upon by the 4-dimensional group $GL(2)$; where does η lie in the 4-dimensional space (or stack to be more precise) of orbits? What examples can we construct? The numerous possible generalizations of this type of question are perhaps better understood in terms of nonabelian cohomology.

The discussion would not be complete without pointing out that the Hodge conjecture is an example (perhaps the first) of the construction problem. In that case, the problem is to construct subvarieties such that the cycle map gives a certain topological class. There are many different formulations, for example one can say that we want subvarieties such that the restriction of some cohomology class is nonzero. The situation for the Hodge conjecture remains much the same as the more general situation we are reflecting on here: we really do not have any good constructions of cycles. Indeed, most cases where it is known are obtained by showing that there are not too many Hodge classes, in other words that there are sufficiently many restrictions on the classes that we do not have to look too far to get the rest. Two notable exceptions are Steenbrink's semiregularity theorem [**182**] and Voisin's construction for Kummer surfaces in [**194**] (ii).

Another side of the story is related to our previous questions for Betti numbers. One of the simplest "standard conjectures" is algebraicity of the Kunneth projectors. But, even to construct varieties for which this question is interesting, we need ones with nontrivial Betti numbers in different degrees. It is not even clear that we really know a very wide collection of such varieties

Let's move onto the non-simply connected case, and more specifically to questions about the fundamental group. One can note that by the existence of Artin neighborhoods, the topology of quasiprojective $K(\pi, 1)$'s can be considered as the basic building block for the topology of any other varieties. (Also a recent result of Beilinson [**23**] and Katzarkov, Pantev and Toen [**114**] refines this to a statement about schematic homotopy types.) In this sense, it is natural to pay what might at first seem to be an inordinate amount of attention to questions about the fundamental group.

The situation is pretty well summed up by the first two sentences of Reznikov's paper [**170**] (vii):

Fundamental groups of complex projective varieties are very difficult to understand. There is a tremendous gap between few computed examples and few general theorems. ...

Reznikov is referring to what is actually a fairly diverse collection of constructions serving to give examples of fundamental groups. These range from uniformization constructions (which are also seen from other viewpoints e.g. in Deligne and Mostow [**67**], Livne [**132**]) to explicit constructions for example of nilpotent fundamental groups [**44**] [**180**]. See the discussion in the book of Amorós *et al* [**7**]. Nonetheless, this collection of examples leaves us with the impression that there are many examples of groups that we do not know about. It is not altogether trivial to formulate precise questions which express this feeling.

We can start with a very general one, which was inspired by Toledo's examples of varieties with non-residually finite π_1 [**191**] [**7**]. Can any finitely presented group be a subgroup of a $\pi_1(X)$ for a smooth projective variety X? Or, on the contrary, are there restrictions on which groups can occur as subgroups? As far as I know, none of the Hodge theory results that we currently have serves to rule out arbitrary subgroups. It is possible that some type of L^2 cohomology techniques, or maybe some Yang–Mills techniques such as in Ni and Ren [**150**], on the (possibly infinite) covering $\tilde{X} \to X$ determined by a subgroup $\Gamma \subset \pi_1(X)$, could give some restrictions. Toledo's example shows that one cannot hope for any results on this question using finite-dimensional Hodge theory.

On the construction side there is something obvious to try: given a finitely presented group Γ, it is the fundamental group of a smooth compact manifold M (say, of dimension 4). This manifold can be embedded in some $\mathbf{R}^N$ and then approximated by a real algebraic manifold. Thus we may assume that M is a connected component of the real points of an algebraic variety X defined over $\mathbf{R}$. We can further replace X by any covering whose ramification locus does not intersect W. We might hope in this way to get an injection $\pi_1(M) \hookrightarrow \pi_1(X)$. I do not see any way of proving that a map of this type is injective, though.

An interesting test case (suggested by M. Larsen) would be $\Gamma = G(n^{-1}\mathbf{Z})$, the group of points in an algebraic group G defined over $\mathbf{Z}$, with coefficients in a finite localization $n^{-1}\mathbf{Z}$. This group is linear, but cannot immediately be embedded in the fundamental group of a Hermitian symmetric domain quotient. One suspects that such Γ cannot itself be a Kähler group, because it would tend to have rigid representations which are not integral, which would go against the conjecture that all rigid representations are motivic. It might, on the other hand, be possible to carry out a real-points construction such as in the previous paragraph, keeping enough control over a linear representation to prove injectivity of the map.

More generally, after Toledo's example and several others in the same direction (these are nicely presented in [**7**]), we can ask what type of non-residually finite groups can occur. A folkloric observation/question[9] is that in all examples so far, there is a nontrivial linear quotient group; do there exist Kähler groups without any subgroups of finite index at all?

A question which was recalled in E. Rees' Math Review MR1396672 of the paper by Amorós [**4**], as well as in [**7**] is whether there are groups which are fundamental groups of Kähler manifolds but not of smooth projective varieties. It might now be possible to give such an example, based on Voisin's construction [**194**].

One of the main invariants of a finitely presented group is its space of representations, say into $GL(n, \mathbf{C})$. It provides a way of measuring how well we can construct examples of fundamental groups. Say that a representation ρ is *non-rigid* if its semisimplification lies in a positive-dimensional component of the moduli space of representations (recall that the moduli space parametrizes Jordan equivalence classes of representations, which is the same as isomorphism classes of semisimple representations). The non-rigidity condition stated this way implies non-vanishing of the deformation space, but the converse is not necessarily true: there can exist representations with nontrivial deformation space but where all the deformations are obstructed so there are no truely integrable deformations (and indeed this seems to be what happens in some Teichmüller examples, see below). The known factorization results leave ample room for existence of interesting non-rigid representations. However, the only examples I know are obtained in the following way: if $f : Z \to X$ is a morphism and if, for some reason we already have a nonrigid local system V over Z, then some higher direct image $R^i f_* V$

[9] I do not remember who said it first but it was not me.

will tend to be nonrigid on X. This construction has the notable property that the birational equivalence class of the irreducible component of the moduli space of representations of $\pi_1(X)$ which is constructed, is not necessarily any different from the one we started with for $\pi_1(Z)$. In other words, no new *moduli spaces* are constructed in this way,[10] even if old moduli spaces are reinterpreted in new ways for new varieties.

This inevitably raises the question, do we already know[11] all of the possible moduli spaces of representations? Or are there varieties having new moduli spaces not on the basic list? A first comment is that by the Lefschetz theorem restricting to a curve, any new moduli spaces are subspaces of moduli spaces of representations on a curve. This includes all of the structure, notably the hyperkähler structure [**99**] [**78**]. So one possible attack on this question would be to try to classify all hyperkähler subspaces of the moduli spaces of representations on curves. The extensive works of Verbitsky, Kaledin [**193**] [**108**] [**109**] are certainly relevant to any such project.

It is probably not easy to prove that two given hyperkähler varieties of the same dimension are distinct. For this part of the question, several possibilities might work: analysis of the singularities, looking at the power-series expansion of the hyperkähler structure at a special point, analysis of the structure at infinity, or calculation of the metric (e.g. curvature) properties. These things are all undoubtedly easier to do in low dimensions. This suggests that an important variant of the question would be, can we construct varieties whose space of representation has components with positive but low dimension? (Recall that the complex dimension is divisible by 2, so the first possibilities would be dimension 2 or dimension 4.) Here we have fallen back into a situation analogous to that for the geography of surfaces: there are many natural examples of rigid representations, but for the "nearby" case of representations with

[10] This is a bit of a misstatement. The moduli spaces are the same, if the collection of representations obtained as higher direct images from representations on Z constitutes an irreducible component of the moduli space of representations on X. It of course remains possible that the representation on X could acquire additional deformations in other directions not coming from deformations on Z. In this case we would get a candidate for a new moduli space (although that would not be a proof that it was really different from the old ones, either ...). Comparing deformation spaces of the local system V on Z and the higher direct image $R^i f_* V$ on X is therefore an interesting question.

[11] The known ones would be products of moduli spaces of representations on curves, and of rank one representations on abelian varieties.

low-dimensional moduli we know only very few[12]. For orbicurves (which also applies to certain fibrations with multiple fibers), the situation is much better: by work of Kostov **[121]** and Crawley–Boevey **[55]** we have a good classification of the irreducible components of the moduli spaces, and in particular there exist components of all possible (even) dimensions. Small components appeared in the "toy example" in **[175]**.

At this juncture, note that cohomology support loci[13] provide natural hyperkähler subvarieties. One might therefore ask whether the irreducible components of support loci occur as components of moduli spaces of representations. There might be geometric constructions relating the situation of a support locus, to a nonrigid representation. For these reasons (as well as because it fits into our general construction question) it would be interesting to have constructions of varieties giving rise to nontrivial cohomology support loci. As mentioned in the previous paragraph, a variant would be to ask for support loci of small positive dimension.

This brings us to a question raised by Hain and/or Looijenga (communicated to me by R.H. but he said it was more due to E.L.). They looked the moduli of k-pointed genus g curves $M_{g,k}$ and looked at spaces of representations of $\pi_1(M_{g,k})$ or perhaps more precisely the Teichmüller group which is the orbifold fundamental group. It is of course well-known that the standard representation is rigid. However, they found other representations (subobjects of tensor powers of the standard one) which were not infinitesimally rigid. The ones they found all had the property that there were enough obstructions that they were globally rigid even if not infinitesimally so. This then raises the question: are there any non-rigid (in our above sense) representations of $\pi_1(M_{g,k})$? Or on the contrary are all representations globally rigid?

Actually if a variety X has the property that all of its representations are globally rigid, then the pro-algebraic completion of $\pi_1(X)$ has a reasonable structure (something like a product of the relative Malcev completions at the various points) and similarly the schematic homotopy type $X \otimes \mathbf{C}$ as defined by Toen **[189]** **[114]** will have reasonable finiteness

---

[12] One should consider the possibility that this phenomenon might be substantial, rather than just a consequence of our lack of suitable constructions, see **[53]** (ii) for example.

[13] By "support loci" we mean the loci $\Sigma_{i,k}$ of local systems V such that $\dim_{\mathbf{C}} H^i(X, V) \geqslant k$. One could introduce variants here, such as loci where certain cup products in cohomology have a given rank.

properties (for example the bad behavior noted at the beginning of Artin and Mazur [**14**] would not occur). This, plus the fact that the representations found so far by Hain and Looijenga turned out to be globally rigid in spite of their infinitesimal non-rigidity, leads us to conjecture that $\Gamma = \pi_1(M_{g,k})$ does not have any globally non-rigid representations.

If the above questions are not enough, one could pose similar ones for higher nonabelian cohomology [**177**]. The reader will readily imagine the details, so it does not seem necessary to stress this further here. The long-range hope is that nonabelian cohomology might help explain or organize what is going on—some hint of this can already be seen in the fact that it was natural to look at cohomology support loci.

We have been concentrating on the question of the pure topology of X. On the other hand, a great deal of research in differential geometry has looked at the classification of differential manifolds and other structures such as symplectic manifolds. A major part of this work has been to consider the place of algebraic or Kähler manifolds in the classification. Thus, what we know about the Donaldson–Seiberg–Gromov–Witten invariants of algebraic varieties, is quite comparable to what we know about them for other types of manifolds (I maintain that, as a whole, this is still not very much.) See [**142**], [**77**] [**72**] [**154**] for example. Lönne has considered an interesting question, which is the group of diffeomorphisms of a variety and its action on cohomology. He was able to calculate this in one case using Seiberg–Witten theory [**133**] (see also [**103**]). An important question is, to what extent will consideration of the new invariants help us for example to construct varieties with given Hodge numbers?

Another possible invariant that we could look to obtain would be intersection cohomology for singular varieties. All of the same types of questions could be posed in that context. One can also ask things like how many (and what kind of) singularities a variety can have, see [**73**] for example.

A general research direction, tangentially related to our present preoccupations, is the question of what happens when we go to "infinity." As became apparent with work of Wentworth and others, there are three natural directions to look at:

— We can go to infinity in an open (say, quasiprojective) variety, and ask for the behavior of representations, vector bundles, cohomology classes and so forth near the divisor at infinity.

— We can go to infinity in the space of representations. This leads to the whole subject of "singular perturbation theory," which is growing very diverse. There are lots of things to investigate. The question of how the harmonic metric degenerates when the representation goes to infinity was raised in [**56**] [**58**], in particular what relationship this has with harmonic maps to buildings occurring on the boundary of the moduli space of representations. There is also the whole question of the Stokes phenomenon for the asymptotic expansion of the monodromy, which actually is probably closely related to the harmonic map question, for example spectral curves play an important role in the Stokes phenomenon [**61**] [**102**] [**9**] [**33**].

— We can look at a degenerating family of varieties, considered as a smooth family over a punctured disc, and go to infinity at the puncture. This amounts to saying that we look at varieties where certain vanishing cycles become infinitely small. The degeneration of cohomology has been extensively studied in this case. The degeneration of the space of representations has been much less studied, see however [**58**] and [**50**] for example. Some time ago, Katzarkov raised the question of whether there is an analogue of the Clemens–Schmid exact sequence for spaces of representations or other nonabelian cohomology.

The above topics suggest their own formulations of the construction question. For example, do open varieties, or degenerating varieties, exist with given topological behavior? This is a vast question, and is also related to the smoothing and branching techniques for construction even of smooth projective examples.

Perhaps the biggest research area in Kähler geometry is the study of special metrics, for example Kähler–Einstein metrics or metrics with other rigidifying curvature properties [**69**] [**186**]. This continues to motivate a large segment of the field, and we could not go into sufficient detail here. This subject probably has or will have important implications for the topological construction problem we are discussing, on the one hand because the existence of metrics with special curvature properties implies strong topological restrictions, and on the other hand because this type of global analysis might contain key techniques for the construction of algebraic objects. A similar question is what new information will be brought by the Ricci flow techniques of Perelman [**161**]: can these be used to construct Hodge cycles by a minimization process, for example? More generally, is there a way in which one can look at "almost Kähler manifolds," then define obstructions and minimization processes allowing us to determine when an "almost Kähler manifold" could be deformed to a true one?

Now it's time to look at some of the main techniques for constructing varieties. The first and most obvious is by writing down equations. It is a nontrivial task just to understand what are the topological properties of a variety which is given by its equations. The other nontrivial task is to find systems of equations which do not reduce to complete intersections,[14] a question which Landsberg has looked at in a theoretical way [**123**].

The equational method becomes very subtle already in codimension two. Perhaps a fundamental example is the determination by Gruson and Peskine of the values of (d, g) for which there exist a smooth curve of degree d and genus g in $\mathbf{P}^3$ (which after all is an early version of the whole construction question we are considering). They succeeded by a very careful construction noticing that you got interesting examples out of the relatively large Picard groups of cubic and singular quartic surfaces. This type of method has not been sufficiently exploited in relationship with more overtly topological questions.

The next technique, which could sometimes be considered as a variant on the equational method, is to construct varieties as branched coverings of other varieties with a given branch locus. This has led to a large and intensive study of some of the first possible cases of branch loci, for example hyperplane arrangements, or curves with a few components of low degree in the projective plane [**130**] [**97**]. Fulton's theorem that π_1 is abelian if the branch curve has only nodes [**80**] shows that the question is subtle, see also [**86**]. The branched variety construction serves as a bridge between the topology of compact smooth varieties, and the topology of the open branch locus complements. In many cases it is easier to envision calculations, for example of fundamental groups, in the open case. This program has already been carried a long way by Moishezon and Teicher [**140**] with Robb [**171**] and others.

Another variant on the equational method is looking at degeneracy loci of vector bundle maps (see Decker, Ein, Schreyer who extensively investigated this in [**60**], see also [**71**]). Eyssidieux's construction of some varieties as base loci of linear systems [**75**] could also fit in here, although

[14] Although it is not even completely clear that we fully understand the topology of complete intersections together with their embeddings, specially in twisted cases such as weighted projective spaces. A few references are [**122**], [**131**] but this is not a complete list.

it has the additional subtlety that the linear system is L^2, taking place on an infinite covering—equivariance says that the base locus descends to a usual subvariety downstairs.

A hitherto underexplored idea is Kapovich and Millson's theorem saying that any variety can be viewed as the moduli space of a linkage [**110**]. It is possible that the "mechanical" properties of a linkage could show up in the topological properties of the moduli space, and in this way we might have a more natural relationship between the presentation of a variety and its topology. It seems urgent to look at this in more detail.

A generalization would be to try to construct varieties as moduli spaces of any kind of geometric object, and to hope that the topological properties of the moduli spaces would be related to the moduli problem. This is of course what happens with well-known moduli spaces. However, up to now the philosophical take on this theory has always been that the "moduli spaces" are natural objects, and we should be interested in their topological properties. A more intentional approach would say that we are interested in constructing given topological types, and we would like to pick and choose our moduli problem in order to get what we want. One place to start might be to look at moduli spaces of group representations (a point of view present in [**110**]). For strange groups Γ which the researcher should carefully choose (probably not Kähler ones!), does the moduli space of representations of Γ have an interesting topological type? Other possibilities might include components of the moduli spaces of space curves [**91**], or (related by [**16**]) components of the Noether–Lefschetz loci [**48**] [**155**].

Something which was mentioned previously is the existence of uniformization constructions for algebraic varieties. The simplest examples are direct construction of the quotient of a symmetric domain by a discrete group action. In recent years this has been considerably extended. Perhaps one of the first extensions was Toledo's construction of examples with non-residually finite fundamental group [**191**]. The idea is that certain natural subspaces of symmetric domains can fit together into interesting configurations in the quotient variety. Then other techniques such as branched covers, or in Toledo's case blowing down a negative subvariety and then taking hyperplane sections which miss the singularities, give new varieties. Lately some more complicated configurations have been considered by Allcock, Carlson and Toledo [**3**]. These types of constructions are good candidates; one of the major difficulties is in calculating what happens.

Another important technique is the notion of fibration, which came up under the notion of Lefschetz devissage above. Whereas one often uses fibrations to study the topology of a variety, it is also possible to try to construct varieties by explicitly constructing a fibration. The easiest example is that of Kodaira surfaces, where the fibration is automatically smooth (a construction which easily generalizes in higher dimensions too). Recall that the basic idea consists of saying that given a pointed curve we can take a covering ramified at the point. As the point moves, this gives a family of curves fibering over the original curve; the construction can be iterated in the fibers. One can write down the group automorphisms determining the monodromy, but the formulae tend to be long and not very enlightening. Things might get more tractable by putting computers to work to do the calculations. Other examples of fibration constructions include the ones of Bogomolov and Katzarkov [**35**]—but here again it is difficult to know what the result is. This theory of course spills over into the symplectic world via Donaldson's notion of almost-Lefschetz fibration. Gompf [**82**], Amorós *et al* [**6**], Reznikov [**170**] (ii) show that the symplectic "almost-Lefschetz" world is more flexible in an essential way, for example any fundamental group can arise. In view of these results, it does not seem clear what implications there are for our present construction problem.

Braid groups come into play when looking at fibrations whose fibers are punctured projective lines [**140**]. In this case, a family is just a moving system of points in $\mathbf{P}^1$. The de Rham version of the devissage for representations of the fundamental group is what is often known as "isomonodromic deformation," or "Painlevé equations." The isomonodromy equations or nonabelian Gauss–Manin, are equations which tell us when we have a family of representations on the fibers which fit together to give a representation of the fundamental group of the total space. So, looking for algebraic solutions of these equations could be a way of constructing quasiprojective varieties with nonrigid representations. Some solutions have been found by Boalch [**33**] and Hitchin [**99**].

Birational transformations enter into algebraic geometry in many ways, and they constitute a way of modifying a variety which changes its topological type in a readily calculable way. These provide a tool for transforming a subvariety into a modification, for example. Also a singularity could become a topological feature after resolution. This type of construction showed up very successfully in Voisin's examples of compact Kähler varieties whose homotopy types cannot be smooth projective [**194**].

One of the major potential techniques for construction of smooth projective varieties is smoothing of degenerate configurations. This was mentioned, for example, in Peters' "Math Review" [**166**] of U. Persson's paper [**162**]. This tool plays an important role in the determination of the "geography of surfaces" [**163**]. While not immediately trivial, the calculation of the homotopy type of the smooth nearby fiber is something which can be done reasonably well using traditional Hodge-theoretic techniques. One of the main difficulties is that not enough research has been done to establish useful criteria stating when a configuration can be smoothed. This has to be a subtle problem [**165**], because all of the Hodge-theoretic restrictions on homotopy types show that you cannot just smooth any configuration. Smoothings are known to exist in some special cases [**90**] [**168**] [**169**] [**186**] [**188**]. There is a wide choice of initial configuration, which could be a union of projective spaces or other easy pieces—in which case the combinatorics of the configuration would determine the topology of the nearby fiber; or it could on the other hand be a variety having a complicated singularity which contributes a lot of topology in the smoothing.

Whatever methods are used for constructing varieties, there is always the problem of calculating what you get. The calculation problem is rendered difficult by the fact that, not only do we need to know how to find the answer, but we need a sufficiently agile understanding of how the calculation works that we can choose the construction accordingly. Indeed, the main point of the whole "construction question" we are asking here is to be able to influence the output topological type as much as possible, by varying the input construction of the variety.[15]

Some recent advances in Hodge theory are likely to be key elements of this computational aspect. Notable is T. Mochizuki's work on non-abelian Hodge theory on open varieties [**138**], see previously [**32**] [**128**]. We will describe in a bit more detail how this is relevant. In the short range, one of the most promising avenues for constructing things remains the consideration of complements of configurations of curves in $\mathbf{P}^2$ (maybe one would blow up at singular points to obtain a normal

[15] Some kind of "engineering" in the wonderful spirit of the title of [**25**].

crossings configuration). The isomonodromy equations constitute one approach to look for representations of the fundamental group of the complement. However, a more global approach would be to look for harmonic bundles over $\mathbf{P}^2$ with singularities along the curves in the configuration. There are three incarnations we could use to look for these: the Dolbeault incarnation would say to look for logarithmic parabolic Higgs bundles; the de Rham incarnation, to look for parabolic bundles with logarithmic connexion; and the Betti incarnation (hitherto less explored) would be to look for "filtered local systems." T. Mochizuki's work describes the properties of harmonic bundles on open varieties and their extensions to the normal crossings compactification. There is still a lot to be done here, but he has crossed the essential step of understanding the local behavior at the singularities. Thus, we can imagine that in the semi-near future we will have the full panoply of tools necessary to analyze the correspondence between the Dolbeault, de Rham and Betti points of view. In each case, there should be a Bogomolov–Gieseker inequality (for the logarithmic Dolbeault case this was mentioned to me recently by Narasimhan). It is the limit case of this inequality which gives rise to representations. The case of filtered local systems is worth highlighting. The easiest possibility would be to look at filtrations on the trivial local system. Then a filtered structure is just specification of a filtration (indexed by real weights) for each divisor in the configuration. The parabolic Chern classes would be polynomials in the weights whose coefficients are determined by the combinatorics of which divisors intersect. The Bogomolov–Gieseker inequality might well give nontrivial constraints on the possible combinatorics of the configuration; and when the limiting case of equality is attained, one would get families of harmonic bundles which, after transformations such as described in [**177**] (which would require the fully worked-out theory to implement), would give representations of the fundamental group on the complement. It would then become a combinatorial problem to find configurations where the limiting case of equality could be attained. We can vividly hope that T. Mochizuki's work will allow us to carry out this type of investigation.

It is really unclear what type of answer we would expect from calculations such as described in the previous paragraph. We would be looking to attain the limiting case of a Bogomolov–Gieseker type inequality, and that would seem hard to do. A lot of negative results formalizing this difficulty would lend credence to the conjecture on Teichmüller groups stated above, and would also increase the hope for some type of strong classification results about representations. Generalizing the rank two classification of [**54**], it might be possible to get say a classification for rank three representations.

If there is any conclusion to be drawn, it is perhaps that even though there is no single fully satisfactory choice of a method of construction, nonetheless there are many methods which have not been fully exhausted yet; and similarly there are so many different variants of the basic construction question, that maybe we should just spend some time trying to see what topological shapes can be constructed with the methods currently at hand. This might direct us towards the most promising construction methods, and might also lead to ideas for improving the underlying Hodge-theoretical tools of calculation.

References

1. D. Abramovich, A.J. de Jong, *Smoothness, semistability, and toroidal geometry*, J. Algebraic Geom. **6** (1997), 789-801.
2. E. Allaud, *Variation de structure de Hodge et systèmes différentiels extérieurs*, Thesis, Université Paul Sabatier (2002);. See `http://fermat.ups-tlse.fr/~webthesards/theses.htm`.
3. D. Allcock, J. Carlson, D. Toledo, *Orthogonal complex hyperplane arrangements*, Symposium in Honor of C. H. Clemens (Salt Lake City 2000), Contemp. Math. **312** (2002), 1–8.
4. J. Amorós, *On the Malcev completion of Kähler groups*, Comment. Math. Helv. **71** (1996), 192–212.
5. J. Amorós, I. Bauer, *On the number of defining relations for nonfibered Kähler groups*, Internat. J. Math. **11** (2000), 285-290.
6. J. Amorós, F. Bogomolov, L. Katzarkov, T. Pantev, *Symplectic Lefschetz fibrations with arbitrary fundamental groups*, Appendix by Ivan Smith. J. Differ. Geom. **54** (2000), 489-545.
7. J. Amorós, M. Burger, K. Corlette, D. Kotschick, D. Toledo, *Fundamental groups of compact Kähler manifolds*, Am. Math. Soc. Mathematical Surveys and Monographs, **44** (1996).
8. Y. André,
   (i) *Pour une théorie inconditionnelle des motifs*, Publ. Math. I.H.E.S. **83**, (1996), 5-49.
   (ii) *Quelques conjectures de transcendence issues de la géométrie algébrique*, Preprint Institut de Mathématiques de Jussieu **121** (1997).
9. T. Aoki, T. Kawai, Y. Takei, *On the exact steepest descent method: A new method for the description of Stokes curves*, J. Math. Phys. **42** (2001), 3691-3713.
10. V. Apostolov, O. Muskarov, *Weakly Einstein–Hermitian surfaces*, Ann. Inst. Fourier **49** (1999), 1673-1692.
11. D. Arapura.

(i) *Hodge theory with local coefficients and fundamental groups of varieties*, Bull. Amer. Math. Soc. **20** (1989), 169-172.

(ii) *Smoothable varieties with torsion free canonical sheaf*, Manuscripta Math. **63** (1989), 17-20.

(iii) *Fundamental groups of smooth projective varieties*, Current Topics in Complex Algebraic Geometry (Berkeley, 1992/93), MSRI Publ. **28**, Cambridge Univ. Press (1995), 1-16.

(iv) *Geometry of cohomology support loci for local systems, I*, J. Algebraic Geom. **6** (1997), 563-597.

12. D. Arapura, P. Bressler, M. Ramachandran. *On the fundamental group of a compact Kähler manifold*, Duke Math. J. **68** (1992), 477-488.
13. D. Arapura, M. Nori, *Solvable fundamental groups of algebraic varieties and Kähler manifolds*, Compositio Math. **116** (1999), 173-188.
14. M. Artin, B. Mazur, *Etale Homotopy*, Lecture Note Math. **100** (1969).
15. D. Auroux, L. Katzarkov, *Branched coverings of* $\mathbf{CP}^2$ *and invariants of symplectic 4-manifolds*, Inventiones **142** (2000), 631-673.
16. S. Azziz, *Exemples de composantes irréductibles et non réduites du schéma de Hilbert des courbes lisses et connexes de* $\mathbf{P}^3$ *(II)*, Preprint, alg-geom/9701018.
17. V. Balaji, I. Biswas, D. Nagaraj, *Principal bundles over projective manifolds with parabolic structure over a divisor*, Tohoku Math. J. **53** (2001), 337-367.
18. S. Bando, Y.-T. Siu, *Stable sheaves and Einstein–Hermitian metrics*, in: Geometry and Analysis on Complex Manifolds, World Sci. Publ., 1994, pp. 39–50.
19. I. Bauer, *Irrational pencils on non-compact algebraic manifolds*, Internat. J. Math. **8** (1997), 441-450.
20. T. Bauer, *Seshadri constants and periods of polarized abelian varieties* (appendix by the author and T. Szemberg), Math. Ann. **312** (1998), 607-623.
21. I. Bauer, F. Catanese, *Some new surfaces with* $p_g = q = 0$, Preprint, math.AG/0310150. [to appear in the Proceedings of the Fano Conference (Torino, 2002).]
22. A. Beauville, *A Calabi–Yau threefold with non-Abelian fundamental group*, in: New Trends in Algebraic Geometry (Warwick, 1996) London Math. Soc. Lecture Note Ser. **264** Cambridge Univ. Press (1999), 13-17.

(ii) *Quantum cohomology of complete intersections*, Mat. Fiz. Anal. Geom. **2** (1995), 384-398.

(iii) *Variétés Kähleriennes dont la première classe de Chern est nulle*, J. Diff. Geom. **18** (1983), 755-782.

23. A. Beilinson.

(i) *Higher regulators and values of L-functions*, in: Current Problems in Mathematics **24** Itogi Nauki i Tekhniki, Akad. Nauk SSSR, Vsesoyuz. Inst. Nauchn. i Tekhn. Inform., Moscow, 1984, pp. 181–238.

(ii) *Notes on absolute Hodge cohomology*, in: Applications of Algebraic K-Theory to Algebraic Geometry and Number Theory (Boulder, 1983) Contemp. Math. **55**, Am. Math. Soc., 1986, pp. 35–68.

(iii) *On the derived category of perverse sheaves*, in: K-Theory, Arithmetic and Geometry (Moscow, 1984-1986), Lecture Note Math. **1289** (1987), 27–41.

24. E. Bierstone, P. Milman, *Canonical desingularization in characteristic zero by blowing up the maximum strata of a local invariant,* Invent. Math. **128** (1997), 207-302.
25. M. Bershadsky, A. Johansen, T. Pantev, V. Sadov, C. Vafa, *F-theory, geometric engineering and $N = 1$ dualities,* Nuclear Phys. B **505** (1997), 153-164.
26. A. Berthomieu, *Proof of Nadel's conjecture and direct image for relative K-theory,* Bull. S.M.F. **130** (2002), 253–307.
27. J. Bertin, J.-P. Demailly, L. Illusie, C. Peters, Introduction à la théorie de Hodge. Panoramas et Synthèses **3**, S.M.F., Paris (1996).
28. F. Bihan, *Asymptotiques de nombres de Betti d'hypersurfaces projectives réelles,* Preprint, math.AG/0312259.
29. I. Biswas,
(i) *Vector bundles with holomorphic connection over a projective manifold with tangent bundle of nonnegative degree,* Proc. Amer. Math. Soc. **126** (1998), 2827–2834.
(ii) *Flat connections on a punctured sphere and geodesic polygons in a Lie group,* J. Geom. Phys. **39** (2001), 129–134.
30. I. Biswas, T. Gómez, *A Torelli theorem for the moduli space of Higgs bundles on a curve,* Q. J. Math. **54** (2003), 159–169.
31. I. Biswas, M. Mitra, S. Nag, *Thurston boundary of Teichmller spaces and the commensurability modular group,* Conform. Geom. Dyn. **3** (1999), 50-66.
32. O. Biquard, *Sur les fibrés paraboliques sur une surface complexe,* J. London Math. Soc. **53** (1996), 302-316.
33. P. Boalch, *The Klein solution to Painleve's sixth equation,* Preprint, math.AG/0308221.
34. H. Boden, K. Yokogawa, *Moduli spaces of parabolic Higgs bundles and parabolic $K(D)$ pairs over smooth curves, I,* Internat. J. Math. **7** (1996), 573-598.
35. F. Bogomolov, L. Katzarkov, *Complex projective surfaces and infinite groups,* Geom. Funct. Anal. **8** (1998), 243-272.
36. F. Bogomolov, T. Pantev, *Weak Hironaka theorem,* Math. Res. Lett. **3** (1996), 299-307.
37. S. Bradlow, *Special metrics and stability for holomorphic bundles with global sections,* J. Differ. Geom. **33** (1991), 169-214.
38. S. Bradlow, G. Daskalopolous. Moduli of stable pairs for holomorphic bundles over Riemann surfaces, II. Int. J. Math **4** (1993), 903-925.
39. S. Bradlow, G. Daskalopolous, R. Wentworth. Birational equivalence of vortex moduli. Topology **35** (1996), 731-748.
40. S. Bradlow, O. García-Prada, P. Gothen. Surface group representations and $U(p,q)$-Higgs bundles. J. Diff. Geom. **64** (2003), 111-170.
41. S. Broughton, *On the topology of polynomial hypersurfaces,* Singularities, Part 1 (Arcata, Calif., 1981) AMS Proc. Sympos. Pure Math. **40** (1983), 167-178,
42. P. Buser, P. Sarnak, *On the period matrix of a Riemann surface of large genus* (appendix by J. Conway and N. Sloane), Inventiones **117** (1994), 27-56.
43. E. Calabi, X. Chen, *The space of Kähler metrics. II.* J. Differ. Geom. **61** (2002), 173-193.
44. F. Campana,

(i) *Remarques sur les groupes de Kähler nilpotents*, Ann. Sci. E.N.S. **28** (1995), 307-316.
(ii) *Ensembles de Green–Lazarsfeld et quotients résolubles des groupes de Kähler*, J. Algebraic Geom. **10** (2001), 599-622.
(iii) *Connexité abélienne des variétés kählériennes compactes*, Bull. S.M.F. **126** (1998), 483-506.

45. J. Carlson, D. Toledo,
(i) *Harmonic mappings of Kähler manifolds to locally symmetric spaces*, Publ. Math. I.H.E.S. **69** (1989), 173-201.
(ii) *Rigidity of harmonic maps of maximum rank*, J. Geom. Anal. **3** (1993), 99-140.
(iii) *Quadratic presentations and nilpotent Kähler groups*, J. Geom. Anal. **5** (1995), 359-377.
(iv) *On fundamental groups of class VII surfaces*, Bull. London Math. Soc. **29** (1997), 98-102.
(v) *Discriminant complements and kernels of monodromy representations*, Duke Math. J. **97** (1999), 621-648.
46. J. Carlson, H. Clemens, J. Morgan, *On the mixed Hodge structure associated to π_3 of a simply connected complex projective manifold*, Ann. Sci. E.N.S. **14** (1981), 323-338.
47. F. Catanese,
(i) *Moduli and classification of irregular Kähler manifolds (and algebraic varieties) with Albanese general type fibrations*, Inventiones **104** (1991), 263-289 (Appendix by A. Beauville).
(ii) *Fundamental groups with few relations*, in: Higher Dimensional Complex Varieties (Trento 1994), De Gruyter (1996), 163-165.
48. E. Cattani, P. Deligne, A. Kaplan, *On the locus of Hodge classes*, J. Amer. Math. Soc. **8** (1995), 483-506.
49. E. Cattani, A. Kaplan, W. Schmid,
(i) *L^2 and intersection cohomologies for a polarizable variation of Hodge structure*, Inventiones **87** (1987), 217-252.
(ii) *Degeneration of Hodge structures*, Ann. Math. **123** (1986), 457-535.
50. J. Chen, G. Tian, *Compactification of moduli space of harmonic mappings*, Comment. Math. Helv. **74** (1999), 201-237.
51. H. Clemens,
(i) *Degeneration techniques in the study of threefolds*, *Algebraic threefolds, Varenna, 1981*, Lecture Note Math. **947** (1982), 93-154.
(ii) *Degeneration of Kähler manifolds*, Duke Math. J. **44** (1977), 215-290.
52. D. Cohen, P. Orlik, *Arrangements and local systems*, Math. Res. Lett. **7** (2000), 299-316.
53. K. Corlette,
(i) *Flat G-bundles with canonical metrics*, J. Differ. Geom. **28** (1988), 361-382.
(ii) *Rigid representations of Kählerian fundamental groups*, J. Differ. Geom. **33** (1991), 239-252.

(iii) *Nonabelian Hodge theory,* in: Differential Geometry: Geometry in Mathematical Physics and Related Topics (Los Angeles, 1990), AMS Proc. Sympos. Pure Math. **54** (1993), 125-144.
(iv) *Nonabelian Hodge theory and square integrability,* Actes du colloque Géométrie et Topologie des Variétés Projectives, Toulouse 1992), Preprint, Toulouse (1992).
54. K. Corlette, C. Simpson, *The classification of rank two local systems on algebraic varieties* [A paper that we are still planning to write.]
55. W. Crawley–Boevey, *On matrices in prescribed conjugacy classes with no common invariant subspace and sum zero,* Duke Math. J. **118** (2003), 339-352.
56. G. Daskalopolous, S. Dostoglou, R. Wentworth,
(i) *On the Morgan–Shalen compactification of the* $SL(2,\mathbf{C})$ *character varieties of surface groups,* Duke Math. J. **101** (2000), 189-207.
(ii) *Character varieties and harmonic maps to* R*-trees,* Math. Res. Lett. **5** (1998), 523-533.
57. G. Daskalopolous, L. Katzarkov, R. Wentworth, *Harmonic maps to Teichmüller space,* Math. Res. Lett. **7** (2000), 133-146.
58. G. Daskalopoulos, R. Wentworth,
(i) *The Yang–Mills flow near the boundary of Teichmüller space,* Math. Ann. **318** (2000), 1-42.
(ii) *Local degeneration of the moduli space of vector bundles and factorization of rank two theta functions I,* Math. Ann. **297** (1993), 417–466.
59. B. de Oliveira, L. Katzarkov, M. Ramachandran, *Deformations of large fundamental groups,* Geom. Funct. Anal. **12** (2002), 651-668.
60. W. Decker, L. Ein, F. Schreyer, *Construction of surfaces in* $\mathbf{P}^4$, J. Algebraic Geom. **2** (1993), 185-237.
61. E. Delabaere, H. Dillinger, F. Pham, *Résurgence de Voros et périodes des courbes hyperelliptiques,* Ann. Inst. Fourier **43** (1993), 163-199.
62. E. Delabaere, J. Howls, *Global asymptotics for multiple integrals with boundaries,* Duke Math. J. **112** (2002), 199-264.
63. P. Deligne,
(i) *Théorie de Hodge II,* Publ. Math. IHES **40** (1971), 5-58.
(ii) *La conjecture de Weil I,* Publ. Math. IHES **43** (1974), 273-307.
(iii) *Théorie de Hodge III,* Publ. Math. IHES **44** (1974), 5-77.
64. P. Deligne, A. Dimca, *Filtrations de Hodge et par l'ordre du pôle pour les hypersurfaces singulières,* Ann. Sci. E.N.S. **23** (1990), 645-656.
65. P. Deligne, P. Griffiths, J. Morgan, D. Sullivan, *Real homotopy theory of Kähler manifolds,* Inventiones **29** (1975), 245-274.
66. P. Deligne, J. Milne, A. Ogus, K. Shih, *Hodge cycles, motives, and Shimura varieties,* Lecture Note Math. **900** (1982).
67. P. Deligne, G. Mostow,
(i) *Monodromy of hypergeometric functions and nonlattice integral monodromy,* Publ. Math. I.H.E.S. **63** (1986), 5-89.
(ii) *Commensurabilities among lattices in* $PU(1,n)$, Ann. of Math. Studies **132**, Princeton University Press (1993).

68. A. Dimca, A. Nemethi, *On the monodromy of complex polynomials*, Duke Math. J. **108** (2001), 199-209.
69. S. Donaldson,
(i) *Symplectic submanifolds and almost-complex geometry*, J. Diff. Geom. **44** (1996), 666-705.
(ii) *Scalar curvature and projective embeddings. I*, J. Differ. Geom. **59** (2001), 479-522.
(iii) *Planck's constant in complex and almost-complex geometry*, in: XIIIth International Congress on Mathematical Physics (London, 2000), Int. Press, 2001, pp. 63–72.
(iv) *Infinite determinants, stable bundles and curvature*, Duke Math. J. **54** (1987), 231-247.
(v) *Anti self-dual Yang–Mills connections over complex algebraic surfaces and stable vector bundles.* Proc. London Math. Soc. **50** (1985), 1-26.
70. L. Ein, R. Lazarsfeld,
(i) *Seshadri constants on smooth surfaces*, J. Géométrie Algébrique d'Orsay, 1992, Astérisque **218** (1993), 177-186.
(ii) *Singularities of theta divisors and the birational geometry of irregular varieties*, J. Amer. Math. Soc. **10** (1997), 243-258.
71. D. Eisenbud, S. Popescu, C. Walter, *Lagrangian subbundles and codimension 3 subcanonical subschemes*, Duke Math. J. **107** (2001), 427-467.
72. G. Ellingsrud, J. Le Potier, S. A. Stromme, *Some Donaldson invariants of* $\mathbf{CP}^2$, Moduli of vector bundles (Sanda, Kyoto, 1994) Lect. Notes Pure Appl. Math. **179** (1996), 33-38.
73. S. Endrass, U. Persson, J. Stevens, *Surfaces with triple points*, J. Algebraic Geom. **12** (2003), 367-404.
74. H. Esnault, M. Nori, V. Srinivas, *Hodge type of projective varieties of low degree*, Math. Ann. **293** (1992), 1-6.
75. P. Eyssidieux,
(i) *La caractéristique d'Euler du complexe de Gauss–Manin*, J. Reine Angew. Math. **490** (1997), 155-212.
(ii) *Un théoréme de Nakai–Moishezon pour certaines classes de type* (1,1), Preprint, math.AG/9811065.
(iii) *Invariants de Von Neumann des faisceaux coherents*, Preprint, math.AG/9806159.
(iv) *Sur la convexité holomorphe des revêtements linéaires réductifs d'une variété projective algébrique complexe*, Prépublication **179**, Laboratoire E. Picard, Toulouse (2000). To appear.
76. B. Feix, *Hyperkähler metrics on cotangent bundles*, J. Reine Angew. Math. **532** (2001), 33-46.
77. R. Friedman, J. Morgan, *Algebraic surfaces and Seiberg–Witten invariants*, J. Algebraic Geom. **6** (1997), 445-479.
78. A. Fujiki,
(i) *An* L^2 *Dolbeault lemma and its applications*, Publ. R.I.M.S. **28** (1992), 845-884.
(ii) *Hyper-Kähler structure on the moduli space of flat bundles*, in: Prospects

in Complex Geometry (Katata and Kyoto, 1989), Lecture Note Math. **1468** (1991), 1-83.

79. A. Fujiki, R. Kobayashi, S. Lu, *On the fundamental group of certain open normal surfaces,* Saitama Math. J. **11** (1993), 15-20.
80. W. Fulton, *On the fundamental group of the complement of a node curve,* Ann. Math. **111** (1980), 407-409.
81. W. Goldman, J. Millson, *The deformation theory of representations of fundamental groups of compact Kähler manifolds,* Publ. Math. I.H.E.S. **67** (1988), 43-96.
82. R. Gompf, *A new construction of symplectic manifolds,* Ann. Math. **142** (1995), 527-595.
83. R. Gompf, T. Mrowka, *Irreducible 4-manifolds need not be complex,* Ann. Math. **138** (1993), 61-111.
84. P. Gothen, *Components of spaces of representations and stable triples,* Topology **40** (2001), 823-850.
85. M. Green, R. Lazarsfeld, *Higher obstructions to deforming cohomology groups of line bundles,* J. Amer. Math. Soc. **4** (1991), 87-103.
86. G.-M. Greuel, C. Lossen, E. Shustin, *Plane curves of minimal degree with prescribed singularities,* Inventiones **133** (1998), 539-580.
87. P. Griffiths, J. Harris, *Principles of Algebraic Geometry,* Wiley-Interscience, New York, 1978.
88. M. Gromov, *Sur le groupe fondamental d'une variété kählérienne,* C.R.A.S. **308** (1989), 67-70.
89. M. Gromov, R. Schoen, *Harmonic maps into singular spaces and p-adic superrigidity for lattices in groups of rank one,* Publ. Math. I.H.E.S. **76** (1992), 165-246.
90. M. Gross, *Deforming Calabi–Yau threefolds,* Math. Ann. **308** (1997), 187-220. [See also the references mentionned in the Math Reviews MR1464900 (98i:14039) of this paper.]
    (ii) *The deformation space of Calabi–Yau n-folds with canonical singularities can be obstructed,* Mirror symmetry, II AMS/IP Stud. Adv. Math. **1** (1997), 401-411.
91. L. Gruson, C. Peskine, *Genre des courbes de l'espace projectif,* Algebraic geometry (Tromso, 1977), Lecture Note Math. **687** (1978), 31-59.
92. D. Guan, *On the Betti numbers of irreducible compact hyperkähler manifolds of complex dimension four,* Math. Res. Lett. **8** (2001), 663-669.
93. R. Hain,
    (i) *The de Rham homotopy theory of complex algebraic varieties, I,* K-Theory **1** (1987), 271-324.
    (ii) *The de Rham homotopy theory of complex algebraic varieties. II,* K-Theory **1** (1987), 481-497.
    (iii) *Infinitesimal presentations of the Torelli groups,* J. Amer. Math. Soc. **10** (1997), 597-651.
    (iv) *The Hodge de Rham theory of relative Malcev completion,* Ann. Sci. E.N.S. **31** (1998), 47-92.

(v) *The rational cohomology ring of the moduli space of abelian 3-folds,* Math. Res. Lett. **9** (2002), 473-491.

94. R. Hain, E. Looijenga, *Mapping class groups and moduli spaces of curves,* Algebraic geometry, Santa Cruz 1995, Am. Math. Soc. Proc. Sympos. Pure Math. **62** (2) (1997), 97-142.
95. T. Hausel, *Compactification of moduli of Higgs bundles,* J. Reine Angew. Math. **503** (1998), 169-192.
96. T. Hausel, M. Thaddeus, *Mirror symmetry, Langlands duality, and the Hitchin system,* Inventiones **153** (2003), 197-229.
97. E. Hironaka, *Alexander stratifications of character varieties,* Ann. Inst. Fourier **47** (1997), 555-583.
98. H. Hironaka, *Resolution of singularities of an algebraic variety over a field of characteristic zero. I, II,* Ann. Math. **79** (1964), 109-203; 205-326.
99. N. Hitchin,
   (i) *Stable bundles and integrable systems,* Duke Math. J. **54** (1987), 91-114.
   (ii) *The self-duality equations on a Riemann surface,* Proc. London Math. Soc. **55** (1987), 59-126.
   (iii) *Poncelet polygons and the Painlevé equations,* Geometry and Analysis (Bombay, 1992), Tata Inst. Fund. Res., Bombay (1995), 151-185.
100. B. Ilic, J.M. Landsberg, *On symmetric degeneracy loci, spaces of symmetric matrices of constant rank and dual varieties,* Math. Ann. **314** (1999), 159-174.
101. Y. Imayoshi, M. Ito, Manabu; H. Yamamoto, *On the Nielsen–Thurston–Bers type of some self-maps of Riemann surfaces with two specified points,* Osaka J. Math. **40** (2003), 659-685.
102. M. Inaba, K. Iwasaki, Masa–Hiko Saito, Moduli of Stable Parabolic Connections *Riemann–Hilbert correspondence and Geometry of Painlevé equation of type VI, Part I,* Preprint, math.AG/0309342.
103. D. Joe, M. Kim, *On the topology of algebraic surfaces and reduction modulo p,* Int. Math. Res. Not. **28** (2002), 1505-1508.
104. F. Johnson, E. Rees, *On the fundamental group of a complex algebraic manifold,* Bull. London Math. Soc. **19** (1987), 463-466.
105. J. Jost, K. Zuo,
   (i) *Arakelov type inequalities for Hodge bundles over algebraic varieties, I. Hodge bundles over algebraic curves,* J. Algebraic Geom. **11** (2002), 535-546.
   (ii) *Representations of fundamental groups of algebraic manifolds and their restrictions to fibers of a fibration,* Math. Res. Lett. **8** (2001), 569-575.
   (iii) *Vanishing theorems for L^2-cohomology on infinite coverings of compact Kähler manifolds and applications in algebraic geometry,* Comm. Anal. Geom. **8** (2000), 1-30.
   (iv) *Harmonic maps into Bruhat–Tits buildings and factorizations of p-adically unbounded representations of π_1 of algebraic varieties, I,* J. Algebraic Geom. **9** (2000), 1-42.
   (v) *Harmonic maps of infinite energy and rigidity results for representations of fundamental groups of quasiprojective varieties,* J. Differ. Geom. **47** (1997), 469-503.

(vi) *Harmonic maps and $SL(r, \mathbf{C})$-representations of fundamental groups of quasiprojective manifolds*, J. Algebraic Geom. **5** (1996), 77-106.

106. J. Jost, Y. Yang, K. Zuo, *The cohomology of a variation of polarized Hodge structures over a quasi-compact Kähler manifold*, Preprint, math.AG/0312145.
107. A. Kabanov, *The second cohomology with symplectic coefficients of the moduli space of smooth projective curves*, Compositio Math. **110** (1998), 163-186.
108. D. Kaledin, *Hyperkaehler structures on total spaces of holomorphic cotangent bundles*, Preprint, alg-geom/9710026.
109. D. Kaledin, M. Verbitsky,
(i) *Hyperkähler manifolds*, Mathematical Physics **12**, International Press (1999).
(ii) *Period map for non-compact holomorphically symplectic manifolds*, Geom. Funct. Anal. **12** (2002), 1265-1295.
110. M. Kapovich, J. Millson,
(i) *On representation varieties of Artin groups, projective arrangements and the fundamental groups of smooth projective varieties*, Publ. Math. I.H.E.S. **88** (1998), 5-95.
(ii) *Universality theorems for configuration spaces of planar linkages*, Topology **41** (2002), 1051-1107.
(iii) *Hodge theory and the art of paper folding*, Publ.R.I.M.S. **33** (1997), 1-31.
111. L. Katzarkov,
(i) *Schematization of homotopy types and realizations*, Comm. in arithmetic fundamental groups, Kyoto 1999-2001 Kurakaisekikenkyusho Kokyuroku **1267** (2002), 147-156.
(ii) *On the Shafarevich maps*, Algebraic geometry, Santa Cruz 1995 AMS Proc. Symp. Pure Math. **62** (2) (1997), 173-216.
(iii) *The geometry of large fundamental groups*, J. Math. Sci. **94** (1999), 1100-1110.
112. L. Katzarkov, T. Pantev,
(i) *Representations of fundamental groups whose Higgs bundles are pullbacks*, J. Differ. Geom. **39** (1994), 103-121.
(ii) *Nonabelian (p,p)-classes*, Motives, Polylogarithms and Hodge Theory, Irvine, 1998 Int. Press Lecture Ser. **3** (II), 2002, pp. 625–715.
113. L. Katzarkov, T. Pantev, C. Simpson,
(i) *Density of monodromy actions on non-abelian cohomology*, Adv in Math. **179** (2003), 155-204
(ii) *Nonabelian mixed Hodge structures*, Preprint, math.AG/0006213.
114. L. Katzarkov, T. Pantev, B. Toen, *Schematic homotopy types and non-abelian Hodge theory I: The Hodge decomposition*, Preprint, math.AG/0107129.
115. L. Katzarkov, M. Ramachandran, *On the universal coverings of algebraic surfaces*, Ann. Sci. E.N.S. **31** (1998), 525-535.
116. M. Kim, A. Sommese, *Two results on branched coverings of Grassmannians*, J. Math. Kyoto Univ. **38** (1998), 21-27.
117. B. Klingler,
(i) *Structures affines et projectives sur les surfaces complexes*, Ann. Inst. Fourier **48** (1998), 441-477.

(ii) *Sur la rigidité de certains groupes fondamentaux. l'arithméticité des réseaux hyperboliques complexes, et les "faux plans projectifs"*, Inventiones **153** (2003), 105-143.

118. H. Konno, *Construction of the moduli space of stable parabolic Higgs bundles on a Riemann surface*, J. Math. Soc. Japan **45** (1993), 253-276.
119. M. Kontsevich, Y. Manin, *Gromov–Wiiten classes, quantum cohomology, and enumeration geometry*, Comm. Math. Phys. **164** (1994), 525-562.
120. N. Korevaar, R. Schoen, *Sobolev spaces and harmonic maps for metric space targets*, Comm. Anal. Geom. **1** (1993), 561-659.
121. V. Kostov,
    (i) *On the Deligne–Simpson problem*, C.R.A.S. **329** (1999), 657-662.
    (ii) *On the Deligne–Simpson problem*, Monodromiya v Zadachakh Algebr. Geom. i Differ. Uravn., Tr. Mat. Inst. Steklova **238** (2002), 158-195.
    (iii) *On some aspects of the Deligne–Simpson problem*, J. Dynam. Control Systems **9** (2003), 393-436.
122. R. Kulkarni, J. Wood, *Topology of nonsingular complex hypersurfaces*, Adv. in Math. **35** (1980), 239-263 [See also the numerous references given in the Math Reviews MR0563926 of this paper.]
123. J. M. Landsberg, *Differential-geometric characterizations of complete intersections*, J. Differ. Geom. **44** (1996), 32-73.
124. B. Lasell, M. Ramachandran, *Observations on harmonic maps and singular varieties*, Ann. Sci. E.N.S. **29** (1996), 135-148.
125. G. Laumon,
    (i) *La transformation de Fourier géométrique et ses applications*, Proceedings of ICM, Kyoto 1990, Math. Soc. Japan (1991), 437-445.
    (ii) *Sur la cohomologie à supports compacts des variétés de Shimura pour* $GSp(4)_Q$, Compositio Math. **105** (1997), 267-359.
    (iii) *Manuscript on the Fourier transform between* $\mathcal{D}$*-modules on an abelian variety and bundles on its universal extension*,
126. G. Laumon, M. Rapoport, *The Langlands lemma and the Betti numbers of stacks of G-bundles on a curve*, Internat. J. Math. **7** (1996), 29-45.
127. R. Lazarsfeld, *Lengths of periods and Seshadri constants of abelian varieties*, Math. Res. Lett. **3** (1996), 439-447.
128. J. Li, M.S. Narasimhan, *Hermitian–Einstein metrics on parabolic stable bundles*, Acta Math. Sin. **15** (1999), 93-114.
129. J. Li, Y.-D. Wang, *Existence of Hermitian–Einstein metrics on stable Higgs bundles over open Kähler manifolds*, Internat. J. Math. **10** (1999), 1037-1052.
130. A. Libgober,
    (i) *Alexander polynomial of plane algebraic curves and cyclic multiple planes*, Duke Math. J. **49** (1982), 833-851.
    (ii) *First order deformations for rank one local systems with a non-vanishing cohomology*, Arrangements in Boston: a Conference on Hyperplane Arrangements (1999). Topology Appl. **118** (2002), 159-168.
131. A. Libgober, J. Wood,
    (i) *Diffeomorphic complete intersections with different multidegrees*, Bull. Amer. Math. Soc. **2** (1980), 459-461.

(ii) *Differentiable structures on complete intersections. II*, Singularities, Part 2 (Arcata, 1981), AMS Proc. Sympos. Pure Math. **40** (1983), 123-133.
132. R. Livné, *Appendix to* [**139**].
133. M. Lönne, *On the diffeomorphism groups of elliptic surfaces*, Math. Ann. **310** (1998), 103-117.
134. J. Lurie, *On Infinity Topoi*, Preprint; math.CT/0306109.
135. E. Markman,
(i) *Brill–Noether duality for moduli spaces of sheaves on K3 surfaces*, J. Algebraic Geom. **10** (2001), 623-694.
(ii) *On the monodromy of moduli spaces of sheaves on K3 surfaces, I and II*, Preprints math.AG/0305042, math.AG/0305043.
(iii) *Algebraic geometry, integrable systems, and Seiberg–Witten theory*, Integrability: the Seiberg–Witten and Whitham equations eds. H. Braden, I. Krichever. Gordon and Breach, Amsterdam (1999), 23-41.
136. E. Markman, E. Xia, *The moduli of flat* $\mathrm{PU}(p,p)$*-structures with large Toledo invariants*, Math. Z. **240** (2002), 95-109.
137. M. Mehta, *Birational equivalence of Higgs moduli*, Preprint, math.AG/0311122.
138. T. Mochizuki,
(i) *Asymptotic behaviour of tame nilpotent harmonic bundles with trivial parabolic structure*, J. Differ. Geom. **62** (2002), 351-559.
(ii) *The Gromov–Witten class and a perturbation theory in algebraic geometry*, Amer. J. Math. **123** (2001), 343-381.
(iii) *Asymptotic behaviour of tame harmonic bundles and an application to pure twistor D-modules*, Preprint, math.DG/0312230.
139. B. Moishezon, *Complex surfaces and connected sums of complex projective planes* (appendix by R. Livne). Lecture Note Math. **603** (1977).
140. B. Moishezon, M. Teicher,
(i) *Braid group technique in complex geometry. I. Line arrangements in* $\mathbf{CP}^2$, Braids (Santa Cruz, 1986), Contemp. Math. **78**, Am. Math. Soc. (1988), 425-555.
(ii) *Braid group technique in complex geometry. II. From arrangements of lines and conics to cuspidal curves*, in: Algebraic Geometry (Chicago, 1989), Lecture Note Math. **1479** (1991), pp. 131-180.
(iii) *Braid group technique in complex geometry. V. The fundamental group of a complement of a branch curve of a Veronese generic projection*, Comm. Anal. Geom. **4** (1996), 1-120.
141. J. Morgan. *The algebraic topology of smooth algebraic varieties*, Publ. Math. I.H.E.S. **48** (1978), 137-204.
142. J. Morgan, T. Mrowka, *On the diffeomorphism classification of regular elliptic surfaces*, Internat. Math. Res. Notices **6** (1993), 183-184.
143. I. Mundet i Riera, *Yang–Mills–Higgs theory for symplectic fibrations*, Preprint, math.SG/9912150.
144. A. Nadel,
(i) *Singularities and Kodaira dimension of the moduli space of flat Hermitian–Yang–Mills connections*, Compositio Math. **67** (1988), 121-128.
(ii) *Invariants for holomorphic vector bundles*, Math. Ann. **309** (1997), 37-52.

145. H. Nakajima, *Hyper-Kähler structures on moduli spaces of parabolic Higgs bundles on Riemann surfaces*, Moduli of vector bundles (Sanda, Kyoto, 1994) Lect. Notes Pure Appl. Math. **179** (1996), 199-208.
146. Y. Namikawa, J. Steenbrink, *Global smoothing of Calabi–Yau threefolds*, Inventiones **122** (1995), 403-419.
147. T. Napier, M. Ramachandran, *The $L^2\overline{\partial}$-method, weak Lefschetz theorems, and the topology of Kähler manifolds*, J. Amer. Math. Soc. **11** (1998), 375-396.
148. V. Navarro Aznar.
149. W. Neumann, P. Norbury, *Vanishing cycles and monodromy of complex polynomials*, Duke Math. J. **101** (2000), 487-497.
150. L. Ni, H. Ren, *Hermitian–Einstein metrics for vector bundles on complete Kähler manifolds*, Trans. Amer. Math. Soc. **353** (2001), 441-456.
151. N. Nitsure,
(i) *Moduli space of semistable pairs on a curve*, Proc. London Math. Soc. **62** (1991), 275-300.
(ii) *Moduli of semistable logarithmic connections*, J. Amer. Math. Soc. **6** (1993), 597-609.
152. M. Nori, *Zariski's conjecture and related problems*, Ann. Sci. E.N.S. **16** (1983), 305-344.
153. K. O'Grady, *The weight-two Hodge structure of moduli spaces of sheaves on a K3 surface*, J. Algebraic Geom. **6** (1997), 599-644.
154. C. Okonek, A. Teleman.
(i) *Master spaces and the coupling principle: from geometric invariant theory to gauge theory*, Comm. Math. Phys. **205** (1999), 437-458.
(ii) *Gauge theoretical equivariant Gromov–Witten invariants and the full Seiberg–Witten invariants of ruled surfaces*, Comm. Math. Phys. **227** (2002), 551-585.
155. A. Otwinowska, *Composantes de petite codimension du lieu de Noether–Lefschetz: un argument asymptotique en faveur de la conjecture de Hodge pour les hypersurfaces*, J. Algebraic Geom. **12** (2003), 307-320.
156. V. Pati, *L^2-cohomology of algebraic varieties in the Fubini metric*, in: Hodge theory (Sant Cugat, 1985), Lecture Note Math. **1246**, 1987, pp. 154-164.
157. T. Pantev, *Comparison of generalized theta functions*, Duke Math. J. **76** (1994), 509-539.
158. S. Papadima, A. Suciu, *Higher homotopy groups of complements of complex hyperplane arrangements*, Adv. Math. **165** (2002), 71-100.
159. R. Pelissier, *Catégories enrichies faibles*, Thesis, Université de Nice-Sophia Antipolis (2002), Preprint math.AT/0308246.
160. O. Penacchio, *Structures de Hodge mixtes et fibrés sur le plan projectif complexe*, Thesis, Université Paul Sabatier, Toulouse (2002), Preprint math.AG/0307156.
161. G. Perelman, *The entropy formula for the Ricci flow and its geometric applications*, Preprint, math.DG/0211159.
162. U. Persson,

(i) *On degenerations of algebraic surfaces*, Mem. Amer. Math. Soc. **11** (1977).
(ii) *An introduction to the geography of surfaces of general type*, in: Algebraic geometry (Bowdoin, 1985), Am. Math. Soc. Proc. Sympos. Pure Math. **46** (1987), 195-218.

163. U. Persson, C. Peters, *Some aspects of the topology of algebraic surfaces*, in: Proceedings of the Hirzebruch 65 Conference on Algebraic Geometry (Ramat Gan, 1993) Israel Math. Conf. Proc. **9**, Bar-Ilan Univ. (1996), 377-392.

164. U. Persson, C. Peters, G. Xiao, *Geography of spin surfaces*, Topology **35** (1996), 845-862.

165. U. Persson, H. Pinkham, *Some examples of nonsmoothable varieties with normal crossings*, Duke Math. J. **50** (1983), 477-486.

166. C. Peters,
(i) *Introduction to the theory of compact complex surfaces*, in: Differential Geometry, Global Analysis, and Topology (Halifax, 1990), CMS Conf. Proc. **12**, AMS, Providence (1991), 129-156.
(ii) *Rigidity for variations of Hodge structure and Arakelov-type finiteness theorems*, Compositio Math. **75** (1990), 113-126.
(iii) *On Arakelov's finiteness theorem for higher-dimensional varieties*, Conference on algebraic varieties of small dimension (Turin, 1985). Rend. Sem. Mat. Univ. Politec. Torino (1987), 43-50.
(iv) Math Reviews MR0466149 (57 # 6030), review of [**162**] (i).

167. D. Quillen,
(i) *Rational homotopy theory*, Ann. of Math. **90** (1969), 205-295.
(ii) *Homotopical Algebra*, Lecture Note Math. **43** (1967).
(iii) *Characteristic classes of representations*, Algebraic K-theory (Northwestern Univ., Evanston, 1976), Lecture Note Math. **551** (1976), 189-216.

168. F. Rahmati, *Déformation des surfaces projectives toriques*, Bull. Soc. Sci. Lett. Lódź Sér. Rech. Déform. **23** (1997), 23-31.

169. Z. Ran,
(i) *Deformations of manifolds with torsion or negative canonical bundle*, J. Algebraic Geom. **1** (1992), 279-291.
(ii) *On the variety of rational space curves*, Israel J. Math. **122** (2001), 359-369.

170. A. Reznikov,
(i) *Continuous cohomology of the group of volume-preserving and symplectic diffeomorphisms, measurable transfer and higher asymptotic cycles*, Selecta Math. **5** (1999), 181-198.
(ii) *Symplectic twistor spaces*, Ann. Global Anal. Geom. **11** (1993), 109-118.
(iii) *Simpson's theory and superrigidity of complex hyperbolic lattices*, C. R. A. S. Math. **320** (1995), 1061-1064.
(iv) *Characteristic classes in symplectic topology*, Selecta Math. **3** (1997), 601-642.
(v) *Rationality of secondary classes*, J. Diff. Geom. **43** (1996), 674-692.
(vi) *All regulators of flat bundles are torsion*, Ann. Math. **141** (1995), 173-186.
(vii) *The structure of Kähler groups. I. Second cohomology*, in: Motives, Polylogarithms and Hodge Theory (Irvine, 1998), Int. Press Lect. Ser. **3** (II) (2002), 717-730.

171. A. Robb, M. Teicher, *Applications of braid group techniques to the decomposition of moduli spaces, new examples,* Special issue on braid groups and related topics (Jerusalem, 1995). Topology Appl. **78** (1997), 143-151.

172. C. Sabbah,
(i) *Frobenius manifolds: isomonodromic deformations and infinitesimal period mappings,* Exposition. Math. **16** (1998), 1-57.
(ii) *Harmonic metrics and connections with irregular singularities,* Ann. Inst. Fourier **49** (1999), 1265-1291.
(iii) *Polarizable Twistor $\mathcal{D}$-Modules,* Preprint, Ecole Polytechnique.

173. M. Saito,
(i) *Mixed Hodge modules and admissible variations,* C.R.A.S. **309** (1989), 351-356.
(ii) *Mixed Hodge modules,* Publ. RIMS **26** (1990), 221-335.

174. W. Schmid, *Variation of Hodge structure: the singularities of the period mapping,* Inventiones **22** (1973), 211-319.

175. A. Schmitt, *Projective moduli for Hitchin pairs,* Internat. J. Math. **9** (1998), 107-118.

176. H. Schoutens, *Bounds in cohomology,* Israel J. Math. **116** (2000), 125-169.

177. C. Simpson,
(i) *Transcendental aspects of the Riemann–Hilbert correspondence,* Illinois J. Math. **34** (1990), 368-391.
(ii) *Products of matrices,* in: Differential Geometry, Global Analysis, and Topology (Halifax, 1990), CMS Conf. Proc. **12** (1991), 157-185.
(iii) *Some families of local systems over smooth projective varieties,* Ann. Math. **138** (1993), 337-425.
(iv) The Hodge filtration on nonabelian cohomology. Algebraic geometry, Santa Cruz 1995 AMS Proc. Symp. Pure Math. **62** (2) (1997), 217-281.
(v) Mixed twistor structures. Preprint, alg-geom/9705006.
(vi) Algebraic aspects of higher nonabelian Hodge theory. Motives, polylogarithms and Hodge theory, Irvine, 1998, Int. Press Lect. Ser. **3** (II) (2002), 417-604.

178. Y.-T. Siu, *Strong rigidity for Kähler manifolds and the construction of bounded holomorphic functions,* Discrete Groups and Analysis, Progr. Math. **67**, Birkhäuser (1987), 124-151.

179. I. Smith,
(i) *Lefschetz pencils and divisors in moduli space,* Geom. Topol. **5** (2001), 579-608.
(ii) *Lefschetz fibrations and the Hodge bundle,* Geom. Topol. **3** (1999), 211-233.

180. A. Sommese, A. Van de Ven, *Homotopy groups of pullbacks of varieties,* Nagoya Math. J. **102** (1986), 79-90.

181. A. Sommese, J. Verschelde, *Numerical homotopies to compute generic points on positive dimensional algebraic sets,* in: Complexity Theory, Real Machines, and Homotopy (Oxford, 1999), J. Complexity **16** (2000), 572-602.

182. J. Steenbrink, *Some remarks about the Hodge conjecture,* Hodge theory, Sant Cugat, 1985, Lecture Note Math. **1246** (1987), 165-175.

183. J. Steenbrink, S. Zucker, *Variation of mixed Hodge structure. I*, Inventiones **80** (1985), 489-542.
184. C. Teleman, *Borel–Weil–Bott theory on the moduli stack of G-bundles over a curve*, Inventiones **134** (1998), 1-57.
185. M. Thaddeus, *Variation of moduli of parabolic Higgs bundles*, J. Reine Angew. Math. **547** (2002), 1-14.
186. G. Tian,
(i) *Smoothness of the universal deformation space of compact Calabi–Yau manifolds and its Petersson–Weil metric*, in: Mathematical Aspects of String Theory (San Diego, 1986), Adv. Ser. Math. Phys. **1** World Scientific (1987), 629-646.
(ii) *Smoothing 3-folds with trivial canonical bundle and ordinary double points. Essays on mirror manifolds*, 458–479, Internat. Press, Hong Kong, 1992.
(iii) *Degeneration of Kähler–Einstein manifolds, I*, Differential geometry: geometry in mathematical physics and related topics (Los Angeles, 1990) AMS Proc. Sympos. Pure Math. **54** (1993), 595-609.
(iv) *Kähler–Einstein metrics with positive scalar curvature*, Inventiones **130** (1997), 1-37.
187. K. Timmerscheidt, *Mixed Hodge theory for unitary local systems*, J. Reine Angew. Math. **379** (1987), 152-171.
188. A. Todorov, *The Weil–Petersson geometry of the moduli space of $SU(n \geqslant 3)$ (Calabi–Yau) manifolds, I*, Comm. Math. Phys. **126** (1989), 325-346.
189. B. Toen,
(i) *Vers une interprétation galoisienne de la théorie de l'homotopie*, Cah. Topol. Géom. Diff. Catég. **43** (2002), 257-312.
(ii) *On motives for Deligne–Mumford stacks*, Int. Math. Res. Notices **17** (2000), 909-928.
(iii) *Champs Affines*, Preprint, math.AG/0012219.
(iv) *Homotopical and Higher Categorical Structures in Algebraic Geometry*, Preprint, math.AG/0312262.
190. B. Toen, G. Vezzosi,
(i) *"Brave New" Algebraic Geometry and global derived moduli spaces of ring spectra*, Preprint, math.AT/0309145.
(ii) *Homotopical Algebraic Geometry I: Topos theory*, Preprint, math.AG/0207028.
(iii) *From HAG to DAG: Derived Moduli Spaces*, Preprint, math.AG/0210407.
191. D. Toledo,
(i) *Representations of surface groups in complex hyperbolic space*, J. Differ. Geom. **29** (1989), 125-133.
(ii) *Examples of fundamental groups of compact Kähler manifolds*, Bull. London Math. Soc. **22** (1990), 339-343.
(iii) *Projective varieties with non-residually finite fundamental group*, Publ. Math. I.H.E.S. **77** (1993), 103-119.
(iv) *Rigidity theorems in Kähler geometry and fundamental groups of varieties*, Several complex variables (Berkeley, 1995-1996) MSRI Publ. **37**, Cambridge Univ. Press (1999), 509-533.

192. K. Uhlenbeck, S.-T. Yau, *On the existence of Hermitian–Yang–Mills connections in stable vector bundles*, Frontiers of the mathematical sciences: 1985 (New York, 1985). Comm. Pure Appl. Math. 39 (1986), no. S, suppl., S257–S293.
193. M. Verbitsky,
(i) *Tri-analytic subvarieties of hyper-Kaehler manifolds*, Geom. Funct. Anal. **5** (1995), 92-104.
(ii) *Deformations of trianalytic subvarieties of hyper-Kähler manifolds*, Selecta Math. **4** (1998), 447-490.
194. C. Voisin,
(i) *Variations of Hodge structure and algebraic cycles*, Proc. ICM (Zürich, 1994), Birkhäuser (1995), 706-715.
(ii) *Remarks on zero-cycles of self-products of varieties*, Moduli of vector bundles (Sanda, Kyoto, 1994), **179** (1996), 265-285.
(iii) Théorie de Hodge et géométrie algébrique complexe. S.M.F. Cours Spécialisés **10** (2002). English translation by L. Schneps: Cambridge Studies in Adv. Math. **76**, **77**, Cambridge University Press (2003).
(iv) A counterexample to the Hodge conjecture extended to Kähler varieties. Int. Math. Res. Not. **20** (2002), 1057-1075.
(v) On the homotopy types of compact kaehler and complex projective manifolds. Preprint, math.AG/0312032.
195. F. Warner. Introduction to manifolds, Scott-Foresman, New York (1971).
196. M. Wolf. Harmonic maps from surfaces to **R**-trees. Math. Z. **218** (1995), 577-593.
197. E. Xia, *The moduli of flat $U(p,1)$ structures on Riemann surfaces*, Geom. Dedicata **97** (2003), 33-43.
198. Y.-H. Yang, *On non-Kählerianity of nonuniform lattices in $SO(n,1)(n \geqslant 4)$*, Manuscripta Math. **103** (2000), 401-407.
199. K. Yokogawa,
(i) *Compactification of moduli of parabolic sheaves and moduli of parabolic Higgs sheaves*, J. Math. Kyoto Univ. **33** (1993), 451-504.
(ii) *Infinitesimal deformation of parabolic Higgs sheaves*, Internat. J. Math. **6** (1995), 125-148.
200. J. Zhang. Threefolds with vanishing Hodge cohomology. Preprint, math.AG/0312239.
201. S. Zucker,
(i) *Variation of mixed Hodge structure. II*, Inventiones **80** (1985), 543-565.
(ii) *Hodge theory with degenerating coefficients: L^2 cohomology in the Poincaré metric*, Ann. Math. **109** (1979), 415-476.
202. K. Zuo,
(i) *Factorizations of nonrigid Zariski dense representations of π_1 of projective algebraic manifolds*, Inventiones **118** (1994), 37-46.
(ii) *Representations of fundamental groups of algebraic varieties*, Lecture Note Math. **1708** (1999).
(iii) *On the negativity of kernels of Kodaira–Spencer maps on Hodge bundles and applications*, Kodaira's issue. Asian J. Math. **4** (2000), 279-301.

Subject Index